MATLAB® for
Electrical and Computer Engineering
Students and Professionals

with Simulink®

MATLAB® for Electrical and Computer Engineering Students and Professionals

with Simulink®

Roland Priemer
University of Illinois at Chicago

SciTECH
PUBLISHING
an imprint of the IET

Edison, NJ
scitechpub.com

Published by SciTech Publishing, an imprint of the IET.
www.scitechpub.com
www.theiet.org

10 9 8 7 6 5 4 3 2

ISBN 978-1-61353-188-4 (paperback)

Typeset in India by MPS Limited
Printed in the USA by Sheridan Books, Inc.
Printed in the UK by Hobbs the Printers Ltd

Contents

Preface

This book combines the teaching of MATLAB® programming skills with the presentation and development of carefully selected electrical and computer engineering (ECE) fundamentals. This is what distinguishes it from many others: it is directed specifically to ECE concerns. Students will see, quite explicitly, how and why MATLAB is well suited to solve practical ECE problems.

For ECE graduates of BS programs, MATLAB programming skills are an increasingly important component in the tool set for competing successfully in the job market. This requires that students start early in their academic studies to learn and apply MATLAB programming skills in their EE and CE curricula.

Audience

This book is intended primarily for the freshman or sophomore ECE major who has no programming experience, no background in EE or CE, and is required to take a MATLAB programming course. It can also be used as the text of an introduction to electrical and computer engineering course where learning MATLAB programming is strongly emphasized. A first course in calculus, usually taken concurrently, is essential. This book can be used in various ways by other instructors and readers. For example, it can be used to accompany any text in a higher level EE or CE course where use of MATLAB is desired. It can be used by junior and senior level students in other fields of engineering who are required to learn fundamentals of EE and CE. Certainly, the professional engineer or scientist in EE, CE or any other field can easily learn (or be refreshed) and apply MATLAB to practical problems and come to appreciate its convenience and powerful computing capabilities. Lastly, it can be used by professional engineers for self-study to learn MATLAB, and I suspect it will. The many in-depth examples and programs included in this book will be welcomed by such readers. MATLAB is useful not only for those with academic or professional needs but also for those who enjoy using a computer to work with, for example, audio (speech, music, and biomedical) signals, images, games, animation, and application development in virtually all fields. Many examples and programs in this book demonstrate this.

Rationale of the Book

The distinguishing feature of this book is that about 15% of this MATLAB book develops ECE fundamentals gradually, from very basic principles. Because these fundamentals are interwoven throughout, MATLAB can be applied to solve relevant, practical problems. The plentiful, in-depth example problems to which MATLAB is applied were carefully chosen so that results obtained with MATLAB *also* provide insights about the fundamentals. With this "feedback approach" to learning MATLAB, ECE students also gain a head start in learning some core subjects in EE and CE curricula. For example, complex numbers and time functions are very important in ECE but are barely touched upon in other MATLAB books, even though MATLAB works conveniently with the complex data type. From basic principles, complex numbers and time functions are introduced such that students learn how to apply MATLAB to solve AC (alternating current) circuit analysis problems and analyze a circuit to find its frequency response.

The effort required to learn how to *actually use* MATLAB quickly pays for itself. Throughout the first eleven chapters of this book, nearly 200 examples and over 80 programs demonstrate this, showing students how solutions of practical problems can be obtained with MATLAB. After using this book, the ECE student will be well prepared to apply MATLAB in all coursework that is commonly included in EE and CE curricula.

Synopsis of Core Chapters 1–11

The core chapters for learning MATLAB are Chapters 1-11. In Chapter 1, the MATLAB environment is presented, including navigation among many of its linked windows. The chapter shows how to do immediate and interactive computing and how to access the MATLAB help facilities. Chapter 2 introduces the Edit Window for program and function, called m-files, development. Electric current and voltage are defined, based on fundamental physical principles. Ohm's Law and the operation of a p-n junction diode are given to apply MATLAB. How to use the MATLAB code analyzer is explained, and students learn how to build their own library of functions, called a toolbox.

Chapter 3 starts by solving two linear equations in two unknowns by repeated substitution, and the solution is written in a form to motivate the definition of a matrix and matrix multiplication, addition, equality, and inversion. Then matrix algebra is presented more thoroughly with the objective of showing how well MATLAB works with matrices. Kirchhoff's Laws are introduced in Chapter 3, and MATLAB is applied to do component, nodal, and mesh analysis of resistive circuits. Several methods to find the solution of N linear equations in N unknowns, written in matrix notation as $AX = Y$, are examined, including Gauss-Jordan elimination and eigenvector and singular value decomposition of the matrix A. After a matrix norm is defined, the condition number of the matrix A is examined to assess the accuracy of a solution.

Chapter 4 addresses MATLAB constructs for program flow control: for loop, if-elseif-else, while loop, and switch-case-otherwise. After describing the MATLAB relational and logical operators, there follows a discussion about probability, random number generators, histograms, median filtering, numerical integration, and optimization, including the method of least squares and the method of steepest descent. Many extensive MATLAB program examples are included to demonstrate program flow control while applying MATLAB to the mentioned topics.

Chapter 5 discusses binary data. Elements of Boolean algebra are introduced, and MATLAB is used to evaluate Boolean functions. Logic gates are defined, and it is shown how to design a combinatorial logic circuit. Algorithms are given for base ten to binary conversion of integers and fractions. Binary arithmetic is discussed. A MATLAB simulation of a serial binary adder is given, which is performed with an animation in Chapter 9. The floating point notation that MATLAB uses to store numeric data is described. A data acquisition system is described, and quantization error that occurs in the process of analog to digital conversion (A/D) is examined. The process of digital to analog (D/A) conversion is illustrated.

Chapter 6 provides an extensive introduction to complex numbers. After Euler's identity is proved, the concept of a phasor of a sinusoidal time function is introduced, and then the Fourier series of a periodic signal is examined. MATLAB is used to obtain and plot the magnitude spectrum of a periodic signal and to reconstruct a periodic signal from its Fourier series. Through Euler's identity, the concept of an impedance is presented, and MATLAB is applied to AC circuit analysis. This leads to the transfer function and frequency response of a linear RLC (resistor, inductor, and capacitor) circuit, and MATLAB is used to calculate and plot the frequency response. The ideal operational amplifier (op-amp) is then defined, and the gain of several op-amp based circuits is obtained, including an active RC (op-amp, resistor, and capacitor) circuit.

Chapter 7 is concerned with the character data type and character string manipulation. This very useful part of MATLAB makes it possible to develop user-friendly applications. Structure and cell arrays that can hold any data type are presented. Structure and cell arrays are utilized for keeping PCB (printed circuit board) inventories. Since MATLAB includes many functions concerned with file management, an example shows how such functions can be used to manage a directory of musical WAV files and retrieve a particular song to play. Some MATLAB functions return a structure array for the symbolic solution of a set of equations. This is applied to find in symbolic form the output of a differential amplifier in terms of its two inputs.

Chapter 8 is about MATLAB support for data input and output (I/O) of binary and text data. Formatted I/O of many data types is discussed. In addition to input of binary and text data, it is also useful to input analog signals for further analysis. An example shows how to record an audio signal.

Chapter 9 is concerned with the MATLAB functions for 2-D and 3-D data visualization. Through colorizing, added annotation, viewpoint control, sizing, multiple plotting, plotting style, animation and much more, many fundamental principles of EE and CE can be better understood. MATLAB functions to create many types of 2-D and 3-D plots are presented. With MATLAB functions or the plot edit GUI, all plot details can be customized. It is also convenient that MATLAB, by using default options, creates useful plots given only the plot data. Plots can be saved or exported to other documents in many image formats. Examples show (1) the output of a digital to analog converter given an input WAV file, (2) the real-time (but slowed down) operation of a serial binary adder, (3) the sweep frequency input and output of a filter, and (4) how to make movies, and more.

Debugging is an unavoidable part of programming, so it is covered in Chapter 10. MATLAB includes a helper program that automatically runs in the background to inform programmers of syntax programming errors and to suggest fixes. By code folding or setting breakpoints, the programmer can single step through any line or segment of a program and check intermediate results. Using an in-depth example throughout the chapter, many MATLAB facilities for debugging MATLAB code are demonstrated.

Chapter 11 is about symbolic computing, where solutions of linear and nonlinear problems can be found in symbolic form instead of numeric form. The chapter starts by describing how to define symbolic objects. Then, throughout the chapter, many built-in functions with symbolic alternatives are discussed, including complex arithmetic, matrix inversion, series summation, integration, and differentiation, to mention only a few. Symbolic integration is applied to find the Fourier series coefficients of a periodic signal. By using rational arithmetic instead of floating point arithmetic, it is shown that with the MATLAB variable precision arithmetic (VPA), one can compute to any precision. With the built-in function **solve**, one can find the solution of sets of linear and nonlinear algebraic equations. The solve function is applied to symbolically find the output in terms of the two inputs of an instrumentation amplifier. It is also applied to symbolically solve Kirchhoff's equations of an RLC circuit to find the circuit transfer function. The built-in function **limit** is useful to find symbolically the limit of a function as its variable approaches a point, and the unit step and unit impulse functions are described and studied with the limit function.

Supplemental Chapter Synopsis

The remaining two chapters cover topics that require the knowledge obtained from the typical first three of four calculus sequence of courses. It would be quite rare to include these chapters in an introductory course syllabus for freshman or sophomore students, but advanced students and professionals would certainly benefit from studying them. After completing Chapter 12, the ECE student will be well prepared to apply MATLAB in a first signals and systems course and beyond.

Chapter 12 starts by presenting fundamentals of signal representation. The discrete Fourier transform (DFT) is developed. Many of its properties are examined with numerous MATLAB examples that employ the built-in function **FFT** (fast Fourier transform). The time domain version of the sampling theorem is presented. MATLAB is applied to do spectral analysis of both stationary and non-stationary signals. Chapter 12 continues by presenting, from basic principles, time domain solution methods of linear differential equations and linear difference equations. Many MATLAB functions are introduced and applied to continuous and discrete time linear system analysis, including finding the impulse response, transfer function, eigenvalues (natural frequencies), frequency response, convolution, and conversion to state variable descriptions. MATLAB also includes several ordinary differential equation (ODE) solvers, which are applied to analyze both linear and nonlinear circuits. Over 25 examples and 20 programs are included to demonstrate the significant role that MATLAB can play in the academic carriers of ECE students. Some examples are concerned with FIR (finite impulse response) and IIR (infinite impulse response) digital filters to demonstrate that a difference equation, which is implemented with a computer, can exhibit frequency selective behavior like an analog circuit.

Chapter 13 is an introduction to Simulink®. Simulink uses a graphical user interface (GUI) for building models of dynamic systems with building blocks from an extensive library of building blocks included in Simulink. The chapter starts by examining the Simulink environment, model editor, and how to navigate among the libraries of building blocks. Then, model building is demonstrated by drag and drop of blocks and interconnecting them. It is shown how one can build their own library of custom blocks and use them in the same way as built-in blocks. Simulink is used to simulate an RLC circuit, and it is used to compare the performance of an open-loop system to a closed-loop control system.

Appendix A describes how to generate hardcopy of MATLAB work.

Chapter Exercises

Each chapter includes numerous end-of-chapter problems to exercise the reader's understanding of the material. Most problems require writing MATLAB programs. Answers to selected problems are given in Appendix C. By adopting instructor's request, a solution manual is available from the publisher.

How the Book Can be Adapted

The coverage of this book (Chapters 1–11) may be too extensive to be completed in a one term course. Course lengths, student abilities, and instructor objectives vary from program to program. An instructor can decide which sections of this book to require, while skipping, if necessary, others or assigning them for self-study. I would offer the following suggestions for trimming the coverage:

- All of Chapter 1 (MATLAB Environment) must be covered.
- In Chapter 2 (Programs and Functions), the section concerned with MATLAB functions need only be used to introduce anonymous and primary functions and perhaps the built-in function **eval**.

- In Chapter 3 (Matrices, Vectors, and Scalars), once MATLAB has been applied to resistive circuit analysis, the remainder of this chapter, which presents the most mathematically challenging material in this book, can be skipped without loss of continuity.
- All of Chapter 4 (Program Flow Control) must be covered.
- Depending on interest, the material about quantization error in Chapter 5 (Binary Data) can be skipped. It would be very practical to return to Chapter 5 later, if time permits.
- All of Chapter 6 (Complex Numbers) should be covered.
- The material in Chapter 7 (Character Data) about structure and cell arrays can be skipped.
- All of Chapter 8 (Input/Output) should be covered.
- All of Chapter 9 (Graphics) should be covered, at least because it is so much fun.
- All of Chapter 10 (Debugging), which is very short, must be covered, perhaps on a self-study basis.
- Chapter 11 (Symbolic Computing) is very interesting, and should be covered if time permits.

Once the assigned material is covered, it is likely that skipped material will be investigated by students on their own, because they will have come to appreciate the convenience and utility of MATLAB for practical problem solving in the entire EE and CE curricula.

Final Notes

There is much literature available about MATLAB. A primary source of material about MATLAB is MathWorks, Inc., the manufacturer of MATLAB. At their website one can find many audio/video tutorials that will help to learn MATLAB. Anyone can also download a 30 day free trial of MATLAB, and a student version of MATLAB that includes many toolboxes, and Simulink can be purchased for less than the average cost of a textbook.

I began to use MATLAB a long time ago, mainly because of its graphic capabilities, ease with which it works with matrix and complex data types and the intuitive syntax of the language. Over the years and now, MATLAB has evolved continuously. As new algorithms are invented in virtually all fields of science and engineering, MATLAB grows to implement these algorithms with new built-in functions. Learning to use MATLAB is a worthwhile investment of time, whether you do it with this book or any other way. However, in all cases, access to MATLAB and learning by trial and error are absolutely necessary.

Roland Priemer

Acknowledgments

After a long career of teaching and working in industry, it is virtually impossible to mention the many colleagues and students who have influenced me. However, a long time ago, as an undergraduate, I took a course taught by Prof L. C. Peach, whom I will never forget. He had an uncanny ability to know what questions must be answered in his beautifully prepared lectures. He was a kind and gentle person, very knowledgeable and always ready to help. As a teacher, he was my role model.

I want to thank Dudley Kay, publisher and editor, of SciTech Publishing who so strongly encouraged me to write this book. If I had known about the considerable effort it would take to produce a manuscript to his and my liking, I am not sure that I would have gotten started. After some thought about the approach I would take to MATLAB, I am extremely glad to have been given the opportunity to present the subject in a new way. I also want to express my appreciation to the many reviewers for their helpful suggestions.

CHAPTER 1

MATLAB® Environment

MATLAB®, which means **Matrix Laboratory**, is a computing tool that is widely used by professionals in Electrical Engineering (EE), Computer Engineering (CE) and many other fields. MATLAB consists of two linked parts. The first part is the programming language, which consists of only a few basic programming constructs. The second part is the library of an enormous number of built-in programs, called functions, that perform operations commonly required in many fields of engineering and science. For this reason, MATLAB will improve an engineering student's academic and professional productivity.

MATLAB can work with a variety of data types, including integers, real numbers, complex numbers, characters, and logic variables, which can all be organized into arrays and other structures. MATLAB is an interactive system, where intermediate results can be accessed as programs are developed. You can develop solutions of problems for application in other MATLAB programs. One of its outstanding features is the capability to produce a variety of two- and three-dimensional graphic outputs that can be exported into documents. Also, MATLAB provides extensive built-in and online tutorials, demonstrations, and documentation to help you learn MATLAB programming.

After you have completed this chapter, you will know how to

- start MATLAB and bring the MATLAB desktop to its default state
- navigate through some of the windows in the MATLAB system
- use MATLAB in its immediate mode of computing for some of your work
- use the extensive MATLAB help facility

1.1   Default MATLAB® Desktop

It is assumed that you have access to a computer that has MATLAB installed in it. This book is based on MATLAB version 7.14 (R2012a), which is a 64-bit MATLAB running on

Windows 7. To launch MATLAB, double-click on the MATLAB icon on your computer's desktop. You can also left-click on the computer's start button, move to the programs menu, and navigate to MATLAB from there.

After launching MATLAB, the MATLAB desktop will appear on the screen. The desktop consists of one or more windows and one or more toolbars positioned above the windows, depending on the appearance of the desktop when MATLAB was last shut down. Through this *Graphical User Interface* (GUI), you can conveniently access all of the resources and features of MATLAB.

To bring the desktop to its default state, left-click the MATLAB desktop button in the **menu bar** located at the top of the desktop and follow the highlighted menus shown in Fig. 1.1. Hereafter, a sequence of button (menu) selections like the sequence shown in Fig. 1.1 will be given by

$$\textbf{Desktop} \ \rightarrow \ \textbf{Desktop Layout} \ \rightarrow \ \textbf{Default}$$

The default MATLAB desktop is shown in Fig. 1.2. The default desktop consists of four windows, the ***Command Window***, the ***Current Folder Window***, the ***Work Space Window***, and the ***Command History Window***. Only one window can be active at a time. To activate or select a window, left-click anywhere within the window. In Fig. 1.2, the *Command Window* is the active window, as indicated by the highlighted **title bar**. Below and associated with the *Current Folder Window* is a window that shows the first comment line that

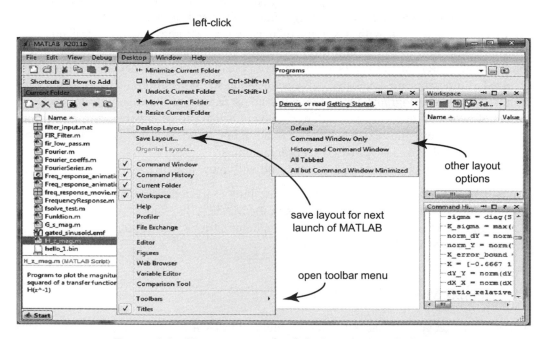

Figure 1.1   How to set up the default MATLAB desktop.

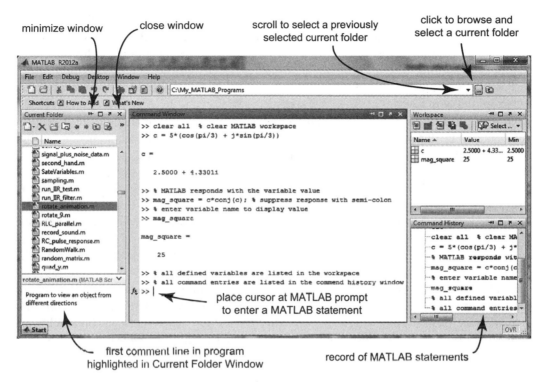

Figure 1.2   The default MATLAB desktop.

appears in the program that is highlighted in the *Current Folder Window*. This window gives you a quick peek at the highlighted program to help you find the program that you want to work on. Soon, these windows will be described in more detail.

MATLAB uses a part of computer memory, called the **Workspace**, where it stores the names of variables that you define while using MATLAB and the results of various kinds of computations that you cause MATLAB to perform. The size of this Workspace can be as large as the available memory that is not used by other applications.

At any time you can quit MATLAB with

File → Exit MATLAB

Or, you can quit MATLAB by typing the command **quit** in the *Command Window*. This command shuts down MATLAB in an orderly manner, and it is one of many built-in MATLAB functions that you can invoke by merely including its name in a MATLAB statement.

You can use MATLAB in two different modes. One mode is the **immediate mode** (also called **Command Line mode**). In this mode, a MATLAB statement is entered in the *Command Window*, and after you depress the keyboard **Enter Key** the action specified by the MATLAB statement is immediately executed. For example, in Fig. 1.2, four executable MATLAB statements were entered in the *Command Window*. The first statement consists of the built-in function **clear** with **all** as its operand, which causes MATLAB to delete from its Workspace all variable names and their associated values. This statement continues with the percent symbol **%**, followed by text. The percent symbol causes MATLAB to ignore the following text until the end of the line. With the percent symbol you can append a **comment** to a MATLAB statement. The second line consists of a MATLAB statement that assigns a complex number to a variable, followed by a comment. Complex numbers will be discussed in Chapter 6. Notice that the result is displayed immediately after you depress the Enter Key. This mode of operation is similar to the way we use a calculator. However, as you will see, MATLAB is much more powerful.

The other mode of MATLAB operation is the **program mode**. In this mode you can write, with an editor, a sequence of MATLAB statements that can become a MATLAB program. A sequence of MATLAB statements is called a **script**. A script can be made into a MATLAB program by saving it as a file with a name that ends in *.m* as a suffix, resulting in a file that is called an **m-file**. MATLAB provides an *Editor Window* in which you can write MATLAB scripts.

In the MATLAB desktop (see Fig. 1.2), there is an option for you to specify a folder as the *Current Folder*, where files can be stored and retrieved without specifying a complete path name. If you save an m-file in the *Current Folder*, then the program can be invoked by typing the program's file name, but without the *.m* suffix, in the *Command Window*. The program mode of operation and the creation of m-files will be discussed in Chapter 2.

1.2   Quick Start

Before we continue our exploration of the MATLAB environment, let us see how easy it is to obtain useful results with MATLAB.

Example 1.1 _____

The following sequence of MATLAB statements is an example of a MATLAB script. The purpose of this script is to create a plot of a function of time.

```
1     % MATLAB statements to plot the function v(t) = A cos(w t + B) for 0 ≤ t ≤ 1 sec
2     % specify parameters
3     t_start = 0.0  % plot start time
4     t_end = 1.0  % plot end time
5     A = 2  % amplitude of v(t)
6     B = pi/12  % phase of v(t) in radians
7     % MATLAB automatically replaces pi with the number: 3.1415926 ···
8     f = 2.0  % frequency in hertz (cycles/sec) of the sinusoidal time function
9     w = 2.0*pi*f  % frequency conversion from cycles/sec to radians/sec
10    T = 0.01  % plot time increment in seconds
```

```
11      % evaluate the function v(t) over the range t = t_start to t = t_end
12      t = t_start : T : t_end  % specify a sequence of time points incremented by T
13      v = A * sin(w*t + B)  % evaluate v(t) for all specified time points
14      vt_plot = plot(t, v, 'k') % use black, 'k', for plot color
15      grid on  %  include a grid in the plot
16      xlabel('time - secs')  % put a label along the abscissa
17      ylabel('voltage')  % put a label along the ordinate
18      title('Example of plotting with MATLAB')  % put a title above the plot
19      saveas(vt_plot, 'plotting_example.emf')  % save the plot named vt_plot to a file
```

Some statements need little explanation, while others cause action that is more complicated. All of these statements were copied from here, without the line numbers, and pasted into the *Command Window*, and MATLAB immediatcly executed them. Let us study these MATLAB statements, and see how they cause the result given in Fig. 1.3.

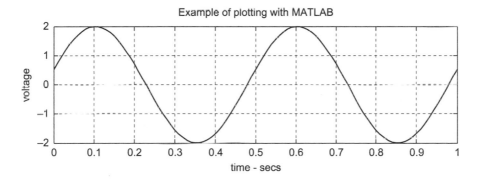

Figure 1.3 Plot obtained with MATLAB.

Line 1 is a comment to explain the purpose of this script. MATLAB ignores all characters following the percent symbol % until the end of the line. Line 2 is another comment line. Line 3, which is called an **assignment statement**, assigns the value given on the right side of the equal sign (**assignment operator**) to the MATLAB variable name t_start, which you choose, given on the left side of the equal sign. In MATLAB, a variable name cannot contain spaces, and to make a variable name that is made out of several terms readable, underscores can be used to connect terms. The variable name and its associated value are placed in the MATLAB Workspace. This is analogous to you writing, t_start = 0.0, on a piece of paper. Then, if at a later time you need to know the value of the variable t_start, you can look for the name t_start on the piece of paper to find its value. Lines 4–6 assign parameter values to variable names that help us to remember the purpose of each parameter. Line 7 is a comment line. Lines 8–10 also assign parameter values to variable names. In lines 6 and 9, MATLAB automatically replaces the character string, "pi", with the number, 3.1415926 ⋯ . Since the variable f has been assigned a value in line 8, MATLAB can perform the multiplication (symbolized with an asterisk) on the right side of the equal sign in line 9 by finding the value of f in the Workspace. Notice that comments are appended to these assignment statements to explain the purpose of the parameters.

To plot the function v(t) versus t, a set of values of t must be specified, and for each value of t in the set, v must be evaluated to obtain a set of corresponding values of v. For example, if t_1 is the first value of t, then $v_1 = v(t_1)$ is the first value of v. A plot over a uniformly spaced set of time points, $t_1 \leq t \leq t_N$, can be made by connecting with straight lines the set of N points $((t_1,v_1), (t_2,v_2), (t_3,v_3), \ldots, (t_N,v_N))$. As the number N of points is increased, the plot will look smoother.

For convenience, a set of N numbers is written as a one-dimensional array of numbers enclosed in brackets, called a **vector**, which can be assigned to a variable name. For example, let x be a vector defined by x = [−1.5, 0.5, 2.5, 4.5]. This vector has N = 4 elements, where x(1) = −1.5 is the first element and x(4) = 4.5 is the last element. We can refer to any element of x with x(n), n = 1, ..., N, where the variable named n is the **element index**. In MATLAB, any valid name, to be described soon, can be used for a vector name and a vector index name. Vectors and other types of arrays will be discussed more thoroughly in Chapter 3.

Line 12 uses the **colon operator** to assign a range of numbers (time points) to the variable name t, which becomes a vector of numbers. The first element in the vector t is t(1) = 0.0, and the colon operator increments each element in t by T = 0.01 to find the next element in t until the last number t_end = 1.0 is reached. Generally, for some vector x the syntax of this assignment statement is given by

```
x = first_value : increment : last_value
```

Other syntax options for defining a vector with the colon operator will be described in Chapter 3. After MATLAB has executed line 12, the vector t is given by

```
t = [ t(1) t(2) t(3) t(4) ... ] = [ 0.0 0.01 0.02 0.03 ... 1.0 ]
```

where the delimiter between vector elements can be a comma or one or more blanks.

Since a vector is a part of the argument of the sine function in line 13, MATLAB automatically evaluates the sine function for each entry (**element**) in t to obtain a corresponding value (element) in the resulting vector named v, and we get

```
v(1) = A*sin(w*t(1) + B)
v(2) = A*sin(w*t(2) + B)
  ⋮
```

All of the variable names defined in this script are listed in the *Workspace Window*.

Line 14 uses the built-in MATLAB function, **plot**, to create a plot of v versus t. This assignment statement associates the resulting plot with the name vt_plot, or whatever other name you might choose. Line 15 uses the built-in MATLAB function, **grid**, with the operand **on** to cause MATLAB to place grid lines in the plot, and notice that MATLAB automatically determines their spacing. The built-in functions **xlabel**, **ylabel**, and **title** make it very easy to define and locate plot labels. When MATLAB executes the plot function, it automatically opens a ***Figure Window*** into which MATLAB inserts the plot and the output of functions concerned with plotting. This window is shown in Fig. 1.4, and within this window you can do much editing of figure attributes. Notice that MATLAB has automatically sized the plot, and inserted x-axis and y-axis tick marks. The last MATLAB statement (line 19) uses the built-in MATLAB function, **saveas**, to save the plot as an **enhanced metafile** (**.emf**) (one of many options), which can be inserted into a Microsoft Word® document file or opened with the Microsoft Paint® application. Many of the figures given in this book, such as Fig. 1.3, were made this way.

Another way to place a figure into a document file is to use the menu sequence

Edit → Copy Figure

in the *Figure Window* shown in Fig. 1.4. Then, you can paste the figure into a document, resulting in Fig. 1.3.

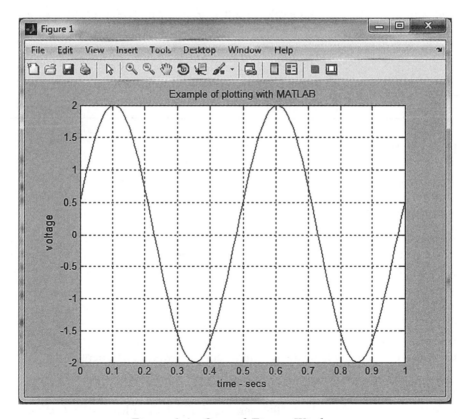

Figure 1.4   Opened Figure Window.

It is understood that at this point in this book the purpose of some of the operations introduced in this example script may not be totally clear. However, by the end of this chapter many of the questions that you may have will be resolved, and you will find that it is easy to obtain useful results with MATLAB. The built-in MATLAB functions introduced in this example will be discussed in more detail in the following chapters.

To place more than one MATLAB statement on a line, separate the statements with a semicolon. With a little experience, you will write the script more efficiently, as follows.

```
% MATLAB statements to plot a sinusoidal time function v(t) = A cos(w t + B) for 0 ≤ t ≤ 1 sec
A = 2; B = pi/12; f = 2.0; w = 2.0*pi*f; % four MATLAB statements define parameters of v(t)
t_start = 0.0; t_end = 1.0; T = 0.01; t = t_start : T : t_end; % specify plot time points
v = A * sin(w*t + B); % evaluate v(t) for all specified time points
% plot v(t) versus t, and use the color black for the plot
vt_plot = plot(t, v, 'k'); grid on; xlabel('time - secs'); ylabel('voltage');
title('Example of plotting with MATLAB'); % put a title above the plot
% save the plot named vt_plot to a file named "plotting_example.emf"
saveas(vt_plot, 'plotting_example.emf')
```

If you want to achieve the same plot result with the least typing effort, then the script can be

```
t = 0.0 : 0.01 : 1.0; v = 2.0*sin(4*pi*t + pi/2); plot(t,v,'k');
grid on; xlabel('time - secs'); ylabel('voltage');
title('Example of plotting with MATLAB');
```

1.3   Default MATLAB® Desktop Continued

Let us now investigate some of the features of the MATLAB desktop.

Command Window

The *Command Window* is the space where you can enter MATLAB statements for immediate execution (or action). Statements are entered at the **command prompt**, ">>". Depending on the kind of MATLAB statement and how you terminate it, the result (or output) is shown immediately following the statement (see Fig. 1.2).

Workspace Window

The *Workspace Window* (see Fig. 1.2) shows the names and types of all variables that you have defined in the immediate or the program mode of MATLAB operation.

Current Folder Window

The *Current Folder Window* shows the names of all of the files in the *Current Folder*. Generally, to save or load (store or retrieve) a file, a complete **path name** must be given

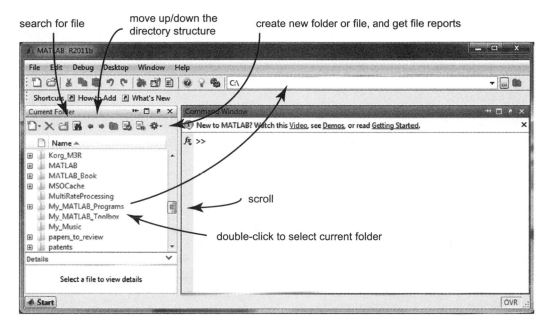

Figure 1.5   Selecting the current folder.

starting with the root directory. For convenience, when using a file directory that is selected as the *Current Folder*, only the file name must be given to store or retrieve a file. The *Current Folder* is the directory in which MATLAB looks first to find and retrieve a file, and it is the default directory where files are saved by MATLAB.

Within the *Current Folder Window*, you can change the *Current Folder* to another folder, such as a folder in a flash drive, to save and load files or results. To do this, make the *Current Folder Window* the active window. Below the highlighted title bar are icons to move up or down one level at a time in the directory structure of your computer, add a new folder, find files, and get directory reports. You can scroll to any folder, and make it the *Current Folder* by double left-clicking on its name. This is shown in Fig. 1.5. The resulting MATLAB desktop is shown in Fig. 1.2. Notice that the *Current Folder* name appears in the title window in the MATLAB desktop toolbar. A *Current Folder* can also be selected with the *Current Folder* button in the desktop toolbar (see Fig. 1.2).

Command History Window

The *Command History Window* (see Fig. 1.2) keeps a running log of MATLAB statements that have been entered in the *Command Window.*

Within the default MATLAB desktop all of the windows are connected (or **docked**). You can free (or **undock**) a window by left-clicking on the small arrow next to the "x" (close window button) located in the upper right-hand corner of a window (see Fig. 1.2). Once undocked, a reverse pointing arrow appears that can be left-clicked to dock the window. To close a window, left-click its close window button "x". You can minimize a window by left-clicking the arrow in the window title bar (see Fig. 1.2). With these options, you can configure your MATLAB desktop with MATLAB windows to suit your preferences. Your preferred layout can be saved by using

Desktop → Save Layout

After MATLAB has been launched, the initial appearance of the desktop will be like the appearance of the desktop just prior to previously shutting MATLAB down. You can clear the default desktop windows with the following menu selections.

Edit → Clear Command Window
Edit → Clear Command History
Edit → Clear Workspace

You can also clear the *Command Window* by right-clicking while the mouse cursor is within the window. A menu will appear that has the clear window option. Yet another way to clear the *Command Window* is by entering the built-in MATLAB function **clc** within the *Command Window*.

Located below the desktop menu bar is the MATLAB **toolbar**. The MATLAB toolbar is populated with a set of icons. The left most icon is the **new script** icon. You can find out the name of an icon by placing the mouse cursor over the icon, in which case, its name will appear. To execute the functionality of an icon, click on it. Many windows have toolbars. You can select the icons that you want to appear on a toolbar with, for example:

File → Preferences → Toolbars

which opens the *Toolbars Preferences Window*. Then, click on the toolbar pull-down menu, and select MATLAB. Check all of the check boxes to fully populate the desktop toolbar, and click the **Apply** button. As you become familiar with the other windows, you can select the icons that you want to appear in the toolbars of these windows. In the preferences list, notice all of the other options you have to control the appearance and operation of the MATLAB environment. As you become familiar with MATLAB, you may want to customize its environment.

If you have access to MATLAB, then it will be very useful to gain some experience about the topics that have been introduced so far by completing Practice 1.1.

Practice 1.1 _____

(a) Launch MATLAB, and bring the MATLAB desktop to its default state. Minimize the *Current Folder Window*.
(b) Select (make activate) the *Workspace Window*. Activate the *Command Window*.
(c) Click the *Edit* button and use the *Edit* menu to clear the *Command Window*, *Command History Window*, and the *Workspace Window*. Notice that within the *Command History Window* the present date and time are kept.
(d) To demonstrate the immediate mode of operation, at the prompt in the *Command Window*, type the MATLAB statement, "arg = pi/2", and then depress the Enter Key.

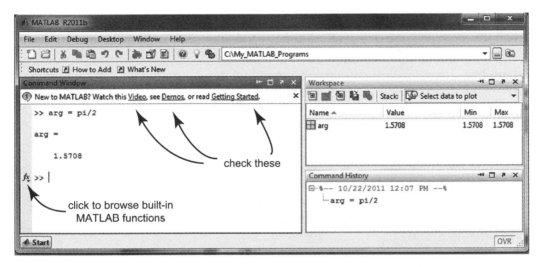

Figure 1.6   Using MATLAB in the immediate mode of operation.

Do not include the double quotes. The resulting MATLAB desktop is shown in Fig. 1.6. The character string, "pi", is a MATLAB **reserved word** that stands for the number π. This MATLAB statement is an assignment statement that assigns to the **character string**, "arg", which is a valid MATLAB variable name, the value resulting from evaluating the expression on the right side of the equal sign.

(e) Activate the *Workspace Window*. Undock the *Workspace Window*, and drag it elsewhere on your computer's desktop. Now dock it. The result should look like Fig. 1.6.

(f) You can activate any window by left-clicking on it. Or, click on the *Window* button in the desktop toolbar. This lists the open windows. To activate the *Command Window* you can also use

     Window $\rightarrow$ **Command Window**

Then, close the *Workspace Window* and *Command History Window*. Now, left-click on the *Window* button to see that only the *Command Window* is listed.

(g) Reinstate the *Workspace*, *Current Folder*, and *Command History Windows* with

     Desktop $\rightarrow$ **Workspace Window**
     Desktop $\rightarrow$ **Current Folder Window**
     Desktop $\rightarrow$ **Command History Window**

(h) Activate the *Command History Window*. You can transfer any previously entered MATLAB statement logged in the *Command History Window* to the *Command Window* by double-clicking the MATLAB statement. Do this with a MATLAB statement logged in the *Command History Window*.

(i) Activate the *Command Window*. Use the built-in function clc to clear the *Command Window*. Notice that previously entered MATLAB statements are still listed in the *Command History Window*.

Several additional windows will be introduced in the following chapters.

1.4 Built-in MATLAB® Functions

A built-in MATLAB function is a program that performs a particular task. You initiate execution of a function (program) by naming (invoking) it in a MATLAB statement. Some built-in functions require that you provide some input information, on which the function operates to obtain a result that is returned to you. For example, consider the following MATLAB statements:

```
>> input_argument = 9.0   % assign to a variable a value
>> sqrt_output = sqrt(input_argument)  % invoke the built-in function sqrt
```

The first MATLAB statement assigns to the variable named input_argument the value 9.0. The second MATLAB statement invokes a built-in function, **sqrt** in this case, which is a program that calculates the square root of the value assigned to input_argument. The program's name is sqrt, and its acronym was arbitrarily chosen by the developers of MATLAB to make it convenient to remember what the function does. To execute the function sqrt, include its name in a MATLAB statement, where you must supply input information. The function is executed, and the result is assigned to the variable name sqrt_output. The names input_argument and sqrt_output were arbitrarily chosen.

Some built-in MATLAB functions require no input information. When invoked (named in a MATLAB statement), such functions perform a particular task. For example, the built-in function **quit** requires no input and does not return an output. When named in a MATLAB statement, this function shuts down MATLAB in an orderly manner.

Some built-in functions perform very complex tasks, and may require much input information and provide much output information. The large number of built-in MATLAB functions makes MATLAB a powerful computational tool.

1.5 MATLAB® Variables

Unlike most other programming languages, MATLAB does not require any variable **type declarations** or **dimension statements** to accommodate variables. When MATLAB encounters a new variable name, such as the variable "arg", in Fig. 1.6, it automatically creates the variable name, which in computer memory is a sequence of binary codes for the characters in the name, and associates with it an appropriate amount of storage space where its value is stored. If the variable name already exists and its value is redefined in another MATLAB statement, then MATLAB changes the content of the associated storage space.

The memory space where variable names and their values are stored is the Workspace, and the content of the Workspace is shown in the *Workspace Window*.

MATLAB variable names can be any character string that consists of a letter, followed by any number of letters, digits, or underscores. Variable names cannot have any blanks in them. MATLAB is case sensitive. For example, the character strings "Arg" and "arg" are not the same variable names. Although the length of a variable name is not limited, MATLAB uses only the first N characters of the name, where N is found by using the built-in function **namelengthmax**, as in the MATLAB statement

```
N = namelengthmax
```

This, with the version of MATLAB used here, produces N = 63.

You can check if a character string is a valid MATLAB variable name with the built-in function **isvarname**. For example, the MATLAB statement

```
chk = isvarname ( 'voltage_1' )
```

produces the result chk = 1 (interpreted as true) to indicate that the character string, "voltage_1", is a valid MATLAB variable name. However, the MATLAB statement

```
chk = isvarname ( '2_current' )
```

produces the result chk = 0 (interpreted as false) to indicate that, "2_current", is not a valid MATLAB variable name.

MATLAB uses a large number of words, characters, or character strings, called **reserved words**, and they are used for operations, built-in MATLAB functions, and pre-defined constants. Some reserved words constitute the behavior of the MATLAB programming language, and therefore these reserved words are called **key words**, which cannot be used for any purpose other than their default MATLAB programming modality. You can check if a character string is a key word with the built-in function **iskeyword**. For example, the MATLAB statement

```
chk = iskeyword( 'end' )
```

produces the result chk = 1 (interpreted as true) to indicate that "**end**" is a key word.

You can use a reserved word as a variable name if it is not a key word. However, then the reserved word no longer means the intended built-in MATLAB function or predefined MATLAB constant. The use of a reserved word as a variable name is reversed with the built-in MATLAB function **clear**, as in the MATLAB statement

```
clear variable_name
```

This removes variable_name from the Workspace, and reinstates it as a reserved word.

All variable names and associated values are placed in the MATLAB Workspace, which is shown in the *Workspace Window*. Recall that all variable names and their values can be removed from the Workspace with the built-in function **clear all**.

All numbers are stored in computer memory using floating point notation with approximately 16 significant digits (called **double precision**) as specified by the IEEE floating-point standard. This finite precision gives a finite range of values from about 10^{-308} to about 10^{+308}. All computing is done in MATLAB using double precision. However, you can control the precision and style in which MATLAB displays numbers with the **format** built-in MATLAB function. For example, to display 16 digits in floating point notation, use the statement

```
format long e
```

The operand options are given in Table 1.1.

Table 1.1   Format function

Function operand	Example of a displayed number
bank	3.14
compact	Suppresses blank lines in output
hex	400921fb54442d18
long	3.141592653589793
long e	3.141592653589793e+000
long g	Will use long or long e
loose	Insert blank lines in output
rat	355/113
short	3.1415
short e	3.1415e + 000
short g	Will use short or short e

Example 1.2 ⎯⎯⎯⎯⎯⎯⎯⎯⎯⎯⎯⎯⎯⎯⎯⎯⎯⎯⎯⎯⎯⎯⎯⎯

The following MATLAB statements were executed in the immediate mode of MATLAB operation and copied from the *Command Window*:

```
>> clear all % delete all variable names and values from the Workspace
>> format compact
>> N = namelengthmax % this function has no input, but returns an output
N =
   63
>> chk = isvarname PowerSource
chk =
   1
>> % the character string, "PowerSource", is a valid MATLAB variable name
```

```
>> % the character string, "1_power", is not a valid variable name
>> chk = isvarname ('1_power')
chk =
    0
>> % the above MATLAB statement could have been: chk = isvarname 1_power
>> arg = pi/4 % MATLAB replaces the character string, "pi", a reserved word, by 3.14159 ···
arg =
    0.7854
>> format short e
>> % just naming a defined variable causes MATLAB to print its value
>> arg
 7.8540e-001
>> format long e
>> x = cos(arg)
x =
    7.071067811865476e-001
>> cos = pi  %  using the reserved word, "cos", as a variable name
cos =
    3.141592653589793e+000
>> x = cos(pi/2)  % this is an error
??? Subscript indices must either be real positive integers or logicals
>> % MATLAB gave the above error message
>> % the character string, "cos", has been used as a variable name
>> % MATLAB has interpreted, "cos(pi/2)", as reference to an element of a
>> % vector named, "cos", where the vector index must be a positive integer
>> % avoid using reserved words as variable names
>> clear cos % reinstate the character string, "cos", as a reserved word
>> x = cos(pi/2)  % ideally, the result should be zero
x =
    6.123233995736766e-017
```

The symbols for elementary arithmetic operations are given in order of precedence in Table 1.2.

Table 1.2   Elementary arithmetic operations

Operation symbol	Operation, example
( )	Control evaluation order, $z = (x+y)^{\wedge}w$
$\wedge$	Exponentiation, $z = x^{\wedge}y$
'	Complex conjugate, $z = x'$
*	Multiplication, $z = x*y$
/	Division, $z = x/y$, right slash
+	Addition, $z = x+y$
−	Subtraction, $z = x-y$

Table 1.3   Special reserved words

Variable	Returned by function
eps	Floating-point relative precision, $\varepsilon = 2^{-52}$
i	Imaginary number, $\sqrt{-1}$
inf	Infinity
j	Same as i
NaN	Not a number
pi	3.14159265...
realmax	Largest floating-point number, $(2 - \varepsilon)2^{1023}$
realmin	Smallest floating-point number, 2^{-1022}

Several special reserved words have predefined meanings or values. Some of these reserved words are given in Table 1.3.

1.6   MATLAB® Statements

The previous examples included a few different kinds of MATLAB statements, one of which is the **assignment statement**. There are only a few additional kinds of MATLAB statements. Some of these kinds of statements will be discussed now. Other kinds of MATLAB statements will be presented in several of the following chapters.

The MATLAB **assignment statement** has an intuitive mathematical structure, as given by

MATLAB_variable_name = MATLAB expression

On the left side of the equal sign (**assignment operator**) is any valid MATLAB variable name. On the right side of the equal sign is a MATLAB expression that can involve any previously defined MATLAB variables, MATLAB functions, or predefined MATLAB constants along with MATLAB operations. There can be any number of blanks within a statement to make it more readable. After the Enter Key has been depressed, MATLAB evaluates the expression and associates the result with the variable name. The result is also displayed in the *Command Window* (see Figs. 1.2 and 1.6). The result of the expression evaluation will determine the data type of the variable name. This information is given in the *Workspace Window*. Table 1.4 gives several basic data types. These and other data types will be discussed in later chapters.

The display of output in the *Command Window* can be **suppressed** by terminating an assignment statement with a **semicolon**, as in

MATLAB_variable_name = MATLAB expression ;

Now, you can reduce the clutter in the *Command Window*, and see only the statement outputs of interest. To aid debugging, you can cause the display of intermediate calculation

Table 1.4   MATLAB data types

Data type	Example
integer	x = int64(5), use 64 binary digits; could be: int8(5), int16(5) or int32(5)
real	x = 3.1415e+001
complex	x = 5 − j*2.6, j = $\sqrt{-1}$ (MATLAB also uses i = $\sqrt{-1}$); could be: x = complex(5,−2.6)
logical	x = logical(0), or x = logical(1)
character	x = 'any character string'; must be enclosed in single quotes
arrays	Matrices, cell arrays, structures
cell arrays	Multidimensional arrays whose elements are arrays
structures	Multidimensional arrays with elements accessed by textual field designators

results in a lengthy algorithm by omitting the semicolon at the end of only some assignment statements.

To conserve the number of lines in a script, multiple statements can be placed on the same line. To suppress the display of output, each statement must be terminated with a semicolon. To display output, use a comma instead of a semicolon to separate statements.

Sometimes a MATLAB statement is too long to fit on one line. Use an **ellipsis** (three periods),..., followed by the Enter Key to indicate that the statement continues on the next line. Use continuation (an ellipsis) to make a long MATLAB statement more readable.

Often it is useful to include explanations of the purpose of a MATLAB statement or set of statements. For this, a **comment statement** should be used. Recall that such a statement begins with the percent symbol, "%", and that MATLAB ignores everything between the percent symbol and the next Enter Key depression.

The value of a variable that has been assigned a value by a previously executed MATLAB statement can be displayed by typing the variable name in the *Command Window*, as in

MATLAB_variable_name

Let us call this kind of a statement a **query statement**. Similarly, the result of evaluating an expression need not be assigned to a MATLAB variable. In this case, by default, MATLAB assigns the result of evaluating an expression to a generic variable named **ans**. For example:

```
>> my_pi = 4*atan(1)  % expression result is assigned to a variable name
my_pi =
3.1416
>> 4*atan(1)  % by default, expression result is assigned to the variable name ans
ans =
3.1416
```

Practice 1.2

The usage of the symbols in Table 1.2 and the special reserved words given in Table 1.3 is demonstrated with the following MATLAB assignment statements. If you have access to a computer running MATLAB, then it will be very useful for you to activate the *Command Window* and enter the following MATLAB statements at the ">>" prompt:

```
>> a = cos(pi/4) % the character string, "pi", is used as a reserved word
a =  7.0711e-001
>> b = sin(pi/4)
b =  7.0711e-001
>> c = a + j*b % the character, "j", is reserved to mean the square root of -1
c =  7.0711e-001 +7.0711e-001i
>> x = 2^-52 % x is assigned the smallest difference between two floating point numbers
x =  2.2204e-016
>> x - eps % the character string, "eps", is a reserved word
ans =    0
>> undefined_big = inf % the character string, "inf", is a reserved word
undefined_big =   Inf
>> 1/undefined_big % MATLAB gives a useful result
ans =    0
>> y = 1/0 % MATLAB gives a useful result
y =   Inf
>> z = 0/0 % the specified computation is indeterminate
z =   NaN
>> cos = pi % using a reserved word as a variable name
cos =   3.1416e+000
>> format hex % require output to be given in hexadecimal notation
>> small = 2^-1022 % assign the smallest positive number to a variable name
small =   0010000000000000
>> % each hexadecimal digit is a 4-bit binary number
>> big = (2 - eps)*2^1023 % assign the biggest positive number to a variable name
big =   7fefffffffffffff
>> format short e
>> small % display small using short floating point notation
small =  2.2251e-308
>> big % display big using short floating point notation
big =  1.7977e+308
>> clear cos  % reinstate the character string, "cos", as a reserved word
>> cos(pi)
ans =    -1
```

A **directive statement** usually uses a built-in MATLAB function to cause MATLAB to perform a housekeeping kind of task. Three examples are **clear all**, **clc**, and **quit**.

You can write MATLAB statements using either a **function format** or a **command format**, as described below. This will also be discussed in the next chapter.

A MATLAB statement in the **function format** consists of the function name followed by open parenthesis, one or more arguments separated by commas and closed parenthesis. This format is

```
functionname(arg1, arg2, ···, argn)
```

You can assign the output of a function to one or more, depending on the function, output variable names separated by commas and all enclosed in square brackets, as in

```
[out1, out2, ···, outm] = functionname(arg1, arg2, ···, argn)
```

With the function format, arguments are passed to the function by value. See the examples below.

A MATLAB statement in the **command format** consists of the function name followed by one or more arguments separated by spaces. This format is

```
functionname arg1 arg2 ... argn
```

Unlike the function format, you cannot assign the output of the function to a variable. Attempting to do so generates an error and a message. Arguments are treated as character strings. See the examples below.

In the function format, arguments are passed by value. In the command format, arguments are treated as character strings. Consider the following examples. The formats are

```
disp(A); % passes the value of variable A to the built-in disp function
disp A; % passes the variable name, "A"
```

For example, let A = pi, and we get

```
>> A = pi;
>> disp(A);  % function format, value of argument is passed
3.1416

>> disp A; % command format, character string is passed
A
>> % instead, you could also have entered disp('A')
```

In the next example, let: str1 = 'one' and str2 = 'one'. We will use the built-in MATLAB function **strcmp**, which compares character strings. With the variables str1 and str2 we get

```
>> strcmp(str1, str2); % function format, compares, "one" and "one"', their values
ans =
    1
>> % true, equal or the same
>> strcmp str1 str2; % command format, compares "str1" and "str2"
ans =
    0
>> % false, unequal or not the same
```

When using the function format to pass a character string to a function, you must enclose the string in single quotes. For example, to display the character string, "Press Enter", use

```
>> disp('Press Enter')
    Press Enter
```

On the other hand, a variable to which a character string has been assigned does not need to be enclosed in quotes. In the function format we get

```
>> directive = 'Press Enter';
>> disp(directive)  % value of argument is passed
    Press Enter
```

However, with the command format we get

```
>> disp directive
    directive
```

It may occur that you have entered and executed several MATLAB statements, and you would like to execute one of them again, perhaps after some editing. While in the *Command Window*, you can scroll up or down (use the up and down arrow keys on the keyboard) through previously entered lines to recall, edit (use left and right arrow keys on the keyboard for position), and execute again previously entered MATLAB statements.

Practice 1.3 _____

The following examples were copied from the *Command Window* and pasted here. It would be useful for you to activate the *Command Window* and enter each of the following statements:

```
>> clear all   % clear the workspace of all variables and values
>> time_now = clock; % get the date and time with the built-in MATLAB function clock
>> % the built-in function clock returns a vector with 6 elements defined as follows
>> % [Year Month Day Hour Minute Second]
>> disp time_now % command format
```

```
time_now
>> disp(time_now)  % function format
 1.0e+003 *
 Columns 1 through 6
  2.0120   0.0050   0.0070   0.0080   0.0030   0.0144
>> time_now(1)  % the first entry in the vector time_now is the year
ans =    2012
>> time_now(2)  % display month
ans =    2
>> time_now(3)  % display day
ans =    21
>> time_now(4)  % display hour
ans =    11
>> time_now(5)  % display minute
ans =    18
>> time_now(6)  % display second
ans =    5.7593e+001
```

1.7   MATLAB® Elementary Math Functions

Some elementary mathematical built-in functions are given in Table 1.5. It is not possible to include in this book an explanation of the usage of even a small fraction of the large number of built-in MATLAB functions. You must become accustomed to using the extensive MATLAB help facility.

Table 1.5   Elementary math functions

Trigonometric	
sin	Sine
sinh	Hyperbolic sine
asin	Inverse sine
asinh	Inverse hyperbolic sine
cos	Cosine
cosh	Hyperbolic cosine
acos	Inverse cosine
acosh	Inverse hyperbolic cosine
tan	Tangent
tanh	Hyperbolic tangent
atan	Inverse tangent
atan2	Four quadrant inverse tangent

(Continues)

Table 1.5   (*Continued*)

Trigonometric	
atanh	Inverse hyperbolic tangent
sec	Secant
sech	Hyperbolic secant
asec	Inverse secant
asech	Inverse hyperbolic secant
csc	Cosecant
csch	Hyperbolic cosecant
acsc	Inverse cosecant
acsch	Inverse hyperbolic cosecant
cot	Cotangent
coth	Hyperbolic cotangent
acot	Inverse cotangent
acoth	Inverse hyperbolic cotangent
hypot	Square root of sum of squares

Exponential	
exp	Exponential
log	Natural logarithm
log10	Common (base 10) logarithm
log2	Base 2 logarithm and dissect floating point number
pow2	Base 2 power and scale floating point number
realpow	Power that will error out on complex result
reallog	Natural logarithm of real number
realsqrt	Square root of number greater than or equal to zero
sqrt	Square root
nthroot	Real n-th root of real numbers
nextpow2	Next higher power of 2

Complex	
abs	Absolute value
angle	Phase angle
complex	Construct complex data from real and imaginary parts
conj	Complex conjugate
imag	Complex imaginary part
real	Complex real part
unwrap	Unwrap phase angle
isreal	True for real array
cplxpair	Sort numbers into complex conjugate pairs

Rounding and remainder	
fix	Round toward zero
floor	Round toward minus infinity
ceil	Round toward plus infinity
round	Round toward nearest integer
mod	Modulus (signed remainder after division)
rem	Remainder after division
sign	Signum

Example 1.3 _____

```
>> arg = pi/3  % MATLAB replaces pi with the number 3.1415926 ···
arg =    1.0472
>> y = cos(arg)
y =    0.5000
>> acos(y)
ans =    1.0472
>> x = 1000;  power_of_ten = log10(x)  % the display of the first assignment is suppressed
power_of_ten =    3
>> x = 6.5;
 >> two_power = nextpow2(x)  % get the smallest power of 2 number greater than x
two_power -    8
>> z = complex(-4,5)  % assign a complex number to z
z =    -4.0000 + 5.0000i
>> conj(z)  % conjugate the complex number assigned to z
ans =    -4.0000 - 5.0000i
>> x = -2.37;
>> fix(x)  % round toward zero
ans =    -2
>> floor(x)  % round to integer less than or equal to x
ans =    -3
>> ceil(x)  % round to integer greater than or equal to x
ans =    -2
>> mod([0:5],3)  % remainder after division by 3 of each element of vector [0 1 2 3 4 5]
ans =    0    1    2    0    1    2
```

Many more built-in MATLAB functions will be introduced in the following chapters.

1.8 Help Facility

MATLAB provides much documentation, audio/video tutorials, and demos to help you learn how to use it. In the *Command Window*, type in the built-in function **demo**, and then select and watch the audio/video tutorial concerned with the MATLAB environment. This is an excellent way to help you get started. If you do not have MATLAB, then you can access this information online at the Mathworks, Inc. website. Mathworks, Inc. is the manufacturer of MATLAB. You can also download from this website a free 30-day trial version of MATLAB.

If you do have MATLAB, then you can access help from the default MATLAB desktop with the F1 key. Or, you can follow

Help → **Product Help** → **Contents**

resulting in the *Help Window* (**help browser**) shown in Fig. 1.7. From this window there are many options.

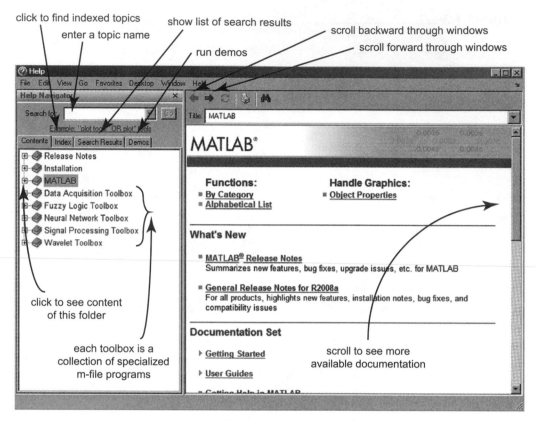

Figure 1.7   Window to start searching for help.

The MATLAB documentation set is extensive. Scroll down to see the many possibilities. I strongly urge you to go online to access audio/video tutorials about MATLAB. Also, you can see the MATLAB folder content by clicking on its plus button, and then click on items of interest. If you have a particular item in mind, but do not know its exact name, then click the "Index" tab for an alphabetical list that you can search. Or, click on the "Alphabetical List" option. To have MATLAB search for a topic, enter a topic name in the "Search for" window, and then click the "Go" button. You can then see all results of the search by clicking the "Search Results" tab. MATLAB includes an extensive set of demos with m-files that you can access by clicking the "Demos" tab. By example, these demos will answer many detailed questions that you may have about MATLAB programming.

MathWorks, Inc. also markets collections of specialized programs and m-files, called toolboxes. Toolboxes were developed by experts in their fields for solving problems in these fields. For electrical and computer engineers there are toolboxes concerned with, for example, communication, control, filter design, image processing, database management, signal processing, and neural networks to mention only a few. Several toolboxes are included in the student version of MATLAB.

Example 1.4 _____

To illustrate finding information, let us find information about the **atan2** built-in function by
letting MATLAB search for it. First, activate the *Help Window*, as shown in Fig. 1.7, and then
type into the "Search for" window the name of the function (or topic). After clicking the
"Go" button, MATLAB returns the atan2 help window, a part of which is shown in Fig. 1.8.
Here we see a description of what the atan2 function does and examples of how to use it.

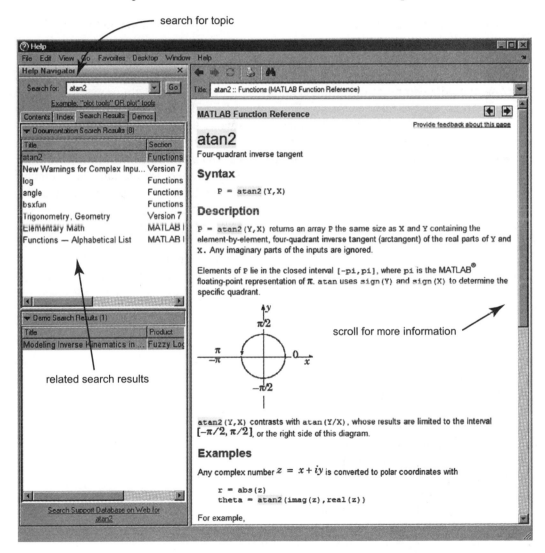

*Figure 1.8   Result of searching for help about the **atan2** built-in function.*

A search result usually includes syntax information, topic description, examples, computing method, and related built-in MATLAB functions. Furthermore, instead of going through the *Help Window*, you can access help directly from the *Command Window* by using the built-in MATLAB function **help** with the command format. If you know the name of the topic for which you want help, then, for example, you can enter

```
>> help atan2  % using command format
```

The help command has many options. To see the possibilities, type **help help** in the *Command Window*. A particularly useful help option is the command **help //**, which lists all of the operators and special characters used by MATLAB.

Another way to access information is with the MATLAB function **doc**. In the *Command Window*, type the command doc without an argument to open the help browser, shown in Fig. 1.7. To access the reference page of a function from the *Command Window*, type the command **doc function_name,** where function_name is the name of the function about which you want more information. Like the command help, the command doc has many options. To see the possibilities, type **help doc** in the *Command Window*.

If you are not sure of the name of a built-in function, then use the **lookfor** command with whatever word or term you think is relevant, for example, enter **lookfor tan**, and then MATLAB will look through its entire set of function descriptions to find and show you every occurrence of tan. In MATLAB, the instant access to help information is particularly convenient.

As you continue to work with this book, regularly browse through the alphabetical list of the built-in MATLAB functions to increase your knowledge about the many tools that are available to help you solve problems.

1.9   Conclusion

In this chapter we explored the MATLAB environment. There is much more to find out. We found it easy to launch MATLAB, get to the default desktop, and navigate in the default MATLAB desktop. You should now know

- how to set up the default MATLAB desktop
- how to dock and undock a window
- the purpose of several MATLAB windows
- how to use MATLAB pull-down menus
- what is meant by a built-in MATLAB function
- how to form valid MATLAB variable names
- the structure of a MATLAB statement
- how to suppress display in the *Command Window*

- how to use some of the built-in functions in the immediate MATLAB operation mode
- what is meant by a script
- how to scroll through previously entered MATLAB statements and how to retrieve statements from the *Command History Log*
- how to utilize the MATLAB help facility
- how to do some useful work in the MATLAB immediate operation mode

Table 1.6 lists the built-in MATLAB functions that were introduced in this chapter. Also see Tables 1.1 through 1.5. You can obtain detailed explanations about all of these built-in functions with the MATLAB **help** function.

In the next chapter, you will learn how to create MATLAB programs and functions.

Table 1.6 Built-in MATLAB functions introduced in this chapter

Function	Brief explanation
clc	Clear *Command Window*
clcar all	Clear Workspace
clear variable_name	Remove a variable from Workspace
clock	Get date and time
demo	Access the many audio/video tutorials
disp	Display text or variables
format	Specify how numbers are displayed
grid on	Insert a grid into a plot
help	Use the MATLAB help facility
iskeyword	Check if a character string is a key word
isvarname	Check if a character string is a variable name
lookfor	Let MATLAB search for a word or phrase
namelengthmax	Get maximum number of characters
plot	Create a plot
quit	Terminate MATLAB session
saveas	Save figure to a file
strcmp	Compare character strings
title	Place a title above plot
xlabel	Place a label along *x*-axis
ylabel	Place a label along *y*-axis

Further reading

Besides using the MATLAB built-in **Help** facility as well as the documentation that comes with MATLAB, the following texts are general references that provide information about the capabilities of MATLAB.

Chapman, S.J., *Essentials of MATLAB programming*, Cengage Learning, Stamford, CT, 2009
Gilat, A., *MATLAB, an introduction with applications*, Wiley, Hoboken, NJ, 2011
Hahn, B.D., Valentine, D.T., *Essential MATLAB for scientists and engineers* (3rd edn.), Elsevier, Burlington, MA 01803, USA, 2007
Moore, H., *MATLAB for beginners*, Pearson Prentice-Hall, Upper Saddle River, NJ, 2012
Pratap, R., *Getting started with MATLAB, a quick introduction for scientists and engineers*, Oxford University Press, New York, NY, 2010

Problems

Section 1.1

1) List and give a one to two sentence description of the purpose of each window in the default MATLAB desktop.
2) In the desktop menu bar, click on the File button. List and click each item in the file menu, and give the title of the window or menu that opens. You need not take any action within these windows.
3) List the name and give a one to two sentence description of the activity of each icon in a fully populated desktop toolbar.

Section 1.2

4) Start with the default MATLAB desktop.
   (a) Enter the following statements in the Command Window:

```
>> clear all; clc; % clear the Workspace and the Command Window
>> f = 2; w = 2*pi*f; % specify a frequency in Hz and convert to rad/sec
>> T = 0.01; % specify a time increment
>> time = 0 : T : 0.5; % specify a vector of time points
>> % evaluate the sine function for each element of the vector time
>> x = sin(w*time);
>> plot(time,x) % plot x versus time
```

A Figure Window will open showing a plot. Activate the Command Window, and enter

```
>> grid on
```

Enter MATLAB statements to add the x-axis label, "seconds", y-axis label "sin (wt)", and title "Plot of a Sinusoidal Function". Activate the Figure Window. In the Figure Window obtain a print of the plot by clicking on the Print Figure icon in the toolbar.

    (b) Give the content of the Workspace and Command History Windows. How many elements does the vector x have?

5) Repeat Prob. 4(a), but replace the statement plot(time,x) with the statements

```
>> y = cos(w*time);
>> plot(x,y)
>> axis equal
```

You will also have to change axes labels and the title. Use help axis, and explain the purpose of the statement, axis equal. Before printing, enter an axis command to make the x and y axes limits −1.1 to +1.1.

Section 1.3

6) (a) In the desktop toolbar, click on the Current Folder pull-down menu, and give all, but no more than three of the complete path names.

    (b) Click on Browse for Folder button in the desktop toolbar. Highlight a folder in the Browse for Folder Window, where you want to save and retrieve your files. You should create a new folder; name it My_Files to use for this purpose. Then, click OK. Repeat part (a).

7) Enter the following statements in the Command Window:

```
>> clear all; clc;
>> arg - -pi:pi/100:pi; x = sin(arg);
>> plot(arg,x);
```

    (a) Give the content of the Command, Workspace, and Command History Windows.

    (b) Dock the Figure Window. Activate the Command Window, and enter commands to place a grid, axes labels, and title on the plot. In the Figure Window, use the print icon to print the figure.

Section 1.4

8) Give an example of a MATLAB function that (a) requires no input and does not return an output, (b) requires an input, but does not return an output, (c) requires no input, but returns an output, or (d) requires an input and returns an output. Explain the activity of each function.

Section 1.5

9) Explain the difference between a reserved word and a key word, and give an example of each. In the Command Window, use appropriate MATLAB functions to find out if the following names are valid MATLAB variable names and key words. (a) For, (b) for, (c) 2volts, (d) current-3, (e) cos, (f) case, (g) pi. Print your statements.

10) For the number x, $x = -23.034786$, use the format function to display x using each of the formats given in Table 1.1. Print your results.

11) In the Command Window, evaluate each of the following expressions. Use format long e. (a) $(x+3)/(x^2-3x+2)$, $x=-1$ and $x=1$, (b) $3e^{-2t}\cos(5\pi t - \tan^{-1}(\sqrt{3}/2))$, $t=0.1$, (c) $(-1)^{1/3}$, (d) $(-2)^{1/2}$, (e) log(0) (natural log), (f) log(-1), (g) realmin, (h) $(x-1)/\log(x)$, $x=1$. Print your results.

12) In the Command Window, evaluate each of the following expressions. Use format long e. (a) pi $= \sqrt{3}$; give a statement to reinstate pi, (b) $(-1)^{1/2}$, (c) 2^{1024}, (d) 2^{-52}, (e) eps, (f) $(\text{pi})^{1/2}$, (g) $(2 - \text{eps})^{1023}$, (h) $\cos^2(\pi/3) + \sin^2(\pi/3)$. Print your results.

Section 1.6

13) Give example MATLAB statements of the following kind: (a) assignment, (b) continuation, (c) query, (d) comment, (e) directive (other than clear all, clc or quit). Explain the difference between assignment statements that are terminated with a comma (or blank) or a semicolon.

14) In the Command Window, set EE = 'Electrical Engineering' and CE = 'Computer Engineering'.
   (a) Enter the statements strcmp(CE,EE) and strcmp EE CE. Do they produce the same result? What is compared in these two statements?
   (b) Give a statement that uses the disp function in the command format to display Electrical and Computer Engineering.

15) Write a one line MATLAB statement that assigns $f = 440$, $w = 2\pi f$, $T0 = 1/f$, $t = T0/4$, and $x = \sin(w\,t)$, where all output, except the output for t, is suppressed.

16) Using the function clock, give an assignment statement that assigns to D_T the date and time. Display all of the elements of D_T.

Section 1.7

17) Enter and explain the difference between atan(2/$-$2) and atan2(2, $-$2). Use the MATLAB help facility.

18) Enter and explain the result of each of the following statements:
   (a) $x = -3 - j*4$, (b) $y = \text{complex}(-3, -4)$, (c) $x\_r = \text{real}(x)$, (d) $x\_i = \text{imag}(x)$, (e) $z = \text{conj}(x)$.

19) Start with the default MATLAB desktop. Enter the following statements in the Command Window:

```
>> clear all; clc;
>> T = 0.05; t = 0 : T : 5;
>> x = exp(-t);
>> plot(t,x); hold on; plot(t,-x);plot(t,1-x);plot(t,x-1); hold off
>> grid on
```

Obtain a print of the plot. Explain the purpose of the statements hold on and hold off.

20) For $x = 6.5$, enter and explain the results of (a) fix(x) and fix$(-x)$, (b) floor(x) and floor$(-x)$, (c) ceil(x) and ceil$(-x)$, (d) round(x) and round$(-x)$.

21) Obtain and explain mod(x, y) for (a) $x = 3$ and $y = 7$, (b) $x = 7$ and $y = 3$, (c) $x = -5$ and $y = 2$.

22) Obtain and explain whether or not mod(x, y) and rem(x, y) produce the same results for (a) $x = 7$ and $y = 3$, (b) $x = -7$ and $y = 3$, (c) $x = 7$ and $y = -3$, (d) $x = -7$ and $y = -3$.

23) Given two numbers x and y, then in MATLAB, $z = x * y$ is the product of the two numbers. Suppose x and y are two vectors, each having N elements, then in MATLAB, $z = x. * y$ is also a vector with N elements, where the kth element of z is $z(k) = x(k) * y(k)$ for $k = 1, 2, \ldots, N$. Here, z is the element by element product of x and y.

(a) Start with the default MATLAB desktop. Enter the following statements in the Command Window.

```
>> clear all; clc;
>> T - 0.05; t = 0 : T : 10;
>> f = 2; w = 2*pi;
>> x = 1 - exp(-t).*sin(w*t+pi/2);
>> % each element by element product is subtracted from 1
>> plot(t,x);
>> grid on
```

Obtain a print of the plot.

(b) Add a statement to obtain a vector of values of $y(t) = e^{-t/2} - e^{-2t}$. Then, using y and values of sin(wt), redefine x to obtain a plot of x that looks like

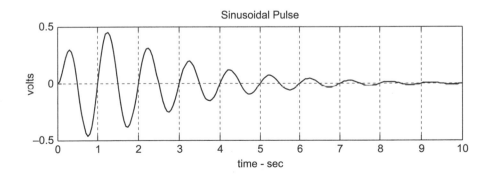

Section 1.8

24) In the Command Window, enter the line

```
>> tic, pause(5), toc
```

Then, get help for each of the three functions in this line with

Help → Product Help

Start by entering tic into the search box to get documentation about the tic function. Use the MATLAB help facility, and explain the operation of each function. You can also get help with Command Window statements such as help tic or doc tic.

25) In the Command Window, enter the command **demo**. A demo window will open. Then, watch the video tutorial, Getting Started with MATLAB. From what you have learned by watching this tutorial, give a MATLAB statement using brackets that assigns to a vector x the elements 5, -2, 3.4, and pi/4. To display the vector, type x in the Command Window.

CHAPTER 2

Programs and Functions

MATLAB® assumes that an **m-file**, a file ending with the suffix, ".m", contains MATLAB statements. There are two kinds of m-files: a **program** and a **function**. A program is created by writing a **script** of MATLAB statements, which is then saved as an m-file. All variables used in a program are stored in the Workspace, and a program can operate on all of the data in the Workspace. A program is executed by entering its file name in the *Command Window* or by including its file name in another program.

A function is created by writing a **function script** of MATLAB statements, which is also saved as an m-file. A function script must start with a function declaration statement. A function can accept inputs from a program or other function that invokes it and return outputs to the program or other function. All variables used within a function can be accessed only from within the function, and they are called **local variables**. A function uses its own work space, which is separate from the Workspace used by a program, to store and retrieve data. A function can be invoked by entering its name in the *Command Window*, in a program, or in another function. Several kinds of functions will be introduced in this chapter.

After you have completed this chapter, you will know how to

- create a program
- create and use functions
- use MATLAB in its program mode of operation
- create your own toolbox of function m-files that you can add to the MATLAB search path

2.1 Current Folder

When you include a program or a function file name in a MATLAB statement, you have the option to not provide a complete path name. To execute the statement, MATLAB must search for the file name. By default, the first place that MATLAB looks for this file name is in the *Current Folder*. The *Current Folder* is the folder with a complete path name given in the *Current Folder* title window located in the toolbar at the top of the MATLAB desktop (see Fig. 1.2). Referring to Fig. 1.2, MATLAB keeps a list of previously used *Current Folders*, and you can scroll through this list to select a path name to be used for the *Current Folder*. Or, you can browse through all of your folders to select a path name to be used for the *Current Folder*.

Another way to select a *Current Folder* is to activate the *Current Folder Window* (see Fig. 1.5). Click on the left/right arrows in the toolbar of the *Current Folder Window* to scroll through the folders to select one as the *Current Folder*. Or, by selecting the new folder option in this window you can create a new folder within the folder of your choice to be used as the *Current Folder*. This allows you to organize the programs and functions that you develop.

If you invoke a program or a function, and MATLAB cannot find the file name in the *Current Folder*, then MATLAB searches a list of *Directories*, called the search path, which was created at the time MATLAB was installed in your computer. These *Directories* contain all of the built-in MATLAB functions. You can add directories to this list. If MATLAB does not find the file name in this list of *Directories*, an error message is generated. This will be discussed later in this chapter.

2.2   Program Development

Before you start to develop a new program, specify a *Current Folder*. To write a new script, start with the default MATLAB desktop and follow the menus given by

File → New → M-file

This will open the ***Editor Window*** as shown in Fig. 2.1, where a few MATLAB statements have been entered. This script is still untitled, because it has not yet been saved as an m-file. In this window you can enter, highlight, copy, cut and paste text, as with any text editor. You can write a script with any conventional word processor and copy and paste it into the

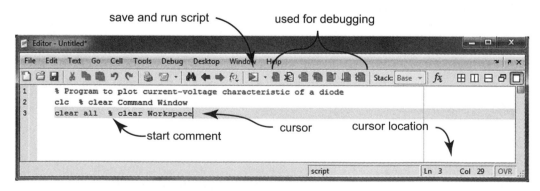

Figure 2.1   Editor Window for writing scripts.

Editor Window for program development, or you can paste the script into the *Command Window* for immediate execution. Notice that MATLAB automatically assigns a number to each line. For convenience, in the event of a programming error, MATLAB uses these line numbers to indicate the error location. As you develop a script, if you are in doubt about the syntax or meaning of any MATLAB operation, you can activate the *Command Window* at any time and use the built-in MATLAB function **help**. Or, you can click on the Help tool in the toolbar of the *Editor Window* (see Fig. 2.1).

2.3   Electric Current and Voltage

Before we continue MATLAB program development, let us examine some fundamentals of electrical and computer engineering. With this background, we will see how MATLAB can be applied to solve a wide variety of problems in these fields.

2.3.1   *Current*

Fig. 2.2 shows a section of a long straight conductor (copper wire, for example), where through a cross-section of the conductor electrons can move from point (b) to point (a) or in the reverse direction. The charge of an electron is $q_e = -1.60217646 \times 10^{-19}$ coulombs, and the charge of a proton is $q_p = -q_e$.

   With respect to points (a) and (b), the movement of negative charge from point (b) to point (a) is equivalent to the movement of positive charge from point (a) to point (b). In the conductor shown in Fig. 2.2, **electric current** $i(t)$ is the rate at which positive charge moves through the conductor cross-section. It is conventional to have current represent the movement of positive charge. The current $i(t)$ is positive (negative) if in a time interval Δt an amount of positive charge Δq has moved through the conductor cross-section in the direction that is the same as (opposite of) the current reference direction, shown by the arrow. The current is defined by

$$i = \frac{\Delta q}{\Delta t}$$

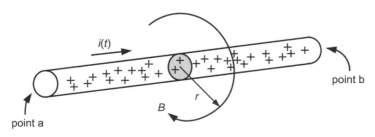

Figure 2.2   Charge moving through a cross-section of a long straight ideal conductor.

From a time interval to the next time interval, the amount and direction of positive charge moving through the cross-section may change, and we use the **instantaneous current** given by

$$i(t) = \lim_{\Delta t \to 0} \frac{\Delta q}{\Delta t} = \frac{dq(t)}{dt} \text{ coloumbs/second} \tag{2.1}$$

One coulomb/second is called an **ampere** (A).

Example 2.1

Fig. 2.3 describes the movement of the net amount of positive charge that has moved through a cross-section of a conductor, as depicted in Fig. 2.2. Prog. 2.1 shows the MATLAB script that generated this piecewise linear plot. Soon, you will see how to produce MATLAB programs.

```
% Plot amount of charge that has passed through a cross-section of a conductor
clear all; clc; % clear Workspace and Command Windows
t = [-4 -2 2 6 10 14]; % specify time points
q = [0 0 3 3 -2 -2]; % specify the amount of charge
plot(t,q) % get plot
grid on   % turn grid on
axis([-4.5, 14.5, -3, 4]) % specify x-axis and y-axis limits
xlabel('time - sec') % place a label along the x-axis
ylabel('charge - coulombs') % place a label along the y-axis
```

Program 2.1   Script for a piecewise linear plot.

Referring to Fig. 2.3, until $t = -2$ secs, no charge has moved through the cross-section. Then, with respect to the current reference direction, charge begins to move steadily from point (a) through the cross-section to point (b), and by $t = 2$ secs, 3 coulombs of charge has

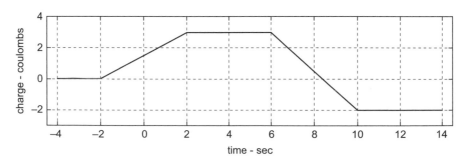

Figure 2.3   Positive charge movement through a conductor cross-section.

moved through the cross-section. From $t = 2$ secs until $t = 6$ secs, the movement of charge stopped. Starting at $t = 6$ secs, the movement of positive charge is reversed, and this decreases the net amount of charge that has moved through the cross-section down to -2 coulombs at $t = 10$ secs, when the movement of positive charge stops.

Fig. 2.4 shows the result of applying (2.1) to the movement of positive charge described in Fig. 2.3. When the amount of positive charge that has moved through the wire cross-section increases (decreases), the current is positive (negative).

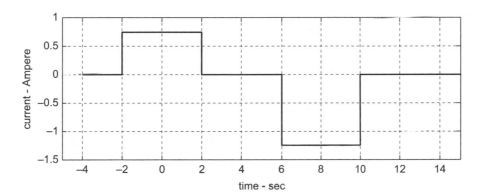

Figure 2.4   Current.

Looking at Fig. 2.4 and given the current reference direction shown in Fig. 2.2, we see that from $t = -2$ secs until $t = +2$ secs positive charge is moving from point (a) to point (b) at the rate of $+0.75$ coulombs/sec (A), and, at an instant of time, say at $t = 0.0$ secs, the current is also $+0.75$ coulombs/sec. Since the current is negative from $t = 6$ secs until $t = 10$ secs, positive charge is moving from point (a) to point (b) at the rate of -1.25 coulombs/sec, or from point (b) to point (a) the rate is $+1.25$ coulombs/sec. If the reference direction is reversed, then from $t = -2$ secs until $t = +2$ secs the current will be negative, still indicating that during this time interval positive charge is moving from point (a) to point (b).

A current (moving charge) is a source of a **magnetic field** B, as shown in Fig. 2.2. According to **Ampere's Law**, the magnitude B of the magnetic field at a distance r along a circular path around a straight conductor with steady current I is given by

$$B = \frac{\mu I}{2\pi r} \text{tesla} \tag{2.2}$$

where μ, called **magnetic permeability**, is a measure of the ability of a material to support a magnetic field within itself. In a vacuum, $\mu = \mu_0 = 4\pi \times 10^{-7}$ newtons/ampere^2. In terms of basic units, 1 tesla $= 1$ newton/((coulomb)(meter/sec)). The tangential direction of the magnetic field is perpendicular to r and the current direction.

Suppose there are two infinitely long parallel conductors that are r meters apart with each conductor (wire) carrying a current I in the same (opposite) direction, then, according to **Ampere's Force Law**, these wires will exert an attractive (repulsive) **magnetostatic force** on each other given by

$$F = \frac{\mu I^2}{2\pi r} \text{ newtons/meter} \qquad (2.3)$$

An important application of this physical property is the design of electric motors, which convert electrical power to mechanical power.

2.3.2   Voltage

An electric charge q_0 is the source of an **electric field** E, as depicted in Fig. 2.5. For a positive (negative) charge q_0 the electric field is directed radially outwards (inwards) in all directions from (to) the charge.

According to the **Lorentz Force Law** the magnitude of the electric field at a distance r from the charge is given by

$$E = \frac{1}{4\pi\varepsilon} \frac{|q_0|}{r^2} \text{ newtons/coulomb} \qquad (2.4)$$

where ε, called **electric permittivity**, is a measure of how much a material resists forming an electric field within itself. In a vacuum, $\varepsilon = \varepsilon_0 = 8.854187\ldots \times 10^{-12}$ (coulombs/meter)^2/newton. It is interesting to note that the speed of light c_0 in a vacuum, μ_0 and ε_0 are related by $c_0^2 = 1/(\mu_0\,\varepsilon_0)$.

Fig. 2.6 shows a charge q in the vicinity of a fixed charge q_0. **Coulomb's Law** gives the **electrostatic force** exerted on a point charge q when it is in the vicinity of another point

Figure 2.5   Electric field created by a point charge.

Figure 2.6   Electrostatic force between two charges.

charge q_0. This is depicted in Fig. 2.6, where the charges are r meters apart. According to Coulomb's Law, the force on the charge q is given by

$$F = \frac{1}{4\pi\varepsilon_0} \frac{q\,q_0}{r^2} \text{ newtons} \tag{2.5}$$

where $F_0 = F$. If the charges q_0 and q have the same (opposite) sign, then F is a positive (negative) repulsive (attractive) force. We have $\lim_{r\to\infty} F = 0$.

Let the charge q_0, which is fixed, and the charge q have the same sign, which makes F positive. The energy required to move the charge q from $r = \infty$ to $r = r_2$ is given by

$$J_2 = \int_{\infty}^{r_2} -F\,dr = \int_{\infty}^{r_2} \frac{-1}{4\pi\varepsilon_0}\frac{q\,q_0}{r^2}\,dr = \frac{q\,q_0}{4\pi\varepsilon_0}\frac{1}{r}\,\bigg|_{\infty}^{r_2} = \frac{q\,q_0}{4\pi\varepsilon_0\,r_2} \text{ joules} \tag{2.6}$$

The energy required to move the charge q from $r = \infty$ to $r = r_1$ is given by

$$J_1 = \frac{q\,q_0}{4\pi\varepsilon_0\,r_1} \text{ joules}$$

If $r_1 < r_2$, then $J_1 > J_2$. To move the charge q from $r = r_2$ to $r = r_1$, we must expend energy given by $J_1 - J_2$ joules, and then, if we move the charge back from $r = r_1$ to $r = r_2$, we get this expended energy back.

An important application of this physical property is the design of electric circuits, which are designed to control the motion of charge.

Voltage is defined as the energy expended (or gained) per unit charge, and it is given by

$$v = \frac{J}{q} = \frac{q_0}{4\pi\varepsilon_0\,r} \text{ joules/coulomb} \tag{2.7}$$

One joule/coulomb is called a **volt** (V). Therefore, at $r = r_1$ the voltage is $v_1 = q_0/(4\pi\varepsilon_0\,r_1)$ volts, which can be positive or negative depending on the sign of the charge q_0, and at $r = r_2$ the voltage is $v_2 = q_0/(4\pi\varepsilon_0\,r_2)$ volts. If q_0 is a positive charge, then as the distance from q_0 changes from $r = r_1$ to $r = r_2$, there is a voltage drop given by $v_1 - v_2$ volts, which is positive, and as the distance from q_0 changes from $r = r_2$ to $r = r_1$ there is a voltage drop given by $v_2 - v_1$, which is negative, that is also said to be a voltage rise given by $-(v_2 - v_1) = v_1 - v_2$, which is positive.

In view of (2.4), voltage is a measure of an electric field over a distance, where the units for the electric field E can also be expressed as volts/meter.

2.3.3 *Resistor*

Electronic circuits are designed to control the motion of charge through them to achieve some desired activity. There are many kinds of electronic devices (components) with different properties that are utilized in a circuit to achieve a desired activity.

For example, an audio amplifier is an electronic circuit where the input voltage is the output voltage of, for example, a microphone, a musical instrument pick-up, or other transducer. However, the microphone produces a low-power signal, which is usually a voltage that varies with time and is proportional to the audio signal applied to the microphone. The output of the amplifier is also a signal, a voltage that is proportional to the amplifier input signal. The amplifier output signal, which has a higher power than the input signal, has enough power to make a speaker, which is connected to the amplifier output, produce a high-power audio signal (sound).

As another example, an adder logic circuit is an electronic circuit with two inputs. Each input is a set of signals. Each signal within a set is either a high voltage, say 5 volts, or a low voltage, say 0 volts, where a high (low) voltage is interpreted to mean the binary digit 1 (0). Each set of input voltages is interpreted to be a binary number. The electronic logic circuit is designed to produce a set of output voltages that is interpreted to be a binary number that is the arithmetic sum of the two input binary numbers. This will be discussed in greater detail in Chapter 5.

One of the many kinds of electronic components is the **resistor**. It is made of carbon material, and, unlike copper, which is used to make wire that does not (almost not) impede the motion of charge through it, a resistor is intended to impede the motion of charge through it. The carbon material of a resistor has a cylindrical shape, and conductors (wires) are attached to the opposite flat faces of the cylinder. Among many other shapes, six-sided shapes are also used, where conductive plates are attached to opposite flat faces.

Fig. 2.7 shows two circuits. The zigzag line is the circuit symbol of a resistor. Each circuit consists of one closed path (or loop), along which charge can move. In circuit (a)

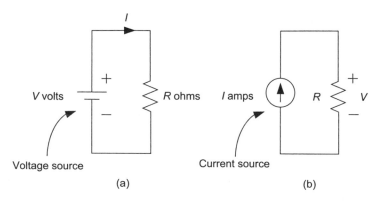

Figure 2.7 (a) Circuit with a battery voltage source. (b) Circuit with a current source.

there is a **battery** (voltage source) connected to a resistor that expends V joules/coulomb (volts) to move charge around the loop (through the resistor). An ideal conductor does not require an expenditure of energy to move charge through it. According to **Ohm's Law**, the rate at which charge moves through the resistor is proportional to the voltage, $I \propto V$, or

$$I = GV \tag{2.8}$$

where G is the proportionality constant. Given V, the current is determined by G, which is called **conductance**. In circuit (b) there is a **current source** that causes charge to move around the loop (through the resistor) at the rate of I coulombs/sec (amps). According to Ohm's Law, the energy/coulomb (joules/coulomb) that is expended to move charge through the resistor is proportional to the current, $V \propto I$, or

$$V = RI \tag{2.9}$$

where R is the proportionality constant. Given I, the voltage is determined by R, which is called **resistance**. We have $R = 1/G$, and the unit for R is **ohms** (symbol is Ω). The unit for G is **siemens** (symbol is S). The degree to which a resistor impedes the motion of charge through it is determined by its resistance value R, which depends on the geometry and chemistry of its carbon material.

The behavior of a resistor is described by (2.9) or $I = V/R$. Notice the relationship between the voltage and current references. If the current reference is reversed or the voltage reference is reversed, then Ohm's Law becomes $V = R\,(-I) = -R\,I$ or $-V = RI$.

The energy delivered by each source in Fig. 2.7 equals the energy absorbed by the resistor, which is lost in the form of heat. The rate (joules/sec) at which energy is delivered to the resistor is called the **power** P delivered to the resistor, which is given by

$$P = V\,I \text{ watts} \tag{2.10}$$

where the unit becomes (joules/coulomb)(coulombs/sec) $\rightarrow$ joules/sec $\rightarrow$ **watt**, and 1 watt = 1 joule/sec.

2.4   Program Development Continued

With MATLAB, we can obtain results to both broaden and deepen our understanding of the fundamental material presented in the previous section. In the remainder of this book, additional fundamental principles will be presented and applied in examples.

Example 2.2 ⎯⎯⎯⎯⎯⎯⎯⎯⎯⎯⎯⎯⎯⎯⎯⎯⎯⎯⎯⎯⎯⎯⎯⎯⎯⎯⎯⎯⎯

This example shows how to write a script of a program that plots the magnetic field due to a steady current in a straight conductor as the magnetic field varies with radius r. To do this, the current must be specified, which can be done with an assignment statement within the script. There will be occasions when a program must be designed to allow a program user to provide information each time the program is executed. This makes the program interactive and flexible. We can achieve this flexibility with the built-in MATLAB function **input**, which has the syntax given by

MATLAB_variable_name = input('character string')

When this statement is executed by MATLAB, the character string, which may give some instruction to the program user, is displayed in the *Command Window*, and then program execution is halted, while MATLAB waits for the program user to enter an input followed by pressing the Enter Key. After the user has pressed the Enter Key, MATLAB assigns the entered input from the program user to the given MATLAB variable name and continues program execution.

Fig. 2.8 shows the script of an interactive program that plots the magnetic field. Notice the use of the built-in MATLAB function **input** in lines 5, 7, 8, and 9. Each input line gives instructions according the meaning of the input information. Line 12 shows a division of a scalar by a vector. The operation, ./(a forward slash preceded by a period) causes MATLAB to divide the numerator by each element of the denominator vector to produce the vector result B, where $B(1) = \mu_0 I / (2\pi r(1))$, ..., $B(N) = \mu_0 I / (2\pi r(N))$. Line 17 uses the built-in MATLAB function **axis** to override the automatic scaling done by MATLAB, and

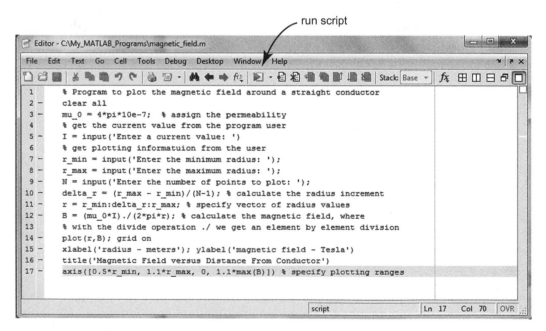

Figure 2.8   Script of an interactive program to plot the magnetic field.

the built-in MATLAB function **max** finds the largest element in the vector *B*. Notice that lines 14 and 15 each have two MATLAB statements separated by a semi-colon.

This script can be made into an m-file from within the *Editor Window* with the menu sequence given by

File → Save as . . .

Save the m-file in the *Current Folder*. You should use a program name that is unique and a reminder of the purpose of the program. **Caution, a file name must never be the same as any variable name in the script**, otherwise, MATLAB will give an error message. Here, the program file name is: magnetic_field.m, which will then appear in the list of files in the *Current Folder Window*. You can get the complete path name of a file with the MATLAB built-in function **which**, as shown below.

```
>> which magnetic_field
        C:\My_MATLAB_Programs\magnetic_field.m
```

After the script has been saved as an m-file in the *Current Folder*, you can run (execute) the program by clicking on the **Run** button in the toolbar of the *Editor Window*. You can edit and re-run the program until it performs as desired. After the m-file has been saved the first time, it is automatically saved by MATLAB every time you run it from the *Edit Window*. You can also run the program from the *Command Window* by entering its name at the prompt. This is shown in Fig. 2.9, where you can see the activity of the input function.

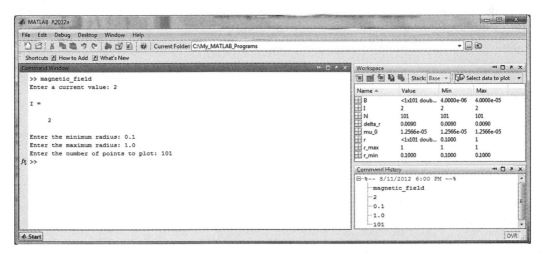

Figure 2.9   Running a program from within the Command Window.

The execution of the built-in MATLAB function **plot** in line 14 of the script causes MATLAB to open a *Figure Window* (e.g., see Fig. 1.4) into which MATLAB inserts the plot and other features specified by lines 15 through 17. Within the *Figure Window* you can edit many plot features, such as the plot line style, width and color; label font type and size; and many others. The plot can be saved as, for example, an enhanced metafile (emf) for insertion into a document (this document, for example), as shown in Fig. 2.10. You can also use

Edit → Copy Figure

and then paste the figure into another document.

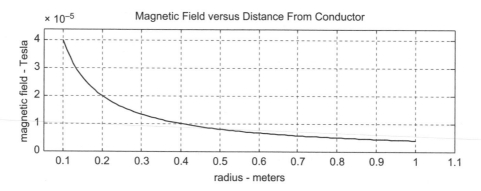

Figure 2.10   Plot obtained with the script given in Fig. 2.8.

Fig. 2.10 shows that the magnetic field decreases much more rapidly as *r* increases when *r* is small than it does when *r* becomes large.

_____

To view an m-file you can use the built-in MATLAB function **type** in the *Command Window*, for example,

```
type magnetic_field
```

Example 2.3 _____

Let us develop a program to plot the current–voltage characteristic of a **p-n junction diode**. The circuit symbol for a diode is shown in Fig. 2.11.

Ideally, a p-n junction diode behaves as follows:

$$v(t) > 0, \text{ short circuit between terminals (a) and (b)}$$
$$v(t) \leq 0, \text{ open circuit between terminals (a) and (b)} \tag{2.11}$$

An ideal diode conducts only if the voltage $v(t)$, with the positive reference at the anode and the negative reference at the cathode, is positive. Then, it is said that the diode is **forward**

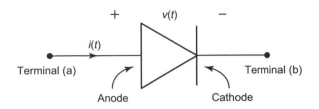

Figure 2.11   A diode with a given current reference and a voltage reference.

biased. If the voltage $v(t)$ is positive, then there is a positive voltage drop from terminal (a) to terminal (b). If the voltage is not positive, then $i(t) = 0$, and the diode is said to be **reverse biased**. A diode permits a current through it only in one direction from the anode (terminal (a)) to the cathode (terminal (b)).

More realistically, the relationship between the current through a p-n junction diode and the voltage across it is given by

$$i(t) = I_S(e^{v(t)/V_T} - 1) \tag{2.12}$$

The current I_S is called the **saturation current** (typically 10^{-12} A), and V_T is the **thermal voltage** given by $V_T = kT/q$, where k is the **Boltzmann constant**, $k = 1.3806503 \times 10^{-23}$ m^2 kg s^{-2} K^{-1} (or joules/K), T is the temperature in degrees kelvin (K), and q is the magnitude of the charge of an electron, $q = 1.60217646 \times 10^{-19}$ coulombs. At $T = 300$ K, $V_T = 25.85$ mV.

We now have information describing the behavior of a p-n junction diode. Prog. 2.2 is a complete script. The m-file diode_characteristic.m was saved in the *Current Folder*. Therefore, the program diode_characteristic can be executed by entering its name in the *Command Window*.

```
% Program to plot the current-voltage characteristic of a diode
%
% Program name: diode_characteristic.m
%
% Purpose:
% This program plots the i-v characteristic of a p-n junction diode.
%
% Program information:
% Date       Programmer      Description
% 3/22/2011   Priemer         Original code, Version 1.1
%
% Define variables
```

```
% I_S         -- saturation current
% V_T         -- thermal voltage at 300 degrees Kelvin
% V_min       -- minimum plot voltage
% V_max       -- maximum plot voltage
% N           -- number of plot points
% delta_v     -- voltage plotting increment
% v           -- vector of voltage values
% i           -- vector of current values
%
clc % clear the Command Window
clear all % clear the Workspace
% specify diode parameters
I_S = 1e-12; % saturation current in amps
V_T = 25.85e-3;  % thermal voltage at 300 degrees Kelvin
% plot the current over a voltage range from -0.25 to +0.75 volts
V_min = -0.25; V_max = 0.75;
N = 101;  % plot N points
delta_v = (V_max - V_min)/(N-1);  % plotting increment
v = V_min:delta_v:V_max; % voltage vector, v(1)=V_min ... v(N)=V_max
i = I_S*(exp(v/V_T) - 1);  % calculate the current vector
plot(v,i)  % open a Figure Window and insert a plot of i versus v
grid on  % insert a grid in the plot
xlabel('voltage - V') % place a label along the x-axis
ylabel('current - A') % place a label along the y-axis
title('i-v characteristic of a p-n junction diode') % place a title
```

Program 2.2   Script of diode_characteristic.m.

The first statement of a program should be a comment statement that identifies the program among the other programs in the *Current Folder*. This will help you to find it in the *Current Folder*, because MATLAB displays this comment in the window associated with the *Current Folder Window* when you highlight the m-file in the *Current Folder Window* (see Fig. 1.2).

As you learn more about MATLAB data types, built-in functions and programming, you will write scripts to solve challenging practical problems. It is likely that such scripts will be developed over an extended time period, and even worked on by others. Over time, you may not remember some of the rationale for some portions of the script. To help long-term script development, it is important to include script comment statements that explain the intent of various parts of the script. Also, the beginning of a script should include comment statements that describe its purpose and give definitions of important variables, if not all of them. It is also useful to include a sample of program inputs and expected program results in comment statements. See Prog. 2.2 for an example of the kinds of information that should be provided.

When writing a script, it is convenient to assign parameter values to variable names, and place all of these assignment statements somewhere toward the beginning of the script. Then, if you want to change a parameter value it will be easy to find the assignment statement, and the parameter value will be changed at all locations in the script where the variable name is used.

In the statement that calculates the current, the built-in MATLAB function **exp** has an argument that is a vector v of voltage values, and MATLAB produces a current vector i given by

$$i(1) = I_S(e^{v(1)/V_T} - 1)$$
$$i(2) = I_S(e^{v(2)/V_T} - 1)$$
$$\vdots$$
$$i(N) = I_S(e^{v(N)/V_T} - 1)$$

This feature of MATLAB is particularly convenient.

Equation (2.12) produces the plot shown in Fig. 2.12. Notice that the current through the diode does not become substantial until the forward-bias voltage across the diode exceeds a threshold voltage of about 0.7 volts.

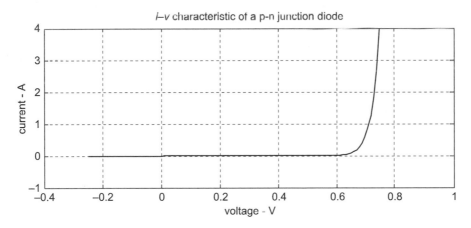

Figure 2.12   Diode characteristic of a p-n junction diode.

For a list of the variables used in this script, use the MATLAB built-in function **who** to get

```
>> who
   Your variables are:
   I_S    V_T    V_min   i
   N      V_max  delta_v v
```

The built-in function **whos** gives more detailed information, for example,

```
>> whos
    Name        Size        Bytes    Class     Attributes
    I_S         1x1             8    double
    N           1x1             8    double
    V_T         1x1             8    double
    V_max       1x1             8    double
    V_min       1x1             8    double
    delta_v     1x1             8    double
    i           1x101         808    double
    v           1x101         808    double
```

2.5 Functions

Consider the following script development situations:

(1) Within a script there is a set of MATLAB statements that implement some particular algorithm (An algorithm is a sequence of computations that accomplish some particular task.), and you want to implement this algorithm at least once again elsewhere in the script.

(2) You want to utilize an algorithm several times within a script, and each time the algorithm is needed, a different value for some parameter in the algorithm must be used.

(3) You have developed a script of some algorithm, and you want to involve this algorithm in some other script.

For each situation, you have the option of repeating the sequence of MATLAB statements that implement some particular algorithm. However, this can be inefficient.

For example, suppose you need to find $y = e^x$ at many places within a script for different values of x, or in any number of other scripts. The exponential function is the power series (Maclaurin series expansion) given by

$$y = e^x = \sum_{k=0}^{\infty} \frac{x^k}{k!} = 1 + \frac{x}{1!} + \frac{x^2}{2!} + \frac{x^3}{3!} + \frac{x^4}{4!} + \ldots, \quad 0! = 1 \tag{2.13}$$

We can write a script to compute the power series, and then include this script wherever the exponential function must be evaluated, or we can develop a function script that uses x as the input and returns y as the output. For our convenience, MathWorks, Inc. has written such a function script and included it in one of the *Directories* (toolboxes) that MATLAB automatically searches whenever it encounters the name **exp** in a MATLAB statement. Any valid variable names can be used for x and y.

In MATLAB, you can define two distinctly different kinds of functions. One kind of function is defined by just one MATLAB statement. There are two types of these functions: **anonymous function** and **inline function**. These functions are defined within a program or in the *Command Window*.

The other kind of function, which generally requires more than one MATLAB statement to define it, is an m-file that must start with a MATLAB function statement that names the function. Following the function statement, you can include in the script of a function any valid MATLAB statements. It can also contain anonymous and inline functions and other types of functions that will be described soon. When you save a function script as an m-file, **the name of the m-file and the function name must be the same**, and the function is called a **primary function**. It is invoked from outside (even inside, a recursive function) its m-file. Like a MATLAB built-in function, the function can accept inputs and return outputs.

A program works with variables and data in the Workspace. A function works with variables and data in its own work space separate from Workspace. However, computation results obtained within a function, like a program, are displayed in the *Command Window*, unless you suppress their display.

2.5.1 Anonymous Function

An **anonymous function** is defined with just one MATLAB statement that you can place anywhere in a program, in a function, or in a command line in the *Command Window* before it is used. The syntax is given by

```
function_name = @(list of input arguments) function_expression
```

where "function_name" identifies the anonymous function. It is called the **function handle**. The symbol @ declares that "function_name" is assigned a **handle**. The **list of input arguments**, which can be any of the MATLAB data types, consists of one or more variables delimited by commas. The **function expression** specifies how the function is to be evaluated. The expression can include variables that are not in the list of input arguments, and **these variables must be assigned values before the function is defined**. MATLAB incorporates the values of these variables when it builds the function.

Example 2.4 _____

Below are MATLAB statements that were entered in the *Command Window*.

```
>> % define an anonymous function to calculate the voltage at a distance r
>> % from a point charge q0
>> permittivity = 8.854e-12; % assign a value to a parameter used in the function
>> voltage = @(q0,r) q0/(4*pi*permittivity*r)  % define the anonymous function
```

```
voltage =
   @(q0,r)q0/(4*pi*permittivity*r)
>> r = 100;  % distance in meters
>> q0 = -1e-7; % charge in Coulombs
>> v = voltage(q0,r)  % use the anonymous function to calculate the voltage
   v =   -8.9877
>> % define an anonymous function for a diode i-v characteristic
>> i_diode = @(x,y,z)  y*(exp(x/z)-1)  % define the anonymous function
i_diode =
   @(x,y,z) y*(exp(x/z)-1)
>> v = 0.7; I_S = 1e-12; V_T = 25.85e-3;
>> diode_current = i_diode(v,I_S,V_T)
   diode_current =   0.5760
>> % this can also be obtained with:  diode_current = i_diode(0.7,1e-12,25.85e-3)
>> v_vector = 0.0:0.2:0.8  % obtain diode current for a set of voltages
v_vector =
   0    0.2000    0.4000    0.6000    0.8000
>> diode_current_vector = i_diode(v_vector,I_S,V_T)
diode_current_vector =
   0    0.0000    0.0000    0.0120   27.5707
```

After an anonymous function has been defined, it can be used within a program or other user-defined function in the same way as built-in MATLAB functions are used. This is convenient for program and function development. If an anonymous function is defined in the *Command Window*, then its definition will be lost when the MATLAB session is terminated. However, since the function definition is also stored in the *Workspace*, its definition is retained even if the *Command Window* is cleared.

2.5.2   Inline Function

An **inline function** is similar to an anonymous function. It is defined with the built-in MATLAB function **inline**, and the syntax is given by

```
function_name = inline('function_expression','var_1','var_2', ...)
```

where **function_name** identifies the inline function. The **function expression**, which must be enclosed in single quotes, specifies how the function is to be evaluated. The expression can include built-in or user-defined functions and involve one or more variables, **var_1**, **var_2**, etc., none of which can be assigned values before the function is defined.

Example 2.5

```
clear all; clc
% define an inline function to calculate the voltage at a distance r
% from a point charge q0
voltage = inline('q0/(4*pi*permittivity*r)','r','q0','permittivity')
p = 8.854e-12; q = -1e-7; r = 100;
v = voltage(r,q,p)  % calculate the voltage v
```

Executing this script gives

```
voltage =
   Inline function:
   voltage(r,q0,permittivity) = q0/(4*pi*permittivity*r)
v =    -8.9877
```

Notice that as MATLAB parses through the character strings that are the arguments of the built-in function inline, it identifies the given independent variables. The result is a function that is used within a MATLAB program or other user-defined function in the same way as built-in MATLAB functions are used.

2.5.3 eval Function

Another way to repeatedly evaluate a MATLAB expression is with the built-in MATLAB function **eval** with syntax given by

eval('MATLAB expression')

where the **MATLAB expression** is a character string within single quotes. Notice that the built-in MATLAB function eval does not create a function with a name, while anonymous and inline functions do create a function with a name. For example,

```
>> I_S = 1e-12; V_T = 25.85e-3;
>> v =0.7;
>> eval('I_S*(exp(v/V_T) - 1)')
ans =   0.5760
>> i = eval('I_S*(exp(v/V_T) - 1)')
i =   0.5760
```

Notice that variables used in the MATLAB expression must be assigned values before the eval function is invoked.

In MATLAB you can assign a character string to a valid MATLAB variable name. For example,

```
>> % assign a character string to a variable
>> diode_characteristic = 'I_S*(exp(v/V_T) - 1)'
diode_characteristic =
   I_S*(exp(v/V_T) - 1)
```

This is useful, because now we do not have to repeat the expression every time we want to evaluate it. For example,

```
>> I_S = 1e-12;  V_T = 25.85e-3;  v = 0.7;
>> i = eval(diode_characteristic)
i =   0.5760
>> v = 0.8;
>> i = eval(diode_characteristic)
i =   27.5707
```

Character strings will be discussed more thoroughly in Chapter 7. By assigning character strings of expressions to variables, preferably near the beginning of a program or function, you can conveniently evaluate them as often as needed for different values of the independent variables.

2.5.4   Primary Function

Before you start to develop a new function m-file, specify a *Current Folder*. To develop a user-defined function, open the *Editor Window* with

File → New → Function

A user-defined function script must start with the primary function definition line that has a syntax given by

function [output arguments] = function_name (input arguments)

where

(1)   the character string, "function" (lower case is required), is a reserved key word
(2)   the **output arguments** are a list of valid MATLAB variable names, separated by commas and enclosed in brackets

(3) the character string **function_name**, which must follow the rules for a valid MATLAB variable name, is a name that you can choose, which must be the same as the file name that you will use to save the function script as an m-file

(4) the **input arguments** are a list of valid MATLAB variable names, separated by commas and enclosed in parentheses

The number of output arguments can be

none, where brackets and the assignment operator (equal sign) are not required, for example,

```
function [ ] = sinusoid_plot(amplitude,frequency,phase,time)
% time is a vector
```

or

```
function sinusoid_plot(amplitude,frequency,phase,time)
```

one, where brackets are optional, for example,

```
function [i] = diode_characteristic(v,temp_F,I_sat)
% if v is a vector, then i is a vector
```

or

```
function i = diode_characteristic(v,temp_F,I_sat)
```

more than one, where brackets are required, for example,

```
function [E v] = electric_field(q0,r)  % E, v and r are vectors
```

The input and output argument lists can include any of the MATLAB data types. A user-defined function is invoked in the same way that a MATLAB built-in function is invoked.

Example 2.6

Suppose that in many places within a script and perhaps other scripts, we want to find the current in a diode given a voltage across the diode and parameter values. It would be efficient to use a function that is an m-file in the *Current Folder* or in some other folder that is included in the search path. A possible MATLAB function definition could be

```
function DiodeCurrent = DiodeCharacteristic_v1(v, temp_F, I_sat)
```

identifies an m-file as a function   causes all comment lines prior to the first
                                         assignment statement to be displayed

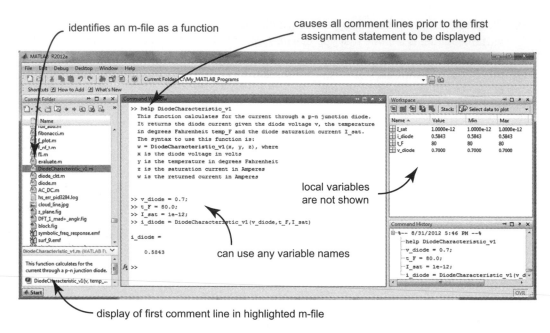

display of first comment line in highlighted m-file

Figure 2.13   Demonstration of using a function m-file.

Here, DiodeCharacteristic_v1.m will be the m-file name of this function. Within the function script, the first few lines should be comment lines that explain what this function does and how to use it. Prog. 2.3 gives a possible script.

```
function DiodeCurrent = DiodeCharacteristic_v1(v, temp_F, I_sat)
% This function calculates for the current through a p-n junction diode.
% It returns the diode current given the diode voltage v, the temperature
% in degrees Fahrenheit temp_F and the diode saturation current I_sat.
% The syntax to use this function is:
% w = DiodeCharacteristic_v1(x, y, z), where
% x is the diode voltage in volts
% y is the temperature in degrees Fahrenheit
% z is the saturation current in Amperes
% w is the returned current in Amperes
%
k = 1.3806503*10e-23; % Boltzmann constant
q = 1.602176646*10e-19; % magnitude of the charge of an electron
% converting input temperature (degrees Fahrenheit) to degrees Kelvin
t_C = (temp_F - 32)*5/9; % temperature in degrees Centigrade
```

```
t_K = t_C + 273.15; % temperature in degrees Kelvin
% compute the diode current
V_T = k*t_K/q; % thermal voltage
DiodeCurrent = I_sat*(exp(v/V_T) -1);
end % this end statement is optional
```

Program 2.3 Function script for DiodeCharacteristic_v1.m.

This function can be used within a program, another function, or in the *Command Window*. Fig. 2.13 shows how this function can be used in the *Command Window*.

Notice the structure of the function m-file shown in Prog. 2.3. The first line is the MATLAB function definition statement, which gives the name of the primary function. The next line, called **H1** (first help line), is a comment statement that appears in the window below the *Current Folder Window* when the function (or program) is highlighted. This is useful to quickly see what the program is about (see Fig. 2.13). The following comment statements explain how to use the program. To view these comment statements, use the MATLAB built-in function **help**, for example,

help DiodeCharacteristic_v1

When the help function is executed, MATLAB displays all of the comment statements until it reaches the first assignment statement (see Fig. 2.13). All of these comment statements are optional. In the function body that follows the comment statements, you can use any MATLAB programming statements, including other functions that you have defined. Furthermore, **within the function body, you must assign values to all variable names included in the function output arguments list**.

All variable names within a function and their values are stored in a work space that is separate from the work space used by the program (or function) from which the function is called (invoked). All variables within a function are called **local variables**. Therefore, you can re-use variable names in a function that were used in the calling program (or other function). The variables in the input arguments list are local variables that receive their values by transferring the values of the variables used as input arguments of the function named in the calling program (or other function). The variables in the output arguments list are also **local variables**, and their values are transferred to the variables used as output arguments of the function named in the calling program (or other function). You can override this method of passing variable values by declaring that some variables are **global variables** with the built-in function **global**, which has the format

global list of variable names delimited by spaces

The global statement must be placed before the variables are used. Any variables can be made global variables.

Example 2.7 ———————————————

Let us write a program that calls the function DiodeCharacteristic_v2, where the Boltzmann constant, the magnitude of the charge of an electron and I_sat are made global variables. A portion of a program script follows.

```
⋮
global  k_Boltzmann  q_electron I_sat  % list of global variables
% assign values to global variables
k_Boltzmann = 1.3806503*10e-23; % Boltzmann constant
q_electron = 1.602176646*10e-19; % magnitude of the charge of an electron
I_sat = 1e-12; % saturation current
% assign values to function input variables
v_diode = 0.7;
t_F = 80.0; % degrees Fahrenheit
i_diode = DiodeCharacteristic_v2(v_diode, t_F);  % get diode current
⋮
```

To access the global variables, the function must also declare them as global variables before they are used in the function. A portion of the revised function script follows.

```
function DiodeCurrent = DiodeCharacteristic_v2(v, temp_F)
⋮
global   k_Boltzmann  q_electron  I_sat
% converting input temperature (degrees Fahrenheit) to degrees Kelvin
t_C = (temp_F - 32)*5/9; % temperature in degrees Centigrade
t_K = t_C + 273.15; % temperature in degrees Kelvin
% compute the diode current
V_T = k_Boltzmann*t_K/q_electron; % thermal voltage
DiodeCurrent = I_sat*(exp(v/V_T) -1);
⋮
```

2.5.5   Sub-Function

Within the body of a primary function there can be the need to execute an algorithm several times. As in a program, we can repeat the MATLAB statements that implement the algorithm. This can be made more efficient by appending to the primary function body a function that starts with a function definition line, which has a format like the format of the primary function definition. The appended function is called a **sub-function**. You can append as many sub-functions as needed. Sub-functions can have the same structure as the primary function, including an H1 line and comments that explain what the sub-function does. The resulting m-file, which still must have the name of the primary function, contains the primary function and all of the sub-functions.

Any of the sub-functions can be invoked from the primary function body and from any other sub-function. However, a sub-function cannot be invoked from outside of the primary function m-file. Furthermore, the primary function and the sub-functions each use their own separate work space, and therefore, within each function all variables are local variables, unless some variables are declared to be global variables. The structure of an m-file is shown below.

```
function [output arguments] = primary_function_name (input arguments)
    ⋮
function [output arguments] = sub_function_1_name (input arguments)
    ⋮
function [output arguments] = sub_function_2_name (input arguments)
    ⋮
```

The definition line of sub_function_1 terminates the body of the primary function, and the definition line of sub_function_2 terminates the body of sub_function_1. You can access the help text of a sub-function with

```
help primary_function_name > subfunction.sub_function_name
```

2.5.6   Private Function

Suppose you want to limit access to a function or you do not want to reveal using it. This can be accomplished by making the function a **private function**. A function is made a private function by saving its m-file in a particular folder. The particular folder must be a sub-folder, which has the special name **private**, of a parent folder. The parent folder can contain m-files and other sub-folders. A private function can only be invoked from function m-files in the parent folder. Private functions cannot be invoked from program m-files in the parent folder, and private functions cannot be invoked from m-files outside of the parent folder. Any function can be made a private function.

Consider the following portion of a directory structure of m-files and folders.

```
folder_1
      ⋮
      program_1.m
      function_1.m
      folder_2
              program_2.m
              function_2.m
              function_3.m
              private
                      function_p.m
                      exp.m
                      function_1.m
      folder_3
      ⋮
```

Here, folder_2, which is a sub-folder of folder_1, is the parent folder of sub-folder private, which contains three private functions. The function function_1 or the program program_1 in folder_1 can invoke function_2, if folder_2 is in the MATLAB search path. However, this function_1 cannot invoke function_p. The functions function_2 and function_3 can invoke a private function. Since private functions are invisible from outside of the parent folder, private functions can use the same names as functions outside of folder_2, such as function_1 in the private sub-folder. The private function exp has the same name as a MATLAB built-in function. MATLAB automatically searches the private folder before looking outside of folder_2 for non-private functions. If function_2 invokes function exp, then MATLAB will use the exp function in the sub-folder private, which is useful if you want to create your own version of some other function, such as exp in this case. Similarly, if function_3 invokes function_1, then MATLAB will use the function_1 in the private sub-folder.

2.5.7   Nested Function

You can define a function within the body of another function. For example, consider the structure of an m-file as shown below.

```
function [out_arg1] = name_1(in_arg1, in_arg2)
   ⋮
var_1 = . . .   % assign some value to var_1
var_2 = name_2(var_1, in_arg2);
   ⋮
      function [out_arg2] = name_2(in_arg4, in_arg5)
```

```
        ⋮
        out_arg2 = . . .  % assign some value to out_arg2
        [var_3  var_4] = name_3(out_arg2)
          ⋮
        end
    ⋮

        function [out_arg3  out_arg4]  = name_3(in_arg6)
          ⋮
        var_5 = . . . % assign some value to var_5
        out_arg3 = . . .  % assign some value to out_arg3
        out_arg4 = . . .  % assign some value to out_arg4
          ⋮
        end
    ⋮
    out_arg1 = . . .  % assign some value to out_arg1
    end
```

Here, the functions name_2 and name_3 are **nested** in the primary function name_1. An **end** statement is required to terminate each nested function and the primary function. You can place nested functions within a nested function.

Like other functions, a nested function uses its own work space, which is separate from the work spaces used by other functions. However, unlike other functions, a nested function and the function in which it is nested can access each other's work space. This provides an alternative to the judicious placement of global statements. For this example, variable var_3 in function name_2 can be accessed from within function name_1. Furthermore, since functions name_2 and name_3 are not nested within each other, var_5 within name_3 cannot be accessed from within name_2.

2.5.8 Function Function

A **function function** is a function where the input arguments include one or more variables with values that are **function handles** of other functions and possibly values or the names of one or more variables. Let us consider the case where the input includes values and variable names and only one variable to which will be passed a function handle, call the variable f_in. The syntax for a function function is

```
function [output arguments] = function_name(f_in, additional inputs)
```

where the **output argument list** can have none or one or more variables, **function_name** is the name of the function function, which must be the same as the name of its m-file. When function_name is invoked, a function handle, call it @f_passed, of some function, called f_passed, must be passed to function_name. When in the body of function_name, **f_in** is invoked, f_passed will actually be used. The functions f_passed and f_in must have the same number of input and output arguments.

Example 2.8

Develop a general purpose function function, call it f_plot, that plots N points of a given function $y = f(x)$ from $x = x_1$ to $x = x_2$. Prog. 2.4 is a function function for f_plot.

```
function f_plot(f_in,N,x1,x2,x_label,y_label)
% General purpose function for plotting a given function
delta_x = (x2-x1)/(N-1); % plotting increment
x = x1:delta_x:x2; % vector of points
y = f_in(x); % evaluate given function and f_in must return a vector
plot(x,y) % get plot
grid on % turn on grid
xlabel(x_label) % label x-axis
ylabel(y_label) % label y-axis
end
```

Program 2.4   General purpose plotting function.

This function function is saved as f_plot.m. With f_plot, plotting details do not have to be repeated each time you want to plot some function.

Let us plot

$$x(t) = A(e^{-a_1 t} - e^{-a_2 t})\sin(\omega t)$$

where a_1 and a_2 control the attack and decay of the sinusoid. The parameters will be passed to the function for $x(t)$ with a global statement. Prog. 2.5 defines the function $x(t)$ to be plotted.

```
function [x] = sinusoidal_pulse(t)
% Function for a sinusoidal pulse
% t is an input vector of time points
global parm_1 parm_2 parm_3 parm_4
% parm_1 and parm_2 receive attack and decay control parameters
a = parm_1; d = parm_2;
% parm_3 receives the amplitude
amp = parm_3;
% parm_4 receives the frequency of the sinusoid in Hz
```

```
freq = parm_4;
w = 2*pi*freq;
x = amp*(exp(-a*t)- exp(-d*t)).*sin(w*t);
end
```

Program 2.5   Function that defines the function to be plotted.

The operation * preceded by a period causes an element by element multiplication of two vectors. For example, let a and b be two vectors, each of length K. Then, $c = a. * b$ produces the vector c, where $c(1) = a(1) * b(1), \ldots, c(K) = a(K) * b(K)$.

This function is saved as sinusoidal_pulse.m. A program to test f_plot is shown in Prog. 2.6. Fig. 2.14 shows the *Figure Window* opened by the built-in function plot.

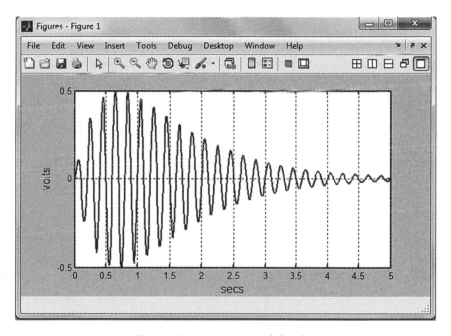

Figure 2.14   A sinusoidal pulse.

```
% Test general purpose plotting program
clear all; % clear Workspace
global parm_1 parm_2 parm_3 parm_4 % used to pass parameters
% specify parameters of function to be plotted
attack = 1; parm_1 = attack; % specify attack and decay control
decay = 2; parm_2 = decay;
amplitude = 2.0; parm_3 = amplitude;
```

```
frequency = 5; parm_4 = frequency;
t1 = 0; t2 = 5; N = 500; x_name = 'secs'; y_name = 'volts';
f_plot(@sinusoidal_pulse,N,t1,t2,x_name,y_name)
```

Program 2.6   Script to test the function function f_plot.

We could have specified which function to use for plotting with a variable that has the function handle as its value, such as

```
x_of_t = @sinusoidal_pulse;
```

and then invoke the function f_plot with

```
f_plot(x_of_t,N,t1,t2,x_name,y_name)
```

Sometimes it is more convenient to pass the handle of an anonymous function to a function function, because the anonymous function is defined within the program that invokes the function function.

Example 2.9

Use f_plot to plot the *i–v* characteristic of a diode. Prog. 2.7 shows how an anonymous function can be used to define the function to be plotted. The resulting plot is shown in Fig. 2.15.

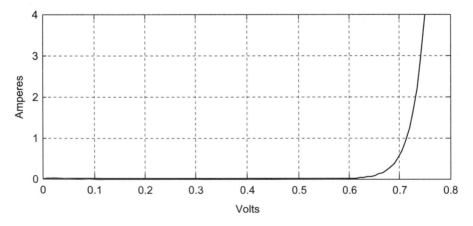

Figure 2.15   Output of the function function f_plot.

```
clear all; clc;
% define an anonymous function for a diode i-v characteristic
I_S = 1e-12; V_T = 25.85e-3;
i_diode = @(v) I_S*(exp(v/V_T)-1);   % define the anonymous function
N = 100;
v1 = 0; v2 = 0.75; % specify the voltage range
x_name = 'Volts'; y_name = 'Amperes';
f_plot(i_diode,N,v1,v2,x_name,y_name)
```

Program 2.7   Program to pass an anonymous function handle to a function function.

MATLAB has several built-in function functions such as **fzero**, for finding where a function is zero, **quadl**, for finding the area under a curve, **fminsearch**, for finding where a function has a minimum value, and others. Use MATLAB help facility to find out more about these function functions.

2.6   Code Analyzer

As you write scripts and functions, syntax errors may occur. Other errors may prevent m-file execution, or may not prevent m-file execution, but cause undesirable results. MATLAB includes a utility, called **M-Lint**, that runs automatically in the background as you write a new script or function or edit a previously created m-file. Fig. 2.16 shows Prog. 2.2 with several intentional errors and warnings.

M-Lint automatically checks for various kinds of syntax and omission errors. It tells you about these errors by underlining their location either in red to indicate an error or in yellow to indicate a warning. Also, in the upper right corner of the *Editor Window* there is a square button, called the **execution status** button, that is either green for no error, yellow for warning, or red for error. Clicking on this execution status button moves the cursor from one error or warning location to the next one. In the column below the status execution button, there can be yellow or red dashes. When you position the mouse pointer on one of these dashes, M-Lint pops up a message that suggests a fix.

Notice the kinds of errors that M-Lint can find. In line 25 the left side of the equal sign uses a minus sign instead of an underscore. In line 26 there is an s instead of an e, and you see the message given by M-Lint. The warning in line 29 occurred because the line is not terminated with a semi-colon. After all errors and warnings have been fixed, the execution status button turns green. While there may still be other kinds of errors that cause undesirable results, M-Lint is a very useful utility.

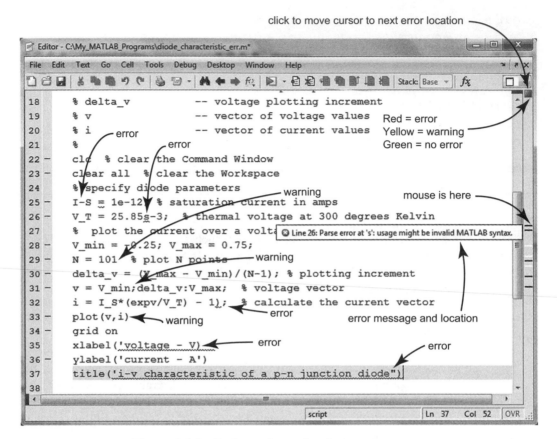

Figure 2.16   Code analyzer showing syntax errors.

You can also apply M-Lint from the *Current Folder Window* to all m-files in the *Current Folder*. Use the help facility to learn more about customizing M-Lint to suit your needs.

2.7   p-Code

There may be occasions when you want to allow someone else to use an m-file that was developed by you, but, for example, for proprietary reasons you do not want someone else to have access to the m-file (the **source code**). This is possible with most other programming languages, for example, the C programming language, because source code written in C is converted (compiled) by a C compiler into **machine code**, which is the binary code that is actually executed by a processor, and it is virtually unreadable.

For execution, MATLAB source code is not directly compiled into machine code. Instead, when MATLAB executes source code, it **interprets** (called **parsing**) each line of the source code and generates protected code, called **p-code**, which is at an intermediate level between source code and machine code. The p-code version of an m-file is also virtually unreadable, and it can be executed. To obtain the p-code version of an m-file use the MATLAB command

```
pcode  source_name.m
```

in the *Command Window*, which produces the file source_name.p. Then, when you invoke the file *source_name* in another m-file or in the *Command Window*, MATLAB will use the p-code.

2.8 Tool Box

When you have used MATLAB to develop many m-files that solve problems in many related and unrelated areas, the folders in which you save these m-files can become difficult to manage to enable convenient access to these m-files at a later time. You could keep all of your m-files in a folder that you always select to be the *Current Folder*, which places this folder first in the MATLAB search path. However, this is not the best way to organize your work product.

MATLAB includes hundreds of built-in functions, many of which are m-files. These functions are stored in many folders, called **toolboxes**, where each toolbox contains functions that solve related problems. For example, there is a MATLAB toolbox that only contains functions concerned with plotting. As another example, there is a MATLAB toolbox that contains functions concerned with signal processing. All of the complete path names of MATLAB toolboxes were placed in the MATLAB search path (the default search path) at the time MATLAB was installed, and all of the complete path names of all toolboxes are stored in a file named **pathdef**, which is stored in the folder named **MATLAB**.

While the signal processing toolbox is included with MATLAB when you purchase it, MathWorks, Inc. has developed toolboxes concerned with problems in many diverse areas, including economics, fuzzy logic, chemical engineering, neural networks, statistics, and many others, too numerous to list here. These toolboxes can be purchased from Math-Works, Inc.

To add a folder and its sub-folders, if there are any, to the MATLAB search path, open the *Set Path Window* with

File → Set Path

which is shown in Fig. 2.17. This window shows the MATLAB search path.

To add a folder and its sub-folders click on the Add with Subfolders button, which opens the *Browse for Folder Window*, where you can select the folders to be added to the search path. Notice that you can edit the sequence in which MATLAB searches toolboxes to find functions that you have invoked. Additions to the MATLAB default search path are removed when you terminate a MATLAB session unless you save the search path in the pathdef file, which MATLAB checks at the start of the next MATLAB session.

You can also manage the search path with built-in functions, some of which are listed in Table 2.1.

click to add folder with sub-folders   highlight folder

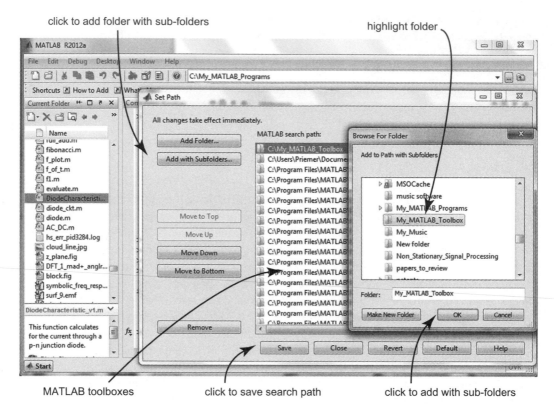

MATLAB toolboxes   click to save search path   click to add with sub-folders

Figure 2.17   Window used to add toolboxes to the MATLAB search path.

Table 2.1   Built-in MATLAB functions to manage the search path

Function	Brief explanation
addpath	Add folders to search path
path	View or change search path
path2rc	Save current search path to pathdef.m file
pathtool	Open set path dialog box to view and change search path
restoredefaultpath	Restore default search path
rmpath	Remove folders from search path
savepath	Save current search path
userpath	View or change user portion of search path

2.9   Conclusion

In this chapter you learned how to create programs and functions from MATLAB
scripts. There are several different kinds of functions, and we saw how to take advantage
of their different properties. Now that you know how to create programs and functions,

you will want to learn more about MATLAB programming to solve challenging and practical problems. Probably, the hardest parts about solving a problem are (1) understanding the problem and (2) developing a method of solution. Then come: (3) developing an algorithm to implement the method of solution and (4) modularizing the algorithm to develop an efficient program with supportive functions. These steps may or may not lead to correct solutions. This will require checking and properly interpreting program output.

Program development requires more knowledge about MATLAB data types, built-in functions and constructs for program flow control to see the possibilities for implementing an algorithm. It may happen that errors, commonly called **bugs**, will be made in steps (1) through (4). MATLAB provides a user-friendly facility for debugging, including M-Lint and much more. Debugging will be discussed in Chapter 10 after you have learned more about MATLAB.

Now, you should know how to

- set up the current folder
- work with electric current and voltage
- create a program
- create several different kinds of functions, including anonymous, inline, primary, private, sub, nested, and function functions
- share variables among programs and functions
- use M-Lint
- produce the p-code of an m-file
- create a toolbox

Table 2.2 gives the built-in MATLAB functions that were introduced in this chapter.

Table 2.2   Built-in MATLAB functions introduced in this chapter

Function	Brief explanation
@	Declares a function handle
axis	Specifies x- and y-axis ranges
eval	Evaluates a function described by a character string
fminsearch	Finds where a function has a minimum
fzero	Finds where a function is zero
global	Enables common access to variables by scripts and functions
input	Gets entry from keyboard
max	Finds largest element in a vector
mlint	Analyzes a script for syntax and omission errors
pcode	Creates a protected version of an m-file
quadl	Finds the area under a curve
type	Display an m-file
who	Lists variables in Workspace
whos	Gives detailed description of variables in Workspace

Use the help facility to find out more about these and related MATLAB built-in functions. You should also see the excellent video tutorials about getting started with MATLAB and program development. To do this, type **demo** in the *Command Window*, and select the audio/video tutorial you want to see.

Problems

In the following programming problems, each program must start with one or more comment statements that explain the purpose of the program. Also, include comments within programs that explain program activity.

Section 2.1

1) (a) When MATLAB is started, what is the default Current Folder path?
   (b) Open the Current Folder pull-down menu in the desktop toolbar, and list all path names, but not more than three, starting at the top of the menu.
   (c) Give the name of the window that opens when you click on the browse for folder icon in the desktop toolbar.
2) Activate the Current Folder Window.
   (a) Give the names of all of the icons in the Current Folder Window toolbar.
   (b) Open the Actions icon pull-down menu, and give the names of all of the menu items.
   (c) In the Actions pull-down menu, click on help, and give the name of the window that opens.

Section 2.2

3) Open the *Editor Window* with: File → New → Script, and list the names of all of the icons in the toolbar.
4) Activate the MATLAB desktop, and specify a current folder where you want to save m-files.
   (a) Open the *Editor Window* to write a new script, and enter the script given in Prog. 2.1. Click on the run icon in the toolbar. Describe what happened. Save the script as an m-file in the Current Folder. Call the m-file charge.m.
   (b) Activate the *Command Window*, and enter help lookfor. Briefly describe the purpose of this function. Then, enter a command to find the character string, "charge". Describe what happened.
5) (a) Use the script given in Prob. 1.19 to create an m-file called multiple_exponentials.m. Provide a program listing and program output.
   (b) Create an m-file, call it sinusoidal_pulse.m, using the script given in Prob. 1.23. Your program should produce the plot given in Prob. 1.23. Provide a program listing and output.

Section 2.3

6) The amount of charge that has moved through a cross-section of a conductor is given by

$$
q(t) = \begin{cases}
0, & t < 0 \\
2t, & 0 < t < 1 \\
0, & 1 < t < 2 \\
-2t + 6, & 2 < t < 3 \\
0, & t > 3
\end{cases}
$$

(a) Manually, give a sketch of $q(t)$ versus t.
(b) Modify the script given in Prog. 2.1 to obtain a plot of $q(t)$ versus t.
(c) Manually, determine and sketch the current $i(t)$ versus t. Label the axes.

7) (a) In terms of basic units (Q – charge, M – mass, L – length and T – seconds) express 1 tesla.
(b) Two infinitely long parallel conductors are 0.001 meters apart. They each carry a current of 10 amperes, but in opposite directions. What is the force/meter that these conductors exert on each other? Is it attractive or repulsive? A force of 4.44822162825 newtons equals a force of 1 pound. Convert your answer to pounds/foot.

8) A negative charge of 0.0001 coulombs is 0.002 meters from another negative charge of 0.005 coulombs. In a vacuum, what force do these charges exert on each other? Give your answer in both newtons and pounds.

9) In a vacuum, a positive point charge $q = 1 \times 10^{-9}$ coulombs is 0.5 meters from a fixed positive point charge $q_0 = 5 \times 10^{-9}$ coulombs. How much energy is required to move the charge q to within 0.2 meters of q_0? What is the change in voltage?

10) A 12 volt battery is connected to a 100 Ω resistor.
(a) What is the current through the resistor?
(b) How much power is delivered to the resistor?
(c) Over an hour, how much energy is expended by the battery?

11) A power source expends 10,000 joules of energy to move 2,000 coulombs of charge through the circuitry of a computer. The charge moves through the computer at a rate of 4 coulombs/sec. How much power does the source deliver to the computer? How much energy does the power source expend in an hour?

12) (a) How much energy is 1 killowatt-hr? Can you find out from an electric bill how much your electric company charges per killowatt-hr?
(b) Assume that when you start your car it starts within 3 seconds of cranking. Assume that the starter motor requires 100 amps to crank the engine. Using a 12 volt battery, how much energy does the battery provide to start the car? How many killowatt-hrs is this energy?

Section 2.4

13) (a) Write a script similar to the script given in Fig. 2.8 that plots the electric field caused by a charge q_0. Your script must input parameter values like the script given in Fig. 2.8. Then, after you have specified a Current Folder, save the script as an m-file, called Electric_Field.m, and run it. Provide a copy of the m-file and input and output. Use the function type to obtain a listing of your m-file.

(b) In the *Command Window*, enter the function name (command) whos. Provide a copy of the response, and explain it. Use help whos.

(c) In the *Command Window*, enter the function (command) what. Provide a copy of the response, and explain it.

14) Write the program Electric_Field.m of Prob. 2.13 in the style illustrated by Prog. 2.2. Provide a copy of the resulting m-file.

15) Use colon notation to define the following vectors.

(a) Vector x with first element (first) equal to –5, element to element increment (delta) equal to 0.7 and last element (last) equal to 2. What is the last element in x? Use help length to find out what the function length does, and use the function length to find out the number of elements in x. What do you get with the statement: x(length(x))?

(b) Vector y has first $= 2.1$, delta $= 0.5$ and last $= 9.0$. What is the last element in y? How many elements does y have? What do you get with $z = y.*y + y + 1$?

(c) Vector z has first $= 5.25$, delta $= -0.3$ and last $= -2.3$. What is the last element of z? How many elements does z have? What do you get with $w = 1./z$?

16) (a) Write a script that (1) uses colon notation to define a vector t with elements that range from –1 to 10 in increments of 0.1, (2) evaluates $y(t) = e^{-0.5t}$, and (3) plots $y(t)$ versus t. Include a grid, axes labels and a title. Save your script as exponential.m, and provide a listing of it.

(b) In the *Command Window*, enter the statement: which exponential.m. What is the response?

(c) In the *Command Window*, enter: exponential. What happened?

17) The voltage source $v_S(t) = 10 \sin(8\pi t)$ volts is connected as shown in Fig. P2.17.

Figure P2.17   Voltage source in series with a diode and a resistor.

(a) Write a script to plot $v_S(t)$ for $0 \le t \le 1$. Use enough plot points to get a smooth looking plot of the sinusoidal function. Include a grid, axes labels and a title. Save the script as sinusoid.m. Provide a program listing and output.

(b) Assume that the diode is an ideal diode as described by (2.11). By hand, sketch $v(t)$. Label the axes.

18) Write a script to evaluate $x(t) = e^{-t}\sin(4\pi t)$ and $y(t) = e^{-t}\cos(4\pi t)$ for $0 \le t \le 5$. Use colon notation to define a vector of values of t with enough time points to obtain smooth looking plots. Use vector multiplication (.*) to multiply the exponential and the sinusoidal functions. Plot y versus x, and include a grid. Save the script as an m-file called spiral.m.

(a) Run the program and provide a copy of the figure.

(b) Add the statement: axis equal, and repeat part (a). How is the result different from part (a)?

(c) Add the statement: axis([−1 1 −1 1]), and repeat part (b). Provide a copy of the final program.

Section 2.5

19) Suppose the power delivered to a resistor with resistance R must be calculated at several places within a program given a current I through a resistor. Write a script that defines an anonymous function named power that has two input arguments R and I and returns the power delivered to the resistor. Save the script as an m-file called power_to_R.m. Your program should use the anonymous function at least two times, and display the inputs and result. Provide a listing of your program and a copy of the function inputs and outputs.

20) Write a script that includes the definition of an anonymous function, which evaluates $x(t) = Ae^{-at}\cos(\omega t + \theta)$ volts, where the input arguments are A, a, $w(\omega)$, $p(\theta)$, and t. The input argument t is a vector of time points. Use colon notation to define a vector of values of t. Your program must produce a figure with a family of plots, where $A = 2$, $a = 0.2$, 1.0, and 3.0, $w = 4\pi$, $p = \pi/2$, and t varies over the range $0 \le t \le 5$. Use the statement: hold on, and your anonymous function several times. Use enough time points to produce smooth looking plots. Provide a program listing and results.

21) Repeat Prob. 2.19, but use an inline function.

22) Repeat Prob. 2.19, but use the function eval.

23) Repeat Prob. 2.20, but use the function eval. For convenience, assign to a variable, for example, x_of_t, a character string of the given function. After assigning values to all input arguments, use, for example, $x = $ eval(x_of_t).

24) Open the *Editor Window* for a new function. Write a script of a function that plots N cycles of $x(t) = A \cos(\omega t + \theta)$ volts, plotting 100 points per cycle. Call the function sinusoid_plot. The inputs are A, $w(\omega)$, $p(\theta)$ and N. There is no output argument. Be sure to include an H1 line. The function must place axes labels and a grid. Write a script that invokes your function. Use $A = 2$, $w = 8\pi$, $p = -\pi/2$, and $N = 3$. Provide program and function listings and results.

25) Open the *Editor Window* for a new function. Write a script of a function, where the only input argument is a vector t of time points, and the only output is a vector x. The function, call it exp_cosine, must evaluate the $x(t)$ given in Prob. 2.20. All other parameters must be received by the function with a global statement. Write a script that utilizes a global statement, invokes the function exp_cosine, and produces a family of plots of $x(t)$ versus t. Use the same parameter values as given in Prob. 2.20.

26) Open the *Editor Window* for a new function. Write a script of a function that returns the voltage drop in a vacuum as the distance from a charge q_0 coulombs is increased from r_1 to r_2 meters. The inputs to the function are q_0, r_1, and r_2, and the output is v_{12}. Be sure to include an H1 comment line. Save the function as v_drop.m. Provide a program that invokes your function to find the voltage drop for (a) $q_0 = 2e\text{-}3$ coulombs, $r_1 = 0.1$ meter and $r_2 = 1.0$ meter and (b) repeat part (a) with $q_0 = -2e\text{-}3$ coulombs. Provide program and function listings and results.

27) Within your Current Folder, create a sub-folder called private. Open the Editor Window for a new function. Write a script of a function that returns the power P delivered to a resistor R, where the input arguments are R and I. Call the function R_power, and save it as an m-file in the subfolder private.

   (a) Write a script, called find_R_power, that assigns values to R and I, invokes the function R_power to obtain the power delivered to the resistor, and save it as an m-file in the Current Folder. Provide program and function listings. Run the program. Describe what happened.

   (b) Open the Editor Window for a new function. Write a script of a function, call it power_to_R, with inputs R and I and returns the power delivered to a resistor. The function power_to_R must invoke R_power. Save power_to_R as an m-file in the Current Folder. Modify find_R_power.m to invoke power_to_R instead of R_power. Provide program and function listings. Run find_R_power. Describe what happened. Discuss the difference between using R_power in parts (a) and (b).

28) Describe the difference between the structure of a function that includes a sub-function and a function that includes a nested function.

29) Open the Editor Window for a new function. Write a script of a function, call it evaluate, that receives the handle of a function m-file and a value of x, and returns the value y of some function $y = f(x)$ defined in some function m-file. For example, start the function evaluate with

```
function y = evaluate(f_in,x)
    ⋮
```

We want to use the function evaluate to find values of functions $f(x)$ defined in other function m-files. For example, let $f1(x) = x^2 + x + 1$ and $f2(x) = e^{-x}$. A function to evaluate f1 could be

```
function y = f1(x)
y = x^2 + x +1;
```

A program can use the function evaluate with statements such as

```
z = evaluate(@f1,3)
```

(a) Complete writing the script of the function evaluate, create m-files for f1 and f2, and write a script that demonstrates using evaluate for f1(x) and f2(x). Provide copies of your program, all functions and results.

(b) How would you modify the m-file for f1 if x is a vector? Provide a demonstration.

30) The debugging tool M-Lint runs automatically in the background as you write scripts. Explain the difference between a warning and an error.

31) (a) Obtain the p-code version of any program m-file that you have written. In the Current Folder Window, rename the resulting p-code version of the m-file to p_code_test.p. This will ensure that there is no m-file with this name. Activate the Command Window, and type p_code_test. Describe what happened.

(b) An m-file can be opened by double clicking its name in the Current Folder Window. Double click the file name p_code_test.p. Describe what happened.

(c) Rename the file p_code_test.p to p_code_test.m. Double click on p_code_test.m. Describe what happened. Change the name back to p_code_test.p.

32) (a) Use the MATLAB help facility, and explain the purpose of the built-in function addpath. Give an example of a MATLAB statement that uses this function, and enter it in the *Command Window*. Verify that this statement worked as intended by clicking the *Current Folder* pull-down menu icon in the MATLAB desktop toolbar. Explain what happened.

(b) In the Command Window, enter the function name pathtool. Describe what happened.

CHAPTER 3

Matrices, Vectors, and Scalars

One of the most important distinguishing features of MATLAB® is how easy it is to work with the matrix data type. We start this chapter with the definition of a matrix. Then, much of the mechanics of working with matrices in MATLAB will be presented. Given these fundamentals, matrix algebra will be applied to several applications, including circuit analysis.

After you have completed this chapter, you will know

- about the origin of the concept of a matrix
- the fundamentals of matrix algebra
- how MATLAB is particularly well suited for matrix algebra
- about many built-in MATLAB functions to find properties of a matrix
- how to use MATLAB for resistive circuit analysis
- how to solve systems of linear and nonlinear algebraic equations

3.1 Matrix Definition

A two-dimensional (2-D) matrix is an organization of numbers into a given number of rows and a given number of columns that is enclosed within brackets. Let A denote some matrix, let N denote the number of rows in the matrix A, and let M denote the number of columns in the matrix A. Then, N and M give the **dimension** of A, and we say that A is an N by M matrix, which is denoted by $(N \times M)$. In MATLAB, any valid variable name can be used to denote a matrix. The structure of the matrix A is shown below.

$$
A = \begin{bmatrix} A(1,1) & A(1,2) & \ldots & A(1,M) \\ A(2,1) & A(2,2) & \ldots & A(2,M) \\ \vdots & \vdots & \ddots & \vdots \\ A(N,1) & A(N,2) & \ldots & A(N,M) \end{bmatrix}
$$

An individual entry in some row n and some column m of the matrix A is called an **element** of the matrix, which is denoted by $A(n, m)$, for $n = 1, 2, \ldots, N$ and $m = 1, 2, \ldots, M$. Here, n is used for the **row index**, and m is used for the **column index** to describe the location of an element in the matrix. It is conventional to give the row index first. In MATLAB, any valid variable name can be used to denote an index.

The term **array** is also used for an organization of elements like a matrix. However, the elements in an array are not necessarily numbers. Array elements can be text, logical variables, or other kinds of information. MATLAB programming that is concerned with array structures will be examined in Chapter 7.

Example 3.1

Let $N = 2$, $M = 3$, and b denote some matrix. Then, b could be, for example, the matrix given by

$$b = \begin{bmatrix} -2.5 & 3.1415926 & 0.1112 \\ 1/3 & 0.0 & -5 \end{bmatrix}$$

Here, b is a (2×3) matrix. In MATLAB, the matrix is assigned to the variable b with the following assignment statement.

```
>> % assign a matrix to the variable b
>> % in MATLAB, the delimiter between the elements in a row must be either
>> % a comma or at least one blank space, and
>> % the delimiter between the rows of a matrix must be a semi-colon
>>% assign a matrix to b
>> b=[-2.5   3.1415926   0.1112;  1/3   0.0   -5]
b =
    -2.5000     3.1416      0.1112
     0.3333          0     -5.0000
>> % MATLAB responds by displaying the defined variable and its elemental values
>> % once the assignment statement has executed, MATLAB knows
>> % the dimension of the matrix and all of the elements in the matrix

>> b(1,2)   %  the element in row 1 and column 2
ans =    3.1416

>> b(2,3)   %  the element in row 2 and column 3
ans =    -5

>> b(3,2)   %  this will cause the following error message
??? Index exceeds matrix dimensions.
>> % in MATLAB a matrix index value CANNOT be zero or negative
>> b(2,-1)
??? Index exceeds matrix dimensions.
>> %  the dimension of a matrix can be found with the built-in function size
```

```
>> % get the dimension of the matrix b
>> size(b)
ans =    2    3
>> % this is the number of rows and columns, respectively
>> % the dimension can be made the elements of a two element row vector with
>> c=size(b)   % place the result in a vector
c =    2    3
>> % the first element of c is the number of rows in b
>> c(1)
ans =    2
```

If $N > 1$, $M > 1$, and $N \neq M$, then A is called a **rectangular matrix**. The matrix b in Example 3.1 is a rectangular matrix. If $N = M$, then A is called a **square matrix**, which has as many rows as it has columns. For an $(N \times N)$ square matrix A, the collection of elements $A(n,n)$, $n = 1, \ldots, N$ is called the **major diagonal** of A. If all elements other than the major diagonal are zero, then the matrix is called a **diagonal matrix**.

If $N = 1$, while $M > 1$, then A has only one row, and it is also called a **row vector**. If $N > 1$, while $M = 1$, then A has only one column, and it is also called a **column vector**. If $N = 1$ and $M = 1$, then A is no longer a matrix, and it is called instead a **scalar**. The elements of a vector B (row or column) having N elements are denoted by $B(n)$, $n = 1, \ldots, N$, where n is the vector index.

For an $(N \times M)$ matrix A, the elements $A(n,n)$, $n = 1, \ldots, N$ can be collected into a vector with the built-in MATLAB function **diag**, as in the MATLAB statement

```
d = diag(A)
```

If $N = M$, then d is the major diagonal of a square matrix. The function diag can also be used to create a diagonal matrix from a vector. If d is an N element vector, then the MATLAB statement

```
B = diag(d)
```

gives an $N \times N$ diagonal matrix B with d as its diagonal.

Example 3.2

The following statement assigns a vector to a variable name.

```
>> % a row vector is defined the same way as one row of a matrix
>> a=[-2.5    1    2^3    5.21+6.3    sin(pi/4)]
a =   -2.5000    1.0000    8.0000    11.5100    0.7071
>> % when MATLAB executes an assignment statement, it finds the values of all
>> % elements
```

```
>> % the built-in function size also gives the dimension of a vector
>> b=size(a)
b =     1     5
>> % here, only the number of columns is of interest
>> % another built-in function, length, gives the number of elements in a vector
>> c=length(a)
c =     5
```

Sometimes, especially for plotting, we want to set up a vector with elements that are incrementally related. The format for a MATLAB assignment statement to define this kind of a vector is given by

$$v = [first value : increment : last value]$$

where the brackets are optional. If we want N points from the first value to and including the last value, then the increment must be

$$increment = \frac{last\ value - first\ value}{N - 1}$$

Example 3.3

The following MATLAB statements illustrate using increments to assign a vector.

```
>> N=6;   %  desired number of points
>> first=0;   %  the first value
>> last=2.5;  %  the last value
>> increment=(last - first)/(N-1)
increment =     0.5000
>> % use the colon operator to define the elements of a vector
>> x=[first : increment : last]
x =     0    0.5000    1.0000    1.5000    2.0000    2.5000

>> % if the increment is not included in the assignment statement,
>> % then it becomes the default value of 1
>> n=[-2:4]
n =     -2    -1    0    1    2    3    4

>> % the increment can be negative
>> r=[5.0: -0.5 : 0]
r =
  Columns 1 through 11
    5.0000    4.5000    4.0000    3.5000    3.0000    2.5000    2.0000    1.5000    1.0000
      0.5000         0
```

With the built-in function **linspace**, we need only to specify the first and the last value and the desired number of points to specify the elements of a vector, because the function finds the increment. The format for the linspace function is

$$\texttt{linspace(first\_value, last\_value, number\_of\_points)}$$

This built-in function is very useful for setting up a vector to hold values of an independent variable to be used for function evaluation or plotting.

Example 3.4 _____

Here are a few examples of using linspace.

```
>> first=-10;  last=10;  N=6; % increment is (10-(-10)/(6-1)=4
>> w=linspace(first,last,N)
w =   -10    -6    -2    2    6    10
>> v=linspace(5,-5,6)  %  the increment can be positive or negative
v =    5    3    1    -1    -3    -5
>> % sometimes a small time range must be segmented into many sub-intervals
>> % segment the time range [0.0, 0.5] into 10^6 - 1 subintervals
>> t=linspace(0,0.5,1+10^6);
>> t(1)   %  the first value
ans =    0
>> t(length(t))  %  the last value
ans =    0.5000
>> t(2)-t(1)  %  time increment between first and second element of t
ans =    5.0000e-007  %  the increment is 0.5 microseconds
>> N=length(t)
N =    1000001
>> t(N)-t(N-1) %  time increment between last and previous element of t
ans =    5.0000e-007
>> % the increment is constant from one time point in t to the next time point
```

It is useful to be able to initialize a matrix (or vector) to have all of its elements equal to the same predetermined value. MATLAB has several built-in functions that are useful for this. With the **zeros** built-in function and the **ones** built-in function the size of a matrix can be specified.

Example 3.5 _____

Here, the zeros and ones functions are used to initialize some matrices. The zeros function is often utilized to allocate storage space for subsequent assignment.

```
>> x=zeros(2,14)  %  preallocating space
x =
     0     0     0     0     0     0     0     0     0     0     0     0     0     0
     0     0     0     0     0     0     0     0     0     0     0     0     0     0
>> y=ones(1,4)  %  create a 4 element row vector
y =   1     1     1     1
>> z=4*y   %  multiply each element of a vector by a scalar
z =   4     4     4     4
```

Table 3.1 gives several built-in MATLAB functions concerned with initializing a matrix and finding basic matrix properties. Use MATLAB help for details and other related functions.

Table 3.1   Some MATLAB functions concerned with initializing a matrix

MATLAB function	Description (See MATLAB help for different forms.)
d = size(X)	Returns the sizes of each dimension of array X in a vector d
N = length(X)	Returns the number of elements N in a vector X
any(X)	Returns 1 if any element of a vector X is a nonzero number
all(X)	Returns 1 if all elements of a vector are nonzero numbers
zeros(N,M)	Initializes an N × M matrix to all zeros
ones(N,M)	Initializes an N × M matrix to all ones
diag(X)	Returns the diagonal elements of a matrix
max(X)	Returns the largest element in a vector; see help max, if X is a matrix
min(X)	Returns the smallest element in a vector; see help min, if X is a matrix
rand(N,M)	Initializes an N × M matrix to uniformly distributed pseudorandom numbers over the range (0,1)
eye(N)	Initializes an N × N matrix to an identity matrix
find(X,K,mode)	Finds at most K indices of the mode =“first” or mode =“last” nonzero elements in a matrix
[]	Null matrix, assigns a matrix with no elements

3.2   Matrix Arithmetic

In MATLAB we can assign a matrix to any valid MATLAB variable name. Let A and B be the variable names of an $(N \times M)$ and an $(L \times K)$ matrix, respectively. Let c and d be the variable names of two column vectors, each having N elements.

The **inner product** (or **dot product**) of two vectors c and d, which have the same number of elements, is denoted by (c, d), and it results in a scalar. Let σ denote the scalar, which is given by

$$\sigma = (c, d) = c(1)\, d(1) + c(2)\, d(2) + \cdots + c(N)\, d(N) = \sum_{n=1}^{N} c(n)\, d(n) \qquad (3.1)$$

This is the sum of the products of corresponding elements in the two vectors. The built-in MATLAB function **dot** obtains the dot product of two vectors. This is shown in the following example.

A common matrix manipulation is the transpose of a matrix. The **transpose** of an $(N \times M)$ matrix A is another matrix F with dimension $(M \times N)$, where the rows of F are the columns of A. In MATLAB, the transpose of A is obtained with $F = A'$, where A' invokes the transpose operation of A. If c and d are two column vectors having the same length, then the inner product is given by $(c, d) = c' * d = d' * c$.

Example 3.6

Let us find the dot product. First, two vectors are defined.

```
>> clear all;  %  remove all variables and values from the Workspace
a=[1;  0;  -2;  3;];  b=[1;  2;  1;  2;];  %  assign values to two column vectors
c=dot(a,b)  %  get the inner product
c =     5
>> d=ones(1,4)  %  preallocate space and assign 1 to all values of a row vector
d =  1   1   1   1
>> e=a' * d'   %  this dot product (could be d*a) sums the elements of the vector a
e =     2
>>  %  this can also be done with the built-in MATLAB function sum
>> f=sum(a)   %  find the sum of the elements of the vector a
f =     2
```

Two matrices A and B can only be equal to each other if they have the same dimension, which means $N = L$ and $M = K$, and the corresponding elements are equal, that is, $A(n,m) = B(n,m)$, $n = 1, \ldots, N$ and $m = 1, \ldots, M$. Similarly, two vectors can only be equal to each other if they have the same length and the corresponding elements are equal to each other.

The addition (subtraction) of any two matrices is only possible if the two matrices have the same dimension, which will also be the dimension of the result. Let the matrix C be given by

$$C = A + B \tag{3.2}$$

The sum is only possible if $N = L$ and $M = K$. The elements of C are found by summing the corresponding elements of A and B, resulting in

$$C(n,m) = A(n,m) + B(n,m), \quad n = 1, \ldots, \ N, \ m = 1, \ldots, M \tag{3.3}$$

In this case we say that A and B are **conformable to addition**.

The mechanics of multiplying two matrices comes from the method of solution of a set of linear equations. We will investigate this in Section 3.5. Now, consider the product of two matrices, written as

$$D = A B \qquad (3.4)$$

which is written as, $D = A * B$, in MATLAB. Here, the matrix A is said to **premultiply** the matrix B, and the matrix B is said to **postmultiply** the matrix A. The product of two matrices is only possible if the premultiplying matrix has as many columns as the postmultiplying matrix has rows. Therefore, the matrices A and B in (3.4) are **conformable to multiplication** only if $M = L$. The dimension of the resulting product matrix D will be $(N \times M)(M \times K) \rightarrow (N \times K)$. Each element $D(n,k)$ in D is given by

$$D(n, k) = \sum_{m=1}^{M} A(n, m) B(m, k), \quad n = 1, \ldots, N, \quad k = 1, \ldots, K \qquad (3.5)$$

which is the dot product of the vector with elements from row n of matrix A and the vector with elements from column k of matrix B. To obtain the product $B A$ requires that $N = K$.

An $(N \times M)$ matrix A can be multiplied by a scalar α resulting in a matrix E, written as $E = \alpha A$, and each element of the matrix E is given by

$$E(n, m) = \alpha A(n, m), \quad n = 1, \ldots, N, \quad m = 1, \ldots, M \qquad (3.6)$$

An important matrix, which is called the **identity matrix** and is denoted by I, is a square $(N \times N)$ matrix defined by

$$I(n, m) = \begin{cases} 1, & n = m \\ 0, & n \neq m \end{cases} \qquad (3.7)$$

In an identity matrix, the **major diagonal** consists of ones, while the remaining elements are zeros. A useful property of the identity matrix is that for any vector or any $(N \times M)$ matrix (square or rectangular) R, we have $R I = R$, with an $(M \times M)$ identity matrix, or $I R = R$, with an $(N \times N)$ identity matrix. In MATLAB, the built-in function **eye** is used to specify an identity matrix according to (3.7).

Example 3.7

Let us use the following matrices in some examples to show how easy it is to use MATLAB for matrix arithmetic. We will use the built-in function eye to size and define an identity matrix.

$$A = \begin{bmatrix} 1 & 2 & 3 \\ -3 & -2 & -1 \end{bmatrix}, \quad B = \begin{bmatrix} 4 & -2 \\ -2 & 2 \end{bmatrix}, \quad C = \begin{bmatrix} 3 \\ 1 \end{bmatrix}, \quad d = 2$$

```
>> A=[1   2   3; -3   -2   -1]
A =
      1      2      3
     -3     -2     -1
```

```
>> A'   %  taking the transpose of A
ans =
     1    -3
     2    -2
     3    -1
>> A'*A  %  a matrix multiplied by its transpose always results in a square matrix
ans =
    10     8     6
     8     8     8
     6     8    10
>> %  a matrix multiplied by its transpose results in a square matrix
>> % that is symmetric about the major diagonal
>> A*A'
ans =
    14   -10
   -10    14
>> B=[4  -2; -2  2];  C=[3;1];  % used in the following examples
>> I=eye(2)  % the built-in function eye is used to specify a (2x2) identity matrix
I =
     1     0
     0     1
>> % A has three columns, and I has two rows
>> % A cannot premultiply I
>> D=A*I  % generates the following error message
??? Error using ==> mtimes
Inner matrix dimensions must agree.
>> D=I*A  %  (2x2)(2x3)  gives a (2x3) matrix, where D is the same as A
D =
     1     2     3
    -3    -2    -1
>> E=B*I  %  E is the same as B
E =
     4    -2
    -2     2
>> F=I*E  %  F is the same as E
F =
     4    -2
    -2     2
>> G=I*C  %  multiplication by I reproduces the vector C
G =
     3
     1
>> H=A*B  %  not conformable to multiplication
??? Error using ==> mtimes
```

```
Inner matrix dimensions must agree.
>> H=B*A  %  conformable to multiplication
H =
    10    12    14
    -8    -8    -8
>> P=B*C   %  conformable to multiplication
P =
    10
    -4
```

In the previous example we see how an identity matrix works as its name implies, and that it is very important to keep track of the order in which matrices are multiplied. While matrix addition is **commutative**, matrix multiplication is **not commutative**.

When we use vectors or matrices in arithmetic operations or arguments of functions, we must be careful about what is to be accomplished. Many built-in functions check the size of their input and process it accordingly. MATLAB includes useful modifications of the basic arithmetic operations. These are given in Table 3.2. Some of these operations are demonstrated in the following example.

In Table 3.3, let a and b designate scalars and let A and B designate vectors or matrices. Then, for example, $\sin(a)$ returns a scalar, while $\sin(A)$ returns a vector or a matrix. This feature of MATLAB is very convenient. Table 3.3 shows some examples of converting

Table 3.2   MATLAB arithmetic operations

Operation	Computation, comment	Function
+	A+B, addition of scalars, vectors, and matrices	plus(A,B)
+	+A, unary plus	uplus(A)
−	A−B, subtraction of scalars, vectors and matrices	minus(A,B)
−	−A, unary minus	uminus(A)
*	A*B, multiplication of scalars, vectors, and matrices	mtimes(A,B)
^	A^B, matrix power, A or B must be a scalar	mpower(A,B)
\	A\B = inv(A)*B, backslash, left divide	mldivide(A,B)
	A*X = B, X and B column vectors, solution is	
	X = A\B	
/	A/B = A*inv(B), slash, right divide	mrdivide
	X*B = A, X and A row vectors, solution is	
	X = Y/B	
.*	A.*B, element-by-element multiply	times(A,B)
.^	A.^B, element-by-element exponentiation	power(A,B)
.\	A.\B = (1./A).*B, element-by-element left divide	ldivide(A,B)
./	A./B = A.*(1./B), element-by-element right divide	rdivide(A,B)
'	A', transpose of A	transpose(A)

Table 3.3   Vectorization for element-by-element results

Scalar version	Possible vector versions
−a	−A
a + b	A + b; A + B
a * b	A * b; A .* B
a/b	a ./B; A/b; A./B
a\b = b/a	a.\B = B/a; A.\B
exp(a); sin(b)	exp(A); sin(B)
exp(a)*sin(b)	exp(A) .* sin(B)
sqrt(a^2 + b^2)	sqrt(A .^ 2 + B .^ 2)
a ^ b	a .^B; A .^ B
1/(a^2 + b^2)	1./(A .^2 + B .^2)

(**vectorizing**) a computation when a scalar is replaced by a vector or a matrix. This can only work if the vectors and matrices have conformable dimensions.

If you are not sure about how to vectorize an expression, then you could try the built-in MATLAB function **vectorize**, for example

```
>> vectorize ('1/(a^2 + b^2)')  % a and b are scalars
ans =
1./(a.^2 + b.^2)  % a and b are vectors or matrices
```

Example 3.8 _____

The following MATLAB statements demonstrate some matrix arithmetic.

```
>> A=[-1  0  2  5  3]; B=[5  4  3  2  1]; % assign two row vectors
>> c=A*B'  % inner product of the vectors A and B, same as a dot product
c =    14
>> C=A'*B  % outer product of the vectors A and B,  (5x1)(1x5) → (5x5)
C =
    -5    -4    -3    -2    -1
     0     0     0     0     0
    10     8     6     4     2
    25    20    15    10     5
    15    12     9     6     3
>> A=[1:5; 6:10]  %  use the colon operator to assign a 2x5 matrix
A =
     1     2     3     4     5
     6     7     8     9    10
```

```
>> B=1 ./ A  %  elemental divide using a scalar and a matrix
B =
    1.0000    0.5000    0.3333    0.2500    0.2000
    0.1667    0.1429    0.1250    0.1111    0.1000
>> D=1 ./ B(1,:)  % elemental divide using a scalar and a vector
D =   1    2    3    4    5
>> E=A .* B  %  element by element multiply of two matrices
E =
    1    1    1    1    1
    1    1    1    1    1
>> F=E./B  %  element by element divide of two matrices
F =
    1    2    3    4    5
    6    7    8    9   10
>> A=[-1  0  2  5  3];
>> C=2 .^ A  %  elemental exponentiation
C =    0.5000    1.0000    4.0000   32.0000    8.0000
>> B=[5  4  3  2  1];
>> D=A .^ B  % element by element exponentiation
D =    -1    0    8    25    3
```

In MATLAB, matrices can be involved in basic arithmetic operations, as demonstrated by the previous examples, and also in arguments of numerous MATLAB functions, both built-in and user defined. These kinds of possibilities allow for very convenient and efficient programming, that is, useful results can be achieved with minimal overhead programming.

3.3 Method of Least Squares

Sometimes a process is described by experimentally obtained data instead of a function relating a dependent variable to an independent variable. We have the data $(x(i), \ y(i))$, $i = 1, \ldots, N$, instead of a function $y = f(x)$ relationship. With a function relationship we can find a y for any given x. However, since N is finite, a particular value of x, for which we would like to know y, may not be included in the given set of data points.

Example 3.9 ————————————————————————————

For example, suppose N students are interviewed and the data obtained from each student is the GPA and hours spent per week studying. A **scatter diagram**, like the diagram shown in

Figure 3.1, gives a perspective about this data. This diagram gives the impression that a useful model for the data is a straight line.

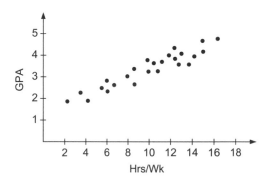

Figure 3.1   Scatter diagram of student data.

Let us model the data $(x(i), \ y(i)), \ i = 1, \ldots, N$ (e.g., the data shown in Fig. 3.1) with a straight line (a first order polynomial) given by

$$y = mx + b \tag{3.8}$$

where m is the slope and b is the intercept. With the model of (3.8) we expect to closely duplicate the given data points and find good estimates of y for values of x not included in the set of data points. Let $\hat{y}$ denote an estimate of y found with the model. For a given $x(i)$ in the data point set, an estimate of $y(i)$ is given by

$$\hat{y}(i) = m \, x(i) + b \tag{3.9}$$

where the estimation error is $e(i) = y(i) - \hat{y}(i)$. The best estimate results by using that slope $m = m_0$ and intercept $b = b_0$ that makes $e^2(i)$ as small as possible, and possibly $e^2(i) = 0$ if the number of data points is $N = 2$. However, when $N > 2$, we cannot expect to find an m_0 and b_0 that makes $e^2(i) = 0$ for all data points. Therefore, find m_0 and b_0 that makes the average error squared $\varepsilon^2(m, b)$ a minimum, where the average error squared is given by

$$\varepsilon^2(m, b) = \frac{1}{N} \sum_{i=1}^{N} e^2(i) = \frac{1}{N} \sum_{i=1}^{N} (y(i) - \hat{y}(i))^2 = \frac{1}{N} \sum_{i=1}^{N} (y(i) - (m \, x(i) + b))^2 \tag{3.10}$$

The average error squared is a quadratic function of m and b, which has a single unique minimum where the derivatives of $\varepsilon^2(m, b)$ with respect to m and b are zero. Using the chain rule of differentiation, these derivatives (partial derivatives) are given by

$$\frac{\partial \varepsilon^2(m,b)}{\partial m} = \frac{1}{N}\sum_{i=1}^{N} 2e(i)\frac{\partial e(i)}{\partial m}$$

$$= \frac{1}{N}\sum_{i=1}^{N} 2(y(i) - \hat{y}(i))\left(-\frac{\partial \hat{y}(i)}{\partial m}\right)$$

$$= \frac{1}{N}\sum_{i=1}^{N} 2(y(i) - (m\,x(i) + b))(-x(i)) \tag{3.11}$$

and

$$\frac{\partial \varepsilon^2(m,b)}{\partial b} = \frac{1}{N}\sum_{i=1}^{N} 2e(i)\frac{\partial e(i)}{\partial b} = \frac{1}{N}\sum_{i=1}^{N} 2(y(i) - (m\,x(i) + b))(-1) \tag{3.12}$$

Setting (3.11) and (3.12) to zero, where $m = m_0$ and $b = b_0$, gives

$$\frac{1}{N}\sum_{i=1}^{N} 2(y(i) - (m_0\,x(i) + b_0))(-x(i)) = 0 \rightarrow$$

$$\frac{1}{N}\sum_{i=1}^{N} y(i)\,x(i) - m_0\frac{1}{N}\sum_{i=1}^{N} x^2(i) - b_0\frac{1}{N}\sum_{i=1}^{N} x(i) = 0$$

and

$$\frac{1}{N}\sum_{i=1}^{N} 2(y(i) - (m_0\,x(i) + b_0))(-1) = 0 \rightarrow \frac{1}{N}\sum_{i=1}^{N} y(i) - m_0\frac{1}{N}\sum_{i=1}^{N} x(i) - b_0 = 0$$

Let

$$\bar{x} = \frac{1}{N}\sum_{i=1}^{N} x(i), \;\; \bar{y} = \frac{1}{N}\sum_{i=1}^{N} y(i), \;\; \overline{xy} = \frac{1}{N}\sum_{i=1}^{N} x(i)y(i), \;\; \overline{x^2} = \frac{1}{N}\sum_{i=1}^{N} x^2(i)$$

and we get

$$\bar{y} = m_0\bar{x} + b_0 \tag{3.13}$$

$$\overline{xy} = m_0\,\overline{x^2} + b_0\,\bar{x} \tag{3.14}$$

Solving (3.13) for b_0 gives

$$b_0 = \bar{y} - m_0\bar{x} \tag{3.15}$$

and after substituting (3.15) into (3.14), we can find m_0, which is given by

$$m_0 = \frac{\overline{xy} - \overline{x}\,\overline{y}}{\overline{x^2} - \overline{x}^2} \tag{3.16}$$

These results were obtained by optimizing the average error squared given in (3.10). This is a quadratic function of the two variables m and b, which has a unique minimum. We found the best values for m and b by setting the derivatives of the average error squared with respect to m and b to zero. This resulted in two linear equations in m_0 and b_0 that were easy to solve. Generally, optimizing a function is not always so easy. For example, a given function may not be a quadratic, there may not be a unique minimum, derivatives of the function to be optimized may be difficult or practically impossible to obtain, and even if the derivatives can be found, the resulting equations from setting the derivatives to zero might be difficult to solve. There are numerical methods to optimize a function, and MATLAB includes built-in functions for optimizing a given function, such as the function described by (3.10). For example, the built-in function **fminsearch** is a function function, which finds the minimum of a multivariable function using a derivative-free method, and it could have been applied to the function defined by (3.10). Or, the built-in function **polyfit** can be applied directly to the given data to fit a first-order polynomial to the data. See the MATLAB help facility to learn more about built-in optimization functions. Use the built-in function **doc**, as in doc polyfit, to find out more about polyfit.

Example 3.9 (continued) _____

Given N data points $(x(i), y(i))$, $i = 1, \ldots, N$, we can use (3.16) and then (3.15) to find the best (in the **least mean error squared sense**) model. To illustrate this, suppose $N = 12$ data points are given by

GPA	1.8	2.1	2.4	2.9	3.1	3.3	3.9	4.1	3.8	4.2	4.7	4.9
Hrs/Wk	2.6	3.3	4.6	5.7	5.9	6.3	8.2	11.7	13.1	13.5	14.9	17.8

A MATLAB script to find and test the model follows.

```
clear all; clc;
% given data
Hrs_per_Wk=[2.6 3.3 4.6 5.7 5.9 6.3 8.2 11.7 13.1 13.5 14.9 17.8];
xdat=Hrs_per_Wk; % will make code easier to follow
GPA=[1.8 2.1 2.4 2.9 3.1 3.3 3.9 4.1 3.8 4.2 4.7 4.9];
ydat=GPA; % will make code easier to follow
stem(xdat,ydat); % open Figure Window and plot given data
xlabel('Hrs/Wk');
ylabel('GPA')
grid on
hold on   % causes next plot to appear in the same Figure Window
```

```
N=length(xdat); % number of data points
%
Av_xdat=sum(xdat)/N; % sum elements of xdat, then divide by N
% the built-in MATLAB function mean finds the average value of the elements
% in a vector, e.g., Av_xdat = mean(xdat)
%
Av_ydat=mean(ydat);  % average GPA
%
Av_xydat=mean(xdat.*ydat);  % first do element by element multiply
Av_xxdat=mean(xdat.*xdat);  % first square the elements
% find the slope
m=(Av_xydat - Av_xdat * Av_ydat)/(Av_xxdat - Av_xdat^2);
% find the intercept
b= Av_ydat - m * Av_xdat;
% plot linear regression (linear model) over a range
x=linspace(0,20,101);
y=m*x+b; % least squared error model of data
plot(x,y)
y_est=m*xdat+b; % estimate of GPA for given Hrs_per_Wk
est_er=ydat-y_est; % estimation error for each data point
est_er_sqr= est_er .* est_er; % estimation error squared
[max_er_sqr,i]=max(est_er_sqr) % data point that gives max estimation error
```

Program 3.1   Program to find the linear regression of a data set.

Figure 3.2 shows the given data and the least mean error squared model, called a **linear regression**, of the given data. From the regression analysis, we may conclude that, on average, the expected GPA of a student that spends 10 hours/week studying is approximately GPA=3.7/5.

The slope is $m = 0.1890$, and the intercept is $b = 1.7385$. With the built-in function **polyfit**, we get

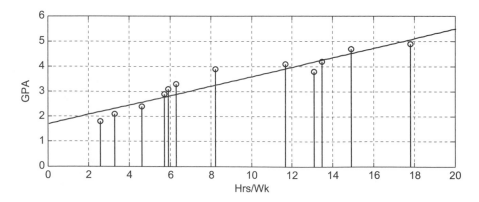

Figure 3.2   Linear regression of a given data set.

```
n = 1; % fit a first order polynomial to the data
coeffs = polyfit(xdat,ydat,n)

coeffs =
    0.1890    1.7385
```

The slope gives us an indication of the dependence of the GPA on the hours per week spent studying. For example, if the slope had been near zero, then the GPA does not depend very much on the hours per week spent studying.

A better indicator of the degree to which two variables are linearly related is the **correlation coefficient** ρ, which is a normalized slope, given by

$$\rho = \frac{\overline{xy} - \overline{x}\,\overline{y}}{\sqrt{\overline{x^2} - \overline{x}^2}\sqrt{\overline{y^2} - \overline{y}^2}} \tag{3.17}$$

For the data used in Example 3.9, (3.17) gives $\rho = 0.9471$. The built-in function **corrcoef** computes the correlation coefficient, which was found with

```
R = corrcoef(xdat,ydat) % compute the correlation matrix
R =
    1.0000    0.9471
    0.9471    1.0000
```

The correlation coefficient range is restricted to $-1 \leq \rho \leq +1$. If $\rho = 0$, then there is no linear relationship between the variables x and y. If $\rho = 1$, then there is a strong linear relationship between the variables, while if $\rho = -1$, then there is a strong but opposite relationship between the variables.

There are many MATLAB built-in functions concerned with data modeling and analysis. Some of these are listed in Table 3.4.

Table 3.4   *Functions concerned with data modeling and data analysis*

Function	Brief description
polyfit	Polynomial curve fitting
corrcoef	Correlation coefficient
spline	Cubic spline interpolation
mean	Average or mean value of a vector or matrix
var	Variance
interp2	2-D data interpolation
mode	Most frequent value in a vector or matrix
std	Standard deviation

3.4   Function of a Matrix

A noteworthy feature of MATLAB is that many of its built-in functions return a matrix when the argument of a function is a matrix, where each element of the returned matrix is the function evaluated for the corresponding element of the function argument matrix.

Example 3.10 _____

The following MATLAB statements illustrate only a few of the many ways a matrix can be used to provide the argument of a function.

```
>> t=linspace(0,1,5);  % a vector with 5 elements evenly distributed from 0 to 1
>> x=2*cos(pi*t)  %  evaluate the cos function for each element in t
x =    2.0000    1.4142    0.0000   -1.4142   -2.0000
>> A=[0 1; -0.5 -0.9]; T=0.1;  % define a 2x2 matrix
>> v=2*exp(A*T)  % exponential function of a matrix
v =
    2.0000    2.2103
    1.9025    1.8279
>> w=linspace(0,4*pi,5);  % 5 points over the frequency range 0 to 4 pi
>> s=j*w   %  s is an imaginary number
s =
  Columns 1 through 5
        0         0 + 3.1416i         0 + 6.2832i         0 + 9.4248i         0 +12.5664i
>> H=1./(1+s)  % the denomiator adds 1 to each element of s
H =
  Columns 1 through 5
   1.0000   0.0920 - 0.2890i    0.0247 - 0.1552i    0.0111 - 0.1049i    0.0063 - 0.0791i
>> % convert H magnitude to decibels (dB), logarithmic scale
>> H_mag_dB=20*log10(abs(H))
H_mag_dB =      0   -10.3621   -16.0722   -19.5340   -22.0116
```

In addition to the built-in functions like **dot** and **sum**, MATLAB has numerous other built-in functions that manipulate the elements of vectors and matrices. Some of these kinds of functions are given in Table 3.5. Use MATLAB help to find out more about these built-in functions.

Table 3.5   Some MATLAB functions that manipulate the elements of a matrix

Operation	Comment
tril(A)	Returns lower triangular part of a matrix A
triu(A)	Returns upper triangular part of a matrix A
fliplr(A)	Flip matrix A left to right
flipud(A)	Flip matrix A up to down
circshift(A,[N M])	Shift circularly rows by N and columns by M

3.5   Solution of a Set of Linear Equations

In this section we will investigate how to solve a set of N linear equations in N unknowns, given by

$$
\begin{aligned}
a_{1,1}\,x_1 + a_{1,2}\,x_2 + & \quad \cdots \quad + a_{1,N}\,x_N = y_1 \\
a_{2,1}\,x_1 + a_{2,2}\,x_2 + & \quad \cdots \quad + a_{2,N}\,x_N = y_2 \\
& \quad \vdots \\
a_{N,1}\,x_1 + a_{N,2}\,x_2 + & \quad \cdots \quad + a_{N,N}\,x_N = y_N
\end{aligned}
\tag{3.18}
$$

With matrix notation, these equations can be written succinctly as

$$
A\,X = Y
\tag{3.19}
$$

where A is an $N \times N$ (square) matrix with $A(n, m) = a_{n,m}$ and Y is an $N \times 1$ column vector with $Y(n) = y_n$, each with known elements, and X is an $N \times 1$ vector of unknown elements. We can expand (3.19) to write it as

$$
X(1)\begin{bmatrix} A(1,1) \\ A(2,1) \\ \vdots \\ A(N,1) \end{bmatrix} + X(2)\begin{bmatrix} A(1,2) \\ A(2,2) \\ \vdots \\ A(N,2) \end{bmatrix} + \cdots + X(N)\begin{bmatrix} A(1,N) \\ A(2,N) \\ \vdots \\ A(N,N) \end{bmatrix} = Y
$$

This shows that the vector Y must be some linear combination of the column vectors of A.

The linear equations in (3.19) are **independent** if none of the equations can be derived algebraically from the others. When the equations are independent, each equation contains new information about the unknown variables. The equations are **dependent** if at least one of them can be derived algebraically from the other equations. The N linear equations are **consistent** if they possess a common solution, and **inconsistent** otherwise. If the equations are inconsistent, then they have no solution. For example, consider the following three sets of equations:

$$
(a)\ \begin{aligned} x_1 + 2x_2 - x_3 &= -2 \\ 2x_1 - x_2 + 5x_3 &= -4 \\ -x_1 + 3x_2 - 6x_3 &= 2 \end{aligned}
\qquad
(b)\ \begin{aligned} -x_1 + 3x_2 &= 2 \\ 2x_1 - 6x_2 &= -5 \end{aligned}
\qquad
(c)\ \begin{aligned} -2x_1 + 3x_2 &= 4 \\ -x_1 + x_2 &= -3 \end{aligned}
$$

Can you determine which set of equations is a dependent, inconsistent, or independent set of equations?

Consider the equation $a\,x = y$, where a, x, and y are scalars, and a and y are given numbers. To solve for x, we divide both sides of this equation by a to get $x = y/a$. Or, let us multiply both sides of the given equation by some number b to get $ba\,x = b\,y$, and then find b such that $b\,a = 1$, resulting in $x = b\,y$. From $b\,a = 1$ we can find b, if $a \neq 0$, and therefore, with $b = a^{-1}$, the inverse of a, we have

$$x = a^{-1}y \tag{3.20}$$

In conventional matrix algebra there is no operation analogous to scalar division. Instead, there is an operation analogous to finding b given a in $b\,a = 1$.

To develop the inverse concept further, let us solve the two equations in two unknowns given by

$$a_{1,1}\,x_1 + a_{1,2}\,x_2 = y_1 \tag{3.21}$$

$$a_{2,1}\,x_1 + a_{2,2}\,x_2 = y_2 \tag{3.22}$$

where x_1 and x_2 are the unknowns.

If we consider the left sides of (3.21) and (3.22) to be the dot product of a row vector and a column vector, we can employ the definition of matrix multiplication given in (3.5). Let us define the matrices A, X, and Y as

$$A = \begin{bmatrix} a_{1,1} & a_{1,2} \\ a_{2,1} & a_{2,2} \end{bmatrix}, \quad X = \begin{bmatrix} x_1 \\ x_2 \end{bmatrix}, \quad Y = \begin{bmatrix} y_1 \\ y_2 \end{bmatrix}$$

Then, through matrix multiplication and matrix equality as defined in Section 3.2, we can write (3.21) and (3.22) more succinctly as

$$AX = Y \tag{3.23}$$

Analogously to (3.20), we want to find a matrix B, such that $B\,A = I$, and then $X = B\,Y$.

To find B we will solve for x_1 and x_2 by repeated substitution. From (3.22) we get

$$x_2 = \frac{y_2 - a_{2,1}\,x_1}{a_{2,2}}$$

and substitution into (3.21) gives

$$a_{1,1}\,x_1 + \frac{a_{1,2}\,y_2 - a_{1,2}\,a_{2,1}\,x_1}{a_{2,2}} = y_1$$

Solving for x_1 results in

$$x_1 = \frac{1}{a_{1,1}\,a_{2,2} - a_{1,2}\,a_{2,1}}\,(a_{2,2}\,y_1 - a_{1,2}\,y_2) \tag{3.24}$$

To solve for x_2, use (3.21) to find x_1 and substitute this x_1 into (3.22), and then find x_2 given by

$$x_2 = \frac{1}{a_{1,1}\, a_{2,2} - a_{1,2}\, a_{2,1}} (-a_{2,1}\, y_1 + a_{1,1}\, y_2) \tag{3.25}$$

By inspection of (3.24) and (3.25) we can write the vector X as

$$X = \begin{bmatrix} x_1 \\ x_2 \end{bmatrix} = \frac{1}{a_{1,1}\, a_{2,2} - a_{1,2}\, a_{2,1}} \begin{bmatrix} a_{2,2} & -a_{1,2} \\ -a_{2,1} & a_{1,1} \end{bmatrix} \begin{bmatrix} y_1 \\ y_2 \end{bmatrix} \tag{3.26}$$

and therefore B must be given by

$$B = \frac{1}{a_{1,1}\, a_{2,2} - a_{1,2}\, a_{2,1}} \begin{bmatrix} a_{2,2} & -a_{1,2} \\ a_{2,1} & a_{1,1} \end{bmatrix}$$

which is called the **inverse matrix** of A, and it is denoted by A^{-1}, so that $X = A^{-1}\, Y$, analogously to (3.20). The scalar, $a_{1,1}\, a_{2,2} - a_{1,2}\, a_{2,1}$, is called the **determinant** of A, which is denoted by $|A|$, and therefore

$$A^{-1} = \frac{1}{|A|} \begin{bmatrix} a_{2,2} & -a_{1,2} \\ -a_{2,1} & a_{1,1} \end{bmatrix} \tag{3.27}$$

Here, the determinant is given by

$$|A| = a_{1,1}\, a_{2,2} - a_{1,2}\, a_{2,1} \tag{3.28}$$

The inverse of a square $(N \times N)$ matrix can be found with the built-in MATLAB function **inv**, and the determinant can be found with the built-in MATLAB function **det**.

Example 3.11 _____

Let us solve the following two equations in the two unknowns i_1 and i_2.

$$6i_1 - 4i_2 = 10$$
$$4i_1 - i_2 = 5$$

Let

$$R = \begin{bmatrix} 6 & -4 \\ 4 & -1 \end{bmatrix}, \quad V = \begin{bmatrix} 10 \\ 5 \end{bmatrix}, \quad I = \begin{bmatrix} i_1 \\ i_2 \end{bmatrix}$$

and the problem is to solve: $R\, I = V$, for I. This is accomplished with the following MATLAB statements.

```
>> R=[6  -4; 4  -1];   % specify the coefficient matrix
>> V=[10  5]'   % using transpose to set up a column vector
V =
   10
    5
>> S=inv(R)    %  the inverse of R could also be found with R^(-1)
S =
   -0.1000    0.4000
   -0.4000    0.6000
>> S*R    %  check that S is the inverse of R
ans =
    1    0
    0    1
>> I=S*V    %  The solution I could be found directly with I=R^(-1)*V
I =
    1.0000
   -1.0000
>> d=det(R)    %  check the determinant
d =   10
>> % the left divide gives another way to solve N equations in N unknowns
>> I = R\V
I =
    1.0000
   -1.0000
```

Another way to write a set of N equations in N unknowns X is

$$X A = Y$$

where X and Y are row vectors. Postmultiplying both sides of this equation by A^{-1} gives

$$X A A^{-1} = Y A^{-1} \rightarrow X = Y A^{-1}$$

In MATLAB, the right divide operation solves for X with $X = Y/A$.

Given a set of N linear equations in N unknowns, we can write them succinctly like (3.19), regardless of how large N may be. If the $(N \times N)$ inverse matrix A^{-1} exists such that

$$A^{-1} A = A A^{-1} = I$$

where I is the $(N \times N)$ identity matrix, then the matrix A is said to be **nonsingular**, and the only solution of (3.19) is given by

$$X = A^{-1} Y \tag{3.29}$$

If the matrix A does not have an inverse, that is, A^{-1} does not exist, then the matrix A is said to be a **singular** matrix, and (3.19) has no solution.

If $Y = 0$, then (3.19) becomes a **homogeneous** equation given by

$$AX = 0 \tag{3.30}$$

If A is nonsingular, then the solution of (3.30) is the **trivial** solution $X = 0$. A necessary and sufficient condition that (3.30) has a solution other than the trivial solution is $|A| = 0$, which means that A must be a singular matrix.

There are many methods to find the determinant and inverse of a square matrix. One of these methods is based on finding the **cofactor** matrix of A, $cofactor(A)$, which is an $(N \times N)$ matrix of cofactors given by

$$cofactor(A) = \begin{bmatrix} \alpha_{1,1} & \cdots & \alpha_{1,N} \\ \vdots & \ddots & \vdots \\ \alpha_{N,1} & \cdots & \alpha_{N,N} \end{bmatrix}$$

The cofactor $\alpha_{n,m}$ in row n and column m of $cofactor(A)$ is a sign adjusted determinant found with

$$\alpha_{n,m} = (-1)^{(n+m)} \left| A \right|_{\substack{\text{row } n \text{ and} \\ \text{column } m \\ \text{removed}}}, \quad n = 1, \ldots, N, \quad m = 1, \ldots, N$$

To find $cofactor(A)$ requires finding N^2 determinants of $((N-1) \times (N-1))$ sub-matrices of A. The **inverse matrix** of A is given by

$$A^{-1} = \frac{1}{|A|} (cofactor(A))' \tag{3.31}$$

A square matrix is nonsingular if and only if its determinant is not zero. The **determinant**, $|A|$, is found with

$$|A| = \begin{cases} \displaystyle\sum_{k=1}^{N} a_{n,k}\, \alpha_{n,k}, & \text{using any row}, \quad n = 1, \ldots, N \\[4mm] \displaystyle\sum_{k=1}^{N} a_{k,n}\, \alpha_{k,n}, & \text{using any column}, \quad n = 1, \ldots, N \end{cases} \tag{3.32}$$

Notice that if A has a row or column of all zeros, then $|A| = 0$. Let $A(n, :)$ denote the nth row of A, where the colon means all columns in the nth row. Using (3.32), it can also be proved that if any row, say the nth row, of A can be written as a linear combination of the other rows of A, which means that we can write

$$A(n,:) = \sum_{\substack{k=1 \\ k \neq n}}^{N} c_k\, A(k,:)$$

for some constants c_k, $k = 1,\dots,N$, $k \neq n$, then $|A| = 0$. Similarly, $|A| = 0$, if any column of A can be written as a linear combination of the other columns of A.

Example 3.12

A set of three equations in three unknowns is given by

$$ZI = V \tag{3.33}$$

where the (3×3) matrix Z and the (3×1) vector V are given by

$$Z = \begin{bmatrix} 0 & 4 & -1 \\ -2 & 1 & 0 \\ 5 & 0 & 3 \end{bmatrix}, \quad V = \begin{bmatrix} 1 \\ 0 \\ -1 \end{bmatrix}$$

Let us find the inverse of the matrix Z. The 9 cofactors are given by

$$\alpha_{1,1} = (-1)^{1+1}\begin{vmatrix} 1 & 0 \\ 0 & 3 \end{vmatrix} = 3, \quad \alpha_{1,2} = (-1)^{1+2}\begin{vmatrix} -2 & 0 \\ 5 & 3 \end{vmatrix} = 6, \quad \alpha_{1,3} = (-1)^{1+3}\begin{vmatrix} -2 & 1 \\ 5 & 0 \end{vmatrix} = -5$$

$$\alpha_{2,1} = (-1)^{2+1}\begin{vmatrix} 4 & -1 \\ 0 & 3 \end{vmatrix} = -12, \quad \alpha_{2,2} = (-1)^{2+2}\begin{vmatrix} 0 & -1 \\ 5 & 3 \end{vmatrix} = 5, \quad \alpha_{2,3} = (-1)^{2+3}\begin{vmatrix} 0 & 4 \\ 5 & 0 \end{vmatrix} = 20$$

$$\alpha_{3,1} = (-1)^{3+1}\begin{vmatrix} 4 & -1 \\ 1 & 0 \end{vmatrix} = 1, \quad \alpha_{3,2} = (-1)^{3+2}\begin{vmatrix} 0 & -1 \\ -2 & 0 \end{vmatrix} = 2, \quad \alpha_{3,3} = (-1)^{3+3}\begin{vmatrix} 0 & 4 \\ -2 & 1 \end{vmatrix} = 8$$

and the cofactor matrix becomes

$$cofactor(Z) = \begin{bmatrix} 3 & 6 & -5 \\ -12 & 5 & 20 \\ 1 & 2 & 8 \end{bmatrix}$$

Using the first row of Z and $cofactor(Z)$, we get

$$|Z| = (0)(3) + (4)(6) + (-1)(-5) = 29$$

and the inverse matrix is given by

$$Z^{-1} = \frac{1}{29} \begin{bmatrix} 3 & 6 & -5 \\ -12 & 5 & 20 \\ 1 & 2 & 8 \end{bmatrix}' = \frac{1}{29} \begin{bmatrix} 3 & -12 & 1 \\ 6 & 5 & 2 \\ -5 & 20 & 8 \end{bmatrix}$$

With the built-in MATLAB function **inv** we get

```
>> Z=[0   4  -1; ...   %  continuation makes it easier to see the matrix
      -2   1   0; ...
       5   0   3]
Z =
     0     4    -1
    -2     1     0
     5     0     3
>> det(Z)   %  checking determinant
ans =    29
>>  Y = Z^-1   %  using built-in MATLAB inverse computation
Y  =
    0.1034   -0.4138    0.0345
    0.2069    0.1724    0.0690
   -0.1724    0.6897    0.2759

>> Y*Z   % checking inverse
ans =
    1.0000         0   -0.0000
    0.0000    1.0000         0
    0.0000         0    1.0000
%  due to finite word length arithmetic a number that is expected to be zero
%  may instead be an extremely small number
>> V = [1 0 -1]'
V =
     1
     0
    -1
>> I = Y*V
I =
    0.0690
    0.1379
   -0.4483
```

3.5.1 Gauss–Jordan Elimination

There are many other methods to solve (3.19) for X, including **Gauss elimination**. This method starts by augmenting the given matrix A with the column vector Y to obtain the $N \times (N + 1)$ **augmented matrix** g given by

$$g = [A|Y] = \begin{bmatrix} A(1,1) & A(1,2) & \dots & A(1,N) & Y(1) \\ A(2,1) & A(2,2) & \dots & \vdots & Y(2) \\ \vdots & \vdots & \ddots & \vdots & \vdots \\ A(N,1) & A(N,2) & \dots & A(N,N) & Y(N) \end{bmatrix}$$

which represents both sides of (3.19). Gauss elimination is based on the properties that the solution of (3.19) cannot change if

1) any row of g is multiplied by a nonzero constant
2) any two rows of g are exchanged
3) any row of g is replaced by the same row minus any other row of g

These operations are called **elementary row operations**, and they can be combined to perform more complicated operations without changing the solution of (3.19).

Gauss elimination applies elementary row operations to bring g to the form given by

$$G = [a|y] = \begin{bmatrix} a(1,1) & a(1,2) & \dots & a(1,N) & y(1) \\ 0 & a(2,2) & \dots & \vdots & y(2) \\ \vdots & \ddots & \ddots & \vdots & \vdots \\ 0 & \dots & 0 & a(N,N) & y(N) \end{bmatrix}$$

The matrix g has been transformed to include an upper triangular $N \times N$ matrix a, where the elements of a below the major diagonal are zero, and (3.19) becomes

$$a X = y$$

The solution of (3.19) is then found by repeated backward substitution, which starts with

$$X(N) = \frac{y(N)}{a(N,N)}$$

and continues with

$$X(N-i) = \frac{1}{a(N-i, N-i)} \left[y(N-i) - \sum_{k=0}^{i-1} a(N-i, N-k) * X(N-k) \right], \quad i = 1, \dots, N-1$$

$$(3.34)$$

The mechanics of (3.34) can be avoided by continuing to apply elementary row operations to transform G to the form

$$
J = [b|z] =
\begin{bmatrix}
b(1,1) & 0 & \cdots & 0 & z(1) \\
0 & b(2,2) & \cdots & \vdots & z(2) \\
\vdots & \vdots & \ddots & 0 & \vdots \\
0 & 0 & \cdots & b(N,N) & z(N)
\end{bmatrix}
$$

where the $N \times N$ matrix b is a diagonal matrix. The process from g to J is called **Gauss-Jordan elimination**. With the matrix J, (3.19) becomes

$$bX = z$$

and the solution X is given by

$$X(i) = z(i)/b(i,i), \quad i = 1, \ldots, N$$

If the ith row of J is divided by $b(i,i)$, $i = 1, \ldots, N$, then we get

$$
K =
\begin{bmatrix}
1 & 0 & \cdots & 0 & X(1) \\
0 & 1 & \cdots & \vdots & X(2) \\
\vdots & \vdots & \ddots & 0 & \vdots \\
0 & 0 & \cdots & 1 & X(N)
\end{bmatrix}
$$

and the last column of K is the solution of (3.19). The structure of K is called a **row reduced echelon form**, and MATLAB has a built-in function, **rref**, that receives the matrix g, and returns the matrix K.

Example 3.12 (continued)

Let us continue Example 3.12 to solve for I given the matrices Z and V. The augmented matrix is

$$
g =
\begin{bmatrix}
0 & 4 & -1 & 1 \\
-2 & 1 & 0 & 0 \\
5 & 0 & 3 & -1
\end{bmatrix}
$$

Starting with g, perform the following steps: to get

(a) exchange the first and second rows
(b) add 5 times the first row to 2 times the third row
(c) add -5 times the second row to 4 times the third row

$$
g \xrightarrow{\text{(a)}}
\begin{bmatrix}
-2 & 1 & 0 & 0 \\
0 & 4 & -1 & 1 \\
5 & 0 & 3 & -1
\end{bmatrix}
\xrightarrow{\text{(b)}}
\begin{bmatrix}
-2 & 1 & 0 & 0 \\
0 & 4 & -1 & 1 \\
0 & 5 & 6 & -2
\end{bmatrix}
\xrightarrow{\text{(c)}}
\begin{bmatrix}
-2 & 1 & 0 & 0 \\
0 & 4 & -1 & 1 \\
0 & 0 & 29 & -13
\end{bmatrix}
= G
$$

With the matrix G, the original matrix equation (3.33) becomes

$$aI = y \rightarrow \begin{bmatrix} -2 & 1 & 0 \\ 0 & 4 & -1 \\ 0 & 0 & 29 \end{bmatrix} \quad I = \begin{bmatrix} 0 \\ 1 \\ -13 \end{bmatrix}$$

which has the same solution as (3.33). Starting with the last element of I, we get $I(3) = -13/29$, and then, with repeated substitution described by (3.34), we can find $I(2)$ and $I(1)$.

Earlier we found I using the built-in MATLAB function **inv** to obtain the inverse matrix of Z. However, our main interest is to solve the set of equations, $ZI = V$, for I, which can be accomplished with Gauss elimination, instead of matrix inversion. MATLAB has a built-in function, called **mldivide** (matrix left divide), that applies Gauss elimination to solve a set of linear equations. This alternative to using matrix inversion requires much less computing time. With mldivide we get

```
>> Z = [0   4   -1; -2   1    0; 5  0    3];
>> V = [1; 0; -1];
>> I = Z\V    %   same as I = mldivide(Z,V)
I =
    0.0690
    0.1379
   -0.4483
>> I = mldivide(Z,V)
I =
    0.0690
    0.1379
   -0.4483
```

Let us continue to do Gauss–Jordan elimination. Starting with G, perform the following steps:

(d) add the third row to 29 times the second row
(e) add the second row to -116 times the first row

to get

$$G \xrightarrow{(d)} \begin{bmatrix} -2 & 1 & 0 & 0 \\ 0 & 116 & 0 & 16 \\ 0 & 0 & 29 & -13 \end{bmatrix} \xrightarrow{(e)} \begin{bmatrix} 232 & 0 & 0 & 16 \\ 0 & 116 & 0 & 16 \\ 0 & 0 & 29 & -13 \end{bmatrix} = J$$

With the matrix J, the original matrix equation (3.33) becomes

$$bI = z \rightarrow \begin{bmatrix} 232 & 0 & 0 \\ 0 & 116 & 0 \\ 0 & 0 & 29 \end{bmatrix} I = \begin{bmatrix} 16 \\ 16 \\ -13 \end{bmatrix}$$

which has the same solution as (3.33). Now, I can be easily found. Or, we can use the built-in function **rref** to get

```
>> g = [0  4 -1  1; -2  1  0  0; 5  0  3  -1];
>> K = rref(g)
K =
    1.0000         0         0    0.0690
         0    1.0000         0    0.1379
         0         0    1.0000   -0.4483
```

3.6 Special Matrix Manipulations

With MATLAB we can manipulate the elements of a matrix in ways that are not a part of conventional matrix algebra. However, these kinds of manipulations are particularly useful and convenient for involving matrices in problem solving.

A useful concept about a matrix is an **empty matrix**. An empty matrix is a matrix with no elements in it. To assign an empty matrix to a variable, for example, X, we write: $X = [\ ]$. This assignment statement makes X a matrix with nothing in it. This is not the same as setting the elements of a matrix to zero or blank. The following examples illustrate some possibilities.

3.6.1 Extracting a Sub-Matrix

Given a matrix, it is possible to extract any part (sub-matrix) of the matrix. The colon operator is used to delimit the part of a matrix to be extracted.

Example 3.13 —————————————————————————

Let us use the colon operator to extract various sub-matrices from a matrix. We start by defining a matrix to use in examples.

```
>> A=[0   1   2   3   4    5   6   7   8    9; ...
     -9  -8  -7  -6  -5   -4  -3  -2  -1    0; ...
```

```
      0   1   0   2   0   3   0   4   0   5; ...
      5   0   4   0   3   0   2   0   1   0]
A =
      0     1     2     3     4     5     6     7     8     9
     -9    -8    -7    -6    -5    -4    -3    -2    -1     0
      0     1     0     2     0     3     0     4     0     5
      5     0     4     0     3     0     2     0     1     0
>> x=A(2,:)  % extracting row 2, where the colon means to use all columns
x =     -9    -8    -7    -6    -5    -4    -3    -2    -1     0
>> y=A(:,3)  % extracting column 3, where the colon means to use all rows
y =
      2
     -7
      0
      4
>> B=A(1:3,6:10)  % get sub-matrix, where colon specifies index ranges
B =
      5     6     7     8     9
     -4    -3    -2    -1     0
      3     0     4     0     5
>> A(2,:)=A(1,:).*A(3,:)
% replacing row 2 with the element by element product of rows 1 and 3
A =
      0     1     2     3     4     5     6     7     8     9
      0     1     0     6     0    15     0    28     0    45
      0     1     0     2     0     3     0     4     0     5
      5     0     4     0     3     0     2     0     1     0
>> s=size(A)
s =   4    10
>> C=A(2:end,7:end)  % end is used to indicate the last element index
C =
      0    28     0    45
      0     4     0     5
      2     0     1     0
>>
```

A vector can be used to specify the rows or columns to be extracted from a matrix.

```
>> f=[3,4,1];   %  define sequence to select rows or columns
>> D=A(:,f)  %  use all rows to extract in sequence columns 3, 4 and 1 from A
```

```
D =
     2     3     0
     0     6     0
     0     2     0
     4     0     5
>> % use columns 2 through 5 to extract rows in sequence 3, 4 and 1 from A
>> E=A(f,2:5)
E =
     1     0     2     0
     0     4     0     3
     1     2     3     4
>>
>> g=[s(2),1:s(2)-1]   % g will be used to specify the columns of A to be shifted
g =    10     1     2     3     4     5     6     7     8     9
>> A=A(:,g)   % use all rows for the shift and rotation
A =
     9     0     1     2     3     4     5     6     7     8
    45     0     1     0     6     0    15     0    28     0
     5     0     1     0     2     0     3     0     4     0
     0     5     0     4     0     3     0     2     0     1
```

A row, column or sub-matrix can be removed from a matrix by replacement with an empty (null) matrix.

```
>> A(1,:)=[]   % removing row 1 from A
A =
    45     0     1     0     6     0    15     0    28     0
     5     0     1     0     2     0     3     0     4     0
     0     5     0     4     0     3     0     2     0     1
>> A(:,2)=[]   % removing column 2 from A
A =
    45     1     0     6     0    15     0    28     0
     5     1     0     2     0     3     0     4     0
     0     0     4     0     3     0     2     0     1
```

3.6.2 *Building a Matrix*

It is also possible to build a vector or matrix from other vectors and matrices in a way that is similar to the way a vector or matrix is defined. Table 3.6 gives several built-in functions for building a matrix.

Table 3.6   Built-In MATLAB functions concerned with matrix construction

Function	Brief explanation
cat(dim,A1,A2, ...)	Concatenate matrices, cat(1,A1,A2)=[A1;A2], cat(2,A1,A2)=[A1,A2]
horzcat(A1,A2, ...)	Concatenate matrices horizontally, horzcat(A1,A2)=[A1,A2]
repmat(A,N,M)	Creates an $N \times M$ tiling replication of A
reshape(A,N,M)	Creates an $N \times M$ matrix whose elements are taken columnwise from A
rot90(A,K)	Rotates A by K times (90 degrees) in the counterclockwise direction
sort(A,dim,mode)	Mode is "ascend" or "descend"; dim=1,sort each column of A; dim=2, sort each row of A
sortrows(A)	Sort rows of A in ascending order
vertcat(A1,A2, ...)	Concatenate matrices vertically, vertcat(A1,A2)=[A1;A2]

Example 3.14

```
>> X = []    % set X to an empty matrix
X   =
     []
>> % define two row vectors
>> A = [1   2   3   4]; B = [5   6   7   8];
>> X = [X; A]   % append a row to X, which can also be found with X=vertcat(X,A)
X =     1     2     3     4    % since X was empty, X now contains only one row
>> X = [X; B]   % append another row to X
X =
     1     2     3     4
     5     6     7     8
>> R = repmat(X,2,3)
>> R = repmat(X,2,3)    %   replicate X into 2x3 tiles
R =
     1     2     3     4     1     2     3     4     1     2     3     4
     5     6     7     8     5     6     7     8     5     6     7     8
     1     2     3     4     1     2     3     4     1     2     3     4
     5     6     7     8     5     6     7     8     5     6     7     8
>> Y = X';  % Y is transpose of X
>> C = A'; D = B';  % form two column vectors
>> % place D in the first column of Y,   can also be found with Y=cat(D,Y)
>> Y = [D, Y]
Y =
     5     1     5
     6     2     6
     7     3     7
     8     4     8
>> % reshape the 12 elements of Y, column wise, into a 2x6 matrix
>> S = reshape(Y,2,6)
```

```
S =

     5     7     1     3     5     7
     6     8     2     4     6     8
>> Y = [Y, C]   % append a column to Y, which can also be found with Y=cat(Y,C)
Y =

     5     1     5     1
     6     2     6     2
     7     3     7     3
     8     4     8     4
>> Y = [Y(1,:); A; Y(3:4,:)]   % replace the second row of Y with A
Y =

     5     1     5     1
     1     2     3     4
     7     3     7     3
     8     4     8     4
>> % combine Y and X column wise, which can also be found with Z=cat(Y,X')
>> Z = [Y, X']
Z =

     5     1     5     1     1     5
     1     2     3     4     2     6
     7     3     7     3     3     7
     8     4     8     4     4     8
```

The functions given in Table 3.6 are useful to construct objects that are not necessarily the result of some algebraic operations.

Example 3.15

Let us build a continuous and piecewise linear function $x(t)$ of time. We want $x(t)$ to behave as follows:

$$x(t) = \begin{cases} 0, & t < 0 \\ m_1 t, & 0 \leq t < t_1 \\ m_2 t + b_2, & t_1 \leq t < t_2 \\ m_3 t + b_3, & t_2 \leq t < t_3 \\ m_4 t + b_4, & t_3 \leq t < t_4 \\ m_5 t + b_5, & t_4 \leq t < t_5 \\ 0, & t \geq t_5 \end{cases}$$

Let $x(t = t_1 = 0.2) = 0.8,$ $x(t = t_2 = 0.4) = 1.0,$ $x(t = t_3 = 1.5) = 1.0,$ $x(t = t_4 = 2.5) = 0.3,$ and $x(t = t_5 = 5.0) = 0.0.$ To use $x(t)$ for plotting, for example, we must sample $x(t)$ at a set of discrete time points to obtain N points given by

$$x(nT) = x(t)|_{t=nT}, \quad n = 0, 1, \ldots, (N-1) \tag{3.35}$$

where T is the time increment. For example, if $T = 0.1$ secs, then $x(t)$ will be sampled at the rate $f_s = 1/T = 10$ samples/sec. The parameter f_s is called the **sampling frequency**. While $x(t)$ is a continuous time signal, $x(nT)$ is a **discrete time signal**, which is known only at the discrete time points $t = nT$, $n = 0, 1, \ldots$. Audio CDs contain number sequences that are samples of music obtained at the rate $f_s = 44,100$ samples/sec. Prog. 3.2 gives a MATLAB program to find and plot $x(nT)$. Fig. 3.3 shows the resulting time function.

```
% Program to find and plot a continuous piecewise linear function
% This function consists of 5 straight line segments
clc; clear all
% assign parameters
t1=0.2; t2=0.4; t3=1.5; t4=2.5; t5=5.0; % interval end times
xt1=0.8; xt2=1.0; xt3=1.0; xt4=0.3; xt5=0.0; % interval end values
fs = 44100; % the sampling rate (frequency)
T = 1/fs;  % sampling time increment
N1=fix(t1/T);N2=fix((t2-t1)/T);N3=fix((t3-t2)/T);N4=fix((t4-t3)/T);
N5=fix((t5-t4)/T); % number of points in each interval
m1=xt1/t1; % slope of line in interval 1
m2 = (xt2 - xt1)/(t2 - t1); b2 = xt1 - m2*t1; % slope and intercept
m3 = (xt3 - xt2)/(t3 - t2); b3 = xt2 - m3*t2;
m4 = (xt4 - xt3)/(t4 - t3); b4 = xt3 - m4*t3;
m5 = (xt5 - xt4)/(t5 - t4); b5 = xt4 - m5*t4;
t = 0:T:(N1-1)*T;  % interval 1 time points
x1 = m1*t; time = t; N=N1; % line segment 1 and initialize build of time
t = N*T:T:(N+N2-1)*T; % interval 2 time points
x2 = m2*t + b2; time = [time t]; N=N+N2; % line segment 2, build time
t = N*T:T:(N+N3-1)*T;
x3 = m3*t + b3; time = [time t]; N=N+N3;
t = N*T:T:(N+N4-1)*T;
x4 = m4*t + b4; time = [time t]; N=N+N4;
t = N*T:T:(N+N5-1)*T;
x5 = m5*t + b5; time = [time t]; % line 5, complete build of time
x = [x1 x2 x3 x4 x5]; % build entire function
plot(time,x); axis([0 5 0 1.1]); grid on
xlabel('time - secs')
ylabel('volts')
title('Continuous Piecewise Linear Function')
save('pulse.mat','x','time') % save the vector x in the file pulse.mat
```

Program 3.2   Script to build a continuous piecewise linear function of time.

The last statement in this program uses the built-in MATLAB function **save** to save the vectors x and time in a file, which has the suffix **.mat**. With the function **save** it is possible to save the entire workspace in a file for retrieval with the built-in function **load**. This is useful

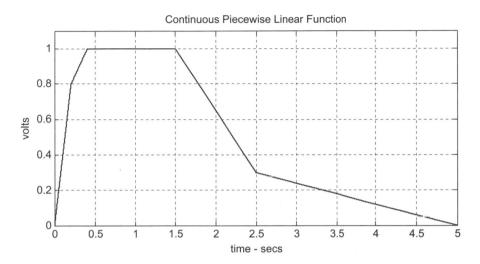

Figure 3.3   Plot of a continuous piecewise linear function using its samples.

to make some or all variables in a program available to another MATLAB program. There are numerous options for doing this, which will be discussed in Chapter 8.

While MATLAB has many built-in functions that can be used to describe the behavior of a signal, there will be occasions when it is necessary to build a signal that has a specific shape.

MATLAB has many built-in functions and operations to find properties about a matrix and to manipulate the content of a matrix (see Tables 3.1–3.3). As you employ MATLAB to solve engineering problems, use the MATLAB help facility to increase your knowledge about the many built-in MATLAB functions.

3.7   Resistive Circuit Analysis

The structure provided by matrix algebra enables a very systematic approach to circuit analysis. Circuit analysis is based on two fundamental principles.

Kirchhoff's Current Law (KCL): *The algebraic sum of the currents leaving a circuit node is zero.* For each of the N nodes in a circuit we can write a KCL equation. For a planar circuit with N nodes, there can be only as many as $(N - 1)$ independent node equations.

Kirchhoff's Voltage Law (KVL): *The algebraic sum of the voltages drops along any closed path in a circuit is zero.* For each closed path (loop) in a circuit we can write a KVL equation. For a planar circuit with M meshes, there can only be as many as M independent loop equations. A **mesh** of a planar circuit is a closed path (loop) that does not enclose any component of the circuit. The M meshes are the closed paths to which KVL will be applied to find M independent loop equations.

3.7.1   Component Circuit Analysis

In component circuit analysis, the unknowns are the voltage across and current through each resistor, the current through each voltage source, and the voltage across each current source.

Example 3.16 ———————————————————————————————————————

Let us see how we can apply matrix notation and the solution method of (3.29) to analyze a circuit. Consider the resistive circuit shown in Figure 3.4, and let $V_1 = 5$ volts, $I_2 = -10$ mA, $R_3 = 100\ \Omega$, $R_4 = 1\text{K}\ \Omega$, $R_5 = 3\text{K}\ \Omega$, $R_6 = 1\text{K}\ \Omega$ and $R_7 = 1\text{K}\ \Omega$. The circuit has $N = 5$ nodes and $M = 3$ meshes.

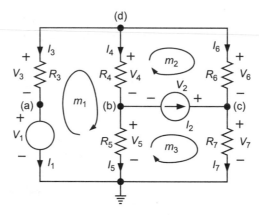

Figure 3.4   A resistive circuit with two sources.

For a component analysis, we label a current and voltage for every component in the circuit. Applying Kirchhoff's current law (KCL) to $N - 1$ nodes (a), (b), (c), and (d) gives

$$
\begin{aligned}
node(a) &: I_1 - I_3 = 0 \\
node(b) &: I_5 - I_4 + I_2 = 0 \\
node(c) &: -I_2 - I_6 + I_7 = 0 \\
node(d) &: I_3 + I_4 + I_6 = 0
\end{aligned}
\tag{3.36}
$$

Applying Kirchhoff's voltage law (KVL) to M meshes gives

$$mesh\ 1: -V_1 - V_3 + V_4 + V_5 = 0$$
$$mesh\ 2: -V_4 + V_6 + V_2 = 0 \tag{3.37}$$
$$mesh\ 3: -V_5 - V_2 + V_7 = 0$$

There are a total of 12 unknown component voltages and currents, and (3.36) and (3.37) are 7 constraints among these unknowns. We can get five more constraints with Ohm's law resulting in

$$V_3 = R_3 I_3$$
$$V_4 = R_4 I_4$$
$$V_5 = R_5 I_5 \tag{3.38}$$
$$V_6 = R_6 I_6$$
$$V_7 = R_7 I_7$$

We have completed applying the physical principles and properties that will enable us to find all voltages and currents in the circuit. Let us define a vector of the unknowns organized as

$$X = [V_2\ V_3\ V_4\ V_5\ V_6\ V_7\ I_1\ I_3\ I_4\ I_5\ I_6\ I_7]' \tag{3.39}$$

Now we can define the matrices A and Y in (3.19), and they are given by

$$A = \begin{bmatrix} 0 & 0 & 0 & 0 & 0 & 0 & 1 & -1 & 0 & 0 & 0 & 0 \\ 0 & 0 & 0 & 0 & 0 & 0 & 0 & 0 & -1 & 1 & 0 & 0 \\ 0 & 0 & 0 & 0 & 0 & 0 & 0 & 0 & 0 & 0 & -1 & 1 \\ 0 & 0 & 0 & 0 & 0 & 0 & 0 & 1 & 1 & 0 & 1 & 0 \\ 0 & -1 & 1 & 1 & 0 & 0 & 0 & 0 & 0 & 0 & 0 & 0 \\ 1 & 0 & -1 & 0 & 1 & 0 & 0 & 0 & 0 & 0 & 0 & 0 \\ -1 & 0 & 0 & -1 & 0 & 1 & 0 & 0 & 0 & 0 & 0 & 0 \\ 0 & -1 & 0 & 0 & 0 & 0 & 0 & R_3 & 0 & 0 & 0 & 0 \\ 0 & 0 & -1 & 0 & 0 & 0 & 0 & 0 & R_4 & 0 & 0 & 0 \\ 0 & 0 & 0 & -1 & 0 & 0 & 0 & 0 & 0 & R_5 & 0 & 0 \\ 0 & 0 & 0 & 0 & -1 & 0 & 0 & 0 & 0 & 0 & R_6 & 0 \\ 0 & 0 & 0 & 0 & 0 & -1 & 0 & 0 & 0 & 0 & 0 & R_7 \end{bmatrix}, \quad Y = \begin{bmatrix} 0 \\ -I_2 \\ I_2 \\ 0 \\ V_1 \\ 0 \\ 0 \\ 0 \\ 0 \\ 0 \\ 0 \\ 0 \end{bmatrix}$$

$$\tag{3.40}$$

Once the vector X of unknowns has been defined, the matrices A and Y are written to combine (3.36), (3.37), and (3.38) into one matrix equation. The inner product of the nth row of A, $A(n,:)$, and X gives the nth element of Y. For example, $(A(1,:),X) = Y(1)$ is the KCL equation written at *node (a)*, $(A(5,:),X) = Y(5)$ is the KVL equation written for *mesh 1* and $(A(8,:),X) = Y(8)$ gives $-V_3 + R_3 I_3 = 0$, which is the first Ohm's law equation.

Since there are so few nonzero elements in the (12×12) matrix A, it is called a **sparse** matrix. It is common in engineering problems to work with very large sparse matrices, and MATLAB has built-in functions that are particularly well suited for solving sparse systems described by (3.19). A MATLAB script that finds X follows.

```
clear all; clc;
V1=5; I2=-0.01; % specify current and voltage sources
R3=100; R4=1000; R5=3000; R6=1000; R7=1000;   % specify resistors
% the vector of unknowns is: X=[V2 V3 V4 V5 V6 V7 I1 I3 I4 I5 I6 I7]'
A=zeros(12,12);Y=zeros(12,1);  % preallocating space
% specify nonzero elements of A
A(1,7)=1;A(1,8)=-1;
A(2,9)=-1;A(2,10)=1;
A(3,11)=-1;A(3,12)=1;
A(4,8)=1;A(4,9)=1;A(4,11)=1;
A(5,2)=-1;A(5,3)=1;A(5,4)=1;
A(6,1)=1;A(6,3)=-1;A(6,5)=1;
A(7,1)=-1;A(7,4)=-1;A(7,6)=1;
A(8,2)=-1;A(8,8)=R3;
A(9,3)=-1;A(9,9)=R4;
A(10,4)=-1;A(10,10)=R5;
A(11,5)=-1;A(11,11)=R6;
A(12,6)=-1;A(12,12)=R7;
% specify nonzero elements of Y
Y(2)=-I2;Y(3)=I2;Y(5)=V1;
% use Gauss elimination to solve AX=Y for X
X=A\Y;  % the backslash stipulates that Gauss elimination will be used
X(7:12)=1000*X(7:12); % converting currents to mA
X  % display X in command window
```

Program 3.3   Program for a component analysis of the circuit in Figure 3.4.

The results of a component analysis are given by

```
X'  = -13.7209  -0.1163  -6.2791  11.1628  7.4419  -2.5581  -1.1628
      -1.1628   -6.2791   3.7209   7.4419  -2.5581
>>
```

An important fundamental physical principle can be demonstrated if we calculate the power delivered to each component in the circuit. Let V be the vector of all component voltages, and let I be the vector of all component currents. The power P_1 delivered to the voltage source is $P_1 = V_1 I_1$, and the power P_2 delivered to the current source is

$P_2 = V_2(-I_2)$. For each resistor, the power delivered to the resistor is $P_k = V_k I_k$, $k = 3, \ldots, 7$. The following MATLAB statements were appended to Prog. 3.3 to find the power delivered to all circuit components.

```
V=[V1 X(1:6)'];  % form the 7 element voltage row vector
I=[X(7) -1000*I2  X(8:12)'];  % form the 7 element current row vector in mA
% -I2 is used to calculate power delivered to the current source
P=V.*I;  % calculate the power vector in mW
disp('Component  Voltage      Current    Power Delivered')
disp(' Number     (volts)       (mA)         (mW)')
K=[1:7];
Table=[K; V; I; P];  % make a table of analysis results
fprintf('   %i      %-9.4f    %-9.4f    %-9.4f \n',Table)
Total_Power = sum(P);  % total power delivered to all circuit components
fprintf('Total power delivered to all components = %9.4f mW \n',Total_Power)
```

The **disp** function could have been used to display the table. However, with disp you have no control of display position. Instead, the built-in function **fprintf**, which will be discussed in Chapter 8, was used. With fprintf, display format can be controlled. This program segment produced the following results:

Component Number	Voltage (volts)	Current (mA)	Power Delivered (mW)
1	5.0000	−1.1628	−5.8140
2	−13.7209	10.0000	−137.2093
3	−0.1163	−1.1628	0.1352
4	−6.2791	−6.2791	39.4267
5	11.1628	3.7209	41.5360
6	7.4419	7.4419	55.3813
7	−2.5581	−2.5581	6.5441

Total power delivered to all components $= -0.0000$ mW

Notice that the power $P_1 = -5.814$ mW delivered to the voltage source is negative. This means that it is actually a power source of 5.814 mW. Similarly, the current source is a power source of 137.2093 mW. The fundamental physical principle is that **the sum of the power sourced equals the sum of the power absorbed**, which is why the total power delivered to all circuit components is zero.

Another fundamental principle in linear circuit analysis is the **superposition principle**. With the matrices given in (3.39) and (3.40) we can find any current or voltage with $X = A^{-1}Y$, where the elements of Y are given by the current and voltage sources in the circuit. Let us split Y into K column vectors, Y_k, $k = 1, \ldots, K$, where K is the total number of sources in the circuit, each Y_k contains elements due to only one of the sources, possibly multiplied by constants, and Y equals the sum of the Y_k, $k = 1, \ldots, K$. In the circuit given in Fig. 3.4, $K = 2$, and we get

$$Y_1 = [0 \ -I_2 \ I_2 \ 0 \ 0 \ 0 \ 0 \ 0 \ 0 \ 0 \ 0 \ 0]'$$
$$Y_2 = [0 \ 0 \ 0 \ 0 \ V_1 \ 0 \ 0 \ 0 \ 0 \ 0 \ 0 \ 0]'$$

Therefore

$$X = A^{-1}Y = A^{-1}(Y_1 + Y_2) = A^{-1}Y_1 + A^{-1}Y_2 = X_1 + X_2 \tag{3.41}$$

where $X_1(X_2)$ gives all the component currents and voltages due to the source $I_2(V_1)$ acting alone. For X_1 the voltage source V_1 is set to zero, which means that it is replaced by a short circuit, and the entire resulting circuit is analyzed to find X_1. Then, for X_2 the current source is set to zero, which means that it is replaced by an open circuit, and the entire resulting circuit is analyzed to find X_2. Finally, X is found by summing the X_k, $k = 1, \ldots, K$.

A circuit with more than one source satisfies the **superposition principle** if a current or voltage of interest can be found by finding the currents or voltages due to each source acting alone and then summing these currents or voltages to obtain the current or voltage of interest.

The **linearity property** of a circuit is another important property, and it is related to the superposition principle. Let us refer to each source in a circuit as an input, and refer to any component voltage or current due to an input as an output. The circuit in Fig. 3.4 has 2 inputs that cause 12 outputs, the elements of X. If the input I_2 is changed to become $\alpha_1 I_2$, where α_1 is an arbitrary constant, then Y_1 becomes $Z_1 = \alpha_1 Y_1$, and like (3.41), X_1 becomes $W_1 = \alpha_1 X_1$. In others words, if you multiply an input by a constant, then the output due to that input is multiplied by that same constant. This part of the linearity property is called the **homogeneity property**. Similarly, if another input, say V_1, is multiplied by an arbitrary constant α_2, then Y_2 becomes $Z_2 = \alpha_2 Y_2$, and like (3.41), X_2 becomes $W_2 = \alpha_2 X_2$. For example, in (3.41) we have the overall input $Y = Y_1 + Y_2$, and the overall output is $X = X_1 + X_2$, where X_1 is due to Y_1 acting alone and X_2 is due to Y_2 acting alone. This part of the linearity property is called the **additivity property**. Furthermore, if $\alpha_1 Y_1 + \alpha_2 Y_2$ is the overall input, then $\alpha_1 X_1 + \alpha_2 X_2$ is the overall output.

The **linearity property** of a circuit states that if $Y_1 \rightarrow X_1$, then the input Y_1 causes the output X_1, and if $Y_2 \rightarrow X_2$, then the circuit satisfies the linearity property if and only if $\alpha_1 Y_1 + \alpha_2 Y_2 \rightarrow \alpha_1 X_1 + \alpha_2 X_2$. If $\alpha_1 = 0$ or $\alpha_2 = 0$, we have the **homogeneity property**, and if $\alpha_1 = \alpha_2 = 1$, we have the **additivity property**. The circuit in Fig. 3.4 is said to be a **linear circuit**, because it satisfies the linearity property. Generally, a set of equations written as $AX = Y$, where Y is a linear combination of inputs, is a linear system of equations, which satisfies the linearity property.

3.7.2 Nodal Analysis

A component analysis is straightforward to apply. However, this method results in many equations. To reduce the number of equations, we can make substitutions that are relatively convenient to do. A modification of component analysis is to utilize KVL and Ohm's law as we apply KCL. We can then immediately generate equations in terms of unknown voltages.

Example 3.16 _____

In the given circuit there are $N - 1 = 4$ nodes where we can define voltages, called **node voltages**, with respect to a common node, designated by the ground symbol. These are the node voltages v_a, v_b, v_c, and v_d. In terms of these node voltages we can express all component voltages. For example, by KVL we get $V_4 = v_d - v_b$. Therefore, if we know the $N - 1$ node voltages we know all component voltages, and then by Ohm's law all component currents can be found. In the given circuit, the node voltage v_a is constrained, and it is given by

$$node(a) : v_a = V_1 \tag{3.42}$$

Now we must find three additional equations constraining the remaining node voltages. Applying KCL to nodes (b), (c), and (d), while applying Ohm's law, gives

$$
\begin{aligned}
node(b) &: \frac{v_b - v_d}{R_4} + I_2 + \frac{v_b}{R_5} = 0 \\
node(c) &: -I_2 + \frac{v_c - v_d}{R_6} + \frac{v_c}{R_7} = 0 \\
node(d) &: \frac{v_d - V_1}{R_3} + \frac{v_d - v_b}{R_4} + \frac{v_d - v_c}{R_6} = 0
\end{aligned}
\tag{3.43}
$$

Collect the unknown node voltages into a vector $v = [v_b \; v_c \; v_d]'$, and (3.42) and (3.43) become

$$Gv = I \tag{3.44}$$

where G and I are given by

$$
G = \begin{bmatrix}
\dfrac{1}{R_4} + \dfrac{1}{R_5} & 0 & \dfrac{-1}{R_4} \\
0 & \dfrac{1}{R_6} + \dfrac{1}{R_7} & \dfrac{-1}{R_6} \\
\dfrac{-1}{R_4} & \dfrac{-1}{R_6} & \dfrac{1}{R_3} + \dfrac{1}{R_4} + \dfrac{1}{R_6}
\end{bmatrix}, \quad
I = \begin{bmatrix}
-I_2 \\
I_2 \\
V_1/R_3
\end{bmatrix}
$$

Equation (3.44) is a linear system of node voltages. A MATLAB script that finds v follows.

```
clear all; clc;
% Program does a nodal analysis
V1=5; I2=-0.01; % specify current and voltage sources
R3=100; R4=1000; R5=3000; R6=1000; R7=1000;   % specify resistors
% the vector of node voltages is: v=[vb vc vd]'
G=zeros(3,3);I=zeros(3,1);   % preallocating space
% specify nonzero elements of G, the conductance matrix
G(1,1)=1/R4 + 1/R5;G(1,3)=-1/R4;
G(2,2)=1/R6 + 1/R7;G(2,3)=-1/R6;
G(3,1)=-1/R4;G(3,2)=-1/R6;G(3,3)=1/R3 + 1/R4 + 1/R6;
% specify nonzero elements of I
I(1)=-I2;I(2)=I2;I(3)=V1/R3;
% use Gauss elimination to solve Gv=I for v
v=G\I;   % the backslash stipulates that Gauss elimination will be used
v' % show the unknown node voltages v in the command window
va=V1;vb=v(1);vc=v(2);vd=v(3);
V2=vc-vb
V3=vd-va
V4=vd-vb
V5=vb
V6=vd-vc
V7=vc
```

Program 3.4   Program for a nodal analysis of the circuit in Figure 3.4.

The results of the nodal analysis are given by

```
v' =   [ 11.1628    -2.5581     4.8837]
```

With the node voltages, the component voltages are given by

```
V2 = -13.7209, V3 = -0.1163, V4 = -6.2791, V5 = 11.1628, V6 = 7.4419, V7 = -2.5581
```

3.7.3   Loop Analysis

An alternative to a nodal analysis is to utilize KCL and Ohm's law as we apply KVL. We can then immediately generate equations in terms of unknown currents.

Example 3.16 (continued) _____

In the given circuit there are $M = 3$ meshes where we can define currents, called **mesh (or loop) currents**. These are the currents i_1, i_2, and i_3. In terms of these mesh currents we can express all component currents. For example, by KCL at *node*(d) we get $-i_1 + I_4 + i_2 = 0$, or $I_4 = i_1 - i_2$. Therefore, if we know the M mesh currents, we know all component currents, and then by Ohm's law all component voltages can be found. In the given circuit, the mesh currents i_2 and i_3 are constrained, since a given current source is common to the corresponding meshes, resulting in

$$meshes\ 2,3: \quad i_3 - i_2 = I_2 \rightarrow i_3 = I_2 + i_2 \tag{3.45}$$

For this circuit we must find two additional equations constraining the three mesh currents. Applying KVL to *mesh* 1 and the combined *meshes* 2 and 3, while including Ohm's law, gives

$$\begin{aligned}
mesh\ 1: & \quad -V_1 + R_3 i_1 + R_4(i_1 - i_2) + R_5(i_1 - (I_2 + i_2)) = 0 \\
meshes\ 2,3: & \quad R_4(i_2 - i_1) + R_6 i_2 + R_7(I_2 + i_2) + R_5((I_2 + i_2) - i_1) = 0
\end{aligned} \tag{3.46}$$

Collect the unknown mesh currents into a vector $i = [i_1 i_2]'$, and (3.45) and (3.46) become

$$Ri = V \tag{3.47}$$

where R and V are given by

$$R = \begin{bmatrix} R_3 + R_4 + R_5 & -R_4 - R_5 \\ -R_4 - R_5 & R_4 + R_6 + R_7 + R_5 \end{bmatrix}, \quad V = \begin{bmatrix} V_1 + R_5 I_2 \\ -R_7 I_2 - R_5 I_2 \end{bmatrix}$$

Equation (3.47) is a linear system of mesh currents. A MATLAB script that finds i follows.

```
clear all; clc;
% Program does a mesh analysis
V1=5; I2=-0.01; % specify current and voltage sources
R3=100; R4=1000; R5=3000; R6=1000; R7=1000;  % specify resistors
% the vector of mesh currents is: [i1 i2 i3]'
R=zeros(2,2); V=zeros(2,1);  % preallocating space
% specify nonzero entries in R, the resistance matrix
R(1,1)=R3+R4+R5;R(1,2)=-R4-R5;
R(2,1)=-R4-R5;R(2,2)=R4+R6+R5+R7;
% you can verify that G equals the inverse of R
% specify nonzero entries in V
V(1)=V1+R5*I2;V(2)=-R7*I2-R5*I2;
%  use Gauss elimination to solve Ri=V for i
i=R\V;   % the backslash stipulates that Gauss elimination will be used
i=1000*i;  I2=1000*I2;  % convert currents to mA
i'   % show the unknown mesh currents i in the command window
i1=i(1);i2=i(2);i3=I2+i(2);
I1=-i1
I3=-i1
```

```
I4=i1-i2
I5=i1-i3
I6=i2
I7=i3
```

Program 3.5 Program for a mesh analysis of the circuit in Figure 3.4.

The results of the mesh analysis are given by

```
i'  =  [1.1628      7.4419]
```

With these mesh currents we get the component currents given by

```
I1 = -1.1628, I3 =  -1.1628, I4 = -6.2791, I5 = 3.7209,  I6 = 7.4419, I7 =  -2.5581
```

Much of the early development of MATLAB was motivated by the need to efficiently and accurately solve (3.19), not only for circuit analysis, but also for linear system analysis in many other fields. Many circumstances (e.g., a very large N, a sparse matrix, or a particular set of matrix elements) can make this a challenging task.

3.8 Linear Transformations

The concept of a matrix and matrix algebra can be very well applied to work with objects in a linear vector space.

3.8.1 Vector Space

An N-dimensional **linear vector space** Ψ is a collection of vectors with an additive operation and scalar multiplication. Associated with the vector space Ψ is a set R of scalars, where R can be, for example, the set of all real or complex numbers. A linear vector space satisfies the following properties:

(a) For any two vectors x and $y \in \Psi$, $x + y \in \Psi$, closure property.
(b) Ψ contains an identity vector 0 (zero vector), such that for any vector x, $x + 0 = 0 + x = x$.
(c) For any vector x, there is another vector y such that $x + y = 0$ (the zero vector).
(d) For any three vectors x, y, and $z \in \Psi$, $(x + y) + z = x + (y + z)$, associative property.
(e) For any vectors x and y and any scalars a and $b \in R$: $ax \in \Psi$, $a(bx) = (ab)x$, $(a + b)x = ax + bx$, $a(x + y) = ax + ay$, there is a multiplication identity $1 \in R$ such that $1x = x$ and a scalar $0 \in R$ such that $0x = 0$ (the zero vector).

Let $x_k \in \Psi$, $k = 1, \ldots, K$ and $\alpha_k \in R$, $k = 1, \ldots, K$. A **linear combination** of the vectors x_k, $k = 1, \ldots, K$ is given by

$$y = a_1 x_1 + a_2 x_2 + \cdots + a_K x_K = \sum_{k=1}^{K} a_k x_k \tag{3.48}$$

where $y \in \Psi$. The vectors x_k, $k = 1, \ldots, K$ are said to be **linearly independent** if the only set of scalars c_k, $k = 1, \ldots, K$ that satisfy

$$c_1 x_1 + c_2 x_2 + \cdots + c_K x_K = 0 \tag{3.49}$$

is the set $c_k = 0$, $k = 1, \ldots, K$. If there are nonzero c_k that satisfy (3.49), then the vectors are said to be **linearly dependent**, and some vector in the set x_k, $k = 1, \ldots, K$ can be written as a linear combination of the other vectors in the set. To be more explicit, it will be assumed hereafter that elements of vectors are either real or complex numbers, and that $K = N$. For the case $K \neq N$, the reader is referred to the literature on linear algebra.

Let us collect the vectors x_n, $n = 1, \ldots, N$ into the columns of an $N \times N$ matrix A to get

$$A = [x_1 | x_2 | \cdots | x_N]$$

Then, (3.49) can be written as

$$AC = 0 \tag{3.50}$$

where C is the column vector given by $C = \begin{bmatrix} c_1 & c_2 & \ldots & c_N \end{bmatrix}'$.

Equation (3.50) is a homogeneous equation. There can be a nonzero solution C if and only if $\det(A) = 0$. If $\det(A) \neq 0$, then $C = 0$, and the columns of A are linearly independent. The MATLAB function **rank** finds the number of linearly independent columns or rows in a matrix, where $\mathrm{rank}(A) \leq N$. For example, let $N = 3$, and then for some given A we have

```
>> A=[1 1 3;2 2 2;3 3 1]    % the second column of A duplicates the first column
A =

     1     1     3
     2     2     2
     3     3     1
>> rank(A)
ans =    2
>> det(A)
ans =    0
```

A square matrix is singular if and only if the columns (rows) are linearly dependent.

It is insightful to use a geometric interpretation of vectors by defining a measure of vector length and direction. The length of a vector x is associated with its **norm**, which is a real scalar denoted by $||x||$. There are many useful definitions of a norm, which must satisfy the following properties:

(a) For any $x \in \Psi$, $||x|| \geq 0$
(b) $||x|| = 0$, if and only if $x = 0$

(c) For any $a \in R$, $||ax|| = |a|\,||x||$
(d) For any vectors x and $y \in \Psi$, $||x+y|| \leq ||x|| + ||y||$, triangle inequality
  A commonly used norm is the l_p-norm (or p-norm), which is defined by

$$l_p - norm : ||x||_p = \left(\sum_{n=1}^{N} |x(n)|^p \right)^{1/p} \tag{3.51}$$

where $p = 1, 2, \ldots, \infty$. If the elements $x(n)$, $n = 1, \ldots, N$ are complex, then $|x(n)| = (x(n)x^*(n))^{1/2}$, the magnitude of a complex number. Some commonly used p-norms are given by

$$l_1 - norm = ||x||_1 = \sum_{n=1}^{N} |x(n)|$$

$$l_2 - norm = ||x||_2 = \left(\sum_{n=1}^{N} |x(n)|^2 \right)^{1/2}, \quad \text{Euclidean norm}$$

$$l_\infty - norm = ||x||_\infty = \max_{n=1,\ldots,N} |x(n)|$$

When a norm is associated with a vector space, the vector space is called a **normed vector space**. It is common practice to define the distance $d(x,y)$ between two vectors x and $y \in \Psi$ with $d(x,y) = ||x-y||$.

The MATLAB function **norm** returns the norm of a vector. There are many norm options, for example:

```
>> x = [3  4]';
>> p = 2;  %  get the 2-norm
>> r = norm(x,p)  %  p can be: 1, 2, ...,  inf
r =    5
```

An N-dimensional **Cartesian space** is a normed vector space. A 2-D Cartesian space is shown in Fig. 3.5. While it is conventional to refer to the two axes as the x and y axes, let us use u_1 and u_2, respectively. In a 2-D Cartesian space, the axes (Cartesian coordinates) are perpendicular (**orthogonal**) to each other. The point where the axes intersect is called the **origin**. Movement along an axis and away from the origin in the direction of the arrow is movement in the positive direction, and in the opposite direction it is movement in the negative direction.

Any point in a 2-D Cartesian space can be located by two numbers, called a **2-tuple**, that specifies movements along the u_1 and u_2 axes. For example, the point labeled v in Fig. 3.5 is the 2-tuple $(-2, 3)$. Another way to denote the location of a point in a 2-D Cartesian space is with a 2-element vector, where the elements $v(1)$ and $v(2)$ are the movements in the u_1 and u_2 directions, respectively, and we write

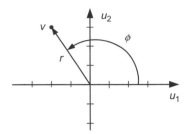

Figure 3.5   A two-dimensional Cartesian space.

$$v = \begin{bmatrix} -2 \\ 3 \end{bmatrix}$$

In Fig. 3.5, the vector v is also considered to be the line pointing from the origin to the point v. Another way to locate the point v is to rotate a line of length r in the counter-clockwise direction from the positive u_1 axis by an angle ϕ. The line length is given by $r = ||v||_2$, called the **magnitude** (or **Euclidian norm**) of v. In this case

$$||v||_2 = (v^2(1) + v^2(2))^{1/2} = \sqrt{13}, \quad \phi = \frac{\pi}{2} + \tan^{-1}\left(\frac{2}{3}\right) = 2.1588 \text{ radians}$$

Let us write (3.19) as

$$y = Ax \tag{3.52}$$

and consider x and y to be vectors in a Cartesian space. Then, given the matrix A, (3.52) is a linear transformation, where A operates on the vector x to obtain the vector y. The matrix A can be designed to achieve different kinds of relationships between the vectors x and y.

Let a_n, $n = 1, 2, \ldots, N$, denote the N columns of A. Then (3.52) can be interpreted to be

$$y = x(1)a_1 + x(2)a_2 + \cdots + x(N)a_N = \sum_{n=1}^{N} x(n)a_n$$

which is a linear combination of the columns of A.

3.8.2   Rotation

Suppose that a particular geometric goal of (3.52) is a matrix A that rotates a given vector x by an angle θ into the vector y. This is depicted in Fig. 3.6. Let us design such a matrix.

By projection, the vectors y and x can be written in terms of the angles θ and ϕ as

$$y = \begin{bmatrix} ||y||\cos(\theta + \phi) \\ ||y||\sin(\theta + \phi) \end{bmatrix} = ||y|| \begin{bmatrix} \cos(\theta)\cos(\phi) - \sin(\theta)\sin(\phi) \\ \sin(\theta)\cos(\phi) + \cos(\theta)\sin(\phi) \end{bmatrix}, \quad x = ||x|| \begin{bmatrix} \cos(\phi) \\ \sin(\phi) \end{bmatrix}$$

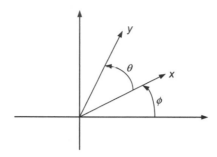

Figure 3.6   Rotation of vector x by θ radians into vector y.

Since we do not want to change the magnitude of x, $||y|| = ||x||$, resulting in

$$y = \begin{bmatrix} \cos(\theta) & -\sin(\theta) \\ \sin(\theta) & \cos(\theta) \end{bmatrix} x = Ax \tag{3.53}$$

which defines a matrix A that operates on a vector to rotate it about the origin by an angle θ. Let us denote this special matrix by $R(\theta)$, a 2×2 rotation matrix. In Chapter 9, rotation in a 3-D Cartesian space will be discussed.

Example 3.17 ———————————————————————————

Prog. 3.6 implements (3.53). This script produces the results given in Fig. 3.7.

```
% Script to rotate a vector
clear all;
% get a vector from the program user
x = input('Enter a 2-element (enclosed in brackets) row vector: ');
x = x'; % convert x to a column vector
plot([0 x(1)],[0 x(2)]) % open a Figure Window and plot the given vector
grid on % include a grid in the figure
x_max = 1.1*(x(1)^2 + x(2)^2)^0.5; % used to override automatic scaling
axis([-x_max x_max -x_max x_max]) % specify range of axes
hold on  % causes MATLAB to place the next plot in the same Figure Window
theta = input('Enter an angle (in radians) of rotation: '); % get angle
R_theta = [cos(theta) -sin(theta); sin(theta) cos(theta)];% rotation matrix
y = R_theta*x; % rotate the input vector
plot([0 y(1)],[0 y(2)]) % plot the rotated vector
```

Program 3.6   A script saved as rotate.m to rotate a vector.

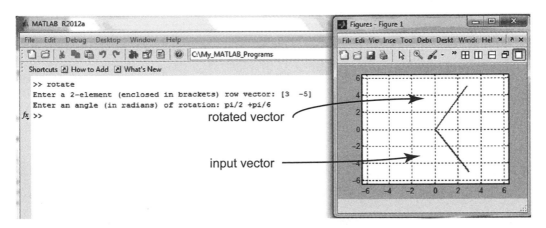

Figure 3.7 Example of a linear transformation designed to rotate a vector.

3.8.3 Eigenvalues and Eigenvectors

In (3.52), assume that a square matrix A is given. Does a vector x exist such that the vector y has the same direction as x? This is an important question in system analysis, especially in finding the solution of linear and time invariant differential equations, which occur when Kirchhoff's laws are applied to linear circuits.

The most that is allowed about y is that it can have a different length from x. This requirement gives

$$\lambda x = Ax \tag{3.54}$$

where λ, a scalar, accounts for the difference between the lengths of x and y. Rearranging (3.54) results in

$$\lambda x - Ax = \lambda Ix - Ax = (\lambda I - A)x = 0 \tag{3.55}$$

where, since A is an $N \times N$ square matrix, I is the $N \times N$ identity matrix and the right side is a column vector of zero elements. Equation (3.55) is a homogeneous equation, which has a nonzero solution if and only if

$$Q(\lambda) = |(\lambda I - A)| = 0 \tag{3.56}$$

The determinant in (3.56) is an Nth order polynomial in λ, called the **characteristic polynomial** $Q(\lambda)$ of A, and

$$Q(\lambda) = 0 \tag{3.57}$$

which is called the **characteristic equation** of A.

Example 3.18 _____

Let $N = 2$, and with A given by

$$A = \begin{bmatrix} 1 & -1 \\ 2 & 4 \end{bmatrix}$$

(3.57) becomes

$$Q(\lambda) = |\lambda I - A| = \left| \begin{bmatrix} \lambda & 0 \\ 0 & \lambda \end{bmatrix} - \begin{bmatrix} 1 & -1 \\ 2 & 4 \end{bmatrix} \right| = \left| \begin{bmatrix} \lambda - 1 & 1 \\ -2 & \lambda - 4 \end{bmatrix} \right|$$
$$= (\lambda - 1)(\lambda - 4) - (-2)(1) = \lambda^2 - 5\lambda + 6 = (\lambda - 2)(\lambda - 3) = 0$$

There are $N = 2$ values of λ that can satisfy (3.57), which are $\lambda = \lambda_1 = 2$ and $\lambda = \lambda_2 = 3$. For each λ there is a vector x that satisfies (3.54). Denote these vectors for λ_1 and λ_2 by x_1 and x_2, respectively, and we have

$$\begin{bmatrix} 1 & -1 \\ 2 & 4 \end{bmatrix} \begin{bmatrix} x_1(1) \\ x_1(2) \end{bmatrix} = 2 \begin{bmatrix} x_1(1) \\ x_1(2) \end{bmatrix}, \quad \begin{bmatrix} 1 & -1 \\ 2 & 4 \end{bmatrix} \begin{bmatrix} x_2(1) \\ x_2(2) \end{bmatrix} = 3 \begin{bmatrix} x_2(1) \\ x_2(2) \end{bmatrix}$$

The vector x_1 must satisfy

$$x_1(1) - x_1(2) = 2x_1(1) \rightarrow x_1(1) = -x_1(2)$$
$$2x_1(1) + 4x_1(2) = 2x_1(2) \rightarrow x_1(1) = -x_1(2)$$

which means that only the direction of x_1 can be found, but not its length, and we write

$$x_1 = \begin{bmatrix} -x_1(2) \\ x_1(2) \end{bmatrix} = x_1(2) \begin{bmatrix} -1 \\ 1 \end{bmatrix} = K_1 \begin{bmatrix} -1 \\ 1 \end{bmatrix} = K_1 e_1$$

where K_1 is an arbitrary number. We can pick $K_1 = 1/\|e_1\| = 1/(e_1^2(1) + e_2^2(1))^{1/2} = 1/\sqrt{2} = 0.7071$ to **normalize** (make $\|x_1\| = 1$) x_1 and get the eigenvector $v_1 = [-0.7071 \ \ 0.7071]'$, where $\|v_1\| = 1$. Similarly, x_2 can be found to be

$$x_2 = K_2 \begin{bmatrix} -\dfrac{1}{2} \\ 1 \end{bmatrix} = K_2 \, e_2, \rightarrow v_2 = \begin{bmatrix} 0.4472 \\ -0.8944 \end{bmatrix}$$

If we use $x = v_1$ in the linear transformation of (3.52), then the resulting y vector will have the same direction as the vector x, and similarly for $x = v_2$.

_____

The N roots of $Q(\lambda)$, λ_n, $n = 1, \ldots, N$ in (3.57) are called **eigenvalues** of A, where in German eigen means belonging to, and the vectors v_n, $n = 1, \ldots, N$ are called **eigenvectors** of A, because they are intrinsic properties of A. We can combine the relationships between the eigenvalues and eigenvectors into one equation by writing

$$A[v_1|v_2|\ldots|v_N] = [v_1|v_2|\ldots|v_N]\begin{bmatrix} \lambda_1 & 0 & \cdots & 0 \\ 0 & \lambda_2 & \ddots & \vdots \\ \vdots & \ddots & \ddots & 0 \\ 0 & \cdots & 0 & \lambda_N \end{bmatrix} \rightarrow AV = V\Lambda \tag{3.58}$$

where the columns of V, called the **modal matrix** of A, are the eigenvectors, and the diagonal elements of Λ are the eigenvalues of A. Therefore, the matrix A can be decomposed into

$$A = V\Lambda V^{-1} \rightarrow A^{-1} = V\Lambda^{-1}V^{-1} \tag{3.59}$$

which requires V to be nonsingular and that all eigenvalues are nonzero. Equation (3.59) gives another method to find A^{-1}.

The high level MATLAB built-in function **eig** finds the eigenvalues and eigenvectors of a square matrix.

Example 3.18 (continued)

The following MATLAB statements demonstrate the application of the function eig.

```
>> A = [1 -1; 2 4];
>> % get two matrices, the eigenvectors and eigenvalues of A
>> [V, Lambda]=eig(A)
V =
    -0.7071    0.4472
     0.7071   -0.8944
Lambda =
     2    0
     0    3
>> V_inv = inv(V)  % get V inverse using the function inv
V_inv =
    -2.8284   -1.4142
    -2.2361   -2.2361
>> Lambda_inv = [1/Lambda(1,1) 0; 0 1/Lambda(2,2)];
>> A_inv = V*Lambda_inv*V_inv   % get A inverse using the modal matrix in (3.59)
A_inv =
     0.6667    0.1667
    -0.3333    0.1667
>> A*A_inv   % check
```

```
ans =
     1.0000     0.0000
     0          1.0000
```

Some or all roots of the characteristic equation can be complex. For example:

```
>> A=[-1 -1;1 -1];
>> % find the eigenvectors and eigenvalues of A with the function eig
>> [V, Lambda]=eig(A)
V =
   0.7071                0.7071
   0 - 0.7071i           0 + 0.7071i
Lambda =
  -1.0000 + 1.0000i            0
         0                  -1.0000  - 1.0000i
%  the complex eigenvalues occur in complex conjugate pairs
%  the corresponding eigenvectors are complex conjugates of each other
>> x=V(:,1)   %  set x to the first eigenvector
x =
   0.7071
   0 - 0.7071i
>> A*x    % apply the linear transformation to an eigenvector
ans =
  -0.7071 + 0.7071i
   0.7071 + 0.7071i
>> Lambda(1)*x    % check to see if (3.54) is satisfied
ans =
  -0.7071 + 0.7071i
   0.7071 + 0.7071i
```

3.9 Singular Value Decomposition

A potential problem with using Gauss–Jordan elimination to solve a set of N equations in N unknowns as described by (3.19) is due to the accumulation of round-off error when N is large. Furthermore, A can have certain properties that can cause a significant error in the solution X, even when N is small.

Another method to find the inverse of A is based on concepts developed in abstract linear algebra. An objective of **singular value decomposition** is to find three $N \times N$ matrices U, Σ, and V such that A can be written as

$$A = U\Sigma V'$$
(3.60)

The matrix U satisfies $U'U = UU' = I$, which means that the column (row) vectors of U are **orthogonal** to each other $((U(:,k), U(:,m)) = 0, (U(k,:), U(m,:)) = 0, k \neq m)$, each with unit length. A matrix with this property is called a **unitary** matrix, and $U^{-1} = U'$. The matrix Σ, a diagonal matrix, has the form

$$\Sigma = \begin{bmatrix} \sigma_1 & 0 & \cdots & 0 \\ 0 & \sigma_2 & \ddots & \vdots \\ \vdots & \ddots & \ddots & 0 \\ 0 & \cdots & 0 & \sigma_N \end{bmatrix}$$

where the σ_n, $n = 1, \ldots, N$, called **singular values**, are nonnegative real numbers arranged such that $\sigma_1 \geq \sigma_2 \geq \ldots \geq \sigma_N$ and V is also a **unitary** matrix. The elements of U and V are real (complex) numbers if the elements of A are real (complex) numbers. With (3.60), (3.19) becomes

$$U\Sigma V'X = Y \tag{3.61}$$

Premultiplying both sides of (3.61) by U', then by Σ^{-1}, and then by V gives

$$X = V\Sigma^{-1}U'Y, \rightarrow A^{-1} = V\Sigma^{-1}U' \tag{3.62}$$

The matrix Σ^{-1} is also a diagonal matrix with diagonal elements $1/\sigma_n$, $n = 1, \ldots, N$. In fact, the matrix A is nonsingular if and only if all singular values are positive real numbers. Furthermore, σ_n^2, $n = 1, \ldots, N$ are the eigenvalues of $A'A$, a symmetric matrix. This method to find X requires more computation than Gauss–Jordan elimination, but the results are less prone to accumulated round-off error.

Let us consider the case when $N = 2$. The matrices U, Σ, and V have the form

$$U = \begin{bmatrix} -\cos(\theta) & \sin(\theta) \\ \sin(\theta) & \cos(\theta) \end{bmatrix} = \begin{bmatrix} -1 & 0 \\ 0 & 1 \end{bmatrix} \begin{bmatrix} \cos(\theta) & -\sin(\theta) \\ \sin(\theta) & \cos(\theta) \end{bmatrix} = \begin{bmatrix} -1 & 0 \\ 0 & 1 \end{bmatrix} R(\theta)$$

$$\Sigma = \begin{bmatrix} \sigma_1 & 0 \\ 0 & \sigma_2 \end{bmatrix}, \quad V = \begin{bmatrix} -1 & 0 \\ 0 & 1 \end{bmatrix} R(\phi) \tag{3.63}$$

Notice that U and V are rotations followed by a reflection (sign reversal) of the first dimension, while Σ causes scaling. Thus, (3.63) shows that the solution of (3.19) involves a rotation, scaling, followed by another rotation to obtain the vector X, a very interesting interpretation of finding the solution of a linear system of equations.

MATLAB includes a large number of built-in functions that perform high-level computations. Another one of them is the built-in function **svd** (singular value decomposition), which performs a singular value decomposition.

Example 3.19 ─────────────────────────────────────

The following MATLAB statements demonstrate the application of the function svd.

```
>> A = [1   -1;2   4];
>> [U,S,V] = svd(A)    % svd returns three matrices
U =
   -0.1091      0.9940
    0.9940      0.1091
S =
    4.4966             0
         0            1.3343
V =
    0.4179      0.9085
    0.9085     -0.4179
>> S_inv = [1/S(1,1)   0; 0   1/S(2,2)];    % inverse of S
>> A_inv = V*S_inv*U'    % get A inverse with (3.62)
A_inv =
    0.6667      0.1667
   -0.3333      0.1667
>> A*A_inv    % check
ans =
    1.0000     -0.0000
         0            1.0000
>> A = [-1 -1;1 -1];   % this matrix has complex eigenvalues
>> [U,S,V] = svd(A)
U =
   -0.7071      0.7071
    0.7071      0.7071
S =
    1.4142             0
         0            1.4142
V =
    1        0
    0       -1
```

3.10   Accuracy of the Solution of $AX = Y$

The accuracy of the solution of $AX = Y$ depends not only on computing accuracy to find A^{-1}, but also on A itself.

As may often be the case, A and Y result from applying physical principles to a practical problem, where the elements of A and Y are estimates of the corresponding parameter values in the problem.

For example, in Example 3.16, a nodal analysis requires that we solve (3.44), which is $Gv = I$ for the node voltages v. An element of I depends on the resistor R_3. Suppose a circuit design calls for $R_3 = 103 \ \Omega$. However, such a resistor value is not commercially available, and a commercially available resistor $R_3 = 100 \ \Omega$ is used instead. Furthermore, due to imprecision in the manufacturing process, the actual resistance may be somewhere in the range $100 \pm 5\% \ \Omega$. The question is, will a small change in the vector I cause a substantial change in the solution, the node voltages v?

Let us model an estimation error of the elements of Y with $\hat{Y} = Y + dY$, where Y is the true value, dY is the estimation error, and $\hat{Y}$ is available to us. Since we only know $\hat{Y}$, we can only find an estimate $\hat{X}$ of X given by

$$\hat{X} = A^{-1}\hat{Y} = A^{-1}(Y + dY) = A^{-1}Y + A^{-1}dY = X + dX, \rightarrow dX = A^{-1}dY \qquad (3.64)$$

Therefore, if the elements of A^{-1} are large, then a small dY can cause a large estimation error dX of X. This does not depend on how accurately A^{-1} is found, but is an intrinsic property of A. We prefer that if there is a small error in estimating Y, then the resulting error dX in the solution $\hat{X}$ is small. Even if we know Y exactly, its value in computer memory can be in error by an amount $dY = $ **eps**, the error due to truncation of Y to a finite number of binary digits. Can a small truncation error in $\hat{Y}$ cause a substantial error in $\hat{X}$?

Example 3.20 _____

The given A and Y result in

$$A = \begin{bmatrix} 3 & 3 \\ 2.001 & 2 \end{bmatrix}, \quad Y = \begin{bmatrix} 1 \\ 0.666 \end{bmatrix}, \rightarrow X = \begin{bmatrix} -0.6667 \\ 1.0000 \end{bmatrix} \qquad (3.65)$$

Now, consider a small change in Y, resulting in

$$A = \begin{bmatrix} 3 & 3 \\ 2.001 & 2 \end{bmatrix}, \quad \hat{Y} = \begin{bmatrix} 0.997 \\ 0.669 \end{bmatrix}, \rightarrow \hat{X} = \begin{bmatrix} 4.3333 \\ -4.0010 \end{bmatrix}$$

Suppose instead that there is a small change dA in A. For example:

$$\hat{A} = \begin{bmatrix} 3.001 & 3.001 \\ 2.002 & 1.999 \end{bmatrix}, \quad Y = \begin{bmatrix} 1 \\ 0.666 \end{bmatrix}, \rightarrow \hat{X} = \begin{bmatrix} -0.0371 \\ 0.3703 \end{bmatrix}$$

We see that small changes in A or Y of the equations described by (3.65) can result in a large change in the solution.

The following A and Y result in

$$A = \begin{bmatrix} 4 & 3 \\ 2 & 1 \end{bmatrix}, \quad Y = \begin{bmatrix} 1 \\ 0.666 \end{bmatrix}, \rightarrow X = \begin{bmatrix} 0.4990 \\ -0.3320 \end{bmatrix} \tag{3.66}$$

and small changes in A or Y give

$$A = \begin{bmatrix} 4 & 3 \\ 2 & 1 \end{bmatrix}, \quad \hat{Y} = \begin{bmatrix} 0.997 \\ 0.669 \end{bmatrix}, \rightarrow \hat{X} = \begin{bmatrix} 0.5050 \\ -0.3410 \end{bmatrix}$$

$$\hat{A} = \begin{bmatrix} 4.001 & 3.001 \\ 2.001 & 0.999 \end{bmatrix}, \quad Y = \begin{bmatrix} 1 \\ 0.666 \end{bmatrix}, \rightarrow \hat{X} = \begin{bmatrix} 0.4978 \\ -0.3305 \end{bmatrix}$$

Here we see that small changes in A or Y of the equations described by (3.66) result in small changes in the solution.

It is said that if a small dY or a small dA causes a small dX, then the matrix A in $AX = Y$ is **well-conditioned**, and if a small dY or a small dA causes a large dX, then the matrix A is **ill-conditioned**.

The **condition number** $K(A)$ of a square matrix A is used to assess the degree to which A is ill-conditioned. To understand its meaning we must understand what is meant by the supremum of a set R of real numbers. The **supremum of R**, denoted by $\sup(R)$, is the least upper bound of the set R, which means that $\sup(R)$ is the smallest real number such that for all $x \in R$, $x \leq \sup(R)$. Similarly, the **infimum of R**, denoted by $\inf(R)$, is the greatest lower bound of R. The supremum (or infimum) of R may or may not be a number contained in R. For example, consider

$$R = \{x : 0 \leq x \leq 1\}$$

Here, $\sup(R) = 1$, $\max_{x \in R}(R) = 1$, and $\sup(R) \in R$. And, $\inf(R) = 0$, $\min_{x \in R}(R) = 0$, and $\inf(R) \in R$. However, consider

$$R = \{x : 0 < x < 1\}$$

Here, $\sup(R) = 1$, $\sup(R) \notin R$, and $\max_{x \in R(R)}$ do not exist. And, $\inf(R) = 0$, $\inf(R) \notin R$, and $\min_{x \in R}(R)$ do not exist. Every nonempty and bounded subset of the set of real numbers has a supremum and an infimum, but not necessarily a maximum or a minimum value.

Also, to understand the meaning of $K(A)$ we must extend the concept of the norm of a vector to the norm of a matrix. In the equation $AX = Y$, the matrix A operates on the vector X with norm $||X||$ to produce the vector Y with norm $||Y||$. Consider the relative change in the norm of X and the norm of Y with $||Y||\,/||X||$, $X \neq 0$. The least upper bound of this ratio is given by

$$||A|| = \sup_{X \neq 0} \left(\frac{||Y||}{||X||} \right) = \sup_{X \neq 0} \left(\frac{||AX||}{||X||} \right) = \sup_{||X||=1} ||AX|| \tag{3.67}$$

which defines the **norm of A**. Therefore, $||A||$ measures the maximum extent that A can change the norm of X, and for all X we have

$$\frac{||AX||}{||X||} \leq ||A|| \rightarrow ||AX|| \leq ||A||\ ||X|| \tag{3.68}$$

The norm of a matrix A has the following properties:

(a) For any $a \in R$, $||aA|| = |a|||A||$
(b) $||A|| \geq 0$; $||A|| = 0$, if and only if $A = 0$
(c) $||A + B|| \leq ||A|| + ||B||$
(d) $||Ax|| \leq ||A||\ ||x||$, if the product Ax of A and a vector x exists
(e) $||AB|| \leq ||A||\ ||B||$, if the product AB of A and another matrix B exists
(f) If A is nonsingular, $||A||\ ||A^{-1}|| \geq 1$
(g) If A is nonsingular, $\frac{1}{||A^{-1}||} = \inf_{x \neq 0} \left(\frac{||Ax||}{||x||} \right)$

An optimization of (3.67) gives the following p-norms $||A||_p$ of A for $p = 1$, $p = 2$, and $p = \infty$.

$$||A||_1 = \max_{1 \leq m \leq N} \sum_{n=1}^{N} |A(n,m)|, \quad \text{largest column sum}$$

$$||A||_2 = \sigma_{max}/\sigma_{min}, \quad \text{ratio of largest to smallest singular value} \tag{3.69}$$

$$||A||_\infty = \max_{1 \leq n \leq N} \sum_{m=1}^{N} |A(n,m)|, \quad \text{largest row sum}$$

Another commonly used norm of a matrix A is the **Frobenius norm** given by

$$||A||_F = \left(\sum_{n=1}^{N} \sum_{m=1}^{N} |A(n.m)|^2 \right)^{1/2}, \quad \text{also called Euclidean norm} \tag{3.70}$$

which is easier to compute than the p-norms given in (3.69). The MATLAB function norm also returns the norm of a matrix A. The syntax is given by

norm_1=norm(A,1)   % get the p-norm for p $= 1$

norm_2=norm(A,2)   % get the p-norm for p=2
norm_inf = norm(A,inf)   % get the p-norm for p=infinity
norm_F = norm(A,'fro')   % get the Frobenius norm

Recall that (3.64) gives the change in the solution X due to a change in Y with $dX = A^{-1}dY$, and like (3.68), where

$$||Y||_p \leq ||A||_p ||X||_p \tag{3.71}$$

we have

$$||dX||_p \leq ||A^{-1}||_p ||dY||_p \tag{3.72}$$

Dividing both sides of (3.72) by $||X||_p$ gives

$$\frac{||dX||_p}{||X||_p} \leq ||A^{-1}||_p \frac{||dY||_p}{||X||_p} \tag{3.73}$$

According to (3.71), $||X||_p \geq ||Y||_p / ||A||_p$, and substituting $||X||_p$ into the right side of (3.73) gives

$$\frac{||dX||_p}{||X||_p} \leq ||A||_p ||A^{-1}||_p \frac{||dY||_p}{||Y||_p} \tag{3.74}$$

which bounds a relative change in X given a relative change in Y. The condition number $K(A)$ of a matrix A is defined by

$$K(A) = ||A|| \, ||A^{-1}|| \tag{3.75}$$

The condition number may vary with the norm that is used. If a change in Y is due to truncation caused by storing Y in computer memory, then

$$\frac{||dX||_p}{||X||_p} \leq K(A) \text{ eps} \tag{3.76}$$

where eps, which is the difference between 1 and the next larger number greater than 1, is returned by the MATLAB function **eps**.

Now consider a change in A, where $\hat{A} = A + dA$. We have $AX = Y$ and $\hat{A}\hat{X} = Y$. Therefore, $AX = \hat{A}\hat{X}$, and

$$\begin{aligned} AX &= (A + dA)(X + dX) = AX + A\,dX + dA\,X + dA\,dX \\ 0 &= A\,dX + dA(X + dX) \rightarrow dX = -A^{-1}dA(X + dX) \end{aligned} \tag{3.77}$$

By property (e) we get

$$||dX|| \leq ||A^{-1}||\ ||dA||\ ||X + dX|| \rightarrow ||A||\ ||dX|| \leq ||A||\ ||A^{-1}|||\ |dA||\ ||X + dX||$$

Therefore

$$\frac{||dX||}{||\hat{X}||} \leq K(A)\frac{||dA||}{||A||} \tag{3.78}$$

which bounds a relative change in the solution $\hat{X}$ given a relative change in the matrix A.

The MATLAB function **cond** returns the condition number of a matrix using the norms given in (3.69) and (3.70). When $K(A)$ is close to its lower bound $(\inf(K(A)) = 1)$, the matrix A is well-conditioned. A relative change in X is nearly the same as a relative change in Y. As $K(A)$ becomes large, the matrix A becomes ill-conditioned. If A is singular, $K(A) = \infty$.

Example 3.20 (continued) _____

For (3.65) we have

```
>> A = [3 3;2.001 2]; Y = [1;0.666]; X = [-0.6667;1.0000];
>> Y_est = [0.997;0.669]; X_est = [4.3333;-4.0010];
>> K = cond(A)   % get the default 2-norm of A
K =  8.6680e+003
% this is a large condition number
>> dX = X_est - X;
>> dX_relative = norm(dX)/norm(X)   % get the relative change in X
dX_relative =    5.8840
>> dY = Y_est - Y;
>> dY_relative = norm(dY)/norm(Y)   % get the relative change in Y
dY_relative =    0.0035
>> K*dY_relative   % get the least upper bound of the relative change in X
ans =   30.6082
```

Here we see that the relative change in X is much larger than the relative change in Y.
For (3.66) we have

```
>> A = [4 3;2 1]; Y = [1;0.666]; X = [0.4990;-0.3320];
>> Y_est = [0.997;0.669]; X_est = [0.5050;-0.3410];
>> K = cond(A)
K =   14.9330
>> dX = X_est - X;
>> dX_relative = norm(dX)/norm(X)
```

```
dX_relative =     0.0180
>> dY = Y_est - Y;
>>  dY_relative = norm(dY)/norm(Y)
dY_relative =     0.0035
>> K*dY_relative  % get the least upper bound of the relative change in X
ans =    0.0527
```

Here we see that the least upper bound of the relative change in X is small.

Example 3.21

Let us apply the concept of a condition number to the circuit analysis given in Example 3.16. The following MATLAB statements were inserted into Prog. 3.3, which does a component circuit analysis, just after the matrix A is defined.

```
[U,S,V] = svd(A); sigma = diag(S); % get singular values of A
sigma_max = max(sigma), sigma_min = min(sigma) % min and max singular values
K = cond(A) % get 2-norm condition number
K_sigma = sigma_max/sigma_min  % check 2-norm condition number
```

The results from these statements are

```
sigma_max =   3.0000e+003
sigma_min =   7.2807e-004
K =   4.1205e+006
K_sigma  =   4.1205e+006
```

The condition number is very large. Let us investigate this by appending the following statement to Prog. 3.3.

```
dX_relative_least_upper_bound = K*eps
```

The function **eps** returns the difference between 1 and the next larger number in double precision. It is used as the relative change in Y. The result from this statement is

```
dX_relative_least_upper_bound =   1.8299e-010
```

Since double precision computation gives 15-digit accuracy, we expect that the last five to six digits in the elements of X will be inaccurate due to the large condition number amplifying the truncation error incurred by storing Y in computer memory. If single precision computation is used, which gives seven-digit accuracy, then the large condition number ($\sim$ 4e6) will cause the solution to be almost useless.

Prog. 3.4 solves the equation $G v = I$ to find the node voltages, and $K(G) = 9.9$, which is small, and Prog. 3.5 solves the equation $R i = V$ to find the mesh currents, and $K(R) = 9.8$, which is also small. In Example 3.16 the results of all three analysis methods are displayed using only four fractional digits, and they appear to be the same.

3.11 System of Nonlinear Equations

The problem is to find a root x of a system of nonlinear equations written as

$$F(x) = 0 \tag{3.79}$$

where x is a vector and $F(x)$ is a function that returns a vector value. Unlike a linear system, where $F(x) = Ax - y$, for a system of nonlinear equations there are no methods that can guarantee to find a solution, if it exists. This depends on the kinds of nonlinearities in the given equations.

The MATLAB built-in function **fsolve** is a methodology that searches for a real root of (3.79), and the search may or may not be successful. A syntax option of fsolve is given by

```
[x,F_val,exit_flag,output] = fsolve(F,x_init,options)
```

which starts at an initial guess x = x_init and tries to find and return an x = x_opt such that the Euclidean norm of F(x_opt) = 0. The minimal syntax is

```
x = fsolve(F,x_init)
```

We will use this syntax and the syntax

```
[x,F_val,exit_flag] = fsolve(F,x_init)
```

where if exit_flag = 1, then fsolve converged to a solution x = x_opt within a default tolerance and F_val = F(x_opt) $\cong$ 0. If exit_flag $\neq$ 1, then fsolve could not find an x_opt using default options, and it could be that merely trying a different x_init may yield a solution. The function F can be specified with a handle of a function m-file, for example, x = fsolve (@my_F, x_init), where my_F is a MATLAB function that can receive a vector input and return a vector. For further details about fsolve, get help with **doc fsolve**.

Example 3.22 _____

Find a root of a quadratic function of a 2×2 matrix x given by $y = Ax^2 + B\,x + C$. Prog. 3.7 sets up the problem, and Prog. 3.8, a function, evaluates the quadratic. Notice that an input argument of fsolve is the handle of the function quad_x. Since fsolve works with vectors, the function **reshape** is used to convert a matrix to a vector and a vector to a matrix.

```
clear all; clc
% Solve for a root of a quadratic function of a 2x2 matrix x
% y = Ax^2 + Bx + C
global A B C
A = [-1 -1;-1 -1]; B = [2 1;1 1]; C = [-1 1;1 -1]; % matrix coefficients
x_init = [1 0;0 1]; % initial 2x2 matrix guess
z_init = reshape(x_init,4,1); % convert x_init to a column vector
z = fsolve(@quad_x,z_init); % find root as a vector
x = reshape(z,2,2) % convert z into a 2x2 matrix
y = A*x*x + B*x + C  % check solution
```

Program 3.7   Set up problem.

```
function F = quad_x(z)
% evaluate a quadratic function of a matrix
global A B C
x = reshape(z,2,2);   % convert vector into a 2x2 matrix
q = A*x*x + B*x + C; % evaluate quadratic function of a matrix
F = reshape(q,4,1);   % convert 2x2 matrix into a vector
end
```

Program 3.8   Evaluate quadratic function of a 2×2 matrix.

Equation solved.
fsolve completed because the vector of function values is near zero
as measured by the default value of the function tolerance, and
the problem appears regular as measured by the gradient.
<stopping criteria details>
x =   y =
   2.0000 −2.0000   1.0e−011 *
  −0.5000 3.0000   0.7266 −0.4718
                            0.7266 −0.4718

Example 3.23 —————————————————————————

Given is the circuit shown in Fig. 3.8. Let the input be $v_s(t) = \sin(\omega t)$, where $f = 1$ Hz and $\omega = 2\pi f$ rad/sec. Find the output voltage $v(t)$.

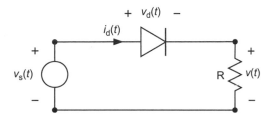

Figure 3.8   A nonlinear diode circuit.

Applying Kirchhoff's voltage law gives

$$-v_s(t) + v_d(t) + Ri_d(t) = 0 \tag{3.80}$$

Recall from Chapter 2 that the diode characteristic is $i_d(t) = I_s(e^{v_d(t)/V_T} - 1)$, and (3.80) becomes

$$-v_s(t) + v_d(t) + RI_s(e^{v_d(t)/V_T} - 1) = 0 \tag{3.81}$$

For each value of t, (3.81) is a nonlinear equation in $v_d(t)$.

Let us find $v_d(t)$ over one period of $v_s(t)$ at the time points $t = nT$, $n = 0, 1, \ldots, N-1$, where $T_0 = 1/f$ sec is the period of $v_s(t)$ and $T = T_0/N$ is the time increment. To put (3.81) into the context of (3.79), let

$$F(x) = -v_s(t) + x + RI_s(e^{x/V_T} - 1) \tag{3.82}$$

and for each value of t, we must find $x = v_d$ to make $F(v_d) = 0$.

Program 3.9 sets up the problem, and uses the function KVL given in Prog. 3.10, which evaluates the KVL function given in (3.82). When finished, the program gives F_mag = 1.2400e–010, which means that the requirement of (3.81) has been achieved for all time points.

```
clear all; clc
% Program to solve for the diode voltage in a nonlinear circuit.
global vs R   % make available in the circuit KVL function
f = 1; w = 2*pi*f; T0 = 1/f; % get results for one cycle
N = 256; T = T0/N; % solve for vd at N time points
n = 0:N-1; t = n*T; % N time points
```

```
vs = sin(w*t); % vector of N values of the sinusoidal input voltage
R = 330; % resistance
vd_init = zeros(1,N); % initial guess of N diode voltages
% apply nonlinear solver to find the diode voltage at N time points
[vd,F_val,exit_flag] = fsolve(@KVL,vd_init);
if exit_flag == 1
    F_mag = norm(F_val) % Euclidean norm
    v = vs-vd; % output voltage across resistor
    plot(t,v,'r'); hold on; plot(t,vs,'k'); plot(t,vd,'b');
    grid on; xlabel('time - sec'); ylabel('volts')
    title('Analysis of a Nonlinear Diode Circuit')
else
    disp('Could not solve for the diode voltages.')
end
```

Program 3.9   Program to find the diode voltage.

```
function F = KVL(x)
% evaluate the KVL equation
global vs R
I_S = 1e-12; % saturation current in amps
V_T = 25.85e-3;   % thermal voltage at 300 degrees Kelvin
id = I_S*(exp(x/V_T)-1); % diode current
F = -vs + x + R*id; % KVL equation at N time points
end
```

Program 3.10   Nonlinear KVL function.

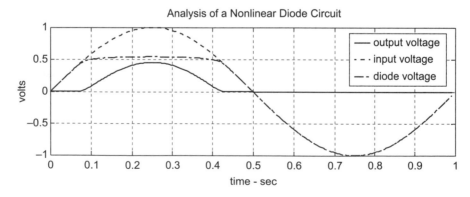

Figure 3.9   Input, output, and diode voltages.

Fig. 3.9 shows how a diode works. While the input is positive over $0 < t < 0.5$ secs, the output is positive over approximately $0.08 < t < 0.42$ secs. The diode does not start to conduct until the input voltage exceeds the forward bias threshold (about 0.5 volts) of the diode at $t \cong 0.08$ secs. Then, the diode voltage drop remains relatively constant, while it is conducting. When the input voltage again goes below the diode threshold voltage at $t \cong 0.42$ secs, the diode stops conducting as indicated by the output voltage becoming substantially zero.

As you apply MATLAB to solve problems in various areas of electrical and computer engineering, you will want to use more of the many high-level MATLAB built-in functions. Browse through a list of functions by category to see the possibilities.

3.12 Conclusion

Matrix algebra has made a tremendous impact on engineering, and with MATLAB we can easily do much of the algebraic and arithmetic work to apply matrix algebra. In this chapter a basic problem (two equations in two unknowns) was used to understand the mechanics of matrix organization and arithmetic. Many special kinds of matrices and matrix properties were introduced. You should now know how

- the mechanics of matrix arithmetic works;
- conveniently MATLAB can be used to work with the matrix data type;
- MATLAB can be used for data analysis;
- Gauss–Jordan elimination can be applied to solve a linear system of equations;
- MATLAB can be used for circuit analysis;
- a linear transformation can be designed to rotate a vector, which is only one of many other kinds of geometric operations;
- a matrix inverse can be found with singular value decomposition or eigenvalue decomposition;
- to assess the accuracy of the solution of a system of linear equations; and
- to solve a system of nonlinear equations.

There is much more to learn about matrix theory that is useful to gain insight into the properties and behavior of data, signals, and systems, and MATLAB has many more built-in functions that are helpful to gain this insight. Table 3.7 gives the MATLAB functions that were introduced in this chapter.

Use the MATLAB help facility, where you will also find many other related built-in functions, to learn more about these built-in functions. You should also use the built-in function **demo** from the *Command Window*, and view an excellent audio/video tutorial about working with matrices and MATLAB.

Table 3.7   Built-in MATLAB functions introduced in this chapter

Function	Brief explanation
: operator	Delimit a range
axis	Specify the minimum and maximum x and y axis range
cond	Returns the condition number of a matrix
doc	Get documentation
size	Get dimensions of a matrix
length	Get number of elements in a vector
diag	Get diagonal elements of a matrix
linspace	Create a vector of equally incremented values
zeros	Create a matrix of all zeros
fix	Round to nearest integer toward zero
ones	Create a matrix of all ones
dot	Compute dot product of two vectors
sum	Sum the elements of a vector, can also be applied to matrices
'	Transpose of a matrix
eye	Create a matrix with ones along the major diagonal and zeros elsewhere
stem	Plot points of data with vertical lines to dependent data points
mean	Get average value of elements in a vector
det	Get determinant of a square matrix
inv	Get inverse of a square matrix
mldivide	Matrix left divide
mrdivide	Matrix right divide
log10	Get log base 10
norm	Returns a measure of the length of a vector or a measure of the amplification factor of a matrix
rank	Finds the number of linearly independent columns(rows) in a matrix
rref	Convert an $N \times N + 1$ matrix to row reduced echelon form
abs	Get magnitude of a vector, can also be applied to complex numbers
svd	Compute a singular value decomposition
eig	Get eigenvalues and eigenvectors of a matrix
fsolve	Solve a system of nonlinear algebraic equations

In the next chapter MATLAB program flow control structures will be introduced, that will make it possible to efficiently execute program segments repeatedly and to make decisions about further program execution based on data and computational results that were obtained within the same program.

Further reading

Anton, H., *Elementary linear algebra (applications version)* (9th edn.), Wiley International, Somerset, NJ, USA, 2005

Lay, D.C., *Linear algebra and its applications* (3rd edn.), Addison Wesley, Reading, MA, USA, 2005

Leon, S.J., *Linear algebra with applications* (7th edn.), Pearson Prentice Hall, Upper Saddle River, NJ, USA, 2006

Problems

Section 3.1

For Probs. 3.1 through 3.4, use the following matrices:

$$A = \begin{bmatrix} 2 & -1 & 2 \\ -2 & 1 & 0 \\ -1 & 1 & -2 \end{bmatrix}, \quad B = \begin{bmatrix} -2 & 1 \\ 1 & -1 \\ 0 & 3 \end{bmatrix}, \quad C = \begin{bmatrix} 3 & 2 & -1 \\ -1 & 2 & 4 \end{bmatrix}, \quad x = \begin{bmatrix} 2 \\ -1 \\ 4 \end{bmatrix}$$

1) Give the dimensions of the matrices A, B, C, x.
2) Give MATLAB statements that define the matrices A, B, C, x.
3) What does MATLAB return with the statements: $a = \text{diag}(A), b = \text{diag}(B), c = \text{diag}(C)$?
4) What does MATLAB return with the statements: $d = \text{size}(B)$, $e = \text{any}(B), f = \text{all}(B)$, $N = \text{length}(x)$, $g = \text{min}(B)$?
5) Give MATLAB statements that define I, a 4×4 identity matrix; Z, a 3×2 matrix of zeros; T, a 100×1 matrix of twos; R, a 4×3 matrix of random numbers uniformly distributed over the range (0,1); a matrix E to be an empty matrix.
6) Give a MATLAB statement using colon notation that defines a vector z, where the first element is $z(1) = -1.5$, the increment from element to element is 0.1, and the last element is 4.3. Then, use the linspace function to define a vector w that is equal to z.

Section 3.2

For Probs. 3.7 through 3.11, use the same matrices as in Prob. 3.1.

7) Manually, obtain the following: inner product (x, x); AB; AA'; $B'B$; Ax; $B + C'$; BI, where I is an identity matrix; IB, where I is an identity matrix; xI, where I is an identity matrix.
8) Manually, obtain $B\,C$; $C\,B$. Is matrix multiplication commutative? Is multiplication of square matrices commutative?
9) Give $F = x\,x'$, and find the dimensions of $C\,A\,B$; $F + A$.
10) Manually, obtain the result of the MATLAB expressions: (a) $1./A$, (b) $B.*C'$, (c) $x.\hat{\ }2$, (d) $2.\hat{\ }x$, (e) $B./C'$, (f) $C'.\backslash B$. Then, in the *Command Window* confirm your results.
11) Assume that scalars q, r, and s satisfy the distributive property $q(r + s) = qr + qs$. Given are three matrices Q, R, and S. Prove that $Q(R + S) = QR + QS$. For $R + S$, R and S must have the same dimension, say $N \times M$, and for multiplication to obtain $Q\,R$ or $Q\,S$, the dimension of Q can be $K \times N$. Hint: let $W = QR$, and write an element of W as in (3.5). Then, let $Q = A$, $R = B$, and $S = C'$, and in the *Command Window* check the distributive property.
12) If $Ax - yx = 0$, where y is a scalar and 0 is a 3×1 vector of zeros, then give the matrix G so that $(A - yG)x = 0$.
13) Vectorize each of the following MATLAB expressions: (a) sqrt$(1 + q\hat{\ }2)$, (b) $1/q$, (c) $\cos(2 * q - \text{pi}/2)$; (d) $\exp(-q)$.

Section 3.3

14) Use the MATLAB help facility and describe what the built-in function **scatter** does. Then, write a MATLAB program that uses the function scatter to obtain a plot of the data used in Example 3.9.

15) Given is the following data:

$$y = (5, 4.3, 3.1, 2.2, 2.5, 1.7, 0.4, -0.3, -2.2, -3.9, -6.1), \text{ dependent variable}$$
$$x = (-2.9, -2.3, -1.5, -1.2, -0.9, -0.7, -0.3, 0.1, 0.5, 0.9, 2.1), \text{ independent variable}$$

Write a MATLAB program that uses the built-in function **polyfit** to fit a straight line to the data. The program must also plot in the same figure the straight line and the given data using the function stem.

16) Using the function **rand**, write a MATLAB program that defines a vector x with $N = 100000$ elements. Then, use the built-in functions **mean** and **var** to find the mean and variance of the elements in x. Let $y = mx + b$, and give m and b such that y ranges over $(-5, 5)$. Use the **min, mean**, and **max** functions to confirm these results.

17) (a) Write a MATLAB program that uses the diode equation given in Prog. 2.2 to find the diode current i_d when the diode voltage is $v_d = 0.7$ volts.

   (b) Modify the program of part (a) to make the diode voltage a vector of random voltage values given by $v_d = 0.7 + v$, where v is a vector with $N = 100000$ elements that are found with the built-in function **randn**. Use the MATLAB statement $v = 0.01*\text{randn}(1,N)$ to assign to v a vector of N random voltages with a standard deviation given by 0.01 volts. Use help randn to find out more about the function randn. Obtain the vector i_d of random currents, and apply the built-in functions min, mean, var, and max to v_d and i_d. Compare the means of v_d and i_d to v_d and i_d from part (a).

   (c) Repeat part (b) with a standard deviation of v given by 0.02 volts. Does the mean of i_d increase significantly as the standard deviation of v is increased?

Section 3.4

18) Write a MATLAB program that plots $N = 101$ points of the signal $x(t) = e^{at}\sin(\omega t)$ over the range $0 \le t \le 3T_0$, where T_0 is the period of the sine function, $a = -2$, and $\omega = 4\pi$ rad/sec.

19) Given is the function $H^2(\omega) = \varepsilon^2 C^2(1/\omega)/(1 + \varepsilon^2 C^2(1/\omega))$, where $C(\theta) = 4\theta^3 - 3\theta$. Write a MATLAB program that plots $N = 201$ points of $H^2(\omega)$ for $-5 \le \omega \le 5$ with $\varepsilon = 0.1$. Note that $H^2(\omega = 0) = 1$. Then, in your program convert $H^2(\omega)$ to dB (decibels) with $H^2(\omega)|_{dB} = 10\log(H^2(\omega))$, and in a second figure, plot $H^2(\omega)|_{dB}$ over the same range of ω. Which plot gives better detail for small ω? And, which plot gives better detail for large ω?

20) Using the matrices given in Prob. 3.1, what does MATLAB return for the statements $a = \text{tril}(A)$, $b = \text{triu}(A)$, $c = \text{fliplr}(A)$, $d = \text{fliplr}(B)$, $e = \text{flipud}(C)$, $f = \text{circshift}(A, [1\ 1])$? Explain each operation.

Section 3.5

21) For $D = \begin{bmatrix} 1 & 0.5 \\ -2 & 3 \end{bmatrix}$ and $E = \begin{bmatrix} 2 & 1 \\ -1 & 0.2 \end{bmatrix}$, manually, find the determinants of D, E, $D\,E$ and $E\,D$. Generally, for two $N \times N$ matrices D and E, $|D\,E| = |D||E|$. Do your results agree with this property of determinants?

22) Given are two equations in two unknowns:

$$3 - 4z + 6\alpha = 5$$
$$-3\alpha + 2 = -2z$$

(a) Define a column vector $X = [z; \alpha]$ of unknowns, and find the matrix A and the vector Y, such that the given equations can be written as $AX = Y$. Manually, by repeated substitution find α and z. Then, in the *Command Window* find A^{-1} and $X = A^{-1}Y$.

(b) Now, redefine X, A, and Y such that the given equations can be written as $XA = Y$, and manually find A^{-1} and X. Does the solution agree with the solution of part (a)? How is the matrix A of part (a) related to the matrix A of part (b)?

23) For the matrix A given in Prob. 3.1, find $|A|$ using (3.32).

24) Suppose every element of any row or any column of a matrix A is multiplied by a scalar k, for example, $B = \begin{bmatrix} a(1,1) & a(1,2) & a(1,3) \\ k\,a(2,1) & k\,a(2,2) & k\,a(2,3) \\ a(3,1) & a(3,2) & a(3,3) \end{bmatrix}$ or

$B = \begin{bmatrix} a(1,1) & k\,a(1,2) & a(1,3) \\ a(2,1) & k\,a(2,2) & a(2,3) \\ a(3,1) & k\,a(3,2) & a(3,3) \end{bmatrix}$ then prove that $|B| = k|A|$. Hint: Use (3.32).

25) Find the solution of $A\,X = Y$ by Gauss elimination and backward substitution, where

$$A = \begin{bmatrix} 1 & -1 & 2 \\ -2 & 1 & 0 \\ -1 & 3 & -2 \end{bmatrix}, \quad Y = \begin{bmatrix} 2 \\ -1 \\ 4 \end{bmatrix}$$

In the *Command Window*, create A and Y, and check your solution with the MATLAB mldivide function.

26) Repeat Prob. 3.25 using Gauss–Jordan elimination.

Section 3.6

For Probs. 3.27 through 3.29, use the following matrices:

$$A = \begin{bmatrix} 2 & -1 & -2 & 3 & 1 \\ -3 & 2 & 0 & -3 & 2 \\ 1 & 3 & -1 & 2 & 4 \\ 4 & 1 & 2 & 5 & 3 \end{bmatrix}, \quad X = \begin{bmatrix} 1 \\ 2 \\ 3 \\ 4 \end{bmatrix}, \quad Y = [5 \quad 4 \quad 3 \quad 2 \quad 1]$$

27) (a) Using continuation notation for each row, give a MATLAB statement that creates the matrix A.

(b) Give MATLAB statements that assign to y the 3rd row of A; assign to x the 4th column of A; create a 2×3 matrix B from the 1st and 2nd row of A and the 3rd through 5th columns of A; remove the 2nd column of B; create a matrix C, where the 1st row comes from row 4, columns 2 through 4 of A and the 2nd row comes from row 1, columns 2 through 4 of A.

(c) Give MATLAB statements that assign to y the elements of Y shifted to the left by one position and make the last element of y the first element of Y, which is called a left rotation; use y to assign to Z a left rotation of A.

28) Give MATLAB statements that add (3/2) times the 1st row of A to the 2nd row of A; and then add $(-1/2)$ times the 1st row of A to the 3rd row of A, and then add (-2) times the 1st row of A to the 4th row of A.

29) Give MATLAB statements that (a) use the function **cat** to create a matrix B by augmenting on the left the matrix A with the vector X, (b) use the function **horzcat** to obtain the matrix B, (c) use the function **vertcat** to create a matrix C by augmenting at the top the matrix A with the vector Y, (d) create the matrix D by sorting the columns of A in descending order, (e) assign to the 10×2 matrix E a reshaped matrix from A', (f) use the function **repmat** to assign to the matrix F the vector Y replicated six times.

30) Write a MATLAB program that creates and plots a piecewise linear function $x(t)$ such that

$$x(t) = \begin{cases} 0, & t < 0 \\ t, & 0 \le t < 1 \\ -t + 2, & 1 \le t < 3 \\ t - 4, & 3 \le t < 4 \\ 0, & t \ge 4 \end{cases}$$

Use a sampling rate of $f_s = 10$ samples/sec.

Section 3.7

For Probs. 3.31 through 3.33, use the following circuit:

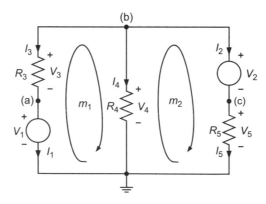

$R_3 = 100\ \Omega$, $R_4 = 1000\ \Omega$, $R_5 = 50\ \Omega$, $V_1 = 5$ volts, $V_2 = -3$ volts

31) (a) To do a component analysis, use the labeled voltages and currents, and (1) apply KVL to nodes (a), (b) and (c); (2) apply KCL to meshes 1 and 2; and apply Ohm's law to each resistor.

   (b) Define a vector of unknowns, such as $X' = [I_1\ I_2\ I_3\ I_4\ I_5\ V_3\ V_4\ V_5]$, and specify a matrix A and a vector Y, such that $AX = Y$ duplicates all of the equations of part (a).

   (c) Write a MATLAB program that assigns to MATLAB variables resistor and voltage source values, creates A and Y, and solves for X and assigns to each component current and/or voltage an element of X, for example, $V_3 = X(6)$. Then, the program should check that all KVL and KCL equations are satisfied.

32) (a) Let i_1 and i_2 be the currents in meshes m_1 and m_2, respectively, with reference directions as shown in the figure. Apply KVL, while also using Ohm's law, to the two meshes, and obtain two equations in two unknowns.

   (b) Define a vector I of unknown mesh currents, a matrix R, and a vector V, such that the equation $R I = V$ duplicates the KVL equations of part (a).

   (c) Write a MATLAB program that solves for the mesh currents and all of the component currents and voltages.

33) Manually, find the sum of the powers delivered to the voltage sources in mW and the sum of the powers delivered to all of the resistors in mW. Find the sum of the powers delivered to all components in the circuit.

For Probs. 3.34 through 3.36, use the following circuits, where $V_1 = 5$ volts, $I_2 = 20$ mA, $R_3 = 300\ \Omega$, and $R_4 - 1000\ \Omega$.

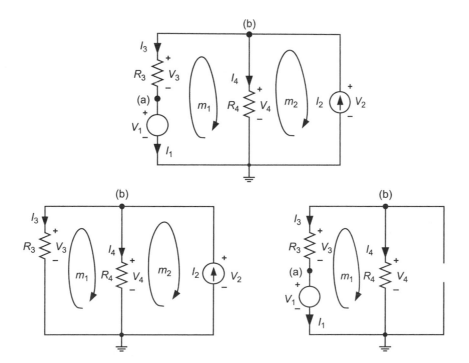

34) For the circuit with two sources, do a component analysis to find the voltage V_4. Follow the steps given in Prob. 3.31.

35) The two lower circuits each have a source removed. In the left circuit, the voltage source was removed, and in the right circuit, the current source was removed. Apply the superposition principle by doing a nodal analysis of the left circuit to find V_4 and doing a loop analysis of the right circuit to find V_4. Then add these two voltages to find V_4 in the top circuit.

36) Repeat Prob. 3.35, but reverse the analyses methods.

Section 3.8

37) Determine whether the system of equations

$$\begin{matrix} 0.5x_1 - 0.75x_2 = 1.5 \\ 1.5x_1 - 2.25x_2 = 2 \end{matrix} \rightarrow AX = Y$$

is consistent and dependent, consistent and independent, or inconsistent. Apply the function **rank** to both A and the augmented matrix.

38) Give the supremum of the following sets of real numbers: (a) the set of negative numbers, (b) the set x such that $x^2 < 3$, and (c) the set $\{1 - e^{-x}\}$, for $x > 0$.

39) The p-norm of a vector x is $||x||_p$. Find the p-norm $||y||_p$ of $y = kx$, where k is a real scalar.

40) Draw the vector $x = \begin{bmatrix} -3 \\ 4 \end{bmatrix}$ in a 2-D Cartesian space. Find its magnitude and angle. Then apply a linear transformation A to x to obtain a vector y, which is x rotated by $-\pi/4$. Give A. Write a MATLAB program that shows both x and y in the same figure.

41) For the matrix A given in Prob. 3.1 write a MATLAB program to find the eigenvalues and eigenvectors of A. For each eigenvalue λ and corresponding eigenvector v verify that $A v = \lambda v$. Then, your program should find the characteristic polynomial of A and the inverse of A, using the modal matrix and the eigenvalues.

Section 3.9

42) For the matrices A and B given in Prob. 3.1 write a MATLAB program to do a singular value decomposition to find U, Σ, and V. The program must verify that U and V are unitary matrices. Then find the inverse of A using the function inv and the matrices U, Σ, and V.

43) Write a MATLAB program to find the determinant and singular value decomposition of the matrix

$$D = \begin{bmatrix} 1 + \varepsilon & 1 - \varepsilon \\ 2 + \varepsilon & 2 - \varepsilon \end{bmatrix}$$

for $\varepsilon = 10^{-n}$, $n = 1, \ldots, 5$. What happens to the singular values as ε decreases?

Section 3.10

44) (a) Use the MATLAB function **norm** to find the l_1, l_2, l_∞, and Frobenius norms of the matrix D given in Prob. 3.43 for each value of ε. Manually find the l_1 and l_2 norms and compare your results with those obtained with the function norm.
   (b) Use the function norm to find the norms of the inverse of each matrix used in part (a).
   (c) Using the results of parts (a) and (b) find the condition number for each ε and for each norm. Tabulate results. Do the condition numbers vary appreciably with different norms?

45) Consider the equation $DX = Y = [\sqrt{2}\ 1]'$, where D is given in Prob. 3.43. To solve this equation with a computer, an estimate of Y must be used, where the estimation error is $dY = [\text{eps}\ 0]'$. Write a MATLAB program to find the effect of decreasing ε on the least upper bound of the relative change in X. For each ε, approximately how many digits in the solution will be inaccurate?

46) Prove that if the product of two matrices A and B exists, then $||AB||_\infty \leq ||A||_\infty ||B||_\infty$. Hint: Start with

$$||AB||_\infty = \max_k \sum_{m=1}^{M} \left| \sum_{n=1}^{N} A(k,n)B(n,m) \right|$$

and use the property that for any two numbers a and b, we have $|a + b| \leq |a| + |b|$.

Section 3.11

47) Write a MATLAB program that uses the function **fsolve** to find the time point where $e^{-t} = t^2$. Use a function m-file to define the function to be solved.

48) Write a MATLAB program and vector function to solve the system of nonlinear equations given by

$$x_1^2 + 4x_2^2 = 8$$
$$x_1 x_2 = 2$$

CHAPTER 4

Program Flow Control

A method to solve a problem may include alternative steps. Which next step to take may depend on results from previous steps. A script that implements a solution method must be able to follow alternative steps based on answers to questions like: Is a variable value different from an allowed set of values? Is the result of a calculation zero? Do two variables have the same value? Which of many variables has a particular value? and other questions. The answers to these kinds of questions can influence what to do next in a script.

After you have completed this chapter, you will know how to

- test conditions and execute alternative script segments
- repeatedly execute a script segment until a condition occurs
- execute one of many script segments depending on input data or intermediate results

4.1 Relational Operators

Relational operators are used to compare variable values and produce a result that is true or false. For example:

```
>> a = 2;  b = 3;  % assign values to two variables
>> c = a < b  % compare a and b using the relational operator <, meaning less than
c =   1
```

Since the value of a is less than the value of b, the comparison, $a < b$, is true, and c is assigned the value $c = 1$, which is interpreted to mean that the comparison result is true. If the value of a is, for example, 4, then the comparison result will be false and c is assigned the value 0. Since the value of c is the result of a relational operation, its value is either 1 (meaning true) or 0 (meaning false), and the variable c is called a **logical variable**, not an arithmetic variable.

Table 4.1   List of relational operators

Relational operator	Description
<	Less than
<=	Less than or equal to
>	Greater than
>=	Greater than or equal to
==	Equal to
~=	Not equal to

Table 4.1 gives the relational operators that can be used in MATLAB®. Notice that the relational operator, $==$, which tests equality, is not an assignment operator. The syntax of a **relational expression** is

left_expression   relational_operator   right_expression

The expressions on each side of a relational operator can be any MATLAB expressions that evaluate to scalars, vectors, or matrices having the same dimension. A relational expression produces a logical result, which can be assigned to a variable.

Example 4.1

```
>> a = 2; b = 3; c = -1; d = 4;  % define four arithmetic variables
>> % arithmetic expressions on each side of a relational operator
>> e = sqrt(a^2+b^2) <= sqrt(c^2+d^2)
e = 1
>> % logical values can be used in arithmetic expressions
>> f = (a > b) +2*(c < d)
f = 2
>> % f, an arithmetic variable, can be 0, 1, 2 or 3
>> w = 2*pi; t = 0:0.1:1; % assign a frequency and a time range
>> % compare two vectors element-by-element
>> X = sin(w*t) <= cos(w*t)  % X is a logical vector
X =   1   1   0   0   0   0   0   1   1   1   1
>> % matrices that have the same dimension can be used on each side
>> % of a relational operator
>> g = exp(c*t) .* sin(w*t); % define an arithmetic vector
```

```
>> % check times at which exponentially decaying sinusoid is positive
>> Z = g > 0
Z =   0   1   1   1   1   1   0   0   0   0   0
>> % Z, a logical vector, results from comparing each element of g
>> % a logical vector can be used in arithmetic expressions
>> g_pos = Z .* g
g_pos =
Columns 1 through 11
0   0.5319   0.7787   0.7046   0.3940   0.0000   0   0   0   0   0
```

Suppose the variables a and b are real scalars. Then, one of the relations

```
a < b
a == b
a > b
```

must be true. However, if a and b are replaced by A and B, which are matrices with the same dimensions, then the relation is applied to corresponding elements of A and B, and the result is a logical matrix with the dimension of A, where some elements are logic 1, while others are logic 0. Since A and B are matrices, then these relations do not produce a scalar logical value. To test the equality of two matrices, use the built-in MATLAB function **isequal**, as in $c = \text{isequal}(A,B)$, where c is a scalar logical variable. Table 4.2 gives some built-in MATLAB functions that operate on matrices and return a logical value.

Table 4.2 Functions that return a logical value

Function	Example, A and B are matrices, c is a logical scalar, and C is a logical matrix
isequal	$c = \text{isequal}(A,B)$, are all corresponding elements equal
islogical	$c = \text{islogical}(A)$, are all elements either 0 or 1
isinteger	$c = \text{isinteger}(A)$, are all elements integers
isfloat	$c = \text{isfloat}(A)$, are all elements floating point numbers
isempty	$c = \text{isempty}(A)$, does A have no elements
true	$C = \text{true}(M,N)$, C is an M $\times$ N logical matrix of all logic 1 values
false	$C = \text{false}(M,N)$, C is an M $\times$ N logical matrix of all logic 0 values
logical	$C = \text{logical}(A)$, convert numeric matrix A to logical matrix C
any	$c = \text{any}(A)$, is any element of A nonzero or logical 1, if A is a vector
	$C = \text{any}(A)$, if A is a matrix, then the function, any, works on each column of A, and C is a vector. To obtain a scalar, use $c = \text{any}(\text{any}(A))$
all	$c = \text{all}(A)$, are all elements of A nonzero, if A is a vector
	$C = \text{all}(A)$, if A is a matrix, then the function, all, works on each column of A, and C is a vector. To obtain a scalar, use $c = \text{all}(\text{all}(A))$

Relational expressions evaluate to a logical value. Relational expressions can be compounded with the logical operators **and, or, xor,** and **not**.

4.2   Logical Operators

Logical operators work on logical variables and produce a logical result. For example, the expression $(a < b)$ and $(c == d)$ can only be true if both relations are true. It is false if either relation or both relations are false. In MATLAB the word "and" is replaced by the symbol &, and a MATLAB assignment statement becomes $e = (a<b)\&(c == d)$. Depending on the values of a, b, c, and d, e can be 0 or 1. Use parentheses to specify the sequence in which the parts of an expression should be evaluated. In a compound expression, MATLAB evaluates the arithmetic parts first, then the relational parts, followed by the logical parts, where logical and has precedence over logical or.

MATLAB uses the symbols **&,** |, and ~ for the logical operators **and, or,** and **not,** respectively, and a built-in function **xor** for the logical operator **xor** (exclusive or). Table 4.3 defines the logical operators, where a, b, and c are scalar logical variables.

Table 4.3   Definition of logical operators

AND			OR			XOR			NOT	
a	b	c = a & b	a	b	c = a | b	a	b	c = xor(a,b)	a	~a
0	0	0	0	0	0	0	0	0	0	1
0	1	0	0	1	1	0	1	1		
1	0	0	1	0	1	1	0	1	1	0
1	1	1	1	1	1	1	1	0		

Logical operators can work with numbers. If a number is 0, then it is used as if it is logical 0, and if it is nonzero, then it is used as if it is logical 1.

Example 4.2 ⎯⎯⎯⎯⎯⎯⎯⎯⎯⎯⎯⎯⎯⎯⎯⎯⎯⎯⎯⎯⎯⎯⎯⎯⎯

```
>> a = 0; b = 1; c = -3;   % assign scalar values
>> a | b   % both operands have logical values
ans =    1
>> b & c   % c does not have a logical value, but it is used as if it is logic 1
ans =    1
>> ~c
ans =    0
>> xor(a,c)  % produces a logic 1 only if one of the two inputs is logic 1
ans =    1
```

MATLAB also has built-in logical functions for the logical operators, that is, **and**$(a,b) =$ a&b, **or**$(a,b) = a \mid b$, and **not**$(a) = \sim a$. There is an option when using the & and | operators. Consider $c =$ expression_a & expression_b. If expression_a evaluates to logic 0 then there is no need to evaluate expression_b, since $c = 0$. To require that MATLAB evaluates expression_b only if expression_a is logic 1, use **&&** instead of &. Similarly, consider $c =$ expression_a | expression_b. If expression_a is logic 1, then there is no need to evaluate expression_b, since $c = 1$. To require that MATLAB evaluates expression_b only if expression_a is logic 0, use || instead of |. These options can reduce execution time.

Logical operators can work with matrices when both operands have the same dimension. The logical operation is done element by element, and the result is a logical matrix. If one operand is a scalar, then the elements of the other operand, a matrix, are used element by element to produce a logical matrix. If A is a matrix of numbers, then $\sim A$ produces a logical matrix where a nonzero element of A results in a logic 0 element in $\sim A$. And, $\sim(\sim A)$ produces a logical matrix, where a nonzero element of A results in a logic 1 element of $\sim(\sim A)$. The built-in functions given in Table 4.2 can be applied to logical matrices.

Another useful built-in function is the function **find** to locate the elements of a matrix that are nonzero. The function returns the indices numbered linearly down each column starting at the left column.

Example 4.3

```
>> A = [1 -5 0; -5 0 4; 0 0 2]    % assign a matrix
A =
   1   -5    0
  -5    0    4
   0    0    2
>> % the indices for the first column are: 1, 2 and 3, and for the second column
>> % the indices are: 4, 5 and 6, and so on
>> find(A)   % find the linear indices of all elements in A that are nonzero
ans =    1    2    4    8    9
>> B = [0  -1.8  -4  5.2  0.6  0  1];
>> I = find(B)   % find the indices of the nonzero elements of a vector
I =    2    3    4    5    7
>> find(B,3,'first')  % find the first 3 indices of the nonzero elements in a vector
ans =    2    3    4
>> find(B,4,'last')  % find the last 4 indices of the nonzero elements in a vector
ans =    3    4    5    7
>> % logical and relational operations can be combined in compound expressions
>> find((B > -1) & (B < 1))  % find indices of the elements in the range: -1 < element < 1
ans =    1    5    6
>> % here, (B > -1) produces a logical vector, which is anded element by element
>> % with the logic vector produced by (B<1) to produce the logical vector that is
>> % the input to the find function
```

Use parentheses to control the sequence in which logical operations are evaluated. MATLAB evaluates an expression within the innermost parentheses first.

Example 4.4

```
>> a = 0; b = 1; c = 1; d = 0;   % assign logical values
>> a & b | c & ~d  % MATLAB evaluates the logical and operations first
ans =    1
>> % this results from: (a&b) | (c & ~d)
>> a & (b | c) & ~d
ans =    0
>> % MATLAB evaluates the term within parentheses first, and then left to right
>> %   parentheses must be carefully used to achieve the desired overall logical
>> % operation
```

4.3   If–Elseif–Else–End

With the **if–elseif–else** structure, a block of statements is executed only if a condition is met. The syntax of the most general form of this structure is

```
if expression_1
     block #1 of statements
elseif expression_2
     block #2 of statements
: insert as many additional elseif and block of statements as needed
else
     block #N of statements
end
```

where each expression must evaluate to a logical value. When MATLAB encounters the **if** key word, it does the following:

1) Evaluate expression_1
2) If expression_1 is true, then execute block #1 statements, and continue execution after the **end** statement
3) If expression_1 is not true, then skip the block #1 statements to the first **elseif** key word and evaluate expression_2
4) If expression_2 is true, then execute block #2 statements, and continue execution after the **end** statement

5)   If expression_2 is not true, then skip the block #2 statements to the next **elseif** key word; if there is one, evaluate its expression, and so on
6)   If all expressions are false, then MATLAB will skip to the **else** key word, if there is one, and execute the block #N statements, after which execution continues after the **end** statement.

It is useful to indent each block of statements to easily see that each **if** key word is matched by an **end** key word. The **elseif** through the **else** sections are optional. Within each block of statements there can be nested **if–elseif–else–end** structures. You can nest these structures as deeply as needed. The simplest **if** structure is

```
if expression
block of statements
end
```

Here, if the expression is true, then the block of statements is executed, and if it is false, then the block of statements is skipped and execution continues after the end statement. If the block of statements can fit on one line, then the if statement could be, for example

```
>> x = -6.8;
>> % limit x to the range: -5 to +5
if (x<-5) | (x>5), x = 5*sign(x); end  % the built-in function sign returns -1, 0 or +1
>> x
x =   -5
```

Example 4.5 _____

Suppose we are given a data point x, where its value can be anywhere in the range from a to b, written as $[a, b]$. The brackets indicate that the end point values are included in this range. If we write (a, b), then both end points are not included in this range. It is desired to know which of N sub-intervals of the range contains x. Let us segment this range into N segments (sub-intervals). Each segment is called a bin. The width of each bin is $w = (b - a)/N$. If we know which bin contains x, then we can say that x is known to be within its bin range. The centers of the bins are given by

$$c(k) = a + (2k - 1)\frac{w}{2}, \quad k = 1, \ldots, N \tag{4.1}$$

Suppose that all we know about x is that it is contained within a particular bin, say bin k. Then, given k only, the best estimate of the value of x is $c(k)$, and the maximum estimation error is $w/2$.

Instead of using bins that all have the same width, a bin center vector c with N elements could be given, where the bin ranges are $[a, (c(1) + c(2))/2]$, $[(c(1) + c(2))/2, (c(2) + c(3))/2], \ldots, [(c(N-1) + c(N))/2, b]$ for a total of N bins over $[a,b]$ that have varying widths. Using bins that all have the same width, Prog. 4.1 finds $k = 1, \ldots, N$, the bin number, given x.

There is another purpose of this program. It illustrates writing a program that is reasonably **robust**. This means that input data or intermediate results within the program do not cause a computer to become nonresponsive or generate error messages. A program may be robust and not give correct (or expected) results, because its programming is based on an incorrect algorithm. If a program executes to completion and the results are incorrect, then the programmer (possibly working with a user) must interpret the results and make corrections.

To find a bin number, the range $[a, b]$ is normalized with

$$y = N\frac{x - a}{b - a}$$

Therefore, if $x = a$, then $y = 0$, and if $x = b$, then $y = N$, and the range of y is $[0, N]$. For a given x, $k = \text{ceil}(y)$, where **ceil** is a built-in function that returns the next higher integer.

```
% Program to find the bin number
clear all; clc
disp('program to find the bin number of x in the range [a,b]')
err = 0; % used for checking if an error occurred
a = input('enter lower range limit: ');
b = input('enter upper range limit: ');
if a < b % a robust program must check inputs
  N = input('enter the number of bins, greater than 0: '); N=round(N);
   if N > 0 % check input
     x = input('enter x: ');
     if x >= a && x <= b % check input
        w = (b-a)/N; % get bin width
        k = ceil((x-a)/w); % round to next higher integer
     else
        err=1; % x is out of range
     end
   else
     err=2;  % N is out of range
   end
else
   err=3; % lower limit not less than upper limit
```

```
end
if ~err % was an error detected
   fprintf('out of N = %i bins, the bin number is: %i \n',N,k)
elseif err == 1, disp('x is out of range')
elseif err == 2, disp('N is out of range')
elseif err == 3, disp('lower limit not less than upper limit')
end
```

Program 4.1 **Program to find the bin number of** x **in the range** $[a,b]$**.**

Notice that all statements within matched **if, else, elseif,** and **end** key words are indented for clarity.

4.4 For Loop

With the **for loop** structure, you can repeat executing a block of statements a predetermined number of times. Its syntax is

```
for loop_index = first_index: index_increment: last_index
     block of statements
end
```

Every **for** key word must be matched by an **end** key word. The block of statements can include additional for loops. For loops can be nested. The block of statements can include if–elseif–else–end structures and any other valid MATLAB statements. It is useful to indent the block of statements to easily see the matched for and end key words. The loop_index must be a MATLAB variable. The first_index value, index_increment value, and last_index value can be expressions that evaluate to a real scalar.

For convenience, denote the loop_index variable by k, first_index by a, index_increment by *incr*, and last_index by b. When MATLAB encounters the **for** key word it does the following:

1) Check incr, and if incr is positive, check that $a <= b$, and if $a <= b$, then set $k = a$.
2) Execute the block of statements.
3) Go back to the **for** statement and set $k = k + incr$, which gives the new loop_index value.
4) If $k <= b$, then execute the block of statements again and go back to step (3).
5) If $k > b$, then skip the block of statements, and continue execution after the end statement.

If $a > b$, then the block of statements is not executed, and execution continues after the end statement.

If in step (1), incr is negative, then check that $a >= b$, and if $a >= b$, then set $k = a$, and do the following:

2) Execute the block of statements.
3) Go back to the **for** statement and set $k = k +$ incr, which gives the new loop_index value.
4) If $k >= b$, then execute the block of statements again and go back to step (3).
5) If $k < b$, then skip the block of statements, and continue execution after the end statement.

If $a < b$, then the block of statements is not executed, and execution continues after the end statement.

You can increment or decrement through a range of loop_index variable values. You can also specify the loop_index values with a vector, as in

```
for k = v
        block of statements
    end
```

where v is a vector and k is first set to $v(1)$ and the block of statements is executed. Then, $k=v(2)$ and the block of statements is executed, and this continues until all elements of v have been used. With this structure you can assign to k any values, including complex values, for example, $v = [1\text{-}j*2 \quad 2* \text{ pi} \quad -1 \quad \exp(j*\text{pi}/4) \quad u]$, where $u = [1/3 \quad j*3]$.

If you do not include an index_increment, as in

```
for k = a:b
        block of statements
    end
```

then a default index_increment value of 1 is used, which requires that $a <= b$.

Example 4.6

Let us take a look at the details of multiplying two matrices with for loops. Let A be an $(N \times K)$ matrix, and let B be a $(K \times M)$ matrix. Consider $C = A*B$ with dimension $(N \times M)$. Prog. 4.2 finds C with for loops and the built-in MATLAB matrix multiply operation.

```
% Program to multiply two matrices (C=A*B) using for loops
clear all; clc
N=3; K=4; M=2; % specify matrix dimensions
A=[1 0 2 -1;2 1 -1 0; 1 -1 1 2]; % specify two matrices
```

```
B=[3 0; -1 5; 1 3; 2 1];
C = zeros(N,M); % preallocating space for C, which decreases execution time
t = tic; % tic, a built-in function, starts a stopwatch timer
for n = 1:N % index for rows of C
   for m = 1:M % index for columns of C
      sum = 0; % initializing the sum of products
      for k = 1:K % get dot product of a row of A and a column of B
         sum = sum + A(n,k)*B(k,m); % accumulate products
      end
      C(n,m) = sum; % save dot product of a row of A and a column of B in C
   end
end
loops_compute_time = toc(t) % toc gets elapsed computing time since tic
C % display result
t = tic % start stopwatch timer
D=A*B; % check
MATLAB_compute_time = toc(t) % get elapsed computing time since tic
```

Program 4.2   Use of for loops to multiply two matrices.

Notice that all statements within matched **for** and **end** key words are indented for clarity. Since the innermost loop has only a few statements, it could have been written as

```
for k = 1:K, sum = sum + A(n,k)*B(k,m);   end
```

This for loop could have been replaced to find $C(n,m)$ with

```
C(n,m) = A(n,:)*B(:,m);   % dot product of row n in A and column m in B
```

The program output is shown below. It was rearranged to take less space.

```
loops_compute_time =   0.0024 secs
C =
   3   5
   4   2
   9   0
MATLAB_compute_time = 1.4848e-005 secs
```

This example demonstrates a programming utility of MATLAB, that is, it works with the matrix data type. Moreover, $D = A*B$ is executed in much less computing time than it took to obtain C, even though space for C was preallocated.

The built-in functions **tic** and **toc** were used in Example 4.6 to measure the execution times of parts of Prog. 4.2. The time required to execute a part of a program depends on the speed of your computer. There may be occasions when, for example, between different program outputs, you want to control the time in between these program outputs. To do this, use the built-in function **pause**(seconds), for example, pause(4.5), which causes a program to stop execution for 4.5 secs. Use **pause** without an argument to stop a program and wait for the user to strike any key before continuing.

If in a program the size of a matrix (or vector) is increased with each pass through a loop, it is best to preallocate space for the largest size of the matrix before getting into the loop. This will save much execution time.

Example 4.7 _____

A signal is $x(t) = A \sin(\omega t)$, and we want to see how the values of $x(t)$ are distributed over one cycle. The signal value varies over the range $-A$ to A, written as $[-A, A]$. Let us segment this range into N bins. The width of each bin is $2A/N$, and using (4.1) the bin centers are given by

$$c(k) = -A + (2k - 1)\frac{A}{N}, \quad k = 1, 2, \ldots, N \tag{4.2}$$

Let $\omega = 2\pi$ rad/sec, and therefore one cycle takes $T_0 = 1$ sec. The signal $x(t)$ will be sampled at the rate $f_s = 44,100$ samples/sec. Prog. 4.3 counts the number of times that the value of $x(t)$ is in each of the N bins. The resulting distribution is called a **histogram**.

```
% Program to find the histogram of a sine wave
clear all; clc
% specify the data
A = 1; freq = 1; T0 = 1/freq; % specify amplitude and freuency
fs = 44100; T = 1/fs; % sampling rate and time increment
t = 0.0:T:T0-T; % time points over one cycle
x = A*sin(2*pi*freq*t); N_data = length(x); % get samples of x(t)
N = 40; w = 2*A/N; k = 1:N; % number of bins, bin width, and bin index
c = -A+(2*k-1)*w/2; % bin centers
bin = zeros(1,N); % preallocate space and initialize bin counts to zero
for n=1:N_data
   k = ceil((x(n)+1)/w);
   if k==0
     k=1; % x=-1 belongs to bin 1
   end
   bin(k)=bin(k)+1; % increment the bin count
end
```

```
bar(c,bin,1) % get a barchart plot
grid on
% the labels were entered using the plot editor
```

Program 4.3 Program to find the histogram of a sine wave.

The histogram is shown in Fig. 4.1. We see that a sine wave spends almost equal amounts of time in segments from -0.8 to $+0.8$, while nearly 20% of the time in a cycle is spent within 5% of the maximum and minimum values.

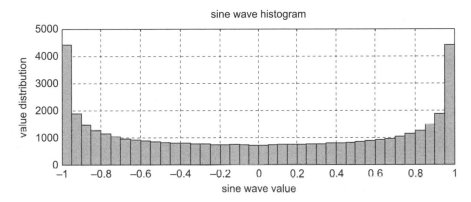

Figure 4.1   Histogram of a sinusoidal signal.

This example introduced the built-in function **bar**, which plots a bar chart. Like the built-in function plot, the first input gives horizontal plotting information, the bar centers, the second input gives the bar heights, and the third input controls spacing between bars in the plot, where a value less than 1 causes a space between bars. More detail will be given in Chapter 9.

In the previous examples, the for loops always executed to completion. Sometimes it is useful to terminate a for loop before it has iterated through its entire index range.

Example 4.8

Evaluate the $\sin(x)$ function using its Taylor series, which is given by

$$\sin(x) = \sum_{k=0}^{\infty} \frac{(-1)^k x^{2k+1}}{(2k+1)!}$$

In a program, the upper summation limit cannot be infinity. We must either set the upper limit to a large integer or use some way of stopping the computation after some

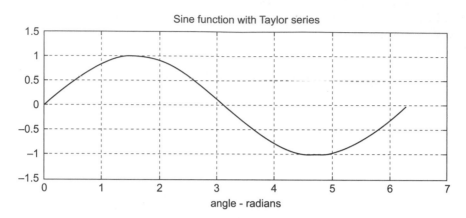

Figure 4.2   Sine function produced with its Taylor series.

predetermined precision has been achieved. When x is incremented by 2π the function $\sin(x)$ repeats itself. Let us evaluate the function for $0 \le x \le 2\pi$. Prog. 4.4 evaluates the Taylor series and plots the result. The resulting approximation of the sin function is shown in Fig. 4.2.

```
% Program to evaluate the sin(x) function using its Taylor series
clear all; clc
N = 101; % N values for x will be used
theta = linspace(0,2*pi,N); % assign values to theta
y = sin(theta); % use the built-in function sin to compare with the series
K = 100;  % K terms in the Taylor series will be used
precision = 1e-6; % this is a precision of 0.0001 percent
y_approx = zeros(1,N); k_value = zeros(1,N); % preallocate space
for n=1:N  % loop for each value in theta
   x = theta(n);
   % no need to evaluate the Taylor series for x=0
   if x == 0.0, y_approx(n) = 0; k_value(n) = 0; continue; end
   sum_T = 0; % initialize Taylor series sum
   factorial_term = 1; % initialize (2k+1)!
   minus_one_term = 1; % initialize (-1)^k
   x_power_term = x;  % initialize x^(2k+1)
   x_square = x^2; % used to get next x^(2k+1), computed only once
   next_term = minus_one_term*x_power_term/factorial_term; % k=0 term
   for k=1:K  % loop to accumulate series terms
      sum_T = sum_T + next_term; % accumulate terms
```

```
   % prepare the next term in the Taylor series
   factorial_term = factorial_term*(2*k)*(2*k+1); % update (2k+1)!
   % uses only two multiplies to find each (2k+1)!
   minus_one_term = - minus_one_term; % update (-1)^k
   x_power_term = x_power_term*x_square; % update x^(2k+1)
   % uses only one multiply to find each x^(2k+1)
   next_term = minus_one_term*x_power_term/factorial_term;
   % compare next term in Taylor series to sum and check precision
   if abs(next_term) >= abs(precision*sum_T);
      continue  % continue this for loop
   else
      sum_T = sum_T + next_term; % include the next term
      break  % jump out of this for loop
   end
  end
  % the size of y_approx is increased with each pass through this loop
  % space for its largest size was preallocated before this loop
  y_approx(n) = sum_T;
  k_value(n) = k; % save index for which this for loop achieved precision
end
% finished evaluating the Taylor series for each element of theta
max_k_value = max(k_value) % get maximum number of for loop iterations
max_error = max(abs(y-y_approx)) % get maximum error
plot(theta,y_approx)
grid on
xlabel('angle - radians')
title('Sine function with Taylor series')
```

Program 4.4 Evaluate the sine function using its Taylor series.

The outer for loop in Prog. 4.4 iterates over the values in the vector theta. The inner for loop in Prog. 4.4 is set up to iterate $K = 100$ times. It computes each term in the Taylor series. Since $K = 100$ terms may not be necessary to achieve a desired precision, each new term is checked to see if it has become small enough to make a negligible contribution to the summation. If it has not become small enough, the built-in function **continue** is used to continue the inner for loop iteration. If it has become small enough, then the built-in function **break** is used to exit the inner for loop in which break occurs. Some program parameters of interest are shown below. While the inner for loop could have iterated up to $K = 100$ times, the maximum number of iterations to achieve the prescribed precision of 0.0001% over a cycle of the sine function turned out to be $k = 22$, and the resulting maximum error when compared to $\sin(x)$ using the MATLAB built-in function is 3.7e-6 percent.

The use of tic and toc will show that the algorithm that MATLAB uses to evaluate the sine function is much faster than a direct application of its Taylor series.

```
max_k_value =   22
max_error =   3.7065e-008
```

Some built-in functions concerned with controlling loop activity are given in Table 4.4.

Table 4.4   MATLAB functions to exit for loops, while loops and user-defined functions

Function	Brief description
continue	When **continue** is encountered in a loop, remaining loop statements are skipped and execution continues with the next iteration of the loop in which it appears. In nested loops, continue causes the next iteration of the loop enclosing it.
break	When **break** is encountered in a loop, it terminates execution of the loop. In nested loops, break cause an exit from the loop enclosing it only.
return	When **return** is encountered in a user-defined function, it causes an exit from the function. Statements in the function after the return statement are not executed. Also, return terminates the keyboard mode which will be discussed in Chapter 10.

While loops will be discussed in the next section.

4.4.1   Probability

To consider some interesting problems, a brief background in probability will now be given.

Suppose we do an experiment, and the outcome is not predictable. Assume that an outcome can be one particular event among a finite number of possible events. The probability of an outcome gives an assessment of how likely it is that a particular outcome will occur compared to all other possible outcomes.

For example, consider tossing a coin. Each coin toss is an experiment, and there are two possible outcomes or events, that is, heads or tails. Suppose you toss a coin N times. Of the N tosses, let n_H be the count of the outcome heads, and let n_T be the count of the outcome tails. The probability p_H of heads is given by

$$p_H = \lim_{N \to \infty} \frac{n_H}{N} \tag{4.3}$$

and for tails, the probability p_T is given by

$$p_T = \lim_{N \to \infty} \frac{n_T}{N} \tag{4.4}$$

Since $N = n_H + n_T$, we have a fundamental property of the probabilities of all possible outcomes, which in this case is

$$p_H + p_T = \lim_{N \to \infty} \frac{n_H}{N} + \lim_{N \to \infty} \frac{n_T}{N} = 1$$

For a fair coin we expect that $p_H = 0.5$ and $p_T = 0.5$.

If there are K possible outcomes of an experiment, and $p_k, k = 1, 2, \ldots, K$ is the probability of each outcome, then

$$\sum_{k=1}^{K} p_k = 1 \tag{4.5}$$

Practically, we cannot do an experiment an infinite number of times to find a probability by counting. However, if we make N large enough, then we might get useful values for the probabilities. The probability of an impossible outcome is zero, while the probability of a certain outcome is one. For the probability p of any outcome we have $0 \leq p \leq 1$. Notice that if an outcome occurs only once in an infinite number of trials of an experiment, then its probability is zero. This means that if the probability of an outcome is zero, we cannot be certain that it will never occur. Similarly, if the probability of an outcome is one, then we cannot be certain that it did not occur at least once.

We can use the MATLAB built-in function **rand** to experimentally find probabilities associated with the outcomes of some experiments. In MATLAB, each **rand** function evaluation returns a number in the range $0 \to 1$ (written as $(0,1)$), which we cannot predict, and where all numbers in this range are equally likely, which means that the function rand is a uniformly distributed random number generator. The MATLAB statement

```
x = rand;   % 0 < x < 1
```

assigns to x a number in the range $(0,1)$. Each use of rand returns a number that does not depend on numbers obtained previously.

A **pseudorandom number generator** is an algorithm for generating a sequence of numbers that approximates the properties of random numbers. The sequence is not truly random. The function **rand** is a pseudorandom number generator. A careful mathematical analysis is required to have any confidence that a pseudorandom number generator produces numbers that are sufficiently random to meet the needs of a particular application. Nevertheless, it is used in simulations to study the behavior of systems in which random events can occur.

The algorithm for the function **rand** is initialized the same way each time a MATLAB session is started. Therefore, when you start a MATLAB session, rand will produce the same random sequence of numbers that it produced the previous time when a MATLAB session was started. During a MATLAB session you can reset the random number generator to the default MATLAB start-up state with the statement **rng('default')**. Use the MATLAB help facility to find out more about the rand function.

Example 4.9 _____

Let us simulate the coin toss experiment to find p_H and p_T. This will be done by counting the number of times each outcome occurs over a large number of coin tosses. Prog. 4.5 simulates the coin toss experiment.

```
clear all; clc;
rng('default'); % initialize random number generator to its default state
total_tosses = 1e6; % specify the total number of coin tosses
heads = 0; tails = 0; % initialize outcome counters
for n = 1:total_tosses
   % do experiment
   x = rand; % get a random number distributed from 0 to 1
   if x < 0.5  % let this result be analogous to heads occurring
      heads = heads + 1;
   else % if not heads, then tails
      tails = tails + 1;
   end
end
disp('probability of heads: ');
pH = heads/total_tosses
disp('probability of tails: ');
pT = tails/total_tosses
```

Program 4.5   Using a random number generator to simulate tossing a coin.

The program output is shown below:

```
probability of heads: pH =  0.4995
probability of tails: pT =  0.5005
```

If we increase the number of tosses, then it is likely that pH and pT will each be closer to 0.5. However, this does not mean that as the total number of tosses N is increased, then the number n_H of heads or the number n_T of tails will ever become exactly $N/2$. Since pH is essentially the same as pT, the outcomes heads and tails are equally likely.

The mean (average) value of the numbers produced by rand can be found experimentally with

```
rng('default') % initialize random number generator
N = 1e7; % specify the number of random numbers to get
% obtain a vector of N random numbers uniformly distributed from 0 to 1
x = rand(1,N);
x_mean = sum(x)/N % find the mean value
% this can be obtained with the MATLAB function mean to get x_mean=mean(x)

x_mean = 0.5000
```

With the function rand we can obtain uniformly distributed random numbers over any range (a,b). The span of this range is $(b-a)$. Since the range of $x =$ rand is $(0,1)$, the range and span of $y = (b-a)x$ are $(0,(b-a))$ and $(b-a)$, respectively. Therefore, the range of $y = (b-a)x + a$ is (a,b).

Example 4.10

To obtain random numbers x with a uniform distribution over the range $-5 < x < +5$, use

```
>> N = 1e7; % specify the number of random numbers
>> % get a vector of N uniformly distributed random numbers between 0 and 1
>> x = rand(1,N);
>> x = 10*x - 5; % multiply by (5-(-5))=10 and translate by -5
>> x_mean = mean(x) % find the mean value
x_mean =   2.9411e-006
>> % ideally, the mean value of x should be x_mean = 0.
>>
```

Let us see the first 100 points in x with

```
>> y = x(1:100); % get first 100 points from x
>> y = y - mean(y); % make the mean of y zero
>> plot(y,'- .'); grid on; xlabel('index'); ylabel('random signal')
>> % within single quotes, the dash causes the plot function to connect the points with
>> % straight lines, and the period causes the plot function to put a dot at each point
```

More about plotting will be given in Chapter 9. Fig. 4.3 shows a plot of some points from x. It looks like a noise signal with a zero mean value.

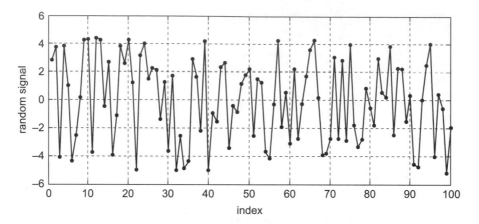

Figure 4.3   A random signal.

An important communications problem occurs due to noise. This is because a received signal $x(t)$ is a noise-corrupted version of a transmitted signal $s(t)$, such as

$$x(t) = s(t) + n(t)$$

where $n(t)$ is additive noise. For example, the signal $x(t)$ could be the playback from an old audio recording, and $s(t)$ is the audio signal that was originally recorded. Or, $x(t)$ could be a received cell phone signal, and $s(t)$ is the signal that was transmitted from a distant location. We are interested to design a process that performs as depicted in Fig. 4.4, where the process output $\hat{s}(t)$ is an estimate of $s(t)$.

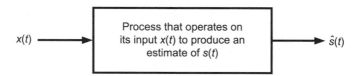

Figure 4.4   Pictorial of a desired operation.

To design an estimation process, it will be useful to do simulation studies with a computer to test the performance of various process designs and noise signals. This will require simulating the noise signal with a model of noise, such as the noise signal shown in Fig. 4.3. Such simulations are often done with MATLAB.

MATLAB also includes the two built-in functions **randi** and **randn**. The function randi produces uniformly distributed random integers. For example, $A =$ randi(imax,M,N) is an $(M \times N)$ matrix of random integers, where each element in A is in the range 1 to imax, while $x =$ randi(imax) is a scalar integer in the range 1 to imax.

Before describing the random numbers produced by **randn**, let us study the random numbers it produces by finding the **mean**, **variance,** and a **histogram**. The variance, denoted by σ^2, of a set of random numbers is a measure of their volatility. The variance is computed with

$$\sigma^2 = \frac{1}{N} \sum_{n=1}^{N} (x(n) - \bar{x})^2 \tag{4.6}$$

where $\bar{x}$ denotes the mean of x, and σ is called the **standard deviation**. More specifically, the variance (and standard deviation) of a set of random numbers is a measure of the degree to which the random numbers in the set deviate away from their mean. MATLAB has a built-in function, **var**, that computes the variance. A histogram of a set of random numbers shows how the numbers are distributed.

The built-in function **hist** returns a vector, the elements of which are counts of the number of times the elements of the vector of random numbers have values within the range of each of the bins. We could have used the hist function in Example 4.7.

Example 4.11 _____

Prog. 4.6 finds the mean, variance, and a histogram of the numbers obtained with randn.

```
% Histogram of numbers produced by the psuedo random number generator randn
clear all; clc
N = 1e7; % specify the number of random numbers to get
x = randn(1,N); % obtain a vector of N random numbers
x_mean = mean(x) % find the mean value
x_var = var(x) % get the variance of the random numbers in x
bin_width = 0.25; % specify a bin width less than 1
N_pos_x_bins = 25; % specify number of bins for positive x(n) values
pos_x_bin_center = N_pos_x_bins*bin_width; % right most bin center
% locate equally spaced bin centers
bin_centers = -pos_x_bin_center: bin_width : pos_x_bin_center;
fprintf('using %i bins \n',length(bin_centers))
x_hist = hist(x,bin_centers); % return a vector with the count in each bin
% prob_x = x_hist/N; % probability of x within each bin
bar(bin_centers,x_hist,1,'w'); % full bar width; and white color
axis([-4 4 0 1.05*max(x_hist)]); grid on
xlabel('random number value'); ylabel('bin count')
title('Distribution');
bins_to_1 = round(1/bin_width); % number of bins accounting for 0<x<1
```

```
index_to_center = N_pos_x_bins + 1; % index of distibution center
kp1 = index_to_center+(bins_to_1 - 1); % index in x_hist for x = 1
km1 = index_to_center-(bins_to_1 - 1); % index in x_hist for x = -1
prob_minus_to_plus_sigma=(x_hist(km1-1)/2 +sum(x_hist(km1:kp1))+ ...
    x_hist(kp1+1)/2)/N
% this is the probability that a value of x will be in the range -sigma to
% +sigma, where sigma is the standard deviation. With randn, sigma = 1
% additional plot editing was done in the Figure Window
```

Program 4.6   Script to find the mean, variance, and a histogram of a set of numbers.

The program output, which is given below, shows that the built-in function **randn** produces random numbers having a zero mean and a variance of one. The standard deviation is 1, and within ±1 standard deviation 67.92% of the returned values occur. The histogram has a bell-shaped appearance. A normally (also called **Gaussian**) distributed random number sequence has these properties.

```
x_mean = -5.9235e-004
x_var = 0.9993
using 51 bins
prob_minus_to_plus_sigma = 0.6792
```

The function randn is designed to be a normally distributed pseudorandom number generator, and produces numbers with zero mean and unity variance. With $y = ax + b$ you can produce a normally distributed random number sequence y having a prescribed mean and variance, where $\sigma_y^2 = a^2$ and $\bar{y} = b$.

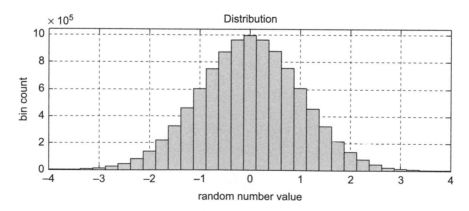

Figure 4.5   Distribution of random numbers obtained with the function randn.

If each bin count is divided by the total of number of samples N used to get the histogram of Fig. 4.5, then, like (4.3) and (4.4), we get the probability that a number will occur in a particular bin. Designate the bin counts for positive (negative) numbers by b_k (b_{-k}), respectively. Then, for example, b_0 is the bin count for $-0.125 \leq x < 0.125$, and the probability that an experiment outcome (invocation of randn) will result in a value of x within the range $-0.125 \leq x < 0.125$ is $p_0 = b_0/N \simeq 10e5/1e7 = 0.1$. The function randn produces numbers in the range $-\infty < x < +\infty$, where a large value of abs(x) is very unlikely, and like (4.5) we have that

$$\sum_{k=-\infty}^{+\infty} p_k = 1 \tag{4.7}$$

4.4.2 Median Filtering

In signal processing, it is often desirable to perform some kind of noise reduction on a signal (one-dimensional or 1-D signal) or an image (2-D signal). **Median filtering** is a digital signal processing algorithm that is easy to apply and often used to reduce noise. For example, an old audio recording, where there is crackling noise, can be improved by median filtering. Median filtering of a speech signal is sometimes a pre-processing step before speech recognition algorithms are applied. Median filtering is widely used in digital image processing.

The median filter concept is actually simple. Let us see how it works with a demonstration. Suppose a signal $s(t)$ is sampled, like in (3.33), to produce the number sequence (discrete time signal) $x = [5\ 6\ 47\ 7\ 6 -29\ 4]$, where $x(1)=s(0)$, $x(2)=s(T)$, $x(3)=s(2T)$, and so on. The third and sixth elements of x appear to be unusual, and they are called **outliers**. Let y denote the number sequence produced by median filtering x. For element n of y, use $x(n-1)$, $x(n)$, and $x(n+1)$. This is called a $K=3$ element **window**, $w = [x(n-1)\ \ x(n)\ \ x(n+1)]$. Sort w into ascending values to obtain a $k=3$ element vector u. Now set $y(n)$ to the middle element of u. Then, increment n by 1, get the next w, and repeat to get the next y. For example:

$n = 1$, $w = [5\ \ 5\ \ 6]$ $\rightarrow$ sort $\rightarrow u = [5\ \ 5\ \ 6]$, $y(1)=u(2)=5$. There is no $x(0)$; $x(1)$ was repeated.

$n=2$, $w=[5\ 6\ 47]$ $\rightarrow$ sort $\rightarrow u=[5\ 6\ 47]$, $y(2)=6$

$n=3$, $w=[6\ 47\ 7]$ $\rightarrow$ sort $\rightarrow u=[6\ 7\ 47]$, $y(3)=7$

$n=4$, $w=[47\ 7\ 6]$ $\rightarrow$ sort $\rightarrow u=[6\ 7\ 47]$, $y(4)=7$

$n=5$, $w=[7\ 6\ -29]$ $\rightarrow$ sort $\rightarrow u=[-29\ 6\ 7]$, $y(5)=6$

$n=6$, $w=[6\ -29\ 4]$ $\rightarrow$ sort $\rightarrow u=[-29\ 4\ 6]$, $y(6)=4$

$n=7$, $w=[-29\ 4\ 4]$ $\rightarrow$ sort $\rightarrow u=[-29\ 4\ 4]$, $y(7)=4$. There is no $x(8)$; $x(7)$ was repeated.

The outliers in x have been removed by median filtering. The signal y is less erratic (noisy) than the signal x. Median filtering works best to replace outliers with a neighboring element when other consecutive elements in x have similar values.

The **window** w is said to slide across the signal x. For this demonstration $K=3$ was used. We could have used $K=4$, in which case w is set to $w=[x(n-2)\ x(n-1)\ x(n)\ x(n+1)]$, and u will not have a middle element. In this case, set $y(n)=(u(n-1)+u(n))/2$. Or, we can apply the $K=3$ median filter again to y, and the result will be even less noisy and most likely too different from x without the noise. The built-in MATLAB function **median** finds the median of a vector w.

Example 4.12

For an 88 key piano, let $n = 1, 2, \ldots, 88$ be the key number from the lowest to the highest frequency tone. The frequency (pitch) f of the tone produced by key n is given by

$$f = 440(2^{1/12})^{(n-49)} \text{ Hz} \tag{4.8}$$

For $n = 40$ the note is middle C, and for $n = 49$, the note played is A440.

Prog. 4.7 consists of two parts. In the first part, a user can enter an integer n, and then the program uses (4.8) to find the frequency of the note. A sinusoid at this frequency is then sampled at the rate of 44,100 samples/sec for 5 sec. To make the sound produced more realistic, the sampled sinusoid is multiplied by the pulse shown in Fig. 3.3, which becomes the envelope of the sound. Therefore, the sound increases quickly and decays slowly. To hear the sound, a two column (two channel stereo) matrix must be set up, and then, the built-in MATLAB functions **sound** or **soundsc** can be used to send the matrix to the audio play device of the computer. You can also use the built-in MATLAB function **wavwrite**, which creates a Windows WAV file that can be played by any multimedia program. The syntax for wavwrite is wavwrite(x,fs,'name.wav'), where x is a two-column matrix and f_s is the sampling rate (also called sampling frequency), and name can be replaced by a name you can choose.

```
% Program to sound an 88-key piano
% Keys are identified by an integer n, n = 1 to 88 from left to right
% Only the tone at the fundamental frequency (pitch) will be sounded
clc; clear all
fs = 44100; % this is the sampling rate used to produce audio CDs.
disp('Play a note on a virtual piano.')
n = input('enter an integer from 1 to 88 of a piano key: '); % using n=40
```

```
n = round(n); % make sure entry is an integer
if n < 1, n = 1; end % keep n in range
if n > 88, n = 88; end
f = 440*(2^(1/12))^(n-49); % the frequency in Hz of the key
w0 = 2*pi*f; % convert Hz to rad/sec
T0 = 5; % the tone will be sounded for 5 secs
delta_t = 1/fs; % the time increment between sound samples
N = round(T0/delta_t); % integer number of samples
time = delta_t*[0:N-1]; % N time points where sound is sampled
p_key = sin(w0*time); % get samples of sound, limit amplitude to +/-1
% get an envelope
load('pulse.mat','x') % retrieve the pulse that was built in Example 3.15
p_key = x .* p_key; % multiply the sound by an envelope, this is optional
stereo_sound = [p_key' p_key']; % form left and right channels
sound(stereo_sound,fs) % send sound to audio play device
```

Program 4.7 (first part)   Sound a piano key.

If the note produced in the first part of Prog. 4.7 comes from an old recording, then it will likely contain some spike-like noise. The second part of Prog. 4.7 simulates the addition of noise and spike noise, which occurs randomly in the sound, and then it applies median filtering to reduce the spike noise contribution to the sound.

```
% Example of median filtering
noise = 0.1*randn(1,N); % using the normal random number generator
p_key = p_key + noise; % add noise to signal
% generating randomly positioned spikes
increment = 100; % add noise spike within every 100 samples of p_key
spikes = zeros(1,N); % zero out and preallocate space
for k = 1:increment:N;
   location = floor(k + rand*increment); % use rand to locate spike
   if rand > 0.5; % use rand to determine sign of the noise spike
      sign = +1.0;
   else
      sign = -1.0;
   end
   spikes(location) = sign*2.0*rand; % use rand to determine spike value
end
p_key = p_key + spikes; % add spike noise to signal
```

```
plot_points = 1500; range = 5e4:5e4+plot_points-1; % plotting few points
t_interval = time(range); y = p_key(range);
plot(t_interval,y);
axis([t_interval(1) t_interval(plot_points) -3 3])
title('Segment of note middle C plus random noise plus random spikes');
xlabel('seconds');
ylabel('amplitude');
grid on
max_value = max(abs(p_key));
% limit sound to +/- 1, required by audio device
normalized_p_key = p_key/max_value;
stereo_sound =[normalized_p_key', normalized_p_key'];
sound(stereo_sound,fs)
% median filtering sound
filtered_p_key = zeros(1,N); % preallocate space
K = 3; % use 3 element median window
for k = 1:N
  if k == 1
    w = [p_key(1) p_key(1) p_key(2)];
    filtered_p_key(1) = median(w);
  elseif k == N
    w = [p_key(N-1) p_key(N) p_key(N)];
    filtered_p_key(N) = median(w);
  else
    w = [p_key(k-1) p_key(k) p_key(k+1)];
    filtered_p_key(k) = median(w);
  end
end
y = filtered_p_key(range); % plot a segment of filtered_p_key sound
figure  % causes function plot to open a new figure window
plot(t_interval,y)
axis([t_interval(1) t_interval(plot_points) -3 3])
title('Segment of median filtered note middle C');
xlabel('seconds');
ylabel('amplitude');
grid on
max_value = max(abs(filtered_p_key));
% limit sound to +/- 1, required by audio device
normalized_filtered_p_key = filtered_p_key/max_value;
```

```
stereo_sound =[normalized_filtered_p_key', normalized_filtered_p_key'];
sound(stereo_sound,fs)
```

Program 4.7 (second part) Median filter piano key sound.

Fig. 4.6 shows note middle C with noise and spike noise, and the sound produced by this signal is awful. Fig. 4.7 shows the result of median filtering to remove spike noise.

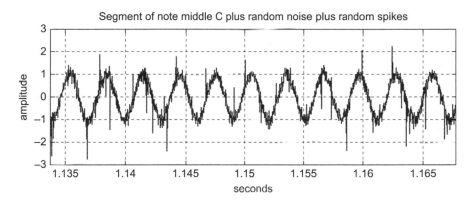

Figure 4.6 Note middle C with noise and spike noise.

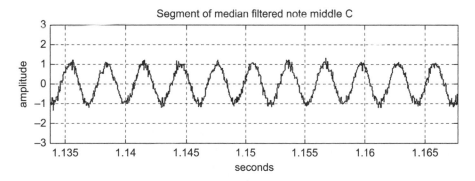

Figure 4.7 Median-filtered note middle C.

The signal shown in Fig. 4.7 still has some noise in it. However, additional digital signal processing can be performed to reduce the noise further.

4.5   While Loop

With the **while loop** structure you can repeat executing a block of statements until an expression is no longer true. Its syntax is

```
while  expression
     block of statements
end
```

Every **while** key word must be matched by an **end** key word. The block of statements can include additional while loops and for loops. While loops can be nested. The block of statements can include if–elseif–else–end structures and any other valid MATLAB statements. It is useful to indent the block of statements to easily see the matched while and end key words. The expression must evaluate to a logical or real scalar. If it evaluates to a real number, then it is treated as a logical value.

When MATLAB encounters the **while** key word, it does the following:

1) Evaluate the expression
2) If the expression is true, then execute the block of statements
3) Go back to the while statement and continue with step 1
4) If the expression is not true, then skip the block of statements, and continue execution after the end statement

A while loop can be made to work like a for loop. For example:

```
n = 1; N = 10;
while n <= N
        :
       n = n +1;
end
```

Here, the statements within the while loop will be executed N times. However, this is not the intention of including a while loop structure in addition to a for loop structure in the programming language. Usually, a while loop structure is used when we do not know ahead of time how many loop iterations are necessary to accomplish some task. For example:

```
err = 0;
while ~err
      block of statements
end
```

Within the block of statements, some condition must be tested that will either leave err unchanged and the loop continues to iterate or set err to err $= 1$, which causes the while loop to terminate.

Example 4.13

The inner for loop in Prog. 4.4 of Example 4.8 could be replaced with

```
k = 1; negligible = 0;
while ~negligible
    sum_T = sum_T + next_term; % accumulate terms
    % prepare the next term in the Taylor series
    factorial_term = factorial_term*(2*k)*(2*k+1); % update (2k+1)!
    % uses only two multiplies to find each (2k+1)!
    minus_one_term = - minus_one_term; % update (-1)^k
    x_power_term = x_power_term*x_square; % update x^(2k+1)
    % uses only one multiply to find each x^(2k+1)
    next_term = minus_one_term*x_power_term/factorial_term;
    % compare next term in Taylor series to sum and check precision
    if abs(next_term) < abs(precision*sum_T);
        sum_T = sum_T + next_term; % include the next term
        negligible = 1;  % no need for additional loop iterations
    end
    k = k + 1; % count iterations required to achieve precision
end
```

The if–end block of statements checks a condition that can change the result of evaluating the while expression.

The **break** and **continue** commands can also be used in while loops. They work the same way as in for loops. If a while loop expression is always true, then, for example, we have

```
while 1
    block of statements
end
```

The block of statements must include a test of some condition that can cause a break, otherwise the loop will execute endlessly. You can terminate a MATLAB program by activating the *Command Window* and depressing Ctrl and C on the keyboard.

4.6   Method of Steepest Descent

Consider the problem of finding the value of x that minimizes a function $f(x)$, assuming that $f(x)$ has a minimum. Denote the value of x that minimizes $f(x)$ by x_{opt}. Therefore, $f(x_{\text{opt}}) \leq f(x)$, for $x \neq x_{\text{opt}}$. A method to find x_{opt} is based on the drawing shown in Fig. 4.8. We can see three minima in Fig. 4.8. At $x = x_{\text{opt}}, f(x)$ has its **global minimum** (also called the **optimizer** of $f(x)$), and the other two minima at $x = x_a$ and $x = x_b$ are called **local minima**.

   Suppose x_n is a guess of x_{opt}. In Fig. 4.8, x_n is to the right of x_{opt}, where the slope of $f(x)$ at $x = x_n$ is positive, as indicated by the tangential line. Therefore, if the slope of $f(x)$ at the guess x_n is positive, then we know that we must decrease this guess. Similarly, if x_n is to the left (but not too far) of x_{opt}, then the slope of $f(x)$ at x_n will be negative, and we know that we must increase this guess. Therefore, given a guess x_n of x_{opt}, adjust the value of x_n to obtain the next guess x_{n+1} with

$$x_{n+1} = x_n - \mu f'(x_n) \tag{4.9}$$

where μ, called the **step size**, is a small positive number that we must pick and $f'(x_n)$ is the derivative of $f(x)$, assuming it exists, evaluated at $x = x_n$.

   To find a value of x that maximizes a function $f(x)$, change (4.9) to

$$x_{n+1} = x_n + \mu f'(x_n) \tag{4.10}$$

   The recursion, (4.9), of the **method of steepest descent** is operated for $n = 0, 1, 2, \ldots$, where x_0 = initial guess of x_{opt}. If x_0 is within the vicinity of x_{opt} and μ is small enough, then $\lim_{n \to \infty} x_n = x_{\text{opt}}$, since as $x_n \to x_{\text{opt}}$, $f'(x_n) \to 0$. Selecting a suitable value of μ is a matter of trial and error, and depends on the behavior of $f(x)$ as well as computation precision. If the step size is too big, then the next value x_{n+1} may be further away from x_{opt} than x_n, or even in the vicinity of a local minimum, such as x_a or x_b in Fig. 4.8, in which case (4.9) may never converge to x_{opt}. If the step size is too small, then (4.9) will require many iterations (much computing time) to converge. The recursion (4.9) (or (4.10)) is operated until $|x_{n+1} - x_n| < \varepsilon$, for some desired precision ε.

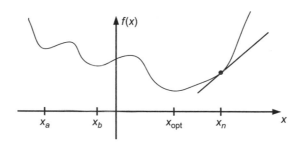

Figure 4.8   A function with a minimum at $x = x_{\text{opt}}$.

We do not have to keep the sequence x_n, $n = 0, 1, 2, \ldots$. Equation (4.9) could be an update of x_{old} to obtain x_{new}, as in

1) set μ to a small positive number and set x_{old} to an initial guess of x_{opt}.
2) compute $x_{new} = x_{old} - \mu f'(x_{old})$.
3) if $|x_{new} - x_{old}| \geq \varepsilon$, then $x_{old} = x_{new}$ and repeat step 2, else $x_{opt} \cong x_{new}$ and stop.

Example 4.14

Find the value of t where $e^{-t} = t^2$. This is an example of a problem where the method of steepest descent, or another kind of optimization method, can be employed, if we convert the given problem to an optimization problem. If it exists, we can find the solution, call it t_{opt}, by minimizing $f(t) = (e^{-t} - t^2)^2$, where $f(t) \geq 0$ for all t, and the minimum value is $f(t_{opt}) = 0$. The square of $(e^{-t} - t^2)$ is used to ensure that $f(t)$ is nonnegative. With $f'(t) = 2(e^{-t} - t^2)(-e^{-t} - 2t)$, Prog. 4.8 finds t_{opt}. The results follow the program.

```
% Application of the method of steepest descent to solve the nonlinear
% algebraic equation: exp(-t) = t^2
clear all; clc; format long e
precision = 1e-8; % desired accuracy
mu = 1e-4; % step size
t_old = 1.0; % initial guess of the optimizer
f_of_t_initial_guess = (exp(-t_old)-t_old^2)^2
f_prime = 2*(exp(-t_old)-t_old^2)*(-exp(-t_old)-2*t_old);
t_new = t_old - mu*f_prime;
n_iterations = 1; % initialize an iterations counter
max_iter = 1e5; err = 0; % limit the number of iterations
while abs(t_new - t_old) >= precision % check for convergence
    n_iterations = n_iterations + 1; % count iterations of algorithm
    if n_iterations > max_iter, err = 1; break; end
    t_old = t_new; % update recursion
    % get new solution
    f_prime = 2*(exp(-t_old)-t_old^2)*(-exp(-t_old)-2*t_old);
    t_new = t_old - mu*f_prime;
end
if err == 0
    t_opt = t_new
    f_of_t_opt = (exp(-t_opt)-t_opt^2)^2
    fprintf('Algorithm required: %6i iterations to converge\n',n_iterations)
else
```

```
    fprintf('Algorithm exceeded: %6i iterations\n',max_iter)
end
```

Program 4.8   Application of the method of steepest descent.

```
f_of_t_initial_guess =   3.995764008937280e-001
t_opt =   7.034812353615486e-001
f_of_t_opt =   6.900847459766987e-010
Algorithm required: 13347 iterations to converge
```

The method of steepest descent has some drawbacks. It is slow to converge, which can be improved by increasing the step size. It requires a derivative of the function to be optimized, which for some problems may be practically impossible to obtain. However, it and its variations are widely applied, because it is a simple algorithm to implement. The required derivative could be approximated with, for example

$$f'(x_n) \simeq \frac{f(x_n + h/2) - f(x_n - h/2)}{h} \tag{4.11}$$

where h is some small positive number, for example, some $h < \mu$. Then, the method requires more function evaluations, which can increase computing time. Also, once x_n is close to the optimizer, (4.11) will give small and erratic $f'(x_n)$ values, making further convergence unlikely. However, this depends on the behavior of $f(x)$ and the desired precision.

The MATLAB built-in function **fminsearch**, which is a function function, is a derivative-free optimization method (Nelder and Mead, 1965), which makes it attractive to use. A syntax is given by

[x_opt,f_opt] = fminsearch(fun,x_initial_guess)

where fun is a function handle. For example, fun could be @myfun, where myfun is the name of a function m-file, or it could be a handle of an anonymous function. Use **doc fminsearch** for more syntax options.

Let us apply fminsearch to the problem of Example 4.14. An anonymous function is given to define $f(t)$ as follows:

```
>> format long e
>> f_of_t = @(t)(exp(-t)-t^2)^2;  % define an anonymous function
>> f_of_t(1)
ans =   3.995764008937280e-001
>> [t_opt f_opt] = fminsearch(f_of_t,1)
t_opt =   7.035156249999998e-001
f_opt =   8.403998856470625e-009
```

While fminsearch is easy to use and can give good results, it can also require many function evaluations or never converge to useful results. This depends on the behavior of the function to be optimized and the initial guess. The function $f(t)$ of Example 4.14 has only one minimum, the global minimum, and both methods of optimization used here worked very well regardless of the initial guess. However, there are no known methods of optimization that can with certainty find the global minimum of a function in finite time (Li, Priemer and Cheng, 2004). This depends on the behavior of the function, and it is important to provide an initial guess in the vicinity of the optimizer.

4.7 Numerical Integration

Given is some function $x(t)$ that is finite for $-\infty < t < +\infty$. To find $y(t)$ given by

$$y(t) = \int_{-\infty}^{t} x(\tau)\, d\tau = \int_{-\infty}^{a} x(\tau)\, d\tau + \int_{a}^{t} x(\tau)\, d\tau, \quad t > a \tag{4.12}$$

is a common problem. We can write (4.12) as

$$y(t) = \int_{a}^{t} x(\tau)\, d\tau + y(a), \quad t \geq a \tag{4.13}$$

where $y(a)$ is called the **integration constant**. If $t = a$, then (4.13) becomes $y(t = a) = y(a)$.

More specifically and for example, recall Fig. 2.3, where

$$\frac{dq(t)}{dt} = i(t)$$

If we integrate this equation, we get

$$q(t) = \int_{-\infty}^{t} i(\tau)\, d\tau = \int_{a}^{t} i(\tau)\, d\tau + \int_{-\infty}^{a} i(\tau)\, d\tau = \int_{a}^{t} i(\tau)\, d\tau + q(a), \quad t \geq a \tag{4.14}$$

The second term $q(a)$ is the amount of charge that has passed through the wire cross-section from $t = -\infty$ until $t = a$, and the first term is the additional amount of charge that has passed through the wire cross-section from $t = a$ until sometime t, where $t > a$.

Depending on $x(t)$ in (4.13), it may be difficult, if not impossible, to find a function for $y(t)$, and we must resort to finding $y(t)$ for a set of discrete time points with some numerical integration method. Sometimes we do not know $x(t)$ for all t. Instead, we only know $x(t)$ for a set of discrete time points.

Without loss of generality, let $a = 0$, and set the upper integration limit to $t = T_0$. Consider the integral

$$A = \int_0^{T_0} x(t)\, dt \tag{4.15}$$

where A is the area under the curve $x(t)$ from $t = 0$ until $t = T_0$.

4.7.1 Euler's Method

Let us segment the time range $[0, T_0]$ into N sub-intervals, where the width of each sub-interval is $T = T_0/N$. This is depicted in Fig. 4.9. The value of $x(t)$ at the beginning of each sub-interval is $x(nT)$, $n = 0, 1, \ldots, N-1$. Euler's method approximates the area under the curve in each sub-interval by a rectangular area given by the width T times the value of $x(t)$ at the beginning of the sub-interval. The integral of (4.15) is approximated by summing all of the rectangular sub-interval areas to get

$$A \simeq \sum_{k=0}^{N-1} x(kT)\, T = T \sum_{k=0}^{N-1} x(kT) \tag{4.16}$$

where $x((N-1)T)$ is the value of $x(t)$ at the beginning of the last sub-interval. The approximation gets better as N is increased, which makes T smaller.

In view of (4.16), let

$$S(n) = \sum_{k=0}^{n-1} x(k\,T)$$

and then $A \simeq T\, S(N)$. For some n, assume that we have $x(kT)$, $k = 0, \ldots, n-1$, which means that we can find $S(n)$. When $x(nT)$ becomes available, we can update $S(n)$ to get $S(n+1)$ with

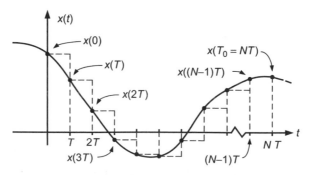

Figure 4.9   Segmentation of $x(t)$ over $[0, T_0 = NT]$.

$$S(n+1) = S(n) + x(nT), \quad n = 0, 1, \ldots, N-1, \quad S(0) = 0 \tag{4.17}$$

This is a recursion in $S(n)$, where

$$S(1) = S(0) + x(0T), \quad \text{set } S(0) = 0$$
$$S(2) = S(1) + x(T)$$
$$S(3) = S(2) + x(2T)$$
$$\vdots$$
$$S(N) = S(N-1) + x((N-1)T)$$

Each time a new value $x(nT)$ becomes available, $S(n)$ is updated to obtain $S(n+1)$. The recursion (4.17) is operated for $n = 0, 1, \ldots, (N-1)$, with $S(0) = 0$.

Example 4.15

Let $x(t)$ be the function defined by

$$x(t) = \begin{cases} 1, & t = 0 \\ \dfrac{\sin(\pi t)}{\pi t}, & t \neq 0 \end{cases} \tag{4.18}$$

This function is called the **sinc** function, that is, $x(t) = \text{sinc}(t)$, and it is a built-in MATLAB function. With

```
>> t = linspace(-10,10,1000);  % specify time points
>> x = sinc(t);  % evaluate x(t) = sin(π t)/πt
>> plot(t,x); grid on
>> xlabel('t - secs'); ylabel('sinc(t)');
```

we get the plot shown in Fig. 4.10. The sinc function oscillates indefinitely in both directions with an amplitude that decays to zero. The integral of $x(t) = \text{sinc}(t)$ from $-\infty$ to $+\infty$ is 1.

An accurate value for the integral of the sinc function from $t = 0$ until $t = 10$ is 0.4898881. Therefore, the integral from $t = 10$ to $t \to +\infty$ is $(0.5 - 0.4898881)$. Prog. 4.9 uses Euler's method to find this integral, and the results for $N = 50, 550, 1050$ follow.

```
% Program to find the integral of a function using Euler's method
clear all; clc
T0 = 10;  % time range is from 0 to T0
```

```
for m=1:3
  N = 50+500*(m-1); % number of sub-intervals
  T = T0/N; % sub-interval width
  S = 0; % initialize the sum
  for n = 0:N-1 % loop to include each sub-interval
    t = n*T; % time point
    S = S + sinc(t); % recursion
  end
  Area(m) = T*S; % area
  N_subs(m) = N;
end
format short
table = [N_subs;Area];
disp('  N     Area')
fprintf('  %i    %f \n',table)
```

Program 4.9   Program to find the integral of sinc(*t*) using Euler's method.

```
   N       Area
   50    0.590224
  550    0.498982
 1050    0.494651
```

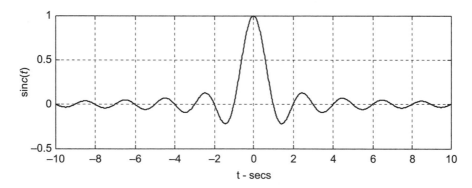

Figure 4.10   The sinc function plotted from t = −10 to t = 10

These results show that N must be large to achieve a small numerical integration error. Nevertheless, depending on the behavior of $x(t)$, Euler's method is often used because of its simplicity.

4.7.2 Trapezoidal Rule

The trapezoidal rule is a slight modification of Euler's method, and gives better results. Instead of using the value of $x(t)$ at the beginning of each sub-interval, the trapezoidal rule uses the average of the values of $x(t)$ at the beginning and end of each sub-interval, and (4.16) and (4.17) become

$$A \simeq \sum_{k=0}^{N-1} \frac{x(kT) + x((k+1)T)}{2} T = \frac{T}{2} \sum_{k=0}^{N-1} x(kT) + x((k+1)T) \tag{4.19}$$

$$S(n+1) = S(n) + x(nT) + x((n+1)T), \quad n = 0, 1, \ldots, N-1, \quad S(0) = 0$$

where $A \simeq (T/2)S(N)$.

Example 4.16 _____

To apply the trapezoidal rule to the integration problem of Example 4.15, the inner for loop in Prog. 4.9 is replaced by

```
xnT = sinc(0); % this is x(nT) for n = 0 and it will be updated in the loop
for n = 0:N-1 % loop to include each sub-interval
    t = (n+1)*T; % next time point
    % evaluating the sinc function just once for each loop iteration
    xn1T = sinc(t);
    S = S + xnT + xn1T; % trapezoidal rule recursion
    xnT = xn1T; % update x(nT)
end
```

and the area is computed with

```
Area(m) = (T/2)*s;
```

The results are

N	Area
50	0.490224
550	0.489891
1050	0.489889

Here, the result for $N = 50$ is better than the result for $N = 550$ using Euler's method.

4.7.3   Built-in Integration Functions

MATLAB has several sophisticated built-in functions for evaluating a definite integral. One of these functions is **quad**. It is a function function, and the syntax to use this function is

```
area = quad(@my_function,a,b, tol)
```

where the name of the m-file that defines the function to be integrated from *a* to *b* must be my_function.m. The fourth argument specifies an error tolerance. If tol is not included in the input argument list, then MATLAB uses a default value of 1e-6.

The function **quad** was used to obtain the accurate result from integrating the sinc function in Example 4.15. The statement that was used is area = quad(@sinc,0,10.0,1e-8).

When MATLAB executes the quad function algorithm, it will want to pass to the user-defined function a vector input. Therefore, the user-defined function must be properly **vectorized** to return a vector. See Table 3.3 for more details about vectorizing expressions within a function that involve function input vectors.

Example 4.17 —————————————————————————————————

Let us find the amount of charge that has passed through a diode over one cycle of a sinusoidal voltage across it. Recall the diode i-v characteristic, for which a function is given below:

```
function current = diode(t)
% current through a diode when the voltage across it is v=Asin(wt)
global w A  % frequency and amplitude
v = A*sin(w*t);
I_S = 1e-12; % saturation current
V_T = 25.85e-3;  % thermal voltage at 300 K
current = I_S*(exp(v/V_T) - 1);
```

Here, the function input variable is *t*, which is used as the input time argument for the sinusoidal voltage. The frequency and amplitude of the voltage are passed to this function as global variables. This function was saved as diode.m. To see how charge moves through the diode over one cycle of the voltage, let us segment one cycle into *N* sub-intervals, and integrate the current over each sub-interval. Prog. 4.10 gives a script to plot charge movement over one cycle.

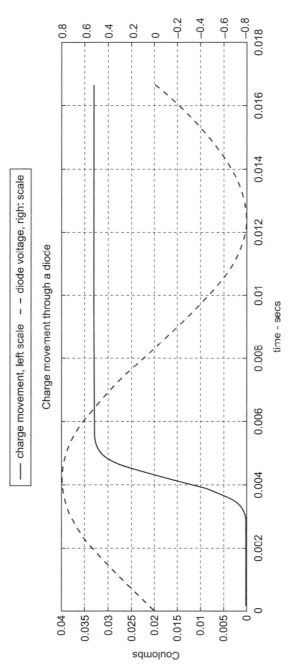

Figure 4.11 Charge movement through a diode over one voltage cycle.

```
% Program to plot charge movement through a diode over one cycle of
% a sinusoidal voltage across it
clear all; clc
global w A
frequency = 60; % specify frequency
w = 2*pi*frequency;
A = 0.8; % amplitude of sinusoid
T0 = 1/frequency; % time of one cycle
N = 100; T = T0/N; % will use N+1 plot points
tol = 1e-8; % specify tolerance
q(1) = 0; % initial charge is zero
v(1) = 0; % initial voltage value
t = 0.0:T:T0; % specify N+1 time points
for k = 1:N
   v(k+1) = A*sin(w*t(k+1));
   q(k+1) = q(k) + quad(@diode,t(k),t(k+1),tol); % accumulate charge
end
plotyy(t,q,t,v); grid on % use right y-axis for v
xlabel('time secs'); ylabel('Coulombs')
title('Charge movement through a diode')
```

Program 4.10   Script to find the amount of charge that has passed through a diode.

Fig. 4.11 shows how charge moves through a diode over one cycle of the voltage across it. Notice that as the voltage increases, there is no appreciable movement of charge through the diode until the voltage reaches a threshold voltage of about 0.7 volts at $t \simeq 2$ msec. Then, once the voltage exceeds this threshold, the amount of charge that has moved through the diode increases quickly until the voltage again goes below the threshold, after which the net amount of charge through the diode no longer increases.

There are several other built-in functions for numerical integration. These functions adaptively adjust the way they work to achieve prescribed error tolerances. For example, **quadl** may be more efficient than **quad**, while also giving more accurate results, especially for smoothly behaving integrands. The built-in function **quad2d** numerically obtains double integrals over a planar surface. See the MATLAB help facility for details and additional built-in functions for numerical integration.

4.8   Switch–Case–Otherwise

The **switch–case–otherwise** structure provides a way to select which block of statements in a set of blocks of statements should be executed. The syntax of this structure is

```
switch expression
      case expression_1
            block #1 of statements
      case expression_2
            block #2 of statements
    : insert as many additional case expression and block of
      statements as needed
      otherwise
            block #N of statements
  end
```

where each **switch** key word must be matched by an **end** key word. The **otherwise** and block #N of statements is optional. The switch expression must evaluate to a scalar or a character string. When MATLAB encounters the **switch** key word, it does the following:

1) Evaluate the switch expression and
2) If the result of evaluating the switch expression matches the case expression_1 value, then execute block #1 statements, and continue execution after the **end** statement
3) If the result of evaluating the switch expression does not match the case expression_1 value, then skip the block #1 statements to the next case
4) If the result of evaluating the switch expression matches the case expression_2 value, then execute block #2 statements, and continue execution after the **end** statement
5) If the result of evaluating the switch expression does not match the case expression_2 value, then skip the block #2 statements to the next case, and so on
6) If the result of evaluating the switch expression does not match any case expression value, then execute the block #N statements after the **otherwise** key word (if there is one) and continue execution after the **end** statement.

It is useful to indent each block of statements to easily see each case block of statements and that each **switch** key word is matched by an **end** key word. Within each block of statements there can be any of the program flow control structures, including nested switch–case–otherwise structures.

Example 4.18

You are approaching an intersection with a traffic light. Different actions are required depending on the color and behavior of the traffic light. Prog. 4.11 is a demonstration of the application of the switch-case-otherwise structure to this situation.

```
% Suggested action when approaching an intersection
% Local conditions may require other actions
clear all; clc
disp('Upon approaching an intersection with working traffic lights,')
```

```
disp('different actions are required.')
behavior = input('Enter a traffic light behavior in single quotes: ');
switch behavior
   case 'red'
      disp('Come to a complete stop; and wait for green light to proceed.')
   case 'yellow'
      disp('Slow down; prepare to stop.')
   case 'green'
      disp('Proceed with caution.')
   case 'flashing yellow'
      disp('Slow down; proceed with caution; be ready to stop.')
   case 'flashing red'
      disp('Come to a complete stop; proceed with caution.')
   otherwise
      disp('Traffic lights may be out of service.')
      disp('Come to a complete stop.')
      disp('Local conditions may require other actions.')
end
```

Program 4.11 Program to suggest action when approaching an intersection with traffic lights.

4.9   Conclusion

Testing a condition, executing alternative parts of a program and repeatedly executing a part of a program are essential for implementing problem solution methods. All programming languages have means to do this. You should now know how

- to use relational operators for formulating condition tests
- to use logical operators to make compounded statements using logical parts
- the if–elesif–else structure is used to execute alternative program parts
- to make a program more robust
- the for loop structure works to repeatedly execute a block of MATLAB statements
- to find the time required to execute a block of statements
- a histogram is obtained
- to determine the probability of an event by counting
- a pseudorandom number generator can be used to simulate random events
- a median filter works
- to use the sound device
- the while loop structure works to repeatedly execute a block of statements depending on a condition

- Euler's method and the trapezoidal rule are used for numerical integration
- the switch–case–otherwise structure is used to execute alternative program parts

Table 4.5 gives the MATLAB functions that were introduced in this chapter. Use the MATLAB help facility to learn more about these built-in functions, where you will also find out about many other related built-in functions.

Table 4.5 Built-in MATLAB functions introduced in this chapter

Function	Brief explanation
isequal(A,B)	Returns logical 1, meaning true, if arrays A and B are the same size and contain the same values, and logical 0, meaning false, otherwise
L = and(A,B)	An element of the output array is set to 1 if both input arrays contain a nonzero element at that same array location, and otherwise, that element is set to 0
L = or(A,B)	An element of the output array is set to 1 if either input array contains a nonzero element at that same array location, and otherwise, that element is set to 0
L = xor(A,B)	An element of the output array is set to 1 if either input array contains a nonzero element at that same array location, but not both, and 0 if both elements are zero or nonzero
L = not(A)	An element of the output array is set to 1 if A contains a zero value element at that same array location, and 0 otherwise
&&	Short logical and
||	Short logical or
I = find(A)	Returns the linear indices corresponding to the nonzero entries of the array A
L = sign(A)	Signum function, returns 1 if the element is greater than zero, 0 if it equals zero and -1 if it is less than zero; for the nonzero elements of complex A, sign(A) = A./abs(A)
tic	Start a stopwatch timer
T = toc	Saves the elapsed time in T (seconds) since the last execution of tic
pause(t)	Pauses for t seconds before continuing, where t can also be a fraction; just pause causes a program or function to stop and wait for the user to strike any key before continuing
bar(X,Y,width)	Draws the columns of the M × N matrix Y as M groups of N vertical bars; bar(Y) uses the default value of X = 1:M. For vector inputs, bar(X,Y) or bar(Y) draws length(Y) bars. There is no space between bars if width = 1
continue	Pass control to the next iteration of for or while loop
break	Terminate execution of for or while loop
return	Causes a return to the invoking program or function or to the keyboard; also terminates the keyboard mode
keyboard	When placed in an m-file, stops execution of the file and gives control to the user's keyboard. The special status is indicated by a K appearing before the prompt. Variables may be examined or changed; all MATLAB commands are valid. The keyboard mode is terminated by executing the command return
R = rand(N,M)	Returns an N × M matrix containing pseudorandom numbers uniformly distributed over the open interval (0,1)
I = randi(N,M,imax)	Returns an N × M matrix containing pseudorandom integer values uniformly distributed over the range 1:imax

(Continues)

Table 4.5   (*Continued*)

Function	Brief explanation
R = randn(N,M)	Returns an N × M matrix containing pseudorandom values from the standard normal distribution
Y = mean(X)	Y is a row vector containing the mean of each column of X
Y = var(X)	Y is a row vector containing the variance of each column of X
I = hist(X,N)	Bins the elements of X into N bins
fprintf	Write formatted data to a display or a file; more about this later
y = median(X)	Returns the median value of the elements in the vector X
sound(X,fs)	Sends the signal in vector X, with sampling frequency f_s to the sound device; X can be a two-column matrix for stereo. Clips values outside the range of -1 to $+1$.
soundsc(X,fs)	Same as sound, but autoscales
wavwrite(X,fs,'file. wav')	Writes data X to a Windows WAV file specified by the file name file.wav, with a sampling rate f_s Hz; stereo data must be specified as a matrix with two columns
Y = sinc(X)	Returns in the matrix Y elements found with sin(pi*x)/(pi*x) of the elements in the matrix X
quad(func,a,b)	Returns an approximation of the integral from a to b of the function specified by the function handle func
plotyy(x1,y1,x2,y2, plot_func)	Plots y1 versus x1 with y-axis labeling on the left and plots y2 versus x2 with y-axis labeling on the right, using the plotting function specified by the function handle plot_func
quadl	Same as quad, but uses a more accurate numerical integration method
quad2d(func,a,b,c,d)	Returns an approximation of the double integral of a function f(x, y) specified by the function handle func, where $a \leq x \leq b$ and $c \leq y \leq d$. See MATLAB help for more options

The essentials of the MATLAB programming language have been presented in the first four chapters of this book. Use demo to watch an excellent audio/video tutorial about if–elseif–else, for, and while loop control structures. In these chapters some special features, for example, plotting, were introduced to make examples more complete with visual displays. In the remaining chapters, special, unique, and particularly useful features of this programming language will be discussed.

In the next chapter the concept of a logical variable will be extended to understand its role in Boolean algebra, logic circuits, and binary arithmetic.

References

Nelder, J.A., Mead, R. 'A simplex method for function minimization', *Comput. J.* 1965; 7:308–313

Li, C., Priemer, R., Cheng, K.H., 'Optimization by random search with jumps', *Int. J. Num. Methods Engg.* 2004;**60**:1301–1315

Problems

In all of the scripts that you are required to write for the following problems, include comments that state the purpose and activity of the script.

Section 4.1

1) Use $a = -2$, $b = 3$, and $c = 5$, and find the result of each of the following relational expressions. (a) $d = a\char94 2 >= c$, (b) $e = (a > b - 4) + c <= b\char94 2$, (c) $f = a + b == c - 4$.
2) For $t = -1:0.1:2$, $w = 2*\text{pi}$ and $x = \cos(w*t)$, use a logical operation to find a vector y, which has the same dimension as x, where all of the positive elements of x have been replaced by zero.
3) $A = [-1\ 9\ 2;\ 0\ -3\ 4]$, $B = []$, $C = \text{ones}(3,2)'$. Give the result of (a) isempty(A), (b) isempty(B), (c) logical(A), (d) any(A), (e) islogical(C), (f) $C < \text{false}(2,3)\ |\ A$, (g) isinteger(sqrt($A$)), (h) isfloat($C$).

Section 4.2

4) For $a = 2$, $b = 0$, and $c = -5$, give the results of (a) $d = a\ |\ (b\ \&\ c)$, (b) $e = \text{and}(a,b)|c$, (c) $f = a\ \&\ c\ |\ (b\ \&\ \mathord{\sim}a)$, (d) $g = a\ \&\ c\ |\ b\ \&\ \mathord{\sim}a$, (e) $h = (b\ \&\&\ c)\ |\ (b\ ||\ a)$, (f) $p = a\ \&\ \text{xor}(\mathord{\sim}b,c)$.
5) (a) Describe the difference between $(a\ \&\ b)$ and $(a\ \&\&\ b)$.
   (b) Describe the difference between $(a\ |\ b)$ and $(a\ ||\ b)$.
6) For $A = [0\ 1;\ 1\ 0]$, $C = [0\ 0;\ 1\ 0]$, and $C = A\ \&\ B$, give all possible B.
7) For $a = -2$, $b = 4$, and $c = 2$, give the results of (a) $d = (a\ \mathord{\sim}= c\ -2)\ \&\ a\char94 2 > b$, (b) $e = \text{sqrt}(a\char94 2 + b\char94 2) < \text{abs}(c)\ |\ (a+b)/2 >= c$, (c) $f = \mathord{\sim}\text{isempty}(\text{find}([a\ b\ 2*c\ \text{sqrt}(a\char94 2 + c\char94 2)] > 4))$.
8) For $A = [-1\ 0.5\ 5;\ 2\ -4\ 0;\ 0\ 0\ 1]$, (a) give $y = \text{find}(A <= -1\ |\ A >= 1)$, and explain the meaning of y. (b) Let $B = \text{reshape}(A,1,9)$ and give $C = B(y)$.

Section 4.3

9) (a) If the inputs to Prog. 4.1 are $a = 0$, $b = 5$, $N = 9$, and $x = 5.6$, then what is the program output?
   (b) If the inputs are the same as in part (a), except that $x = 2.7$, then what is the program output?
10) (a) Convert Prog. 4.1 into a function, call it bin_number. The function input arguments are a, b, x, and N, and the output arguments are k and err. Include comments that explain the purpose of the function and the meaning of the err codes. Provide a copy of the function.
    (b) Write a script that demonstrates using the function bin_number. Provide a program listing and demonstration of program operation.

11) Write a script that implements the flowchart given in Fig. P4.11. Use an if–else–end control structure. Provide a program listing and a demonstration of program operation. Does the current have to be a positive number?

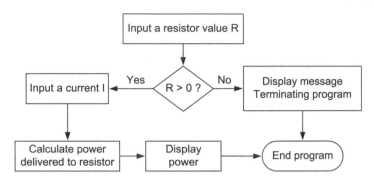

Figure P4.11   Flowchart to calculate power delivered to a resistor.

12) Write a script that inputs an integer N in the range 0–9. Your script must check that the input is an integer and that it is within the required range. Otherwise, display an error message for each input error type. Then, the script must determine if N is an odd or even integer and display a message about the result. Provide a program listing and a demonstration of program operation.

Section 4.4

13) Describe what is wrong with each of the following for statements: (a) for $k = 0$:–1:10, (b) for $n = 10$:2:3, (c) for $2k = 0$:5, (d) for Number $= -1$:–5. In each case, what will MATLAB do when the for loop is completed with an end statement?

14) (a) Write a function script, call it inner_prod, that uses a for loop to find the inner product A_dot_B of two vectors A and B with dimension $1 \times N$. The function input arguments must be A and B, and the function returns A_dot_B.
   (b) Write a script to demonstrate the operation of your function where $t = 0$:T:$T0$-T, $T = T0/N$ sec, $T0 = 2\pi/\omega_0$ sec, $A = \cos(\omega_0 t)$ and $B = \cos(K\omega_0 t)$ for K=1, 2, and 5. You choose N and ω_0. For each K, what is A_dot_B? What do you think A_dot_B will be for any integer $K \neq 1$?
   (c) Repeat part (b) for $B = \sin(K\omega_0 t)$.
   (d) When the inner product of two vectors is zero, it is said that the vectors are orthogonal. Here, we find that this concept can be extended to functions, for example, $\cos(M\omega_0 t)$ and $\cos(K\omega_0 t)$, $M \neq K$, are orthogonal functions over a period $T0$. What other sinusoidal functions are orthogonal over a period?

15) Write a script that uses for loops to evaluate the power series given by (2.13) of an exponential function $x(t) = e^{-2t}$, for $t = 0$:0.1:2. Break out of the for loop that does the

power series computation when the next term in the series contributes less than 0.0001% of the sum of terms. For example, see Prog. 4.4. Provide a figure of a plot of your result. Include in the figure a stem plot that uses the MATLAB built-in function exp. The figure should include a grid, axes labels and a title.

16) Given is the circuit shown in Fig. P4.16.

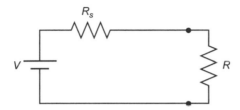

Figure P4.16   Battery with internal resistance R_s connected to a load R.

(a) In terms of V, R_s and R, obtain an expression for the power P delivered to R.
(b) Write a script that receives as input the values of V and R_s, and makes a vector R with element values that range from $0.1*R_s$ to $10*R_s$ in increments of $0.1*R_s$. Use a for loop to create a vector of power delivered to each element value of R. Then, plot the power P versus R. Do this for $V = 10$ volts, $R_s = 10$ Ohms and $R_g = 100$ Ohms For each case, from inspection of the plots, give a value of R such that V delivers the maximum power to R. Provide a program listing and plots, including axes labels and titles.
(c) Vectorize your expression for P, and repeat part(b) without using a for loop.
(d) In general, for a given value of R_s, what value of R enables the resistor R to receive the most power from the voltage source V?

17) Write a script that starts by obtaining a vector x of $N = 1e6$ normally distributed random numbers with the function randn.
(a) Obtain the mean, variance and histogram of x. Use 51 bins for the histogram. Use the MATLAB built-in functions mean, var, and hist to obtain these results.
(b) Let $y = ax + b$, where $a = 3$ and $b = 2$. Use a for loop to find the mean of y. Then, use another for loop to implement (4.6). How does your result compare with using the MATLAB function var to find the variance of y?
(c) Obtain a histogram of y, using 51 bins. How are the parameters a and b related to the variance and mean of y?

18) Suppose that an experiment consists of tossing two die. Each die will show an integer from 1 to 6. This can be simulated with die_1 = ceil(6*rand) and die_2 = ceil(6*rand), where each invocation of rand produces a random number from 0 to 1. Write a script that uses a for loop to do N_exps = 1e7 experiments, and within the loop uses an

if–elseif–end structure to count the number of times die_1 + die_2 = 7 or 11. Begin your script with the statement: rng('default') to bring the random number generator to its default initial state. By counting, find the probability that an experiment outcome will be 7 or 11.

19) Let us study the performance of the algorithm given by

$$y(k\,T) = \frac{1}{4}\left(x(kT) + 2\,x((k-1)T) + x((k-2)T)\right)$$

for $k = 0, 1, \ldots$, where $x(k\,T)$ comes from sampling a sound given by $s(t) = \sin(\omega_1\,t) + 0.5\cos(\omega_2\,t)$ that is corrupted by additive noise $n(t)$. Let $\omega_1 = 2\pi(262)$ and $\omega_2 = 2\pi(440)$. The sampling rate is $f_s = 8000$ samples/second, and $T = 1/f_s$. The data $x(kT)$ for this simulation is given by $x(kT) = s(kT) + a\,n(kT)$, where $n(kT)$ is found by using a normal random number generator and a controls the noise amplitude. To help you get started, use

```
% This is an example of digital signal processing
% Program to implement a digital low-pass filter
clear all; clc;
fs =8000; T = 1/fs;
w1 = 2*pi*262; w2 = 2*pi*440;
a = 0.5;  % used to see how noise affects the result
k = 0; t = k*T; time(k+1) = t;  % initialize sample time
x(k+1) = sin(w1*t) + 0.5*cos(w2*t) + a*randn;  % first sample
% since there is no x(-T) and x(-2T) use
y(k+1) = x(k+1)/4;
k = k+1; t = k*T; time(k+1) = t;  % save time for plotting
x(k+1) = sin(w1*t) + 0.5*cos(w2*t) + a*randn;  % second sample
% since there is no x(-T) use
y(k+1) = (x(k+1)+2*x(k))/4;
N = 100;  % process N-2 more samples of the input
for k = 2:N-1
        t = k*T; time(k+1) = t;  % save time for plotting
        x(k+1) = sin(w1*t) + 0.5*cos(w2*t) + a*randn;  % kth sample
        y(k+1) =
```

Complete the for loop with the given algorithm, and then plot x and y versus time. What does the algorithm appear to do?

Section 4.5

20) Do Prob. 4.11 using a while loop. The program should allow for repeated execution of the program until a program user enters an R value less than or equal to zero. For example:

```
clear all; clc;
while 1
        R = input('Enter a resistance value: ')
        if R <= 0
                disp('Terminating program')
                break
        end
```

Complete the script. For a nice looking output, use, for example, fprintf('Power delivered = %6.2f watts\n', P). Provide a program listing, and demonstrate its operation.

21) Do Prob. 4.15 using a while loop inside a for loop. With the for loop index ranging from $n = 1$ to $n = N$, where $N = \text{length}(t)$, then, for example, the while loop could be

```
x = -2.0*t(n);
exp_sum = 1.0; % first term in power series
next_term = x; k = 1; % second term in power series
while abs(next_term) > 1e-6*abs(exp_sum)
        exp_sum = exp_sum + next_term;
        k = k+1;
        next_term = next_term*x/k;
end
exp_func(n) = exp_sum + next_term; % use last term
```

Explain why the abs function must be used. Complete the script, and provide results as described in Prob. 4.15.

Section 4.6

22) Consider again Prob. 4.16. Write a script that applies the method of steepest ascent, (4.10), to find the value of R that maximizes the power delivered to it by the voltage source V. Use input statements to obtain the values of V and R_s, and a while loop to implement an update method to find R_{new} given R_{old} similar to the update method of (4.9). See Example 4.14. Find a suitable value of the step size by trial and error. Use $R_{\text{old}} = R_s/2$ as the initial guess of R_{opt}. Provide a program listing and a demonstration of its operation.

23) Given a number x, find $y = x^{1/2}$. Of course, $y = $ sqrt(x). However, an algorithm must be designed to find y. This can be done by converting the given problem to an optimization problem. Consider minimizing the function $f(y) = (y^2 - x)^2$, where $f'(y) = 2(y^2 - x)(2y)$. Write a script that uses a while loop within a while loop to apply the method of steepest descent. The outer while loop must contain an input statement to receive a value of x, and if the input is not greater than zero, then break out of this loop, which terminates the program. If the input x is greater than zero, then the inner while loop must apply the method of steepest descent to find y. Implement an update method to find y_{new} given y_{old} like the update method of (4.9). Use $y_{old} = x$ as an initial guess of y_{opt}. See Example 4.14. Find a suitable step size by trial and error. Provide a program listing and a demonstration of its operation for three positive values of x.

24) Repeat Prob. 4.23, but replace the inner while loop with an invocation of the MATLAB built-in function fminsearch. Create a function m-file to define the function to be optimized by fminsearch. Provide listings of your program, the function m-file, and a demonstration of program operation.

Section 4.7

25) Given is the function $x(t) = e^{-at} - e^{-bt}$. Let $a = 1$ and $b = 2$.
   (a) Manually, find an expression for the area A under $x(t)$ from $t = 0$ to $t = 5$.
   Write a script to:
   (b) Plot $x(t)$ for $t = 0$ to $t = 5$. Use $N = 501$ values of t in the given time range
   (c) Calculate A using the expression found in part (a)
   (d) Apply Euler's method to find an approximation of A for $N = 11$, 111, and 211
   (e) Output a table of the difference between A and its approximation for each N.

   Use format long e for all program output. Provide a listing of your program and the results.

26) Repeat Prob. 4.25, but use the trapezoidal rule.

27) For the $x(t)$ given in Prob. 4.25, use the built-in MATLAB function quad to find the area A under $x(t)$ from $t = 0$ to $t = 5$. Use an anonymous function to define $x(t)$. Compare the result to A found in Prob. 4.25, part (a). Use format long e for all program output. Provide a listing of your program and results.

Section 4.8

28) Write a function script, call it linear_eqs_solver, where the inputs are a square matrix A, a column vector Y and a character string named solution_method, and the outputs are the condition number K of A and the solution X of $AX = Y$, assuming that the determinant of A is not zero. First, check the determinant of A, and if it is zero, set solution_method = 'none'. The possible solution methods you should use are inv (the function), left matrix divide, eigenvalues and eigenvectors or singular value decomposition, as designated with solution_method = 'inv', 'left_mat_div', 'eigen', or

'singular_decomp'. Use the switch-case-otherwise-end structure to select the method to be used. Use the otherwise option to return an empty vector X and $K=$ inf in the event that the character string solution_method is not one of the allowed solution method options. Write a script that demonstrates the application of your function. Provide program and function listings and program output.

29) Repeat Prob. 4.18, where within the for loop a switch-case-end structure is used to count the number of times that an experiment outcome is 7 or 11. Use die_1 + die_2, for the switch expression and case 7 and case 11.

CHAPTER 5

Binary Data

A digital circuit is an electronic circuit that receives binary voltage inputs and produces binary voltage outputs. Digital circuits are designed to perform various logical operations, and they are the building blocks of digital systems, which are employed in virtually all electronic products. In digital systems, a data bit is associated with a binary voltage. A data bit can be logic 1 or logic 0, where logic 0 is associated with one of the binary voltages, and logic 1 is associated with the other voltage. A commonly used binary voltage pair and data bit association is: 0 volts $\leftrightarrow$ logic 0 and 5 volts $\leftrightarrow$ logic 1. This is called positive logic, where logic 1 corresponds to the high voltage, which may be a voltage other than 5 volts. Many other binary voltage pairs are in common use. A group of data bits can have different meanings. In this chapter some of the possibilities will be examined.

After you have completed this chapter, you will know

- the elements of Boolean algebra
- about the binary number system and binary arithmetic
- about basic electronic logic gates
- how to formulate and implement Boolean functions
- how MATLAB® can be used to simulate logic circuits
- about the operation of an analog to digital converter and quantization error

5.1   Boolean Algebra

In Boolean algebra, scalar variables can assume only one of two values: 0 or 1, also called **false** or **true**, respectively. Boolean variables are logical variables.

Boolean algebra is based entirely on three basic operations: **AND, OR**, and **NOT**, also called **conjunction**, **disjunction**, and **negation** or **complement**, respectively. These operations were defined in Table 4.3, which is repeated here, for convenience, as Table 5.1.

In Boolean algebra, x **and** y is written as $x \bullet y$ or simply xy, where x and y are logical variables, x **or** y is written as $x+y$, and **not** x is written in several different ways, but we will use $\bar{x}$ to mean not(x). Recall that MATLAB uses $x\&y = xy$, $x \,|\, y = x+y$ and $\sim x = \bar{x}$. Instead, MATLAB also uses $x\&\&y = xy$ and $x \,||\, y = x+y$, which will be explained later.

Table 5.1   Basic operations of Boolean algebra (x and y are logical scalars)

AND			OR			NOT	
x	y	x and $y = x \bullet y$	x	y	x or $y = x+y$	x	not(x) $= \bar{x}$
0	0	0	0	0	0	0	1
0	1	0	0	1	1		
1	0	0	1	0	1	1	0
1	1	1	1	1	1		

The and(or) operation can be extended over many input variables, and the result is logic 1 (0) only if all inputs are logic 1 (0).

The fundamental laws or **axioms** of Boolean algebra are given in Table 5.2. Parentheses are used to specify the order of evaluation. Recall that MATLAB evaluates logical expressions from left to right, and therefore, in MATLAB parentheses must be used to specify the order of evaluation.

Table 5.2   Axioms of Boolean algebra (x, y, and z are logical scalars)

Axiom (or property)	AND version	OR version
Cummutative	$x \bullet y = y \bullet x$	$x+y = y+x$
Associative	$x \bullet (y \bullet z) = (x \bullet y) \bullet z$	$x+(y+z) = (x+y)+z$
Idempotence	$x \bullet x = x$	$x+x = x$
Absorption	$x \bullet (x+y) = x$	$x+(x \bullet y) = x$
Distributive	$x \bullet (y+z) = (x \bullet y)+(x \bullet z)$	$x+(y \bullet z) = (x+y) \bullet (x+z)$
Complement	$x \bullet \bar{x} = 0$	$x+\bar{x} = 1$
Annihilation	$x \bullet 0 = 0$	$x+1 = 1$
Identity	$x \bullet 1 = x$	$x+0 = x$

Any of the axioms can be verified (proved) by an exhaustive search, that is, evaluate the expression on each side of an equality for every binary assignment to the variables, and if the expressions are equal for all possible binary assignments to the variables, the axiom is proved.

Example 5.1 ⎯⎯⎯⎯⎯⎯⎯⎯⎯⎯⎯⎯⎯⎯⎯⎯⎯⎯⎯⎯⎯⎯⎯⎯⎯⎯⎯⎯⎯

Let us prove the distributive property. This property involves $k = 3$ variables, and therefore, there are $2^k = 8$ different binary assignments from all three variables equal to logic 0 to all three variables equal to logic 1. Prog. 5.1 evaluates the expression on each side of the equal sign of the distributive property for each binary assignment to x, y, and z, and it generates a table, called a **truth table**.

```
% Program to prove the distributive property: x(y + z) = xy + xz, or
% x + yz = (x + y)(x + z) by exhaustive search
clear all; clc
x = logical([0 0 0 0 1 1 1 1]); % all assignments to x, y and z
y = logical([0 0 1 1 0 0 1 1]);
z = logical([0 1 0 1 0 1 0 1]);
for k = 1:8
   L_side_and(k) = x(k)&(y(k)|z(k)); % left side AND version
   R_side_and(k) = (x(k)&y(k))|(x(k)&z(k));
   L_side_or(k) = x(k)|(y(k)&z(k)); % left side OR version
   R_side_or(k) = (x(k)|y(k))&(x(k)|z(k));
end
% the for loop could be replaced by
% L_side_and = x&(y|z); R_side_and = (x&y)|(x&z);
% L_side_or = x|(y&z); R_side_or = (x|y)&(x|z);
disp('Truth Table for the Distributive Property')
disp('        AND version   OR version')
disp('x  y  z  left right    left right')
disp('        side side     side side')
table = [x;y;z;L_side_and;R_side_and;L_side_or;R_side_or];
fprintf(' %i  %i  %i    %i  %i      %i  %i \n',table)
% the fprintf function prints a column of a table in a row of output
```

Program 5.1 Program to prove the distributive property.

Truth Table for the Distributive Property						
			AND version		OR version	
x	y	z	left side	right side	left side	right side
0	0	0	0	0	0	0
0	0	1	0	0	0	0
0	1	0	0	0	0	0
0	1	1	0	0	1	1
1	0	0	0	0	1	1
1	0	1	1	1	1	1
1	1	0	1	1	1	1
1	1	1	1	1	1	1

Since the left side column matches the right side column of both versions of the distributive property for all binary assignments to the variables x, y, and z, the property is proved.

Table 5.3 gives several additional commonly used logical operations. Notice that the **eq**uivalence operation (denoted by $\odot$) is the complement of the exclusive **or** operation (denoted by $\oplus$).

Table 5.3   Additional basic operations of Boolean algebra

NAND = NOT($x \bullet y$)			NOR = NOT($x+y$)			XOR			EQ = $\overline{\text{XOR}}$		
x	y	$\overline{(x \bullet y)}$	x	y	$\overline{(x+y)}$	x	y	$x \oplus y$	x	y	$x \odot y$
0	0	1	0	0	1	0	0	0	0	0	1
0	1	1	0	1	0	0	1	1	0	1	0
1	0	1	1	0	0	1	0	1	1	0	0
1	1	0	1	1	0	1	1	0	1	1	1

Another important property is called DeMorgan's theorem, which consists of two parts given by

$$\overline{(x \bullet y)} = \bar{x} + \bar{y}$$
$$\overline{(x + y)} = \bar{x} \bullet \bar{y} \tag{5.1}$$

Let us prove DeMorgan's theorem with the truth table given below, which can be obtained with a program similar to Prog. 5.1. Since columns 6 and 7 match and columns 8 and 9 match for all assignments to x and y, DeMorgan's theorem is proved.

Truth table for proof of DeMorgan's theorem

x	y	$\bar{x}$	$\bar{y}$	$x \bullet y$	$\overline{(x \bullet y)}$	$\bar{x} + \bar{y}$	$x+y$	$\overline{(x + y)}$	$\bar{x} \bullet \bar{y}$
0	0	1	1	0	1	1	0	1	1
0	1	1	0	0	1	1	1	0	0
1	0	0	1	0	1	1	1	0	0
1	1	0	0	1	0	0	1	0	0

A major application of Boolean algebra is the design of electronic systems for general purpose computing and the computing required in many kinds of devices, such as cell phones, digital cameras, monitor of patient vital signs, and much more. Such systems are composed of interconnections of a variety of logic circuit modules. Each module performs a particular task, for example, addition or comparison of numbers. Each input and output of a logic circuit module can have only one of two allowed values, logic 0 or logic 1.

The building blocks of these modules are electronic circuits that perform the basic logical operations that have been described in this section. Physically, the input(s) and output(s) of the electronic circuits are each a voltage that can have only one of two allowed voltage values. To utilize this kind of behavior of an electronic circuit requires that all data (text and numeric) be represented in binary form, where logic 0 is associated with one of the allowed voltages and logic 1 is associated with the other allowed voltage. In the following sections, we will see how data can be represented in binary form.

5.2 Binary Numbers

We use the base ten number system for numeric data, an alphabet for text data, and other special symbols (data such as !, /, #, ~, and others) for various kinds of intentions and operations. To process data with an electronic system requires that each kind of data must be represented by a set of binary voltages, which correspond to a set of logic values. Regardless of its meaning, a set of logic values can be considered to be a binary number. Let us see the kinds of meaning that we can attribute to a binary number.

5.2.1 Base Ten to Binary Conversion

A base ten integer is written as

$$N = \quad d_4\, d_3\, d_2\, d_1\, d_0 \tag{5.2}$$

where the digits d_k, $k = 0, 1, 2, \ldots$ can each be one of $0, 1, 2, \ldots, 9$. N gives a count implicitly. Each digit in N gives the number of times each corresponding power of ten contributes to the count that N represents. The value of N is given by

$$N = \cdots + d_4 \times 10^4 + d_3 \times 10^3 + d_2 \times 10^2 + d_1 \times 10^1 + d_0 \times 10^0 \tag{5.3}$$

and here we see that d_k specifies the number of times that 10^k contributes to the value of N. While N in (5.3) gives the actual count, N in (5.2) conveys the same information.

Similar to the base ten number system, in the base two (or binary) number system an integer N is written like N in (5.2), where instead, the binary digits (also called **bits**) d_k, $k = 0, 1, 2, \ldots$, can each be 0 or 1, and still N gives a count implicitly. The value of N is given by

$$N = \cdots + d_4 \times 2^4 + d_3 \times 2^3 + d_2 \times 2^2 + d_1 \times 2^1 + d_0 \times 2^0 \tag{5.4}$$

Here, since d_k can only be 0 or 1, it determines whether or not 2^k contributes to N. For example, if $N = 101101$, then we know that N does not contain 2^4 and 2^1, and it does contain 2^5, 2^3, 2^2, and 2^0. With (5.4), we can convert the binary version of N to its base ten version, which is $N = 32 + 8 + 4 + 1 = 45$. Table 5.4 gives the range of values of 3-bit binary numbers.

Table 5.4 Range of 3-bit binary numbers from 0 to 2^3-1

Base ten	0	1	2	3	4	5	6	7
Base two	000	001	010	011	100	101	110	111

Notice that in the truth table of Example 5.1, all of the binary assignments to the three variables x, y, and z are listed as a binary count from 0 to 7.

To convert a base ten integer N into binary, start with (5.4), denote N by N_0, and

1) set $k = 0$
2) divide N_k by 2 to get

$$\frac{N_k}{2} = \cdots + d_4 \times 2^{3-k} + d_3 \times 2^{2-k} + d_2 \times 2^{1-k} + d_1 \times 2^{-k} + d_0 \times 2^{-1-k}$$

3) if $N_k/2$ is an integer, then $d_k = 0$, otherwise $d_k = 1$
4) let $N_{k+1} = \frac{N_k}{2} - \frac{d_k}{2}$
5) if $N_{k+1} \neq 0$, then $k = k+1$ and go to step 2, otherwise, stop, process has terminated

For example, if $N = 67$, then the following table is a demonstration of the given algorithm:

$N_0 = 67$	d_k
$N_1 = 33$	$d_0 = 1$
$N_2 = 16$	$d_1 = 1$
$N_3 = 8$	$d_2 = 0$
$N_4 = 4$	$d_3 = 0$
$N_5 = 2$	$d_4 = 0$
$N_6 = 1$	$d_5 = 0$
$N_7 = 0$	$d_6 = 1$

Thus, $N = 67 = d_6\, 2^6 + d_1\, 2^1 + d_0\, 2^0$, and in binary $N = 1\,0\,0\,0\,0\,1\,1$.

In binary, a fraction is written as

$$F = .d_{-1}\, d_{-2}\, d_{-3}\, d_{-4}\, d_{-5} \cdots$$

where the digits d_{-k}, $k = 1, 2, \cdots$ are either 0 or 1. The base ten value of F is given by

$$F = d_{-1} \times 2^{-1} + d_{-2} \times 2^{-2} + d_{-3} \times 2^{-3} + d_{-4} \times 2^{-4} + \cdots \tag{5.5}$$

To convert a base ten fraction F into binary, start with (5.5), denote F by F_{-1}, and

1) set $k = 1$
2) multiply F_{-k} by 2 to get

$$2F_{-k} = d_{-1} \times 2^{-1+k} + d_{-2} \times 2^{-2+k} + d_{-3} \times 2^{-3+k} + d_{-4} \times 2^{-4+k} + \cdots$$

3)   if $2\,F_{-k}$ is a fraction, then $d_{-k} = 0$, otherwise $d_{-k} = 1$
4)   let $F_{-(k+1)} = 2\,F_{-k} - d_{-k}$
5)   if $F_{-(k+1)} \neq 0$, then $k = k + 1$ and go to step 2, otherwise, stop, process has terminated

Depending on F, this process may or may not terminate. For example, if $F = 0.67$, then the following table is a demonstration of the given algorithm.

$F_{-1} = 0.67$	d_{-k}
$F_{-2} = 0.34$	$d_{-1} = 1$
$F_{-3} = 0.68$	$d_{-2} = 0$
$F_{-4} = 0.36$	$d_{-3} = 1$
$F_{-5} = 0.72$	$d_{-4} = 0$
$F_{-6} = 0.44$	$d_{-5} = 1$
$F_{-7} = 0.88$	$d_{-6} = 0$
$F_{-8} = 0.76$	$d_{-7} = 1$
$\vdots$	$\vdots$

Thus, $F = 0.67 = d_{-1}2^{-1} + d_{-3}2^{-3} + d_{-5}2^{-5} + d_{-7}2^{-7} + \ldots,$ and   in   binary $F = .1\,0\,1\,0\,1\,0\,1\ldots$.

To store a fraction in the memory of a computer, where only a finite number of bits can be stored, its binary representation must be truncated, and then the value of the binary fraction stored in memory is an approximation of the given fraction. Assuming that the binary fraction becomes involved in arithmetic operations, subsequent results will be inaccurate. Such errors are referred to as **finite word length effects**. To retain as much accuracy as possible, by default, all computation in MATLAB is done in double precision, which will be explained later.

Bits are commonly grouped into 4 bits (called a **nibble**) and 8 bits (called a **byte**). A group of bits that together represent some unit of information is called a **word**. For example, if 24 bits are used for a binary fraction, then the word length of a binary fraction is 24 bits or 3 bytes. The left most bit of a word is called the **most significant bit** (**MSB**), and the right most bit is called the **least significant bit** (**LSB**). In a computer system all information is stored and processed using bits, nibbles, and bytes, which are interpreted to mean different things, as we shall see in this chapter.

5.2.2   ASCII Codes

Each entry from a keyboard is encoded with a 1-byte binary number. The most commonly used code is **ASCII** (American Standard Code for Information Interchange). See Appendix B for a complete table of ASCII codes. The printable characters have ASCII codes from 32 to 127. To get them, use the **char** built-in MATLAB function, as in

```
>> ASCII_codes = 32:127; % a base ten numeric vector of printable ASCII codes
>> % convert the numeric vector into a character vector, the codes remain unchanged
```

```
>> characters = char(ASCII_codes)
characters =
 !"#$%&'()*+,-./0123456789:;<=>?@ABCDEFGHIJKLMNOPQRSTUVWXYZ[\]^_`abcdefghijklmnop
qrstuvwxyz{|}~
>> characters(1) % this element holds the ASCII code 32
ans =        % this is a space or blank, as when you depress the space bar on the keyboard
 >> characters(2) % this element holds the ASCII code 33
ans = !     % this is the exclamation mark
```

For example, the MATLAB key word **for** is stored in computer memory using three ASCII codes, which are $f \rightarrow 01100110$, $o \rightarrow 01101111$, and $r \rightarrow 01110010$. In base ten we can write $f \rightarrow 102$, $o \rightarrow 111$, and $r \rightarrow 114$. When MATLAB begins to process a statement and it first encounters these three binary numbers, then this context causes the three numbers to be interpreted as the key word **for**, in which case MATLAB sets up further processing to execute a for loop. When a program is entered from a keyboard (using the MATLAB editor or some word processor), it is stored as a sequence of characters using ASCII codes, like in a text file.

When MATLAB executes a statement in an m-file, a sequence of characters (ASCII codes) that represents an actual number is converted into a floating point binary number using 64 bits (double precision), by default.

Example 5.2

```
>> A = [-397 0.183 pi sin(pi/4) exp(-2.5) 0]  % the default display format is short
A =  -397.0000   0.1830   3.1416   0.7071   0.0821      0
>> % the elements of A are stored using 64-bit floating point notation, double precision
>> % for display, the entire line is a string of ASCII codes
>> whos   % the elements of A are stored using 8 bytes per number
 Name     Size        Bytes    Class    Attributes
   A       1x6          48      double
>> format long  % the format function only controls how numbers are displayed
>> A
A =  Columns 1 through 6
-3.97000000000000   0.001830000000000   0.031415926535898   0.007071067811865
 0.000820849986239   0
>> format long e % floating point notation
>> A
A =  Columns 1 through 6
 -3.970000000000000e+002   1.830000000000000e-001   3.141592653589793e+000
  7.071067811865475e-001   8.208499862389880e-002   0
format short
>> whos   % numbers remain in double precision
 Name     Size        Bytes    Class    Attributes
   A       1x6          48      double
```

5.2.3 Storage Allocation

You can control the amount of memory space to be used, depending on the class and range of a number. Table 5.5 lists built-in MATLAB functions with which to do this.

Table 5.5   Built-in functions concerned with numeric data (scalar, vector, and matrix)

Function	Output range (base ten) (activity)	Output type (class)	Bytes per element
X = int8(N)	-128 to $127 \rightarrow -2^7$ to $2^7 - 1$	Signed 8-bit integer	1
int16	$-32,768$ to $32,767 \rightarrow -2^{15}$ to $2^{15} - 1$	Signed 16-bit integer	2
int32	$-2,147,483,648$ to $2,147,483,647$	Signed 32-bit integer	4
int64	$-9,223,372,036,854,775,808$ to $9,223,372,036,854,775,807$	Signed 64-bit integer	8
intmin	-2147483648	Signed 32-bit integer	4
intmax	2147483647	Signed 32-bit integer	4
uint8	0 to $255 = 2^8 - 1$	Unsigned 8-bit integer	1
uint16	0 to $65,535 = 2^{16} - 1$	Unsigned 16-bit integer	2
uint32	0 to $4,294,967,295 = 2^{32} - 1$	Unsigned 32-bit integer	4
uint64	0 to $18,446,744,073,709,551,615 = 2^{64} - 1$	Unsigned 64-bit integer	8
double	Convert to double precision	64-bit floating point	8
single	Convert to single precision	32-bit floating point	4
isnumeric	Test class	Logical	1
isfloat	Test class	Logical	1
isinteger	Test class	Logical	1
islogical	Test class	Logical	1

Example 5.3 _____

```
>> a = 40000; % out of range of a 16 bit signed integer
>> int16(a) % will result in biggest possible positive signed integer
ans = 32767
>> b = uint16(a) % a is within the range of an unsigned 16 bit number
b = 40000
>> c = -98; % c is within the range of a signed 8 bit number
>> int8(c)
ans = -98
>> % c is out of range of an unsigned 8 bit number
>> uint8(c) % will result in smallest possible unsigned integer
ans = 0
>> int8(-37) + int16(2304) % both numbers are within their respective ranges
??? Error using +
Integers can only be combined with integers of the same class, or scalar doubles.
>> d = [-68 57 121]; % each integer element of d uses 8 bytes of memory
>> e = int8(d)
```

```
e = -68  57  121 % each integer element of e uses 1 byte of memory
>> d + e % this will not work, because d and e are not of the same class
??? Error using +
Class of operand is not supported.
>> whos % check to see the class of each variable
Name     Size      Bytes   Class    Attributes
  a       1x1          8    double
  ans     1x1          1    uint8
  b       1x1          2    uint16
  c       1x1          8    double
  d       1x3         24    double
  e       1x3          3    int8
```

If you know the class and range of numbers you will use in a program, then it is useful to use the least amount of memory required for these numbers, as this will also decrease execution time. Table 5.6 gives a list of the data classes that MATLAB supports.

Table 5.6   MATLAB data classes

Class name	Class description
double	Double precision floating point numeric matrix
logical	Logical matrix
char	Character array
single	Single precision floating point numeric matrix
float	Single or double precision floating point numeric matrix
intn	n-Bit signed integer matrix, replace n with 8, 16, 32, or 64, as in int8
uintn	n-Bit unsigned integer matrix, replace n with 8, 16, 32, or 64, as in uint8
integer	Any of the eight integer matrix classes
numeric	Integer or floating point matrix
cell	Cell array (described in Chapter 7)
struct	Structure array (described in Chapter 7)
function_handle	Function handle

So far, a variety of data types have been introduced, and in the following chapters additional data types will be introduced. A MATLAB variable can be a scalar, vector, matrix, or array of any of the data types mentioned in Table 5.6 and more, as we will see. In a programming environment, it is useful to have a term that is nonspecific, but means any variable of any data type. The term **object** is commonly used for this purpose. Thus, for example, a matrix of floating point numbers, a character string, a plot, the handle of a function, etc. are referred to as objects. The built-in function **class** returns the class of an object, and its syntax is given by

```
X = class(object)
```

where the variable X is assigned a character string that is one of the class names given in Table 5.6. For example:

```
>> A = [2.5 pi;atan(pi/4) -3]; % the matrix A is an object
>> X = class(A)
X =   double
>> class(X) % X is an object
ans =   char
```

Another commonly used number base is hexadecimal (base 16). Table 5.7 gives the digits used in this base. Hexadecimal uses the base ten digit symbols for the first ten hexadecimal digits and the remaining hexadecimal digit symbols are the letters a through f. It is easy to convert a binary integer to hexadecimal by starting at the right, group every four binary digits, get the base ten value, but write it in hexadecimal. For example, N in base 2 = 1011101001011110111 = 101, 1101, 0010, 1111, 0111 → 5, 13, 2, 15, 7 in base 10 › 5d2f7 in base 16. For a binary fraction, start at the left and make groups of four binary digits. In reverse it is easy to convert a base 16 number to binary. Or, go directly from base 2 to base 16 using 4 bits at a time. Base 16 is used to reduce output print space. Here, 19 characters are required to print N in binary, while only 5 characters are required to print N in base 16. Base 16 is often used to print (communicate) binary numbers.

Table 5.7   Hexadecimal (base 16) digits

0	1	2	3	4	5	6	7	8	9	10	11	12	13	14	15
0000	0001	0010	0011	0100	0101	0110	0111	1000	1001	1010	1011	1100	1101	1110	1111
0	1	2	3	4	5	6	7	8	9	a	b	c	d	e	f

Note: First row is base ten, second row is base 2, and third row is base 16.

5.2.4   Binary Arithmetic

MATLAB does not explicitly support arithmetic with numbers given in binary notation. In MATLAB a binary number is specified with a character string.

Example 5.4 _____

```
>> a = '10010111'  % assign a character string, which represents a binary number, to a
a =  10010111
>> isfloat(a) % a is not a floating point number
ans =   0
>> isnumeric(a)  % a is not a number
```

```
ans =   0
>> islogical(a)   % a is not a logical variable
ans =   0
>> whos
  Name      Size      Bytes    Class     Attributes
    a        1x8         16     char
   ans       1x1          1     logical
>> a(1)   % a is an 8 element character vector
ans =   1
>> a(2)
ans =   0
>> % convert a into an integer vector of ASCII codes
>> b = uint8(a)
b =   49  48  48  49  48  49  49  49
>> % 49 and 48 are the ASCII codes for 1 and 0, respectively
>> isnumeric(b)   % b is a numeric vector
ans =  1
```

Table 5.8 gives a list of the many built-in MATLAB functions that work with binary character strings.

Table 5.8   MATLAB functions concerned with binary
*   character strings*

Function	Activity
bin2dec	Convert binary string to decimal number
bitand	Bitwise AND
bitcmp	Bitwise complement
bitget	Get bit at specified position
bitmax	Maximum double precision floating point integer
bitor	Bitwise OR
bitset	Set bit at specified position
bitshift	Shift bits specified number of places
bitxor	Bitwise XOR
dec2base	Convert decimal to base N number in string
dec2bin	Convert decimal to binary number in string
dec2hex	Convert decimal to hexadecimal number in string
int2str	Convert integer to string
num2str	Convert number to string
str2num	Convert string to number
strcmp	Compare strings, case sensitive
strcmpi	Compare strings, not case sensitive

Example 5.5 _____

As you go through this example, keep in mind that most of these functions expect unsigned integer inputs and they return unsigned integer outputs. To see a result in binary, the integer result must be converted to a binary character string. Prog. 5.2 exercises some of the functions given in Table 5.8, and the results follow.

```
clear all; clc
a_bin = dec2bin(47) % convert integer to a binary character string, a vector
b_bin = '10011010' % assign a binary character string to b_bin, a vector
% you can also use b_bin = int2str(10011010)
c_int = bin2dec(b_bin) % binary to decimal conversion
d_int = bin2dec(a_bin) + c_int % add integers and get integer result
d_bin = dec2bin(d_int) % convert to binary character string
d_bin_first_bit = bitget(d_int,1) % get first bit
% Notice that for the vector b_bin, the first element is on the left, while
% the first bit of d_int is on the right. See the program output.
d_bin_second_bit = bitget(d_int,2) % get second bit
d_bin_complement = dec2bin(bitcmp(d_int,8)) % complement first 8 bits
d_bin_bit_six_set = dec2bin(bitset(bin2dec(d_bin),6,1)) % set bit 6 to 1
d_2_byte = uint16(bin2dec(d_bin)) % make a 2 byte integer
d_shifted_left = dec2bin(bitshift(d_2_byte,2)) % multiply by 4
% if 2 is replaced by -2, a right shift occurs, which is a divide by 4
whos
```

Program 5.2 Demonstration of various MATLAB binary operations.

```
a_bin =   101111  % 47 in decimal
b_bin =   10011010
c_int =   154
d_int =   201  % this is 47 + 154
d_bin =   11001001  % 201 in binarry
d_bin_first_bit =   1  % least significant bit of d_bin
d_bin_second_bit =   0  % next bit
d_bin_complement =   110110  % bitwise complement
d_bin_bit_six_set =   11101001  % 6th bit of d_bin set to 1
d_2_byte =   201  % convert d_bin to a 2 byte integer to make room for a shift to the left
d_shifted_left = 1100100100 % a shift to the left by 2 bits is like multiplying d_bin by 4
```

Name	Size	Bytes	Class	Attributes
a_bin	1x6	12	char	
b_bin	1x8	16	char	
c_int	1x1	8	double	
d_2_byte	1x1	2	uint16	
d_bin	1x8	16	char	

```
d_bin_bit_six_set      1x8        16      char
d_bin_complement       1x6        12      char
d_bin_first_bit        1x1         8      double
d_bin_second_bit       1x1         8      double
d_int                  1x1         8      double
d_shifted_left         1x10       20      char
```

Notice the various class types. Even 1 bit is stored using 8 bytes. If memory space is critical, then use **uint8** to save space. For example, d_bin_first_bit = uint8(bitget(d_int,1)) uses 1 byte of memory.

5.2.5   *Floating Point Notation*

MATLAB uses two floating point formats as specified by the IEEE-754 standard. They are **single precision** with 32 bits (word length is 4 bytes) and **double precision** with 64 bits (word length is 8 bytes).

5.2.5.1   **Single Precision**

Denote the 32 bits of single precision with d_k, $k = 0, 1, \ldots, 31$, where d_{31} is the MSB and d_0 is the LSB. These 32 bits, which can be written with 8 hexadecimal digits, are used to represent a floating point number. In binary, a normalized 32-bit floating point number is given by

$$N = (-1)^S \times 1.f_{-1}f_{-2}f_{-3}\cdots f_{-21}f_{-22}f_{-23} \times 2^G \tag{5.6}$$

where S is the **sign bit**, 0 for positive or 1 for negative, which is assigned to d_{31}, $1.f_{-1}f_{-2}f_{-3}\cdots f_{-21}f_{-22}f_{-23}$, called the **significand**, are 24 bits, and the 23 fraction bits, denoted by F, are assigned to $d_{22}, \ldots, d_0$, and G, a signed integer, is restricted to the range from -126 to $+127$. The 23 fraction bits F of the significand determine the fraction given by

$$f_{-1} \times 2^{-1} + f_{-2} \times 2^{-2} + \cdots + f_{-22} \times 2^{-22} + f_{-23} \times 2^{-23}$$

Since the digit 1 in the significand of (5.6) is always assumed in a normalized number, it is not assigned to any d_k. This bit is called the **hidden bit**, and it gives the 32-bit format an extra bit of precision. The hidden bit is inserted when N is reconstructed from d_k, $k = 0, 1, \ldots, 31$. With the 24 bits of the significand, 32-bit floating point notation gives a precision of 24 bits, which is equivalent to about 7.2 base ten digits.

For example, $X = -379.46 \rightarrow -101111011.01110101110000101\ldots$, where the binary fraction does not terminate. Normalizing X gives $-1.01111011011110101110000101\cdots \times 2^8$, where 25 bits of the fractional part are shown. The fraction is reduced to 23 bits by **rounding**. There are many rounding methods. One possibility is to **truncate** the fraction (delete bits) after 23 bits, which is rounding toward zero. The IEEE-754 standard uses rounding to the nearest value. To demonstrate this, suppose we have a fraction, and move the binary point to the right of the 23rd bit to get, for example, $Y = 01111011011101011100001.01\ldots$, and here the fractional part is less than 0.1. Now, round Y to the nearest integer resulting in 01111011011101011100001. After rounding, the normalized 32-bit floating point number becomes $N = (-1)^1 \times 1.011110110111$ 01011100001×2^8. Suppose instead that for a different X, $Y = 01111011011101$ $011100001.11\ldots$, where the fractional part is greater than 0.1. Rounding this Y to the nearest integer gives 01111011011101011100010. In the event that the fractional part of Y is exactly 0.1, called a tie, then Y is rounded to the nearest even integer. Also, if the integer part of Y is all 1s, then Y is rounded to an integer by deleting the fractional part.

Regarding the exponent G, there are several commonly used ways to write a positive or negative integer in binary. One way is called **sign magnitude** notation, where the MSB is set to 0 or 1 if the number is positive or negative, respectively, and the remaining bits are used to denote the magnitude of the number in binary. For example, using 8 bits, the negative integer -15 is written as 10001111, where 7 bits are used for the magnitude. Note that 8 bits would not be enough bits to represent, for example, -317 in signed magnitude notation, where instead, at least 10 bits are required.

Given the allowed range of G, 8 bits are enough to write G in sign magnitude notation. However, according to the IEEE-754 standard, the 8 bits $d_{30} \cdots d_{23}$, which account for the signed exponent G, are determined by $E = G + b$, where $b = +127$. Thus, E ranges from 1 to 254, which can be written in binary with 8 bits $e_7 e_6 \cdots e_1 e_0$ that are assigned to $d_{30} \cdots d_{23}$. The number b is called a **bias**. $E = 0$ and $E = 255$ are reserved (to be explained soon) for a purpose other than representing N with (5.6). A single precision floating point number is stored in memory as

$$N \equiv \begin{cases} d_{31} & d_{30} & d_{29} \cdots d_{23} & d_{22}\, d_{21}\, d_{20} \cdots d_1 & d_0 \\ S & e_7 & e_6 \cdots e_0 & f_{-1}\, f_{-2}\, f_{-3} \cdots f_{-22} & f_{-23} \end{cases} \qquad (5.7)$$

With (5.6) stored according to (5.7), there is no way to represent the number 0. Therefore, the rules to get the value of N are as follows:

- If $E = 0$ and F is nonzero, then $N = (-1)^S \times (0.F) \times 2^{(-126)}$, unnormalized values.
- If $E = 0$, $F = 0$, and $S = 1$, then $N = -0$, the hidden bit is replaced by 0.
- If $E = 0$, $F = 0$, and $S = 0$, then $N = +0$, the hidden bit is replaced by 0.
- If $0 < E < 255$ then $N = (-1)^S \times (1.F) \times 2^{(E-b)}$, recall that the 1 in $1.F$ is implicit.
- If $E = 255$ (all 1s) and F is nonzero, then $N = $ NaN ("Not a number").
  This serves as notation for the case: 0/0, which is indeterminate.
- If $E = 255$, $F = 0$ (all 0s), and $S = 1$, then $N = -$ infinity.
  This serves as notation for the case: negative number/0.
- If $E = 255$, $F = 0$, and $S = 0$, then $N = +$infinity.
  This serves as notation for the case: positive number/0.

For example:

$0\ 11111110\ 11111111111111111111111 \rightarrow +1 \times 1.11111111111111111111111 \times 2^{(254-127)}$

$\qquad\qquad = 1.11111111111111111111111 \times 2^{127}$, largest positive value

$0\ 10000000\ 00000000000000000000000 \rightarrow +1 \times 1.0 \times 2^{(128-127)} = 2$

$0\ 01111111\ 00000000000000000000000 \rightarrow +1 \times 1.0 \times 2^{(127-127)} = 1$

$0\ 00000001\ 00000000000000000000000 \rightarrow +1 \times 1.0 \times 2^{(1-127)} = 2^{(-126)}$

$0\ 00000000\ 10000000000000000000000 \rightarrow +1 \times 0.1 \times 2^{(-126)} = 2^{(-127)}$

$0\ 00000000\ 00000000000000000000001 \rightarrow +1 \times 0.00000000000000000000001 \times 2^{(-126)} =$

$\qquad\qquad\qquad\qquad\qquad\qquad\qquad\qquad 2^{(-149)}$, smallest positive value

5.2.5.2 Double Precision

In (5.6), the significand of double precision consists of 53 bits and G ranges from -1022 to $+1023$. Double precision is more precise than twice the precision of single precision. In (5.7), $b = +1023$, and E, which consists of 11 bits, ranges from 1 to 2046. $E = 0$ and $E = 2047$ are reserved, similar to single precision. The remaining rules for finding N are similar to the single precision rules.

5.3 Logic Gates

Electronic circuits that behave as described by Table 5.1 are called **logic gates**. These circuits are fabricated with resistors, diodes, and transistors using integrated circuit (IC) technology. Fig. 5.1 shows how inputs and outputs are connected to a logic gate. The constant voltage source supplies the power, $P = V\,i(t)$ watts to make the electronic circuit work. There are two input voltages $v_x(t)$ and $v_y(t)$ that can be logic 0 (0 volts) or logic 1 (5 volts). The design of the electronic circuit must satisfy an important requirement. Regardless of the value of the input voltages, the currents $i_x(t)$ and $i_y(t)$ must be very small, for example, less than a few µA. Then, the power supplied by the input voltage sources $v_x(t)$ and $v_y(t)$ will also be very small. Suppose the circuit is designed to perform the **AND**

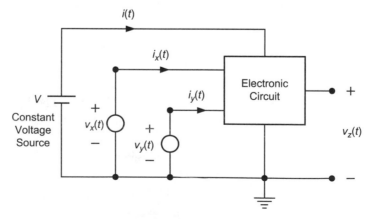

Figure 5.1   Circuit diagram for connection to a logic gate.

operation. Then, $v_z(t) = 5$ volts (logic 1) only if $v_x(t) = 5$ volts (logic 1) and $v_y(t) = 5$ volts (logic 1), and otherwise, $v_z(t) = 0$ volts (logic 0).

It is conventional that logic circuit diagrams do not include the details shown in Fig. 5.1. Instead, only the symbol for a gate is given, with the names of the inputs and outputs. This is shown in Fig. 5.2. In these drawings it is presumed that a constant voltage source supplies power to the circuits, that the inputs are voltages from the input terminals to the negative side of the constant voltage source, and that the output is a voltage from the output terminal to the negative side of the constant voltage source. Furthermore, the inputs and outputs are interpreted to be logical variables.

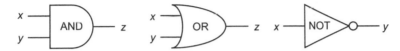

Figure 5.2   Symbols for the and, or, and not gates.

The input voltages to a logic circuit can change with time. For example, Fig. 5.3 shows two input voltages applied to an **OR** gate and the output voltage as the input voltages change with time. Notice that the output does not change exactly at the same time that the inputs change. Ideally, an output should change exactly when an input changes. However, for physical reasons the output of a real circuit cannot change instantaneously with its inputs. The time required for the output to respond to inputs is called **propagation delay**. This and other kinds of propagation delays limit the speed of a digital system.

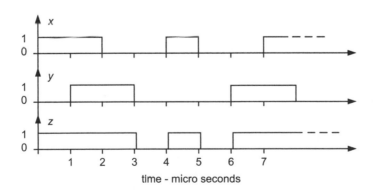

Figure 5.3   Timing diagram of inputs to and output from an OR gate.

In addition to the basic gates depicted in Fig. 5.2, which perform the logic operations given in Table 5.1, there are basic gates that perform the logic operations given in Table 5.3. The NAND operation gives $z = \overline{(x \bullet y)}$. This can be achieved by connecting the output of an AND gate to the input of a NOT gate, which outputs z. The NOR operation gives $z = \overline{(x + y)}$. This can be achieved by connecting the output of an OR gate to the input of a NOT gate, which outputs z. The XOR operation gives $z = x \oplus y$, and soon we will see how this can be achieved with an interconnection of basic gates. The symbols for the NAND, NOR, and XOR gates are shown in Fig. 5.4. The small circle included in the NAND and NOR gates is called a **bubble**, and it is notation for the complement (NOT) operation.

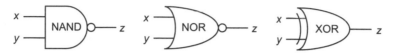

Figure 5.4   Symbols for the NAND, NOR, and XOR gates.

A bubble can be used by itself to indicate a complement (NOT) operation. For example, the equivalent logic circuits shown in Fig. 5.5 have the same output given by

$$y = \overline{(\bar{a}b) + \bar{c}}$$
$$= \overline{(\bar{a}b)}\,c, \quad \text{by DeMorgan's theorem}$$
$$= (a + \bar{b})c, \quad \text{by DeMorgan's theorem}$$

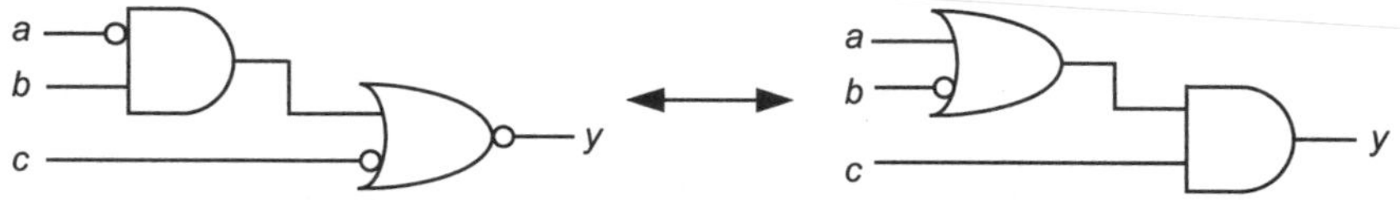

Figure 5.5   Example of using a bubble to complement a variable.

Notice how the logic circuit determines how to write the Boolean expression for y. This also shows that axioms and theorems of Boolean algebra can be useful to modify and possibly simplify a logic circuit.

5.4   Boolean Functions

A Boolean function f of K logic variables is a rule that describes how to assign a logic value to f given a logic assignment to the K variables. There are 2^K different logic assignments to the K logic variables. However, f need not be defined for all possible logic assignments. A Boolean function can be given by a logic statement involving AND, OR, and NOT operations with parentheses to provide precedence or by a truth table.

Example 5.6

Given is the Boolean function

$$f(a, b, c) = \bar{a}(b + \bar{c}) + a\,\bar{b}\,c$$

This is a Boolean function of $K = 3$ logic variables. There are $2^K = 8$ possible combinations (binary assignments) of a, b, and c, which are $abc = 000, 001, 010, 011, 100, 101, 110,$ and 111.

Notice that the binary assignments are a count in binary from 0 to $7 = 2^K - 1$. For each binary assignment the function rule describes how to find the function logic value. Another way to describe the relationship between the three logic variables and the function value is with the truth table shown below, where each binary assignment and the corresponding function value are listed.

Truth Table

a	b	c	$f(a,b,c)$
0	0	0	1
0	0	1	0
0	1	0	1
0	1	1	1
1	0	0	0
1	0	1	1
1	1	0	0
1	1	1	0

A MATLAB script to generate this table follows.

```
clear all; clc
a = logical([0 0 0 0 1 1 1 1]); % all assignments to a, b and c
b = logical([0 0 1 1 0 0 1 1]);
c = logical([0 1 0 1 0 1 0 1]);
for k = 1:8
   f(k) = ~a(k)&(b(k)|~c(k))|(a(k)&~b(k)&c(k));
end
% the for loop could be replaced with f = ~a&(b|~c)|(a&~b&c)
disp('Truth Table')
disp(' a  b  c  f(a,b,c)')
table = [a;b;c;f];
fprintf(' %i  %i  %i    %i \n',table)
```

It is easy (maybe tedious if you do this manually) to find a truth table given a Boolean function. Soon, we will see how to find a Boolean function given a truth table. In practice, however, we usually start with a description of some problem and a desired resolution of the problem. With this starting point we must find a truth table or a Boolean function that resolves the problem.

Example 5.7 ——————————————————————————————

Given are two 4-bit binary numbers a and b. Find a Boolean function $f(a,b)$ that is logic 1 if the binary numbers a and b are equal and logic 0 if the binary numbers are unequal.

Denote the binary number a with $a_3\ a_2\ a_1\ a_0$. In MATLAB, we can write a = '$a_3\ a_2\ a_1\ a_0$', where, for example, a = '1 0 0 1', $a(1) = a_3 = 1$, which defines a as a character string. Define b in a similar way. The numbers a and b are equal if $a(1) = b(1)$, and $a(2) = b(2)$, and $a(3) = b(3)$, and $a(4) = b(4)$.

If instead, a and b are defined as logical vectors given by

$$a = \text{logical}([a_3\ a_2\ a_1\ a_0])$$
$$b = \text{logical}([b_3\ b_2\ b_1\ b_0])$$

then the logical operations EQ (equivalence) and AND can be used, and the Boolean function $f(a,b)$ is given by

$$f = (a(1) \odot b(1))(a(2) \odot b(2))(a(3) \odot b(3))(a(4) \odot b(4)) \tag{5.8}$$

Here, f is a Boolean function of $K = 8$ logic variables, and a truth table would require $2^8 = 256$ rows of binary assignments to a and b. This logical expression for f specifies how to write a MATLAB function, where the inputs a and b are defined as logical binary numbers. A MATLAB function is given in Prog. 5.3.

```
function equal = binary_equality(a,b)
% Boolean function to compare two binary numbers for equality
% a and b can be logical or numeric binary vectors
a_logic = logical(a); b_logic = logical(b); % logical vectors
J = length(a_logic); K = length(b_logic); % get lengths
N = J; % assume length J
if J > K
   b_logic = [false(1,J-K),b_logic]; % add leading logic 0s to b_logic
elseif J < K
   N = K;
   a_logic = [false(1,K-J),a_logic]; % add leading logic 0s to a_logic
end
equal = true; n = 1; % initially assume equal
while equal && (n <= N)
   equal = equal & (~xor(a_logic(n),b_logic(n))); % EQ equals ~XOR
   n = n+1;
end
```

Program 5.3   Function to check if two logical vectors are equal.

With this function we get

```
>> x = logical([1 0 1 1]);
>> y = logical([1 1 0 0]);
>> z = [0 0 1 0 1 1];  % z is a numeric vector
>> x_y = binary_equality(x,y)  % x and y are not equal binary vectors
x_y =   0
>> x_z = binary_equality(x,z)  % leading zeros in z do not change the binary value of z
x_z =   1
>> whos
  Name     Size       Bytes      Class     Attributes

  x         1x4          4        logical
  x_y       1x1          1        logical
  x_z       1x1          1        logical
  y         1x4          4        logical
  z         1x6         48        double
```

The Boolean expression for f given in (5.8) also specifies how to design (build) a logic circuit for f, which is shown in Fig. 5.6. This is called a **realization** of the Boolean function, and Prog. 5.3 simulates the logic of this circuit.

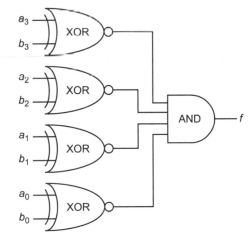

Figure 5.6 Realization of the equivalence Boolean function given in (5.8).

If a and b are defined as character strings, then the function f is given by $f =$ strcmp(a,b). However, this form for f provides no insight about designing a logic circuit.

Any Boolean function can be realized with only AND, OR, and NOT gates. Let us look at the truth table for an XOR gate, which is shown below.

x	y	$x \oplus y$	$x \odot y$
0	0	0	1
0	1	1	0
1	0	1	0
1	1	0	1

In the second row where $x = 0$, $y = 1$, and $x \oplus y = 1$, the term $\bar{x}\,y$ is logic 1. For all other logic assignments to x and y, the term $\bar{x}y$ is logic 0. In the third row where $x = 1$, $y = 0$, and $x \oplus y = 1$, the term $x\bar{y}$ is logic 1. For all other logic assignments to x and y, the term $x\bar{y}$ is logic 0. Therefore, Boolean functions for $x \oplus y$ and $x \odot y$ are given by

$$\begin{aligned} x \oplus y &= \bar{x}y + x\bar{y} \\ \overline{x \oplus y} &= x \odot y = \bar{x}\,\bar{y} + x\,y \end{aligned} \tag{5.9}$$

where the rows in which $x \odot y = 1$ were used to write it in (5.9). With (5.9), the circuit in Fig. 5.6 can be built using only AND, OR, and NOT gates.

With Prog. 5.3 as an example, it should not be difficult to write MATLAB functions that do bitwise AND, NAND, OR, NOR, and more, where the inputs are logical vectors. This will lead to the kinds of logic circuits that are realized within the CPU (central processing unit) of a computer.

Now consider binary addition. Given are two binary numbers a and b, for example, $a = [1\ 1\ 0\ 1]$ and $b = [1\ 1\ 1]$. To add them, line them up in two rows, and start addition in the right most (LSB, least significant bit) column, as shown in Fig. 5.7. Here, $a_0 + b_0 = 10$, which is 2 in binary, and therefore, the sum bit is $s_0 = 0$ and the carry bit into the column to the left is $c_1 = 1$. In the next column, we have $c_1 + a_1 + b_1 = 10$; $s_1 = 0$ and $c_2 = 1$. In the next column, we have $c_2 + a_2 + b_2 = 11$, which is 3 in binary, giving $s_2 = 1$ and $c_3 = 1$. This continues until we get the 5-bit sum. The mechanics of binary addition are the same as the mechanics of base ten addition.

Given a_k and b_k, let us consider addition in the kth column, but without a carry bit from the $(k-1)^{\text{th}}$ column. The table in Fig. 5.8 shows the possibilities. By inspection of this

$$\begin{array}{ccccc} c_4 & c_3 & c_2 & c_1 & \\ & a_3 & a_2 & a_1 & a_0 \\ + & b_3 & b_2 & b_1 & b_0 \\ \hline s_4 & s_3 & s_2 & s_1 & s_0 \end{array} \longrightarrow \begin{array}{ccccc} 1 & 1 & 1 & 1 & \\ & 1 & 1 & 0 & 1 \\ + & 0 & 1 & 1 & 1 \\ \hline 1 & 0 & 1 & 0 & 0 \end{array}$$

Figure 5.7 Addition of two binary numbers.

table, the sum bit is given by $s_k = a_k \oplus b_k$, and the carry bit into the next column is $c_{k+1} = a_k b_k$. This results in the logic circuit shown in Fig. 5.8, which is called a **half-adder** (HA).

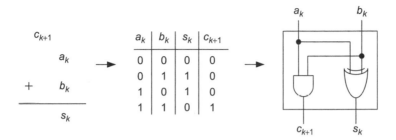

Figure 5.8   Binary addition in the kth column with $c_k = 0$.

For full addition in the kth column, we must take into account the carry bit c_k from the $(k-1)^{th}$ column. Given a_k, b_k, and the carry bit c_k into the kth coulmn from the $(k-1)^{th}$ column, Table 5.9 shows the sum bit s_k and the carry bit c_{k+1} into the $(k+1)^{th}$ column. This is full addition.

Table 5.9 k^{th} column binary addition

a_k	b_k	c_k	s_k	c_{k+1}
0	0	0	0	0
0	0	1	1	0
0	1	0	1	0
0	1	1	0	1
1	0	0	1	0
1	0	1	0	1
1	1	0	0	1
1	1	1	1	1

From the rows where s_k is logic 1, the Boolean function for s_k is given by

$$s_k = \bar{a}_k \bar{b}_k c_k + \bar{a}_k b_k \bar{c}_k + a_k \bar{b}_k \bar{c}_k + a_k b_k c_k \tag{5.10}$$

Here, each of the four terms that are ORed comes from a row where $s_k = 1$, and they can be formed by an inspection of the binary assignment in each of these rows. Furthermore, each term looks as if the variables or their complements are being multiplied. Each term is called a **product term** (also called a **minterm**). The product terms are ORed to form the function

s_k, and this looks like addition. Therefore, this structure of the function s_k is called a **sum-of-products** form, which can be written by inspection of the truth table. Applying the distributive property to the second and third minterms and also to the first and fourth minterms gives

$$s_k = (\bar{a}_k b_k + a_k \bar{b}_k)\bar{c}_k + (\bar{a}_k \bar{b}_k + a_k b_k)c_k \tag{5.11}$$

In view of (5.11), let X_k be defined by $X_k = \bar{a}_k b_k + a_k \bar{b}_k = a_k \oplus b_k$, which is one of the outputs of a half-adder, and then (5.11) becomes

$$s_k = X_k \bar{c}_k + \bar{X}_k c_k = c_k \oplus X_k \tag{5.12}$$

which can be obtained with a second half-adder. From the rows where c_{k+1} is logic 1, a sum-of-products form of a Boolean function for c_{k+1} is given by

$$\begin{aligned} c_{k+1} &= \bar{a}_k b_k c_k + a_k \bar{b}_k c_k + a_k b_k \bar{c}_k + a_k b_k c_k \\ &= (\bar{a}_k b_k + a_k \bar{b}_k)c_k + a_k b_k (\bar{c}_k + c_k) \\ &= X_k c_k + a_k b_k \end{aligned} \tag{5.13}$$

With (5.12) and (5.13), a logic circuit can be designed, where the inputs are a_k, b_k, and c_k, and the outputs are s_k and c_{k+1}. The logic circuit, which is called a **full-adder**, is shown in Fig. 5.9.

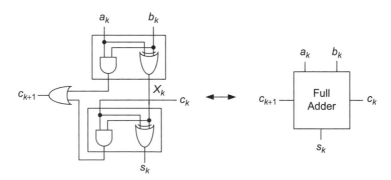

Figure 5.9   Logic circuit for addition obtained with Table 5.8.

In view of (5.9), a full adder can be fabricated using AND, OR, and NOT gates.

Example 5.8 ──────────────────────────────────

Design a logic circuit that can add two 4-bit numbers. Using full-adders, Fig. 5.10 gives a logic circuit that adds two 4-bit numbers. Since $c_0 = 0$, the least significant bit (LSB) full-adder can be replaced by a half-adder. This adder design is called a **serial adder**. Assume that all inputs are applied at the same time. Due to propagation delay, the carry bit c_1 cannot be valid immediately. This means that the sum bit s_1 cannot be valid until after c_1 has become valid. This delay action propagates serially through each full adder until finally s_4 becomes valid.

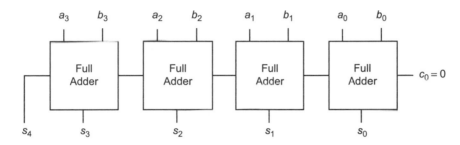

Figure 5.10   A 4-bit serial adder logic circuit using AND, OR, and NOT gates.

The logic of this circuit is simulated by Prog. 5.4, which is a function that includes a sub-function. Some tests of this function follow the program.

```
function binary_sum = binary_adder(a,b)
% This function does binary addition of the logic vectors a and b
% The output is the logic vector binary_sum
% Uses a subfunction to perform the operation of a half-adder
a_logic = logical(a); b_logic = logical(b); % logical vectors
J = length(a_logic); K = length(b_logic); % get lengths
N = J; % assume length J
if J > K
   b_logic - [false(1,J-K),b_logic]; % add leading logic 0s to b_logic
elseif J < K
   N = K;
   a_logic = [false(1,K-J),a_logic]; % add leading logic 0s to a_logic
end
% binary_sum will have N+1 bits
bin_sum = false(1,N); % preallocate space for the binary sum
ck = false; % first carry in bit
```

```
for k = N:-1:1 % processing a and b from right to left
  % use first half-adder in a full-adder
  [ak_and_bk Xk] = half_adder(a_logic(k),b_logic(k));
  % use second half-adder in a full-adder
  [ck_and_Xk bin_sum(k)] = half_adder(ck,Xk);
  ck = ak_and_bk | ck_and_Xk; % carry out bit
end
if ck % check if there is a final logic 1 carry out bit
  binary_sum = [ck, bin_sum]; % including the most significant bit
else
  binary_sum = bin_sum; % suppress a logic 0 final carry out bit
end
%
function [x_and_y x_xor_y] = half_adder(x,y)
% Evaluates the logic of a half-adder; has two outputs
x_and_y = x & y;
x_xor_y = (~x&y) | (x&~y);
```

Program 5.4   A function that does binary addition using full-adders.

```
>> a = logical([1 0 1 0]);
>> b = logical([0 1 0 1 1]);
>> c = logical([1 0 1]);
>> binary_adder(a,b)
ans =    1    0    1    0    1
>> binary_adder(a,c)
ans =    1    1    1    1
```

A MATLAB program that produces an animation of a serial binary adder is given in Chapter 9.

5.5   Quantization Error

Real world phenomena vary continuously with time over continuous ranges. An **analog (analogous) signal** $v(t)$, a voltage or possibly a current, is called an analog signal, because it varies analogously with some physical phenomenon. An analog signal can have any value in a continuous range, for example, $-V \leq v(t) \leq +V$, $0 < V < 1$.

An audio signal produced by a microphone, a voltage that varies over a continuous range continuously with time, is an example of an analog signal. The voltage is analogous to the variation of the pressure $p(t)$ of the air at the surface of the microphone's diaphragm. This analog signal can be amplified to obtain another analog signal within a range like, for example, the range mentioned above. The pressure $p(t)$ is also an analog signal, as it is analogous to the physical phenomenon. The microphone is a **transducer** that converts the analog signal $p(t)$, which is expressed in psi, to another analog signal expressed as a voltage.

To process an analog signal with a computer, it must be sampled, as described by (3.33). Then, each sampled value must be converted into a binary number. An **analog to digital converter** (**ADC**) is an electronic device that receives a sample value (a voltage) of an analog signal and outputs a binary number representing the sampled signal value. This is depicted in Fig. 5.11, where $T = 1/f_s$ is the time between samples of $v(t)$, and f_s is the sampling rate. This entire process is referred to as **data acquisition**.

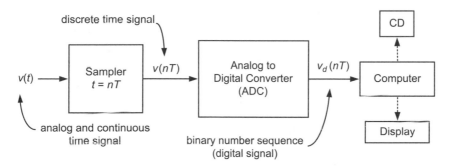

Figure 5.11   Data acquisition (DAQ) system.

If, for example, the goal of a data acquisition system is to produce a music CD, then f_s = 44,100 samples/sec, and each $v_d(nT)$ is a K = 16 bit binary number, where the ADC output consists of K = 16 binary voltages. For CD production, a K = 24 bit ADC is commonly used with a sampling rate of 88,200 samples/sec and even higher. To process the incoming data with algorithms that modify the nature of the sound, each K-bit number received by the computer is converted into a 32-bit floating point number. Modifications such as adding reverberation, enhancing different frequency ranges of the sound, and adding special effects such as flanging are typical, and MATLAB is well suited for program development to do this kind of processing. After processing with 32-bit floating point arithmetic, the resulting data is converted back to 16-bit or 24-bit data at 44,100 samples/sec for recording in a CD.

Computers are used to process all kinds of analog signals, for example, the outputs of sensors in a car, electrocardiograms, microphones in a cell phone, thermometers in a building heating/cooling system, pH meters in a chemical process, and many others. Each of these different applications use an appropriate transducer that converts the behavior of some physical phenomenon to a voltage as it varies continuously in value and continuously with time.

There are many advantages in using a computer to process an analog signal. Two advantages are programming flexibility and the possibility to execute operations that are impossible to do with an electronic circuit that operates directly on $v(t)$. Among others, there is an important concern. Since each binary number in the sequence of numbers $v_d(nT)$ must use a finite number of bits, it is likely that the value of each number in the sequence $v_d(nT)$ is never exactly equal to each number in the sequence of numbers $v(nT)$. The difference is called **quantization error**, and for each n there will be a quantization error, which is given by

$$q(nT) = v(nT) - v_d(nT) \tag{5.14}$$

For example, if for some time point $t = nT$, $v(nT) = +0.86$ V, a finite number of digits, then in sign magnitude notation $v(nT) = 0110111000011010001111\ldots$, (0 for +), which requires an infinite number of bits that must be truncated (or possibly rounded) to obtain $v_d(nT)$. Here, if $K = 16$ bits are used, then $v_d(nT) = 0110111000011010$, a sign bit and a 15-bit fraction. The bits that were discarded to obtain $v_d(nT)$ are the quantization error at the time $t = nT$. With $K = 16$ bits, $v_d(nT)$ can be any one of $2^{16} = 65{,}536$ possible binary numbers.

Let us denote the sequence of numbers $v(nT)$ by $\{v(nT)\}$, and similarly for the number sequences $v_d(nT)$ and $q(nT)$. Since $\{v_d(nT)\}$ results from truncating the binary representation of $\{v(nT)\}$, $\{q(nT)\}$ is considered to be a random number sequence. Since $v(nT)$ can be positive or negative, it is preferred that the mean of $\{q(nT)\}$ is substantially zero and that $\{q(nT)\}$ has a small variance. This depends mainly on the number K of bits used to obtain $\{v_d(nT)\}$. While increasing K reduces the variance of $\{q(nT)\}$, this may not mean that $\{v_d(nT)\}$ will be a better representation of the physical phenomenon to which $v(t)$ is analogous, as $v(t)$ may include electrical noise and the influence of other unwanted physical phenomena.

Let us rearrange (5.14) to become

$$v_d(nT) = v(nT) - q(nT) \tag{5.15}$$

Here, we can think of $v_d(nT)$ as a distorted version of $v(nT)$, and $q(nT)$ is the source of the distortion. If $v(t)$ is perfectly analogous to the physical phenomenon of interest, such as, for example, the sound made by a singer in a studio, then $q(nT)$ is the only source of distortion. However, this depends on an ideally functioning ADC, which may not be the case.

We must know how to assess the degree of distortion caused by quantization error to select an appropriate K value for a particular application. It could be that any one of $K = 4$, 8, 10, 12, 16, or 24 is suitable for a particular application, which depends on how accurately $\{v_d(nT)\}$ must represent some physical phenomenon. For example, in the production of a music CD, it is preferred to use $K = 24$ for acquiring and processing the data. Recall that in single precision the significand consists of 24 bits, and therefore, it is preferred to use all of this precision. Then, a smaller K is preferred for recording data in a CD to increase the amount of music that can be stored in the CD.

Essentially, an ADC performs the activity discussed in Example 4.5, where a MATLAB program is used to determine which of N sub-intervals of a voltage range contains a given voltage $v(nT)$. Here, an ADC is used to do this, which is illustrated in Fig. 5.12 for $N = 2^K$ and the case $K = 3$. V^+ and V^- are positive and negative voltages from voltage sources that supply the power required by the ADC to work. $+V_{\mathrm{Ref}}$ and $-V_{\mathrm{Ref}}$ are reference voltages that specify the voltage range that is subdivided by the ADC, where the width of each sub-interval is given by

$$Q = \frac{+V_{\mathrm{Ref}} - (-V_{\mathrm{Ref}})}{2^K} = \frac{2V_{\mathrm{Ref}}}{2^K} = \frac{V_{\mathrm{Ref}}}{2^{K-1}} \tag{5.16}$$

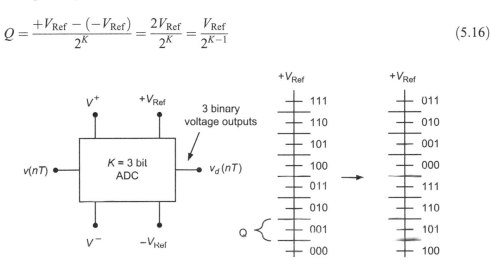

Figure 5.12   Connections of voltage sources to an ADC and the binary output.

The input to the ADC is the sampled voltage $v(nT)$, which is in the range $-V_{\mathrm{Ref}} \leq v(nT) \leq +V_{\mathrm{Ref}}$. V_{Ref} is selected to satisfy $V_{\mathrm{Ref}} \leq V$. It is desired to know which of $N = 2^K$ sub-intervals of the range $[-V_{\mathrm{Ref}}, +V_{\mathrm{Ref}}]$ contains $v(nT)$. Each sub-interval is associated with one of N possible binary numbers, which become the output of the ADC. In Fig. 5.12, where $K = 3$, the output of the ADC consists of three binary voltages grouped together as $v_d(nT)$, which is a binary assignment to each sub-interval. Two kinds of binary assignments are given in Fig. 5.12, a binary count from the most negative to the most positive voltage sub-interval and a modified binary assignment where the MSB indicates a correspondence to negative and positive voltage sub-intervals. A binary assignment of a sub-interval is called a **code** for the sub-interval, and both coding methods are **linear** coding methods. The sub-intervals next to $-V_{\mathrm{Ref}}$ and $+V_{\mathrm{Ref}}$ do not have the same widths as the remaining $N - 2$ sub-intervals, and the reason will be made clear in a moment. These $N - 2$ sub-intervals have a width given by (5.16). The relationship between $v_d(nT)$ and $v(nT)$ is easier to see in Fig. 5.13.

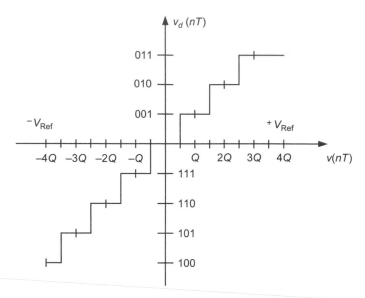

Figure 5.13   Ideal operation of a K = 3 bit analog to digital converter.

The signed binary output of the ADC is shown in Fig. 5.13, and it is given in Table 5.10. There is a one-to-one correspondence between each sub-interval and its binary assignment. If $v(nT)$ is near zero, either positive or negative, then the binary assignment is 000. If the mapping from $v(nT)$ to $v_d(nT)$ is shifted to the right by $Q/2$, then all sub-intervals will have the same width. However, then a small positive $v(nT)$ will produce a different binary assignment than a negative $v(nT)$ near zero, which means that if $|v(\mathrm{nT})|$ is near zero, the binary assignment is not unique, and this is not preferred. Notice that the signed binary output given in Fig. 5.12 distinguishes negative $v(nT)$ from positive $v(nT)$ with the bit on the left, the most significant bit (MSB), which makes it a sign bit. If the binary assignment is instead a straight binary count C from 000 to 111 going up in Fig. 5.12, the binary assignments given in Table 5.10 can be found with $C + 2^{K-1}$, and drop the left most carry bit.

Table 5.10   ADC analog input $\rightarrow$ binary output

Analog input, $v(nT)$	Binary output, $v_d(nT)$
$5Q/2 \leq v(nT) \leq +V_{\mathrm{Ref}}$	011
$3Q/2 \leq v(nT) < 5Q/2$	010
$Q/2 \leq v(nT) < 3Q/2$	001
$-Q/2 \leq v(nT) < +Q/2$	000
$-3Q/2 \leq v(nT) < -Q/2$	111
$-5Q/2 \leq v(nT) < -3Q/2$	110
$-7Q/2 \leq v(nT) < -5Q/2$	101
$-V_{\mathrm{Ref}} \leq v(nT) < -7Q/2$	100

Now that we know what an ADC does, there is the question of how to produce a continuous time signal given a recording of a digital discrete time signal $\{v_d(nT)\}$, a number sequence. For example, a CD player reads a number sequence at the rate of 44,100 samples/sec from a CD and produces a continuous time signal that becomes the sound that you hear. Let us continue to work with $K = 3$, in which case, the binary number sequence from a CD consists of binary numbers from Table 5.10. Suppose that at some time point the number $v_d(nT) = 010$ is retrieved from the CD. From Table 5.10 we know that this occurred because at that time point $3Q/2 \leq v(nT) < 5Q/2$, which does not tell us the actual value of $v(nT)$. Therefore, choose a value for $v(nT)$ to be in the middle of the sub-interval, and denote it by $\hat{v}(nT)$, which is an estimate of $v(nT)$, that is, $\hat{v}(nT) = 2Q$. Since $v(nT)$ could have been anywhere within the sub-interval, the worst estimation error is $\pm Q/2$, which according to (5.14) makes the **maximum quantization error** magnitude, $q_{max} = Q/2$. There is only one sub-interval where the maximum quantization error magnitude is Q instead of $Q/2$, and this occurs in the sub-interval next to $+V_{Ref}$, where if $v_d(nT) = 011$, then $\hat{v}(nT) = 3Q$, where $v(nT)$ could have been near $+V_{Ref}$ (see Fig. 5.13). This quantization error is of no concern, because if, for example, $K = 16$, then it is one sub-interval in 65,536 sub-intervals, and this quantization error is considered to rarely occur.

Based on Fig. 5.13, the quantization error behaves as shown in Fig. 5.14. For example, if $Q/2 \leq v(nT) < 3Q/2$, then $v_d(nT) = 001$, and if $v_d(nT) = 001$, then $\hat{v}(nT) = Q$. Furthermore, if $v(nT) = Q$, then $v_d(nT) = 001$, and $\hat{v}(nT) = Q$, and the quantization error is zero for this particular value of $v(nT)$, as well as any $v(nT) = kQ$, $k = -4, -3, \ldots, 2, 3$

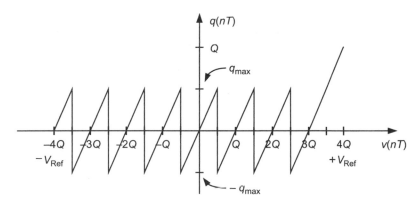

Figure 5.14   Quantization error versus the ADC input for the case $K = 3$.

If $v(nT)$ occurs anywhere within a sub-interval, then its value must change by Q to change the LSB of $v_d(nT)$. The **voltage resolution** of an ADC is given by Q. In Table 5.10, for example, we see that if $v(nT)$ changes in value by Q, then $v_d(nT)$ changes in the LSB. K is called the **resolution** of an ADC, since the range $[-V_{Ref}, +V_{Ref}]$ is subdivided into 2^K sub-intervals. The **dynamic range** of an ADC is the ratio of the largest input signal level change to the smallest input signal level change that the ADC can resolve. Given any input level, the input $v(nT)$ must change by Q to change the LSB of $v_d(nT)$, and the largest input level change is $2^K Q$. Therefore, the dynamic range (DR), which by convention is expressed on a logarithmic scale, is given by

$$DR = 20 \log\left(\frac{2^K Q}{Q}\right) = 20 \log(2^K) \text{ dB}$$

where **dB** is an abbreviation for decibels, a unit used to indicate a logarithmic scale. The **signal variance to noise variance ratio** (SNR) of an ADC is another parameter that is commonly used to assess the performance of an ADC. In (5.15), the signal is $v(nT)$, and the noise is $q(nT)$. As a standard, $v(t) = A \sin(\omega t)$ is used to find the SNR, where $f = 1000$ Hz and $\omega = 2\pi f$.

Example 5.9

Determine the resolution of an ADC such that the maximum quantization error is less than 0.00001% of the full range of the ADC. The full range (also called full scale) is $2V_{Ref}$. Therefore, it is required that

$$q_{max} = \frac{Q}{2} < 10^{-7} \times 2V_{Ref}, \quad Q = \frac{2V_{Ref}}{2^K}, \quad \rightarrow \frac{V_{Ref}}{2^K} < 2 \times 10^{-7} \times V_{Ref} \rightarrow 2^{K+1} > 10^7$$

Since $2^{20} = 2^{10} \times 2^{10} = 1024 \times 1024$, the requirement is satisfied if $K + 1 \geq 24$, where $2^{24} = 2^4 \times 2^{20}$. Therefore, $K \geq 23$. Since a 23-bit ADC is not commercially available, a 24-bit ADC must be used. Suppose that $V_{Ref} = 5$ volts. Then, the voltage resolution of the ADC is given by $Q = 10/2^{24} = 0.596$ μvolts. It is likely that in a real-world situation, electrical noise and other sources of distortion will have amplitudes comparable to this Q, and cause the LSB and possibly bits with higher weightings to not accurately represent the physical phenomenon of interest.

Example 5.10

Suppose that it is known that an analog signal $v(t)$ varies over the range $[-V, +V]$, where $V = 2$ volts. The signal must be sampled and converted into a binary sequence such that the signal resolution is not greater than 1 mV. What ADC resolution is necessary to achieve this requirement? Using (5.16) gives

$$Q = \frac{4}{2^K} \leq 10^{-3} \rightarrow 2^K \geq 4000 \rightarrow K = 12$$

To achieve the required signal resolution, V_{Ref} must be set to $V_{Ref} = 2$ volts, and then the ADC voltage resolution will be the same as the desired signal resolution.

However, if, for example, $V_{Ref} = 5$ volts is used, then the 12-bit ADC voltage resolution is $10/2^{12} = 2.4$ mV. This means that $v(nT)$ must change by 2.4 mV to change the LSB of $v_d(nT)$. Therefore, a change in $v(nT)$ by 1 mV will not necessarily change $v_d(nT)$, and the signal resolution requirement is not satisfied. With $V_{Ref} = 5$ volts, a $K = 14$ ADC must be used to achieve the signal resolution requirement. This means that either V_{Ref} is set to $V_{Ref} = 2$ volts or $v(t)$ is amplified by a factor of 2.5, which changes the signal resolution requirement from 1 mV to an equivalent 2.5 mV.

A continuous time signal $\hat{v}(t)$, which is an estimate of $v(t)$, is produced from $v_d(nT)$ with a digital to analog converter (**DAC**). A DAC is an electronic device that receives a binary number input and it outputs a constant voltage until it receives the next binary number input. For $K = 3$, Table 5.11 lists the activity of a 3-bit DAC. Therefore, $\hat{v}(t)$ is a piecewise constant continuous time signal. This is illustrated in Fig. 5.15.

Table 5.11 DAC digital input → analog output

Digital input, $v_d(nT)$ at $t = nT$	Analog output, $\hat{v}(t)$ over $nT \leq t < (n+1)T$
011	3Q
010	2Q
001	Q
000	0
111	−Q
110	2Q
101	3Q
100	−4Q

Strictly speaking, $\hat{v}(t)$ is not an analog signal, because within each time range $nT \leq t < (n+1)T$ it can have only one of a finite number of values. However, for example, every CD player uses a DAC and additional electronic circuits to smooth out the sudden jumps in $\hat{v}(t)$.

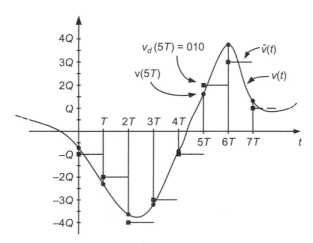

Figure 5.15 Illustration of data acquisition and data reconstruction with $K = 3$.

An inspection of Fig. 5.15 shows that $\hat{v}(t)$ is a poor reconstruction of $v(t)$. K must be increased to reduce the quantization error, and T must be reduced (f_s must be increased) to better follow the rapidly changing $v(t)$.

Example 5.11

Find the dynamic range and signal to noise ratio of an ADC for resolutions given by $K = 4$, 8, 10, 12, 16, and 24 bits. The signal $v(t) = A\sin(\omega t)$, where $f = 1000$ Hz and $\omega = 2\pi f$, will be sampled at the rate $f_s = 44{,}100$ samples/s for 5 secs. To obtain $v_d(nT)$, the binary representation of $v(nT)$ will be truncated with the function given by Prog. 5.5. This function implements the algorithm for converting a base ten fraction to a binary fraction. The built-in function **var** will be used to find a variance.

```
function y = truncate(x,nF)
% Convert a base ten signed fraction x to an nF-bits binary fraction y
% x must be a scalar
nB = nF-1; % the MSB in nF bits is the sign bit
if length(x) > 1, disp('error, input not a scalar'); y = x; return; end
if x == 0, y = 0; return; end
if nB <= 0, disp('error, specified less than 2 bits'); y = x; return; end
mag_x = abs(x); % strip off sign
if x == mag_x, s = 1; else s = -1; end % s is the sign of the input
if mag_x > 1, disp('warning, input greater than one'); y = s; return; end
if mag_x == 1, y = s; return; end
F_x = mag_x; % work with fractional part
y = 0; half_power = 1.0; % initialize truncated fraction
for n = 1:nB % find truncated fraction
   half_power = 0.5*half_power; % used for repeated addition of 2^-n
   two_F_x = 2*F_x; % multiply fraction by 2
   if two_F_x >= 1 % check if not a fraction
      y = y + half_power; % add 2^-n to truncated fraction
      F_x = two_F_x - 1; % get fractional part
   else
      F_x = two_F_x; % two_F_x is a fraction
   end
end
y = s*y; % restore sign
```

Program 5.5   Function to truncate a base ten fraction to an nF-bits binary fraction.

Program 5.6 uses the function truncate to find the SNR for each resolution. A table of results follows the program.

```
% Program to find the SNR and DR of an ADC
clc; clear all
fs = 44100; T = 1/fs; % sampling frequency and time increment
K = [4 8 10 12 14 16 24]; KR = length(K); % KR ADC resolutions
T0 = 5.0; % time range
t = 0.0:T:T0; % time points
N = length(t); % number of samples used for SNR
f = 1000; % frequency of v(t) = A sin(wt)
w = 2*pi*f; A = 0.999; % frequency and amplitude
v = A*sin(w*t); % signal samples
var_v = var(v); % variance of the signal
vd = zeros(1,N); % preallocate space for vd(nT)
for k = 1:KR % loop for resolutions
   for n = 1:N % loop to find truncated fraction of each sample
     vd(n) = truncate(v(n),K(k)); % truncate v(nT)
   end
   q = v - vd; % quantization error
   var_q = var(q); % variance of the quantization error
   SNR(k) = 10*log10(var_v/var_q); % SNR in dB
   DR(k) = 20*log10(2^K(k)); % DR in dB
end
table = [K;SNR;DR]; % table of results
disp('Resolution    SNR(dB)    DR(dB)')
fprintf('   %i      %6.2f    %6.2f   \n',table)
% the format 6.2f means to use 6 print characters with 2 digits
% after the decimal point
```

Program 5.6   **Program to find the SNR and DR of an ADC for $v(t) = A \sin(\omega t)$.**

Resolution	SNR(dB)	DR(dB)
4	19.02	24.08
8	43.74	48.16
10	56.28	60.21
12	67.94	72.25
14	79.79	84.29
16	92.16	96.33
24	140.45	144.49

To put these results into perspective, an SNR = 100 dB means that if the signal is a sound, then the noise sound level is 1/100,000 of the signal sound level.

5.6   Conclusion

The logical data type was introduced in Chapter 4 for decision making in program flow control. In this chapter, the fundamentals of Boolean algebra, which work with logical variables, were presented. It was shown that a Boolean function corresponding to a truth table can be written by inspection of the table. MATLAB was used to evaluate Boolean functions. Boolean algebra was applied to design a logic circuit for binary addition, and with MATLAB the logic circuit was simulated.

Many built-in MATLAB functions were introduced that bridge the different interpretations (numeric, logical, and character) of binary data. Numeric data is structured in integer and floating point formats, both of which are supported in MATLAB. MATLAB has many built-in functions to control the allocation of storage space for binary data. Built-in MATLAB functions for manipulating character strings will be introduced in Chapter 7.

To use MATLAB for the study of real-world signals requires analog to digital conversion. The analog to digital conversion process was studied to understand the impact of quantization error. There is much more to learn about the conversion process. For example, there are ADC encoding methods other than a linear encoding method, which was described here, that can give performance with an 8-bit output code that is almost equivalent to a 13-bit linearly encoded output. This reduces storage space and decreases communication time.

We found that a binary number can have many different meanings.

Now, you should know

- about the fundamentals of Boolean algebra
- how to write truth tables that describe a logical relationship
- how to write a Boolean function from a truth table and use MATLAB to evaluate a Boolean function
- how to realize a Boolean function with basic logic gates
- how to convert a base ten integer or fraction to binary and hexadecimal notation
- about binary arithmetic
- about the built-in MATLAB functions for decimal to binary character string conversion
- how to write a MATLAB program for base ten to binary conversion
- about floating point notation
- about analog to digital conversion and quantization error

Table 5.12 gives the MATLAB functions that were introduced in this chapter. Use the MATLAB help facility to learn more about these built-in functions, where you will also find out about many other related built-in functions.

Table 5.12   Built-in MATLAB functions introduced in this chapter

Function	Brief explanation
&	MATLAB notation for the logical AND operation, as in a & b
|	MATLAB notation for the logical OR operation, as in a | b
~	MATLAB notation for the logical NOT (complement) operation, as in ~a
S = char(X)	Converts the array X that contains nonnegative integers representing character codes into a MATLAB character array S
class(obj)	Returns the name of the class of object obj

A very important kind of number in electrical and computer engineering is the complex number. It is the subject of the next chapter, where MATLAB will be applied for AC (alternating current) circuit analysis.

Further reading

http://www.math.grin.edu/~stone/courses/fundamentals/IEEE-reals.html

ANSI/IEEE Standard754-1985, Standard for binary floating point arithmetic

Problems

Section 5.1

1) Write a MATLAB script that produces a truth table for the AND, OR, EXOR, and EQ operations.
2) Write a MATLAB script that proves both versions of DeMorgan's theorem.

Section 5.2

3) (a) Convert the base ten integers: 99 and 250 to 8-bit binary numbers.
   (b) What is the largest base ten integer that can be converted to an 8-bit binary number?
4) (a) Convert the base ten fractions: 0.2, 0.21, and 0.91 to 8-bit binary fractions.
   (b) What is the smallest base ten fraction that can be converted to an 8-bit binary fraction without error?
5) (a) Give the ASCII codes for the integers: 0 to 9 in base ten and in binary.
   (b) Give the base ten ASCII codes for the name MATLAB. Then, use the function char to obtain the character string, MATLAB, given a vector of its base ten ASCII codes.
6) The standard ASCII code uses 7 bits per code. However, each code is stored as a byte, leaving the eighth bit unused. A useful purpose for the eighth bit is transmission error checking. One possibility is to count the number of ones in a given ASCII code, and if it

is odd(even), make the eighth bit logic one(zero). Therefore, every byte will have an even number of ones. This is called **even parity**, and the eighth bit is called the **parity bit**, denoted by P. When this byte is transmitted, the receiver can count the number of ones, and if the number of ones is not even, then a transmission error must have occurred, prompting the receiver to transmit to the sender an ASCII code which means to request a retransmission of the previously transmitted byte. An alternative to even parity is **odd parity**, where the eighth bit is logic one(zero) to make the total number of ones in a byte an odd number. The parity bit method of transmission error detection assumes that the probability of two transmission errors is very small compared to the probability of one transmission error. To both generate a parity bit and detect correct parity, the circuit given in Fig. P5.6 can be used.

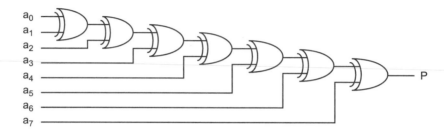

Figure P5.6   Parity bit generator/checker.

Given is a 7-bit ASCII code $d_6d_5d_4d_3d_2d_1d_0$. To generate a parity bit, set the inputs of the parity generator $a_6a_5a_4a_3a_2a_1a_0$ to the ASCII code. Use the input a_7 to select either even or odd parity. Then, P is used as the eighth data bit d_7. For parity detection, another parity generator is used to receive the eight data bits $d_7 \cdots d_0$, and the output P detects correct parity. Write a MATLAB program to investigate the operation of this circuit for parity bit generation. Try different ASCII codes, and find out how to select a_7 for even or odd parity. Then, your program should check parity by finding P given a byte that contains an ASCII code and a parity bit. Explain how the parity generator works as an even or odd parity checker. Give P to check odd and even parity.

7) In a MATLAB program, a matrix X is known to contain elements that are integers with values over the range -9999 to $+9999$. Give a MATLAB statement that allocates the least amount of memory space necessary to store the matrix X. What will happen if the program contains the statement $Y = X + 50000$?

8) (a) Convert the following base ten numbers: 15, 255, 1023, and 2000 to hexadecimal. Hint: First convert each number to binary.

   (b) Convert the hexadecimal numbers 1e, ab, and 7ff to binary and base ten.

9) Given are the unsigned integers $a = 517$ and $b = 9872$. Assign each integer to a variable using 16 bits for each number. Then, in computer memory the numbers are stored as $d_{15}d_{14} \ldots d_3d_2d_1d_0$.

(a) Give MATLAB statements that use the function bitget to find d_8 of each number.
(b) What numbers result from applying the function bitcmp to each number?
(c) Use the function bitand to find the and of each number with $c = 255$.
(d) Use the function bitshift to multiply the number a by two.
(e) Obtain a binary string for each number.
(f) Use the function dec2base to convert each number to base 8 (octal) and hexadecimal.

10) (a) Give the number -17.375 in 32-bit floating point notation.
(b) Give the 32-bit floating point notation for NaN.
(c) Give the 32-bit floating point notation for: $-$infinity.
(d) To the nearest value, round the fraction 0.1011001011101111110111001 to a 16-bit fraction.

Section 5.3

11) Fig. 5.3 gives the inputs and output timing diagram for an OR gate. Using the same inputs, give the timing diagram of the output of an AND gate.

12) Repeat Prob. 5.11 with an exclusive OR gate.

13) Use the signals x, y, and z given in Fig. 5.3 as the three inputs a, b, and c, respectively, in Fig. 5.5. Find the timing diagrams of the output y for both circuits shown in Fig. 5.5. Assume there are no propagation delays in z and the gates shown in Fig. 5.5.

14) (a) Give the truth table for a NAND gate.
(b) Draw a NAND gate, and connect the two inputs together. In view of the truth table of a NAND gate, does this connection convert a NAND gate into a NOT gate?
(c) Using a NAND gate and a NAND gate converted into a NOT gate give a circuit that works as an AND gate.
(d) DeMorgan's theorem states that $\overline{(a + b)} = \bar{a}\,\bar{b}$. Complement both sides, and then use NAND gates converted into NOT gates and a NAND gate to make a circuit that works like an OR gate.

This problem shows that any Boolean function can be realized with only NAND gates, which is called a **universal gate**.

15) (a) Give the truth table for a NOR gate.
(b) Draw a NOR gate, and connect the two inputs together. In view of the truth table of a NOR gate, does this convert a NOR gate into a NOT gate?
(c) Using a NOR gate and a NOR gate converted into a NOT gate, give a circuit that works as an OR gate.
(d) DeMorgan's theorem states that $\overline{ab} = \bar{a} + \bar{b}$. Complement both sides, and then use NOR gates converted into NOT gates and a NOR gate to make a circuit that works like an AND gate.

Like a NAND gate, a NOR gate is also called a **universal gate**.

Section 5.4

16) Write a MATLAB program that generates the truth table for the Boolean function
$f(a,b,c,d) = a\,\overline{b} + (\overline{a}\,bd + \overline{c}\,d)(\overline{b}\,cd + abd)$.

17) Find a Boolean function $f(a,b)$, where a and b are each 3-bit numbers, that is logic 1 if $a > b$, and logic 0 otherwise. This is a Boolean function of six logic variables. Hint: Start by checking if $a_2 = 1$ and $b_2 = 0$, which gives the first term $a_2\,\overline{b}_2$ in f. Then, if $a_2 = b_2$, check if $a_1 = 1$ and $b_1 = 0$ with $(a_2\,b_2 + \overline{a}_2\,\overline{b}_2)(a_1\,\overline{b}_1)$, and so far we have $f = a_2\,\overline{b}_2 + (a_2\,b_2 + \overline{a}_2\,\overline{b}_2)(a_1\,\overline{b}_1)$. Now, continue adding term(s) to f. With the completed function f, write a MATLAB program that generates a truth table for the Boolean function. Check the truth table to confirm that the Boolean function works as specified.

18) Using the basic gates AND, OR, and NOT, give a realization of the function obtained in Prob. 5.17.

19) A decoder is a device that outputs a logic signal to indicate that the inputs are a particular binary assignment. It usually has another input that enables it to function. A 3×8 decoder has three inputs and eight outputs, where each output is a logic signal that indicates that the three inputs have a particular one of the eight possible input binary assignments. A block diagram is shown in Fig. P5.19, where a design of the decoder has been started.

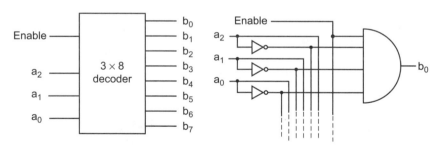

Figure P5.19   A 3 × 8 decoder.

If the enable input is logic 0, then all outputs are logic 0. If the enable input is logic 1, then only one decoder output can be logic 1. So far the circuit works to make b_0 logic 1 if and only if the input binary assignment is $a_2a_1a_0 = 000$. Complete the design of the decoder.

20) Write a MATLAB program that receives from the program user an input D that is any digit, 0–9. Then, the program should obtain its 4-bit binary code, called **binary coded decimal** (BCD), and assign the 4 bits to logic variables w, x, y, and z. Define a matrix of codes, and use the input to make an index to access a code in the table. To display the digit with 7 LEDs (light emitting diodes) arranged as shown in Fig. P5.20, the 4-bit digit code must be the input to seven Boolean functions to determine the on or off state of each diode in the 7-segment LED display.

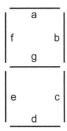

Figure P5.20   Names and configuration of a 7-segment LED display.

The on or off state of each LED is the value of a Boolean function of the logic variables w, x, y and z, for example, a(w,x,y,z), b(w,x,y,z), ... , g(w,x,y,z). Complete the truth table, shown below, for each BCD code. The table shows which LEDs must be on to display the digit corresponding to the given BCD code.

BCD code				Seven segment LED input						
w	x	y	z	a	b	c	d	e	F	g
0	0	0	0	1	1	1	1	1	1	0
0	0	0	1	0	1	1	0	0	0	0
0	0	1	0	1	1	0	1	1	0	1
0	0	1	1	1						
0	1	0	0	0						
0	1	0	1	1						
0	1	1	0	0						
0	1	1	1	1						
1	0	0	0	1						
1	0	0	1	1						

With a completed table, you can find each of the seven Boolean functions to control the on or off state of each diode. For example, the function for diode a is a(w,x,y,z) given by

$$a(w, x, y, z) = \overline{w}\,\overline{x}\,\overline{y}\,\overline{z} + \overline{w}\,\overline{x}\,y\,\overline{z} + \overline{w}\,\overline{x}\,y\,z + \overline{w}\,x\,\overline{y}\,z + \overline{w}\,x\,y\,z + w\,\overline{x}\,\overline{y}\,\overline{z} + w\,\overline{x}\,\overline{y}\,z$$

Another way to write a Boolean function given a truth table is to complement the function, which changes logic 0(1) to logic 1(0). Then, write a sum of products form for the complement of the function. For $\overline{a(w, x, y, z)}$ and then $a(w, x, y, z)$ we get

$$\overline{a(w, x, y, z)} = \overline{w}\,\overline{x}\,\overline{y}\,z + \overline{w}\,x\,\overline{y}\,\overline{z} + \overline{w}\,x\,y\,\overline{z}$$

$$a(w, x, y, z) = \overline{(\overline{w}\,\overline{x}\,\overline{y}\,z + \overline{w}\,x\,\overline{y}\,\overline{z} + \overline{w}\,x\,y\,\overline{z})} = \overline{(\overline{w}\,\overline{x}\,\overline{y}\,z)}\,\overline{(\overline{w}\,x\,\overline{y}\,\overline{z})}\,\overline{(\overline{w}\,x\,y\,\overline{z})}$$

$$= (w + x + y + \overline{z})(w + \overline{x} + y + z)(w + \overline{x} + \overline{y} + z)$$

This will require less hardware to realize a logic circuit, because there are fewer zeros than ones in the truth table for a. However, $a(w, x, y, z)$ can be simplified even further. There are methods, which are beyond the scope of this book, to obtain the most simplified version of a Boolean function.

Give the Boolean functions for the remaining six LEDs in the display.

Continue the MATLAB program, where now the logic variables w, x, y, and z have been assigned values, and evaluate and output each of the seven Boolean functions a through f for the 7-segment LED display.

To repeatedly get a digit input D and find the 7-segment LED display input, place all of the program activity in a while loop, and if the program user enters a number other than an integer in the range 0−9, then terminate the program.

21) (a) Manually, obtain the sum S of the binary numbers A = 10.111011101 and B = 1001100.011.

(b) Define appropriate vectors for A and B, and give MATLAB statements to obtain S.

22) Write a MATLAB function that performs the operation of a half-adder, call it half_add. The function should have two inputs, a and b, and return two outputs, X and c. Then, write another function, call it full_add, that invokes half_add to perform the operation of a full adder. The function full_add should have three inputs, a, b, and c_in, and return two outputs, s and c_out. Finally, write a program that uses the function full_add to find and display s and c_out for each possible combination of a, b, and c_in.

Section 5.5

23) For the case $K = 3$, Fig. 5.13 gives the characteristic of a 3-bit ADC. Write a MATLAB function, call it ADC, that receives V_ref and a number in the range −V_ref to +V_ref, and returns a 3-bit code as given in the figure.

24) Like Fig. 5.13, draw a characteristic of a 4-bit ADC. Then, write a MATLAB function, call it DAC, that receives a 4-bit code and V_ref, and returns a number in the range −V_ref to +V_ref that would be produced by a digital to analog converter.

25) An analog signal varies over the range −1 volt to +1 volt. Assume that $V_{Ref} = 1$ volt. Specify the resolution of an ADC such that the maximum quantization error is less than 0.0001% of full-scale. What will be the resulting voltage resolution of the ADC?

26) For the case $K = 3$, Table 5.11 gives the characteristic of a 3-bit DAC. Write a MATLAB function, call it DAC, that receives a 3-bit logical vector vd and V_ref, and returns a number in the range −V_ref to +V_ref.

CHAPTER 6

Complex Numbers

The complex number is a very useful and important concept in many fields of science and engineering and especially in almost all areas of electrical engineering. An important distinguishing feature of MATLAB® is its ability to work with the complex data type.

After you have completed this chapter, you will know

- about the origin of complex numbers
- fundamental properties of complex numbers
- how MATLAB is particularly well suited for complex number computations
- how complex numbers are used for signal representation
- about the role that complex numbers play in circuit analysis

6.1 Origin of Complex Numbers

Complex numbers were invented as a resolution of the dilemma that can arise when we want to find the solution of

$$f(x) = ax^2 + bx + c = 0 \tag{6.1}$$

where a, b, and c are real numbers and $f(x)$ is a second-order polynomial in x. The values of x for which $f(x) = 0$ are called the **roots** of $f(x)$.

Assuming that a is not zero, (6.1) can be written as

$$x^2 + \frac{b}{a}x + \frac{c}{a} = 0$$

and adding and subtracting $(b/2a)^2$ gives

$$x^2 + \frac{b}{a}x + \left(\frac{b}{2a}\right)^2 - \left(\frac{b}{2a}\right)^2 + \frac{c}{a} = 0$$

This equation can be written as

$$\left(x + \frac{b}{2a}\right)^2 = \left(\frac{b}{2a}\right)^2 - \frac{c}{a} = \frac{b^2 - 4ac}{4a^2} \tag{6.2}$$

The steps from (6.1) to (6.2) are called **completing the square**. Taking the square root of both sides of (6.2) yields

$$x = -\frac{b}{2a} \pm \sqrt{\frac{b^2 - 4ac}{4a^2}} = \frac{-b \pm \sqrt{b^2 - 4ac}}{2a} \tag{6.3}$$

The formula in (6.3) gives the two solutions (roots) of (6.1).

The factor, $b^2 - 4ac$, is called the **discriminant**, and it may be positive, in which case there are two distinct real roots of (6.1). If the discriminant is zero, then there are two real and equal roots. By denoting the two roots by x_1 and x_2, we can write the polynomial $f(x)$ as a product of two first-order factors given by

$$f(x) = a(x - x_1)(x - x_2) = a(x^2 - (x_1 + x_2)x + x_1 x_2)$$

If the discriminant is negative, can $f(x)$ still be written as a product of two first-order factors? In this case, let us write (6.3) as

$$x = \frac{-b \pm \sqrt{b^2 - 4ac}}{2a} = \frac{-b \pm \sqrt{(-1)(4ac - b^2)}}{2a} = \frac{-b \pm \sqrt{-1}\sqrt{4ac - b^2}}{2a}$$

Since $\sqrt{-1}$ cannot have a real value, it is taken into account with a symbol, such as

$$j = \sqrt{-1} \tag{6.4}$$

and $j^2 = -1$. Therefore, if the discriminant is negative, then the two roots of (6.1) are given by

$$x = \frac{-b \pm j\sqrt{4ac - b^2}}{2a} \tag{6.5}$$

and x is called a **complex number**. It is also common (especially in the mathematics literature) to use i for $\sqrt{-1}$. However, in electrical and computer engineering, i is commonly used to mean electric current.

In general, a **complex number** x is given by

$$x = \alpha + j\beta \tag{6.6}$$

where α, a real number, is called the **real part** of x, denoted by $\alpha = \text{Re}(x)$, and β, a real number, is called the **imaginary part** of x, denoted by $\beta = \text{Im}(x)$. This way of writing a complex number is called the **rectangular form**. A complex number can be a real number, in which case the imaginary part is zero, and a complex number can be an **imaginary number**, in which case the real part is zero. A useful operation on a complex number is to change the sign of the imaginary part. If two complex numbers differ only in the sign of their imaginary parts, then these two numbers are said to be complex conjugates of each other. The **complex conjugate** of x in (6.6) is denoted by x^*, and it is given by

$$x^* = (\alpha + j\beta)^* = \alpha - j\beta \tag{6.7}$$

Notice that $(x^*)^* = x$. Also notice that the two complex roots given in (6.5) of a quadratic polynomial with real coefficients occur as a complex conjugate pair. Let us obtain the product of two first-order factors using a pair of complex conjugate roots to get

$$\begin{aligned}(x - (\alpha + j\beta))(x - (\alpha - j\beta)) &= x^2 - x\alpha + jx\beta - \alpha x + \alpha^2 - j\alpha\beta - j\beta x + j\beta\alpha + \beta^2 \\ &= x^2 - 2\alpha x + (\alpha^2 + \beta^2)\end{aligned}$$

As for real roots, the result is a polynomial with real coefficients.

Example 6.1

Let us find the roots of (6.1) for some specific cases.
(a) Let $a = 2$, $b = -2$, and $c = -12$. With (6.3) we get

$$x = \frac{2 \pm \sqrt{4 + 96}}{4} = \frac{2 \pm 10}{4} = 3, -2$$

Let x_1 and x_2 denote the roots obtained with the plus sign and minus sign, respectively. Therefore, $x_1 = 3$ and $x_2 = -2$. With the roots, we can factor the second-order polynomial into a product of two first-order factors given by

$$f(x) = 2x^2 - 2x - 12 = 2(x - x_1)(x - x_2) = 2(x - 3)(x + 2)$$

Since $x = x_1$ and $x = x_2$ make $f(x)$ equal to zero, x_1 and x_2 are also called the **zeros** of $f(x)$.

For another point of view, let us plot $f(x)$ versus x, which is shown in Fig. 6.1. Here we see that the real zeros of a polynomial are the values of x where the polynomial crosses the abscissa, the x-axis, where $y = 0$. The MATLAB script that was used to obtain the plot in Fig. 6.1 is given in Prog. 6.1.

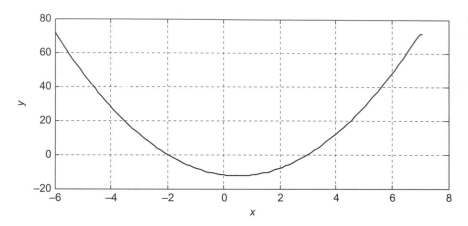

Figure 6.1   Plot of a quadratic function that has two real zeros.

```
% MATLAB program to plot a quadratic function of the form
% y = f(x) = a x^2 + b x + c
% versus x
clear all; clc;
a = 2.0; b = -2.0; c = -12.0; % specifying coefficients
x_begin = -6.0; % specifying the first value of x
x_end = +7.0; % specifying the last value of x
N = 101; % specifying the number of points to be plotted
% using a built-in MATLAB function to specify a vector of points
x = linspace(x_begin,x_end,N);
y = a*x.*x + b*x + c; % using element by element multiply
plot(x,y)
grid on
xlabel('x')
ylabel('y')
```

Program 6.1   MATLAB program to plot a quadratic function of x.

(b) Let $a = 1$, $b = 2$, and $c = 5$. With (6.3) we get

$$x = \frac{-2 \pm \sqrt{4 - 20}}{2} = \frac{-2 \pm 4\sqrt{-1}}{2} = -1 + j2, -1 - j2$$

where the symbol j is used to mean $\sqrt{-1}$. Let us verify that $x_1 = -1 + j2$ and $x_2 = -1 - j2$ are the zeros of $f(x)$. With $x = x_1$ we get

$$x_1^2 + 2x_1 + 5 = (-1 + j2)(-1 + j2) + 2(-1 + j2) + 5$$
$$= 1 - j2 - j2 + (j2)(j2) - 2 + j4 + 5$$
$$= 1 - j4 - 4 - 2 + j4 + 5 = 0$$

Similarly, we can verify that x_2 is a zero of $f(x)$. As with real zeros of $f(x)$, $f(x)$ can be factored into a product of two first-order factors given by

$$f(x) = (x - x_1)(x - x_2) = (x + 1 - j2)(x + 1 + j2)$$
$$= x^2 + x + j2x + x + 1 + j2 - j2x - j2 - (j2)(j2)$$
$$= x^2 + 2x + 5$$

Fig. 6.2 shows $f(x)$ versus x, and we see that $f(x)$ does not cross the abscissa, because it has complex zeros.

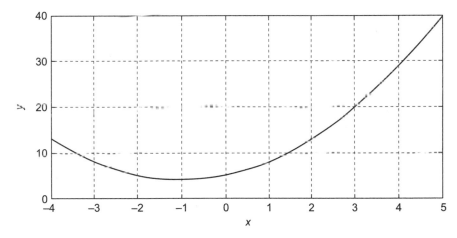

Figure 6.2   Plot of a quadratic function that has a pair of complex conjugate zeros.

6.2   Rectangular Form and Complex Arithmetic

Arithmetic with complex numbers is similar to arithmetic with real numbers. Let $x_1 = \alpha_1 + j\beta_1$ and $x_2 = \alpha_2 + j\beta_2$ be two complex numbers. Table 6.1 lists several arithmetic properties.

In MATLAB, it is very easy to work with complex numbers. MATLAB uses the symbols i or j for $\sqrt{-1}$.

Table 6.1   Properties of rectangular form complex numbers and arithmetic

Equal	$x_1 = x_2$, if and only if $\alpha_1 = \alpha_2$ and $\beta_1 = \beta_2$
Addition	$x_1 + x_2 = (\alpha_1 + \alpha_2) + j(\beta_1 + \beta_2)$
Subtraction	$x_1 - x_2 = (\alpha_1 - \alpha_2) + j(\beta_1 - \beta_2)$
Multiplication	$x_1 x_2 = (\alpha_1 + j\beta_1)(\alpha_2 + j\beta_2) = (\alpha_1\alpha_2 - \beta_1\beta_2) + j(\alpha_1\beta_2 + \beta_1\alpha_2)$
Division	$\dfrac{x_1}{x_2} = \dfrac{x_1 x_2^*}{x_2 x_2^*} = \dfrac{(\alpha_1 + j\beta_1)(\alpha_2 - j\beta_2)}{(\alpha_2 + j\beta_2)(\alpha_2 - j\beta_2)} = \dfrac{\alpha_1\alpha_2 + \beta_1\beta_2}{\alpha_2^2 + \beta_2^2} + j\dfrac{-\alpha_1\beta_2 + \beta_1\alpha_2}{\alpha_2^2 + \beta_2^2}$
Conjugation	$(x_1 + x_2)^* = x_1^* + x_2^*$, $(x_1 x_2)^* = x_1^* x_2^*$ and $\left(\dfrac{x_1}{x_2}\right)^* = \dfrac{x_1^*}{x_2^*}$
Polynomial with real coefficients	$f^*(x) = f(x^*)$, for $x = \alpha + j\beta$

Example 6.2

Using the MATLAB command window, the following statements demonstrate how convenient it is to use MATLAB to work with complex numbers written in rectangular form.

```
>> clear all
>> a=j*2   %  a is a purely imaginary number
a =     0 + 2.0000i
>> % MATLAB replaces the symbol j  with the symbol i
>> b=2-j*3
b =    2.0000 - 3.0000i
>> c=a+b   %  MATLAB follows the rules of arithmetic given in Table 6.1
c =    2.0000 - 1.0000i
>> d=c*j   %  this is a multiplication
d =    1.0000 + 2.0000i
```

MATLAB has several built-in functions that work with complex numbers. The following statements demonstrate some of these functions.

```
>> e=complex(2,3)  % built-in MATLAB function complex assigns a complex number to e
e =    2.0000 + 3.0000i
>> % built-in function isreal checks if e is real, and then sets f to a logic value
>> f=isreal(e)
f =    0
>> g=conj(e)   %  built-in function conj takes the complex conjugate
g =    2.0000 - 3.0000i
```

```
>> h=real(g)   %  built-in function real gets the real part
h =    2
>> p=imag(g)   %  built-in function imag gets the imaginary part
p =    -3
>> q=d/e  % division with complex numbers
q =    0.6154 + 0.0769i
>> r=(d*conj(e))/(e*conj(e))   % the denominator is a real number
r =    0.6154 + 0.0769i
>> %  a complex number multiplied by its complex conjugate always produces
>> %  a real number
```

In MATLAB, complex numbers can be elements of matrices. The following statements illustrate some of the possibilities.

```
>> clear all
>> A=[j   j*2*(1+j)   -1+2*j]  % the vector A has 3 complex elements
A =     0 + 1.0000i   -2.0000 + 2.0000i   -1.0000 + 2.0000i
>> A(2)  % obtain the second element in the vector
ans =  -2.0000 + 2.0000i
>> % the transpose of a complex matrix includes conjugating every element
>> B=[-1   -2+j*2   4-j*5]'  %  B becomes a column vector
B =
  -1.0000
  -2.0000 - 2.0000i
   4.0000 + 5.0000i
>> c=conj(A)*B  %  element by element conjugation of A and then matrix multiply
c -   -11.000 + 6.0000i
>> d=A.*conj(A)  %  each element of A is multiplied by its complex conjugate
d =    1     8     5
```

Example 6.3

In this example we will use a MATLAB script to apply the formula given in (6.3). Most of the programs that we will write serve to illustrate various MATLAB features and operations. However, when we write an application program that is intended to be used by others or the program writer at a later time, it is useful to document the program so that its purpose and methods can be easily understood. The following MATLAB program illustrates preferred program organization. You must always strive to write programs that provide the user an opportunity to test for the proper functioning of the program and a means to terminate the program within the program development environment.

```
% EXAMPLE OF PROGRAM ORGANIZATION
%
% Program calc_roots.m
%
% Purpose:
% This program solves for the roots of a quadratic equation
% of the form: a*x^2 + b*x +c = 0.  It finds real roots
% and complex roots of the equation.
%
% Program information
% Date        Programmer        Description
% 2/22/10     Priemer           Original code, version 1.0
%
% Define variables
% a              --coefficient of x^2 term
% b              --coefficient of x term
% c              --constant term
% disc           --discriminant
% i_part         --imaginary part
% r_part         --real part
% x1             --first root
% x2             --second root
%
clear all; clc;
disp('This program solves for the roots of a quadratic equation');
disp('of the form: a*x^2 + b*x + c = 0.');
disp('To terminate this program,');
disp('enter zero when prompted for the coefficient a');
while 1  % this causes the while loop to execute endlessly
    % Prompt user for equation coefficients
    a=input('Enter the coefficient a: ');
    if a == 0.0
        break  % user has terminated the while loop and the program
    end
    b=input('Enter the coefficient b: ');
    c=input('Enter the coefficient c: ');
    % Find the discriminant
    disc=b^2-4*a*c;
    % Check the discriminant
```

```
    if disc > 0   %   There are two real roots
        d=sqrt(disc);
        x1=(-b+d)/(2*a);   % first root
        x2=(-b-d)/(2*a);   % second root
        disp('The equation has two real roots:');
        fprintf('x1= %f \n', x1); % Using floating point formatted print
        fprintf('x2= %f \n', x2);
    elseif disc == 0 % There are two real roots
        x1=(-b)/(2*a);   % x2 = x1
        disp('The equation has two repeated real roots:');
        fprintf('x1=x2= %f \n', x1);
    else % There are two complex conjugate roots
        r_part=(-b)/(2*a);
        i_part=sqrt(abs(disc))/(2*a);
        disp('The equation has two complex conjugate roots:');
        fprintf('x1= %f +j %f \n', r_part, i_part);
        fprintf('x2= %f -j %f \n', r_part, i_part);
    end
end
disp('Program terminated');
```

Program 6.2   Program to find the roots of a quadratic polynomial.

The method used for formatted print of the results from this program is discussed in Chapter 8. The execution of this program gives the following results.

```
This program solves for the roots of a quadratic equation
of the form: a*x^2 + b*x + c = 0.
To terminate this program,
enter zero when prompted for the coefficient a
Enter the coefficient a: 2
Enter the coefficient b: 4
Enter the coefficient c: 6
The equation has two complex conjugate roots:
x1= -1.000000 +j 1.414214
x2= -1.000000 -j 1.414214
Enter the coefficient a: 1
Enter the coefficient b: 2
Enter the coefficient c: 1
```

```
The equation has two repeated real roots:
x1=x2= -1.000000
Enter the coefficient a: 0
Program terminated
```

Let us write an Nth order polynomial (a polynomial of degree N) as

$$f(x) = x^N + \sum_{k=1}^{N} a_{N-k} x^{N-k} \qquad (6.8)$$

where the coefficients $a_{N-k}, k = 1, \ldots, N$ are real numbers. The Fundamental Theorem of Algebra states that an Nth order polynomial has N roots, where the roots may not be distinct, can be complex and then occur in complex conjugate pairs, and include at least one real root if N is an odd integer.

For polynomials with degree higher than 4, there are no formulas to find the roots. However, there are numerical methods for root finding, and MATLAB has a built-in function **roots** to find the **roots of a polynomial** and another built-in function **poly** to find the coefficients of a polynomial given its roots.

Example 6.4 ────────────────────────────────────

The following script shows how to use the functions: **roots** and **poly**.

```
% Find the roots of a polynomial
clear all; clc;
disp('To find the roots of: x^N + a_N-1 * x^(N-1) + ... + a_1 * x^1 + a_0')
disp('enter the polynomial degree and then the coefficients.');
disp('To terminate the program, enter zero degree.'); disp(' ');
while 1;
    N = input('Enter polynomial degree:  ');
    if N > 0
        a(1)=1.0;  %  coefficient of highest power of x
        for n=1:N;
            disp('Power of x is:'); disp(n-1);
            a(N+2-n) = input('Enter coefficient:  ');
          % roots expects polynomial coefficients in descending power order
        end
        r = roots(a);  %  using built-in function roots to find the N roots
```

```
            disp(' '); %  skip a line
            disp('The roots of the polynomial are:');
            disp(r);
            % find polynomial coefficients given the roots
            b = poly(r);
            disp(' '); disp('The coefficients of the polynomial are:');
            disp(b);
        else
            break;   % causes termination of the while loop
        end
         clear all;  %  clearing a, r and N from previous while loop execution
    end
    clear all;  %  cleaning up work space
    disp('Terminated program')
```

Program 6.3   MATLAB program to find the roots of a polynomial.

For the polynomial: $x^5 + x^4 + 2x^3 + 3x^2 + 4x^1 + 5x^0$, Prog. 6.3 gives

```
To find the roots of: x^N + a_N-1 * x^(N-1) + ... + a_1 * x^1 + a_0* x^0
Enter the polynomial degree and then the coefficients.
To terminate the program, enter zero degree.

Enter polynomial degree:   5
Power of x is:      0
Enter coefficient:   5
Power of x is:      1
Enter coefficient:   4
Power of x is:      2
Enter coefficient:   3
Power of x is:      3
Enter coefficient:   2
Power of x is:      4
Enter coefficient:   1

The roots of the polynomial are:
   0.7145 + 1.3076i
   0.7145 - 1.3076i
  -1.2663
  -0.5814 + 1.2001i
  -0.5814 - 1.2001i
```

```
The coefficients of the polynomial are:
    1.0000    1.0000    2.0000    3.0000    4.0000    5.0000
Enter polynomial degree:   0
Terminated program
```

Notice that complex roots appear in complex conjugate pairs. With the roots $r(n)$, $n = 1, 2, \ldots, N$, we can write a polynomial as a product of N first-order factors given by

$$f(x) = \prod_{n=1}^{N} (x - r(n)) \tag{6.9}$$

6.3   Polar Form and Complex Arithmetic

There is another way, called the **polar form**, to arithmetically define a complex number, which in many circumstances is more convenient to work with than the rectangular form.

A complex number is described by two real numbers. Instead of writing $x = \alpha + j\beta$, we could write $x = (\alpha, \beta)$ to communicate the real part and the imaginary part of x. However, $x = (\alpha, \beta)$ is not an arithmetic expression for x. Another way to write an arithmetic expression for a complex number is based on locating x in a plane, called the **complex plane**, where α is the distance along the abscissa, called the **real axis**, and β is the distance along the ordinate, called the **imaginary axis**. This is illustrated in Fig. 6.3.

The point x can also be located by rotating a line of length, denoted by $||x||$, in the counterclockwise direction from the positive abscissa by an angle, denoted by $\angle x$. By projection we have

$$\alpha = ||x||\cos(\angle x) \tag{6.10}$$

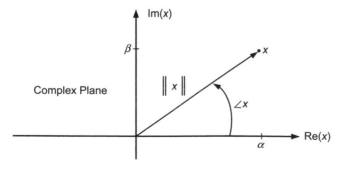

Figure 6.3   Complex number located in the complex plane.

$$\beta = ||x||\sin(\angle x) \tag{6.11}$$

Summing the squares of (6.10) and (6.11) gives

$$\alpha^2 + \beta^2 = ||x||^2$$

Therefore

$$||x|| = \sqrt{\alpha^2 + \beta^2} \tag{6.12}$$

and $||x||$ is called the **magnitude** of x. Dividing (6.11) by (6.10) gives

$$\frac{\beta}{\alpha} = \frac{\sin(\angle x)}{\cos(\angle x)}$$

Therefore

$$\angle x = \tan^{-1}\left(\frac{\beta}{\alpha}\right) \tag{6.13}$$

and $\angle x$ is called the **angle** of x. Substituting (6.10) and (6.11) into the rectangular form of (6.6) gives

$$x = ||x||\cos(\angle x) + j||x||\sin(\angle x) = ||x||(\cos(\angle x) + j\sin(\angle x)) \tag{6.14}$$

Equation (6.14) is a rectangular form expression for x, except that the real and imaginary parts of x are given in terms of the magnitude and angle of x.

Now, let us consider the Maclaurin series expansion (Taylor series expansion about $c = 0$) of the exponential function e^{c}, for any real or complex number c. The power series is given by

$$e^c = \sum_{k=0}^{\infty} \frac{c^k}{k!}, \quad 0! = 1$$

Let $c = j\theta$, and we get

$$e^{j\theta} = \frac{(j\theta)^0}{0!} + \frac{(j\theta)^1}{1!} + \frac{(j\theta)^2}{2!} + \frac{(j\theta)^3}{3!} + \frac{(j\theta)^4}{4!} + \frac{(j\theta)^5}{5!} + \cdots$$
$$= 1 + j\frac{\theta^1}{1!} - \frac{\theta^2}{2!} - j\frac{\theta^3}{3!} + \frac{\theta^4}{4!} + j\frac{\theta^5}{5!} - \cdots$$

Table 6.2   Properties of polar form complex numbers and arithmetic

Equal	$x_1 = x_2$, if and only if $\|x_1\| = \|x_2\|$ and $\angle x_1 = \angle x_2$
Addition	$x_1 + x_2$, must convert to rectangular form
Subtraction	$x_1 - x_2$, must convert to rectangular form
Multiplication	$x_3 = x_1 x_2 = \|x_1\|\|x_2\|e^{j(\angle x_1 + \angle x_2)}$, $\|x_3\| = \|x_1\|\|x_2\|$ and $\angle x_3 = \angle x_1 + \angle x_2$
Division	$x_3 = \dfrac{x_1}{x_2} = \dfrac{\|x_1\|}{\|x_2\|}e^{j(\angle x_1 - \angle x_2)}$, $\|x_3\| = \dfrac{\|x_1\|}{\|x_2\|}$ and $\angle x_3 = \angle x_1 - \angle x_2$
Conjugation	$x = \|x\|e^{j\angle x}$, $x^* = \|x\|e^{-j\angle x}$, $(x_1 x_2)^* = x_1^* x_2^*$ and $\left(\dfrac{x_1}{x_2}\right)^* = \dfrac{x_1^*}{x_2^*}$
Polynomial with real coefficients	$f^*(x) = f(x^*)$, for $x = \|x\|e^{j\angle x}$

Gathering real and imaginary parts gives

$$e^{j\theta} = \left(1 - \frac{\theta^2}{2!} + \frac{\theta^4}{4!} - \cdots\right) + j\left(\frac{\theta^1}{1!} - \frac{\theta^3}{3!} + \frac{\theta^5}{5!} - \cdots\right) \tag{6.15}$$

The real part of (6.15) is the power series (Maclaurin series expansion) for $\cos(\theta)$, and the imaginary part of (6.15) is the power series (Maclaurin series expansion) for $\sin(\theta)$. Thus, (6.15) becomes

$$e^{j\theta} = \cos(\theta) + j\sin(\theta) \tag{6.16}$$

which is called **Euler's identity**.

With Euler's identity, the expression for the complex number given in (6.14) becomes

$$x = \|x\|e^{j\angle x} \tag{6.17}$$

This way of writing a complex number is call the **polar form**. With (6.12) and (6.13), a complex number can be converted from rectangular form to polar form, and with (6.10) and (6.11), a complex number can be converted from polar form to rectangular form. Table 6.2 lists several arithmetic properties of complex numbers written in polar form.

Example 6.5 ———————————————————————————————————

The following statements illustrate how convenient it is to work in MATLAB with complex numbers written in polar form.

```
>> clear all
>> a=3*exp(j*pi/2) % the angle of a is pi/2 radians, which places a on the imaginary axis
a =
   0.0000 + 3.0000i
```

```
>> b=exp(j*pi)  %  the angle of b is pi, which places b on the negative real axis
b =
  -1.0000 + 0.0000i
>> c=a*b
c =
  -0.0000 - 3.0000i
>>% magnitude of c equals the magnitude of a times the magnitude of b
>> % angle of c is pi/2 plus pi, or -pi/2
>> angle_c = angle(c)  %  using the MATLAB built-in function angle
angle_c =
  -1.5708
>> magnitude_c = abs(c)  %  using the MATLAB built-in function abs
magnitude_c =
   3
>> v=[j*pi/2  j*pi  j*2*pi]
v =
        0 + 1.5708i        0 + 3.1416i        0 + 6.2832i
>> w=exp(v)  %  a complex matrix can be the argument of a built-in function
w =
   0.0000 + 1.0000i  -1.0000 + 0.0000i   1.0000 - 0.0000i
```

6.4 Euler's Identity

Euler's identity in (6.16) can be written in several different ways. If θ in (6.16) is replaced by $-\theta$, then we get

$$e^{-j\theta} = \cos(-\theta) + j\,\sin(-\theta) = \cos(\theta) - j\,\sin(\theta) \tag{6.18}$$

Comparing (6.16) and (6.18) shows that conjugating a complex number changes the sign of its angle. Adding (6.16) and (6.18) gives

$$\cos(\theta) = \frac{e^{j\theta} + e^{-j\theta}}{2} \tag{6.19}$$

With (6.19), cos (θ) can be written as the sum of two complex conjugate exponential functions. Subtracting (6.18) from (6.16) gives

$$\sin(\theta) = \frac{e^{j\theta} - e^{-j\theta}}{j2} = \frac{e^{j\theta} - e^{-j\theta}}{2e^{j\pi/2}} = \frac{e^{j(\theta - \pi/2)} + e^{j\pi}e^{j(-\theta - \pi/2)}}{2}$$
$$= \frac{e^{j(\theta - \pi/2)} + e^{-j(\theta - \pi/2)}}{2} = \cos\left(\theta - \frac{\pi}{2}\right) \tag{6.20}$$

With (6.20), $\sin(\theta)$ can be written as the sum of two complex conjugate exponential functions. Equations (6.18)–(6.20) are also called **Euler's identity**.

Example 6.6

Let us verify Euler's identities with the following MATLAB statements.

```
>> theta=pi/3;   % this is 60 degrees
>> a=cos(theta)
a =
0.5000
>> b=sin(theta)
b =
0.8660
>> c=(exp(j*theta)+exp(-j*theta))/2   % this gives cos(theta)
c =
0.5000
>> d=(exp(j*theta)-exp(-j*theta))/(j*2)   % this gives sin(theta)
d =
0.8660
```

Consider the sine function given by

$$x(\phi) = \sin(\phi)$$

where ϕ, the independent variable, is expressed in radians. When ϕ goes through a change of 2π radians, then x will repeat itself, or

$$x(\phi) = x(\phi + r2\pi)$$

for all integers r and any ϕ. We say that 2π is the **period** of $x(\phi)$.

Now, let us work with a sinusoidal function of time (time is expressed in seconds) given by

$$x(t) = A\cos(\omega_0 t + \theta) \tag{6.21}$$

where ω_0 (expressed in radians/sec) is called the **frequency** of $x(t)$; A, a positive number, is called the **amplitude** of $x(t)$; and θ is called the **phase angle** of $x(t)$. Like $\sin(\phi)$, this function will repeat itself when $(\omega_0 t + \theta)$ goes through a change of 2π, where t is now the variable. The time interval, T_0, over which $x(t)$ repeats itself is called the **period** of $x(t)$, and we write

$$x(t) = x(t + T_0) \tag{6.22}$$

This means that

$$A\cos(\omega_0 t + \theta) = A\cos(\omega_0(t + T_0) + \theta) = A\cos(\omega_0 t + \omega_0 T_0 + \theta)$$

This requires that $\omega_0 T_0 = 2\pi$, and we can get ω_0 from knowing T_0 with

$$\omega_0 = \frac{2\pi}{T_0}$$

We also express the frequency with $f_0 = 1/T_0$ cycles/sec (1 cycle/sec = 1 hertz, abbreviated to Hz), and $\omega_0 = 2\pi f_0$.

Example 6.7

Below is a script to plot the signal given in (6.21). This example also shows how to place more than one plot in a figure with the built-in MATLAB function **subplot**. To find out more about the function subplot, use help subplot. Fig. 6.4 shows the signal for two frequencies.

```
clear all; clc;
amplitude = 2.0; % specify amplitude
phase = pi/4.0; % specify phase angle
f0 = 2.0; % specify frequency in Hz
w0 = 2.0*pi*f0; % convert frequency to rad/sec
T0 = 1.0/f0; % find the period, the time of one cycle
N = 101; % plot N points of the signal
total_time = 2*T0; % plot over two periods
t = linspace(0.0,total_time,N); % use linspace to set up time points
x = amplitude*cos(w0*t + phase); % evaluate signal for each time point
% organizing two plots into 2 rows and 1 column
subplot(2,1,1); plot(t,x)   %  placing the first plot in a 2X1 matrix of plots
xlabel('time - seconds')
ylabel('signal unit')
grid on
x = amplitude*cos(2*w0*t + phase); % double the frequency
subplot(2,1,2); plot(t,x)   %  placing the second plot in a 2X1 matrix of plots
xlabel('time - seconds')
ylabel('signal unit')
grid on
```

Program 6.4 **Program that demonstrates using subplot.**

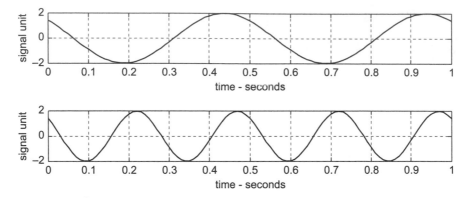

Figure 6.4   Placement of two plots in one figure.

The behavior of most machines, devices, and systems that engineers design is inherently cyclical, which is usually described using sinusoidal functions. However, it is much more convenient to work with exponential functions. This is possible through Euler's identity. Applying (6.19) to (6.21) gives

$$x(t) = A\cos(\omega_0 t + \theta) = A\frac{e^{j(\omega_0 t+\theta)} + e^{-j(\omega_0 t+\theta)}}{2} \tag{6.23}$$

$$x(t) = \frac{A}{2}e^{j\theta}e^{j\omega_0 t} + \frac{A}{2}e^{-j\theta}e^{-j\omega_0 t} = \frac{X}{2}e^{j\omega_0 t} + \frac{X^*}{2}e^{-j\omega_0 t} \tag{6.24}$$

where $X = Ae^{j\theta}$ is called the **phasor** of $x(t)$. Therefore, $x(t)$ has been written as the sum of two complex conjugate exponential functions.

Example 6.8

The following MATLAB script and table verify (6.23).

```
>> A=2.0; % specify the amplitude
>> w=2*pi;  % specify a frequency of 1 cycle/second
>> phase=pi/4; % specify a phase
%
>> %  evaluate the sinusoid over just a quarter of a cycle
%
>> t=[0:0.025:0.25]; % evaluate only 11 points
>> x=A*cos(w*t+phase); % evaluate trigonometric function
```

```
>> X=(A)*exp(j*phase); % specify phasor
>> x_exponential=(X/2)*exp(j*w*t)+(conj(X))/2*exp(-j*w*t);
>> table=[t' x' x_exponential']   % give results in three columns
>> % the first column is time
>> % the second column is the trigonometric expression
>> % the third column is the complex exponential expression
table =
        0           1.4142     1.4142
        0.0250      1.1756     1.1756
        0.0500      0.9080     0.9080
        0.0750      0.6180     0.6180
        0.1000      0.3129     0.3129
        0.1250      0.0000     0.0000
        0.1500     -0.3129    -0.3129
        0.1750     -0.6180    -0.6180
        0.2000     -0.9080    -0.9080
        0.2250     -1.1756    -1.1756
        0.2500     -1.4142    -1.4142
```

6.5   Fourier Series

Practical periodic signals may not have a function representation. However, periodic signals can be represented with a series sum of complex exponential functions, called a **Fourier series**.

Let $s_1(t) = \sin(\omega_0 t)$. The frequency ω_0 is given by $\omega_0 = 2\pi/T_0$, where T_0 is the period of $s_1(t)$. Now, consider the time function

$$s_2(t) = \sin(2\omega_0 t)$$

This time function will go through two cycles over the time range T_0, and the period of $s_2(t)$ is given by

$$\frac{2\pi}{2\omega_0} = \frac{T_0}{2}$$

Although the period of $s_2(t)$ is $T_0/2$, $s_2(t)$ also repeats itself every T_0 sec, over which it goes through two cycles. In general, the function of time given by

$$s_k(t) = b_k \sin(k\omega_0 t)$$

for $k = 1, 2, \ldots, \infty$, where b_k is the amplitude, repeats itself every T_0/k secs, and over the time range T_0, $s_k(t)$ goes through k cycles. Similarly, the cosine function given by

$$c_k(t) = a_k \cos(k\omega_0 t)$$

for $k = 1, 2, \ldots, \infty$, where a_k is the amplitude, repeats itself every T_0/k secs, and over the time range T_0, $c_k(t)$ goes through k cycles.

Therefore, if we sum all sine and cosine functions as defined above, we get

$$x(t) = a_0 + \sum_{k=1}^{\infty} [a_k \cos(k\omega_0 t) + b_k \sin(k\omega_0 t)] \tag{6.25}$$

and then $x(t)$ will also repeat itself every T_0 secs, because for each k the cosine and sine functions go through an integer number of cycles over the time range T_0 secs. Since the average value of the sine and cosine terms is zero, a_0 is included to account for the average value of $x(t)$. The signal $x(t)$ given in (6.25) is a periodic function, since for all t we have

$$x(t) = x(t + T_0)$$

and (6.25) is called a **trigonometric Fourier series** of $x(t)$. The frequency given by $\omega_0 = 2\pi/T_0$ is called the **fundamental frequency** of $x(t)$. If the Fourier series (sum of sinusoidal functions) is to have the same period as $x(t)$, then all sinusoidal functions in the Fourier series must have frequencies given by $k\omega_0$ rad/sec, where k is an integer.

Given a practical periodic signal $x(t)$, we can find the a_k and b_k coefficients such that $x(t)$ can be written as a Fourier series. These coefficients, called the **trigonometric** Fourier series coefficients, are found with

$$a_k = \frac{2}{T_0} \int_0^{T_0} x(t)\cos(k\omega_0 t)dt = \frac{2}{T_0} \int_{t_0}^{t_0+T_0} x(t)\cos(k\omega_0 t)dt, \quad a_0 = \frac{1}{T_0} \int_{t_0}^{t_0+T_0} x(t)dt \tag{6.26}$$

$$b_k = \frac{2}{T_0} \int_0^{T_0} x(t)\sin(k\omega_0 t)dt = \frac{2}{T_0} \int_{t_0}^{t_0+T_0} x(t)\sin(k\omega_0 t)dt \tag{6.27}$$

for any t_0.

Example 6.9

Let us apply (6.25) to represent the sawtooth wave $x(t)$ shown in Figure 6.5. The period of $x(t)$ is $T_0 = 1$ sec. The fundamental frequency is $\omega_0 = 2\pi/T_0 = 2\pi$ rad/sec. A period of $x(t)$ can be any part of $x(t)$ over a time range of $t_0 \leq t < t_0 + T_0$ sec for any time point t_0. Over the time range $-0.25 \leq t < 0.75$, the signal $x(t)$ can be written as $x(t) = t + 0.25$. Let us use this part of $x(t)$ to evaluate (6.26) and (6.27). With (6.26) and (6.27) we get

$$a_0 = \frac{1}{1}\int_{-0.25}^{0.75}(t+0.25)dt = \frac{1}{2}$$

$$a_k = \frac{2}{1}\int_{-0.25}^{0.75}(t+0.25)\cos(2k\pi t)dt = -\frac{\sin(k\pi/2)}{k\pi}, \quad k = 1,2,\ldots$$

$$b_k = \frac{2}{1}\int_{-0.25}^{0.75}(t+0.25)\sin(2k\pi t)dt = -\frac{\cos(k\pi/2)}{k\pi}, \quad k = 1,2,\ldots$$

These coefficients are used in the following MATLAB script to evaluate the Fourier series and plot the signals shown in Figs. 6.6 and 6.7.

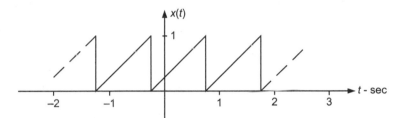

Figure 6.5   A sawtooth wave.

```
% This program uses a given set of trigonometric Fourier series coefficients
% to evaluate a Fourier series over three periods of the function.
T0=1.0, % the period of the function
w0=2*pi/T0; % the fundamental frequency
N=3001; % evaluate function for 3001 time points
t=linspace(-T0,2*T0,N);
K=10; % using only ten terms in the Fourier series
k=1:K; % a vector for the Fourier series coefficient index values
a0=0.5; % the average value of the function
% find all trigonometric Fourier series coefficients
ak=-sin(k*pi/2)./(pi*k); % element by element division of two vectors
bk=-cos(k*pi/2)./(pi*k); % element by element divide
% ak and bk are row vectors
w = k*w0;   % row vector of all frequencies
for n=1:N   % loop to evaluate x(t) at each time point
    x(n)=a0+ak*cos(w*t(n))'+bk*sin(w*t(n))'; % using inner product to do summation
end
plot(t,x)
grid on
xlabel('time - sec')
ylabel('signal value')
```

Program 6.5   Program to construct a signal with a Fourier series.

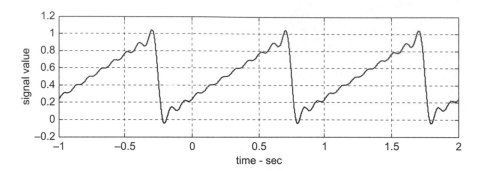

Figure 6.6   Reconstructed signal using K = 10 terms in the Fourier series.

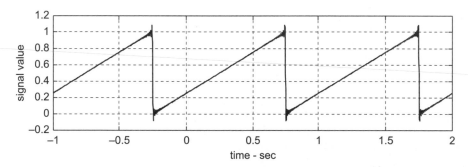

Figure 6.7   Reconstructed signal using K = 100 terms in the Fourier series.

The oscillation about the discontinuity is called **Gibbs' oscillation**. The amplitude of the oscillation does not go to zero as the number of terms used in the Fourier series increases. However, the time duration of Gibbs' oscillation does go to zero as the number of terms increases. This means that the Gibbs' oscillation contribution in the Fourier series of $x(t)$ has no energy, and practically it can be ignored.

The Fourier series expression for $x(t)$ can be written in another way. Based on Euler's identity we have

$$\cos(k\omega_0 t) = \frac{e^{jk\omega_0 t} + e^{-jk\omega_0 t}}{2}$$

and

$$\sin(k\omega_0 t) = \frac{e^{jk\omega_0 t} - e^{-jk\omega_0 t}}{j2}$$

Therefore, the Fourier series in (6.25) for $x(t)$ becomes

$$x(t) = a_0 + \sum_{k=1}^{\infty} \left[a_k \frac{e^{jk\omega_0 t} + e^{-jk\omega_0 t}}{2} + b_k \frac{e^{jk\omega_0 t} - e^{-jk\omega_0 t}}{j2} \right]$$

$$= a_0 + \sum_{k=1}^{\infty} \frac{a_k + jb_k}{2} e^{-jk\omega_0 t} + \sum_{k=1}^{\infty} \frac{a_k - jb_k}{2} e^{jk\omega_0 t}$$

Now, let X_k be defined by

$$X_k = \frac{a_k - jb_k}{2}, \quad k = 1, 2, 3, \ldots \tag{6.28}$$

Since according to (6.26), $a_{-k} = a_k$, and according to (6.27), $b_{-k} = -b_k$, we have

$$X_{-k} = \frac{a_{-k} - jb_{-k}}{2} = \frac{a_k + jb_k}{2} = X_k^*, \quad k = 1, 2, 3, \ldots \tag{6.29}$$

Thus, $a_k = 2\text{Re}(X_k)$, and $b_k = 2\text{Im}(X_k^*)$. Now, $x(t)$ can be written as

$$x(t) = a_0 + \sum_{k=1}^{\infty} X_k^* e^{-jk\omega_0 t} + \sum_{k=1}^{\infty} X_k e^{jk\omega_0 t} = a_0 + \sum_{k=-1}^{-\infty} X_k e^{jk\omega_0 t} + \sum_{k=1}^{\infty} X_k e^{jk\omega_0 t} \tag{6.30}$$

which becomes

$$x(t) = \sum_{k=-\infty}^{\infty} X_k e^{jk\omega_0 t} \tag{6.31}$$

where $X_0 = a_0$ and the X_k, called the **complex Fourier series** coefficients, are found by substituting (6.26) and (6.27) into (6.28) to get

$$X_k = \frac{1}{T_0} \int_0^{T_0} x(t) e^{-jk\omega_0 t} dt = \frac{1}{T_0} \int_{t_0}^{t_0+T_0} x(t) e^{-jk\omega_0 t} dt \tag{6.32}$$

for any t_0 and $k = -\infty, \ldots, -1, 0, 1, \ldots, \infty$. The Fourier series in (6.31) is called the **complex exponential** Fourier series.

The idea of writing a periodic time function or signal in terms of a sum of sinusoidal functions is a very fundamental concept that is widely applied in engineering. In view of (6.28), the X_k are in general complex numbers, which can be written in polar form to get

$$X_k = ||X_k|| e^{j\angle X_k}$$

Since $X_{-k} = X_k^*$, then $||X_{-k}|| = ||X_k||$, which makes $||X_k||$ an even function of k, and $\angle X_{-k} = -\angle X_k$, which makes $\angle X_k$ is an odd function of k. Substituting the polar form for X_k into (6.31) gives

$$x(t) = \sum_{k=-\infty}^{\infty} ||X_k|| e^{j\angle X_k} e^{jk\omega_0 t}$$

$$= \cdots + ||X_2|| e^{-j\angle X_2} e^{-j2\omega_0 t} + ||X_1|| e^{-j\angle X_1} e^{-j\omega_0 t} + X_0 + ||X_1|| e^{j\angle X_1} e^{j\omega_0 t} + ||X_2|| e^{j\angle X_2} e^{j2\omega_0 t} + \cdots$$

Applying Euler's identity to pairs of terms from corresponding positive and negative k gives

$$x(t) = ||X_0|| + 2\sum_{k=1}^{\infty} ||X_k|| \cos(k\omega_0 t + \angle X_k) \tag{6.33}$$

This shows that $2||X_k||$ gives the amplitude and $\angle X_k$ gives the phase of the sinusoidal contribution to $x(t)$ at the frequency $\omega = k\omega_0$ rad/sec. To assess the nature of a signal $x(t)$, the amplitude and phase of each sinusoidal function that contributes to $x(t)$ are plotted versus the frequency of the sinusoidal function.

Example 6.10 _____

For the periodic signal $x(t)$ given in Example 6.9, (6.32) gives the complex Fourier series coefficients, which results in

$$
\begin{aligned}
X_k &= \frac{1}{1}\int_{-0.25}^{0.75} (t+0.25)e^{-jk2\pi t} dt = \frac{1}{2k\pi} e^{j(k+1)\pi/2}, \quad k \neq 0 \\
&= \frac{1}{2}, \quad k = 0
\end{aligned} \tag{6.34}
$$

Integration by parts was applied to obtain this result. Fig. 6.8 shows the amplitudes of the sinusoidal contributions to $x(t)$ versus the frequency of each sinusoid. This plot is called the **magnitude spectrum** of $x(t)$. The plot shows that the average value of $x(t)$ is 0.5, and that the amplitude of the sinusoid at the fundamental frequency ω_0 is almost twice the amplitude of the sinusoid at the frequency $2\omega_0$. Also, we see that at higher frequencies the amplitudes of sinusoids contributing to $x(t)$ decrease.

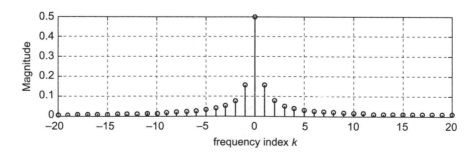

Figure 6.8   Magnitude spectrum of the sawtooth wave.

The following script uses the complex Fourier series coefficients to plot the magnitude spectrum of the sawtooth wave.

```
clear all; clc;
K=20; %
freq_index=[-K:1:K];
% find complex Fourier series coefficients
X0=0.5
for k=1:K
    X_pos_k(k)=exp(j*(k+1)*pi/2)/(k*2*pi);
    X_neg_k(k)=conj(X_pos_k(k));
end
% organize Fourier series coefficients for k = -K, ..., -1, 0, 1, ..., K
X=[fliplr(X_neg_k), X0, X_pos_k]; % flipping X_neg_k
X_mag=abs(X); % get magnitude of complex Fourier series coefficients
stem(freq_index,X_mag); % draw a stem plot
grid on
xlabel('frequency index k')
ylabel('Magnitude')
```

Program 6.6 Using complex Fourier series coefficients to plot the magnitude spectrum.

In this section we saw how convenient it is to use the MATLAB facility to work with complex numbers for representing a periodic signal. In several of the following chapters this approach to signal representation will be extended to include aperiodic continuous time signals, discrete time periodic signals, and aperiodic discrete time signals.

6.6 Energy

Consider the circuits shown in Fig. 6.9, where a voltage source $v(t) = A\cos(\omega t + \theta)$, $(T_0 = 2\pi/\omega)$, is connected to a resistor with value $R = 1\ \Omega$, and a constant voltage source V is connected to another $1\ \Omega$ resistor.

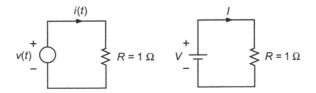

Figure 6.9 Voltage sources connected to 1Ω resistors.

The power delivered by $v(t)$ to the 1 Ω resistor is given by

$$p(t) = v(t)i(t) = \begin{cases} Ri(t)i(t) = i^2(t) \text{ watts} \\ v(t)\dfrac{v(t)}{R} = v^2(t) \text{ watts} \end{cases} \tag{6.35}$$

and the energy E delivered to the resistor by the voltage source $v(t)$ over one cycle is given by

$$E = \int_0^{T_0} p(t)dt = \int_0^{T_0} v^2(t)dt = \int_0^{T_0} A^2\cos^2(\omega t + \theta)dt$$

$$= \frac{A^2}{2}\int_0^{T_0} (1 + \cos(2(\omega t + \theta)))dt$$

$$= \frac{A^2}{2}\int_0^{T_0} dt + \frac{A^2}{2}\int_0^{T_0} \cos(2(\omega t + \theta))dt = \frac{A^2}{2}T_0 \text{ joules} \tag{6.36}$$

The average power P delivered to the resistor is $P = E/T_0 = A^2/2$ watts. In the other circuit, the power and the average power delivered by the battery to the 1 Ω resistor is given by $P = VI = V^2 = I^2$ watts.

Now, consider the question, what constant voltage source delivers the same average power to the resistor as does $v(t)$? If $V = A/\sqrt{2}$ volts, then the voltage sources $v(t)$ and V deliver the same average power to the resistor. In view of (6.36), to assess the power delivery capability of a sinusoidal voltage source, it is common practice to give its **root mean square** (RMS) value given by

$$V_{RMS} = \left(\frac{1}{T_0}\int_0^{T_0} v^2(t)dt\right)^{1/2} = \sqrt{\left(\frac{1}{T_0}\left(\frac{A^2}{2}T_0\right)\right)} = \frac{A}{\sqrt{2}} = V \tag{6.37}$$

It is also common practice to state that the voltage available at a wall outlet is 110 volts. However, this is its RMS value. The voltage available at the wall outlet is actually $v(t) = 110\sqrt{2}\cos(2\pi ft + \theta)$, where the amplitude is $110\sqrt{2} = 155$ volts and the frequency is $f = 60$ Hz.

Again, in view of (6.36), it is common practice to associate with any periodic signal $x(t)$, with period T_0, an energy E given by

$$E = \int_{t_0}^{t_0+T_0} x^2(t)dt \tag{6.38}$$

for any t_0. For example, if $x(t)$ represents temperature in degrees fahrenheit (F) as it varies periodically with time in hours (H), then the unit of E is F^2H, which has no physical meaning.

There is another way to compute E. Substitute (6.31) into (6.38), and we get (with $t_0 = 0$ for convenience)

$$\int_0^{T_0} x^2(t)dt = \int_0^{T_0} \sum_{k=-\infty}^{\infty} X_k \, e^{jk\omega_0 t} \sum_{l=-\infty}^{\infty} X_l \, e^{jl\omega_0 t} dt$$

$$= \sum_{k=-\infty}^{\infty} \sum_{l=-\infty}^{\infty} X_k X_l \int_0^{T_0} e^{j(k+l)\omega_0 t} dt \qquad (6.39)$$

where the integral is given by

$$\int_0^{T_0} e^{j(k+l)\omega_0 t} dt = \int_0^{T_0} (\cos((k+l)\omega_0 t) + j\sin((k+l)\omega_0 t)) dt = \begin{cases} T_0, \ l = -k \\ 0, \ l \neq -k \end{cases}$$

Therefore, (6.39) reduces to

$$P = \frac{1}{T_0} \int_0^{T_0} x^2(t)dt = \sum_{k=-\infty}^{\infty} X_k X_{-k} = \sum_{k=-\infty}^{\infty} ||X_k||^2 \qquad (6.40)$$

which is called **Parseval's relation**. This means that we can calculate the average power P of a periodic signal using the signal or its complex Fourier series coefficients.

Example 6.10 (continued) ──────────────────────────

Apply Parseval's relation to the sawtooth signal of Example 6.10, where for one period $x(t) = t + 0.25, -0.25 \le t < 0.75$ and $T_0 = 1.0$ secs. Let us use the built-in function function **quad** to numerically evaluate (6.38). The signal is passed to quad as an anonymous function, as follows.

```
>> T0 = 1.0;
>> % must use vectorized form of the square operation
>> x_squared = @(t) (t+0.25).^2;
>> E = quad(x_squared,-0.25,0.75);
>> P = E/T0
P =    0.3333
```

This is the same as the integral of t^2 from $t = 0$ to $t = 1$.

Equation (6.34) gives $||X_k||^2 = 1/(2k\pi)^2$ and $||X_0||^2 = 0.25$; therefore, with (6.40) we get

$$\sum_{k=-\infty}^{\infty} ||X_k||^2 = \sum_{k=-\infty}^{-1} \frac{1}{(2k\pi)^2} + 0.25 + \sum_{k=1}^{\infty} \frac{1}{(2k\pi)^2} = 0.25 + \frac{2}{4\pi^2} \sum_{k=1}^{\infty} \frac{1}{k^2}$$

The infinite series is a special case of the Riemann zeta function given by

$$\varsigma(s) = \sum_{k=1}^{\infty} \frac{1}{k^s}, \ s = \alpha + j\beta$$

which converges if and only if $\alpha > 1$. With Parseval's relation we have proved that

$$\sum_{k=1}^{\infty} \frac{1}{k^2} = 2\pi^2 \left(P - \frac{1}{4} \right) = \frac{\pi^2}{6} \tag{6.41}$$

6.7 Impedance

The complex exponential function also plays an important role in linear system analysis and design. To see how readily MATLAB can be involved in this, Euler's identity will be applied to AC circuit analysis. Then, using MATLAB, we will analyze circuits for the sinusoidal response to sinusoidal inputs. Before we proceed, let us consider some electrical devices used to build circuits. Fig. 6.10 shows, from left to right, the commonly used symbols for the components: **resistor**, **capacitor**, and **inductor**. These devices are said to be **passive devices**, because they are not sources of power. A voltage source is said to be an **active device**, because it is a power source.

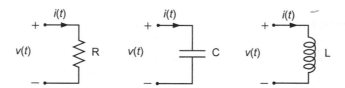

Figure 6.10   Current and voltage references of electrical circuit components: R, C, and L.

An **inductor** is made with a wire (conductor) wound into a coil to increase the magnetic field inside the coil when there is a current. The degree to which the magnetic field in the coil is increased depends on the coil geometry, the number of turns of wire, and the coil core material. These factors determine the value of the parameter L, called **inductance**. The unit for inductance is the **henry** (H).

A time-varying current through the inductor coil causes a time-varying magnetic field in and around the coil, which then causes a voltage across the inductor coil that opposes the change in current. With respect to the current $i(t)$ and voltage $v(t)$ references given in Fig. 6.10, if the current through a 1 H inductor increases at the rate of 1 A/sec, then the voltage across the inductor is 1 volt.

For the inductor, the voltage–current relationship is

$$v(t) = L\frac{di(t)}{dt} \tag{6.42}$$

For example, if the current through an inductor is a constant, then the voltage across the inductor is zero. Furthermore, if the current is increasing (decreasing), then the voltage across the inductor is positive (negative).

Suppose the current is some sinusoidal function, $i(t) = A\cos(\omega t + \theta)$. Then, the phasor I for the current is given by

$$I = Ae^{j\theta}$$

so that

$$\frac{I}{2}e^{j\omega t} + \frac{I^*}{2}e^{-j\omega t} = \frac{Ae^{j\theta}}{2}e^{j\omega t} + \frac{Ae^{-j\theta}}{2}e^{-j\omega t} = A\left[\frac{e^{j(\omega t+\theta)} + e^{-j(\omega t+\theta)}}{2}\right] = A\cos(\omega t + \theta)$$

According to (6.42), the voltage across the inductor becomes

$$v(t) = -A\omega L\sin(\omega t + \theta) = A\omega L\cos\left(\omega t + \theta + \frac{\pi}{2}\right)$$

and therefore, the phasor V for the inductor voltage $v(t)$ is given by

$$V = A\omega L e^{j(\theta+\pi/2)} = j\omega L A e^{j\theta}$$

With the voltage and current phasors we have

$$V = j\omega L I = Z_L I \tag{6.43}$$

The term $Z_L = j\omega L$, which relates the inductor current phasor to the inductor voltage phasor is called the **impedance of an inductor**. If we use instead the current–voltage relationship given by

$$i(t) = \frac{1}{L}\int v(t)dt$$

then we get the phasor relationship given by

$$I = \frac{1}{j\omega L}V = \frac{1}{Z_L}V \tag{6.44}$$

For AC circuit analysis, we have changed the time-domain derivative relationship between the inductor voltage and inductor current into a frequency-domain algebraic relationship between the inductor voltage phasor and the inductor current phasor.

A **capacitor** is made with two conducting plates in close proximity that are separated by a layer of insulating material. A wire is attached to each conducting plate. To reduce package size, long strips of metal foil separated by a long strip of insulating material are wound from one end to the other end of the strips into a cylindrical shape. When a voltage difference is applied across the conducting plates, negative charge is drawn from the plate connected to positive side of the voltage difference, which effectively makes this plate hold a positive charge that attracts negative charge to the other plate. The accumulated positive and negative charges on each plate cause an electric field between the plates. The magnitude of the electric field depends on the area of the plates, the distance between the plates, and the material between the plates. These factors determine the value of the parameter C, called **capacitance**. The unit of capacitance is the **farad** (F).

A time-varying voltage across the plates of a capacitor causes a time-varying electric field between the plates and a current, called **displacement current**. With respect to the current $i(t)$ and voltage $v(t)$ references given in Fig. 6.10, if the voltage across a 1 F capacitor increases at the rate of 1 V/sec, then the displacement current of the capacitor is 1 ampere. Even though there is no conductive path from one plate of a capacitor to the other plate, externally, the displacement current of a capacitor is measured like a conductive current through other electronic devices.

For the capacitor, the current–voltage relationship is

$$i(t) = C\frac{dv(t)}{dt} \tag{6.45}$$

For example, if the voltage across a capacitor is a constant, then the capacitor current is zero. It is said that a capacitor blocks current due to a constant voltage. Furthermore, if the voltage is increasing (decreasing), then the current is positive (negative).

Suppose the voltage is some sinusoidal function, $v(t) = A\cos(\omega t + \theta)$. Then, the phasor V for the voltage is given by

$$V = Ae^{j\theta}$$

According to (6.45), the capacitor current becomes

$$i(t) = -A\omega C\sin(\omega t + \theta) = A\omega C\cos\left(\omega t + \theta + \frac{\pi}{2}\right)$$

and therefore, the phasor I for the capacitor current $i(t)$ is given by

$$I = A\omega Ce^{j(\theta+\pi/2)} = j\omega CAe^{j\theta}$$

With the current and voltage phasors we have

$$I = j\omega CV \tag{6.46}$$

and

$$V = \frac{1}{j\omega C}I = Z_C I \tag{6.47}$$

The term $Z_C = 1/j\omega C$, which relates the capacitor current phasor to the capacitor voltage phasor, is called the **impedance of a capacitor**.

For the resistor, the voltage–current relationship is $v(t) = Ri(t)$, and, since this is an instantaneous relationship, the relationship between the resistor voltage phasor and the resistor current phasor is given by

$$V = RI = Z_R I \tag{6.48}$$

where $Z_R = R$ is called the **impedance of a resistor**.

In (6.43), (6.47), and (6.48), which are structured like Ohm's law, we see that multiplying a current phasor by an impedance produces a voltage phasor. These equations show that if we are only interested to solve the integral–differential equations that result from applying Kirchhoff's laws to linear circuits for the steady-state response to sinusoidal inputs, then we can immediately convert these equations to complex algebraic equations in terms of current phasors, voltage phasors, and impedances.

6.8 AC Circuit Analysis

AC circuit analysis is concerned with finding the steady-state sinusoidal response of a linear circuit to a sinusoidal input. Since MATLAB can work with the complex data type, applying MATLAB to AC circuit analysis is similar to applying MATLAB to resistive circuit analysis.

As a general example, consider the RLC circuit shown in Fig. 6.11, where the input is the voltage given by

$$v_s(t) = A\cos(\omega t + \theta)$$

The amplitude A and phase θ determine the phasor $V_s = Ae^{j\theta}$ of $v_s(t)$, and the frequency ω will be assigned values over a range to see how the circuit responds differently, depending on the frequency of the input.

To find the steady-state sinusoidal current, let us do a mesh analysis of the circuit, and apply Kirchhoff's voltage law. Summing the voltage drops around the mesh gives

$$-v_s(t) + \frac{1}{C}\int i(t)dt + L\frac{di(t)}{dt} + Ri(t) = 0 \tag{6.49}$$

Methods to solve this integral–differential equation will not be considered until Chapter 12. However, since $v_s(t)$ is a sinusoid, then in steady-state $i(t)$ must also be a sinusoid given by $i(t) = ||I||\cos(\omega t + \angle I)$, where $I = ||I||e^{j\angle I}$ is the phasor of $i(t)$. Replacing $v_s(t)$ and $i(t)$ by their Euler identity equivalents results in

$$-\frac{1}{2}(V_s e^{j\omega t} + V_s^* e^{-j\omega t}) + \frac{1}{2C}\left(\frac{1}{j\omega}Ie^{j\omega t} - \frac{1}{j\omega}I^* e^{-j\omega t}\right) +$$
$$\frac{L}{2}(j\omega Ie^{j\omega t} - j\omega I^* e^{-j\omega t}) + \frac{R}{2}(Ie^{j\omega t} + I^* e^{-j\omega t}) = 0$$

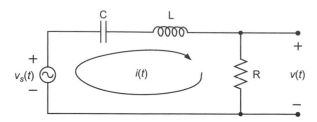

Figure 6.11   Series connected RLC circuit.

Rearranging this equation gives

$$\left(-V_s + \frac{1}{j\omega C}I + j\omega LI + RI\right)e^{j\omega t} + \left(-V_s + \frac{1}{j\omega C}I + j\omega LI + RI\right)^* e^{-j\omega t} = 0$$

which must be true for all t. This requires that

$$-V_s + \frac{1}{j\omega C}I + j\omega LI + RI = 0 \tag{6.50}$$

The time-domain sum of voltage drops in (6.49) has been replaced by the sum of phasor voltage drops in (6.50) resulting in an algebraic equation, which we could have obtained by inspection of the circuit. Solving for I gives

$$I = \frac{1}{\dfrac{1}{j\omega C} + j\omega L + R}V_s = \frac{j\omega C}{1 - \omega^2 LC + j\omega RC}V_s = Y(j\omega)V_s$$
$$= ||Y(j\omega)||e^{j\angle Y(j\omega)}Ae^{j\theta} = ||Y(j\omega)||Ae^{j(\angle Y(j\omega)+\theta)} \tag{6.51}$$

Notice that when the frequency of the input voltage $v_s(t)$ is $\omega = 1/\sqrt{LC}$ rads/sec, then $I = V_s/R$, as if the capacitor and inductor connected in series are equivalent to an ideal conductor. With (6.51) we can find how the circuit modifies the amplitude and phase of the input voltage $v_s(t)$ to obtain the amplitude and phase of the current $i(t)$, and

$$i(t) = ||Y(j\omega)||A\cos(\omega t + \angle Y(j\omega) + \theta)$$

With (6.48) the phasor V for the voltage $v(t)$ is given by

$$V = RI = \frac{j\omega RC}{1 - LC\omega^2 + j\omega RC}V_s = H(j\omega)V_s$$
$$= ||H(j\omega)||Ae^{j(\angle H(j\omega)+\theta)} \tag{6.52}$$

where $H(j\omega)$ is a complex function of the real variable ω and the frequency of the input voltage $v_s(t)$. Since $H(j0) = 0$, $H(j\infty) = 0$, and $H(j/\sqrt{LC}) = 1$, then the circuit behaves like a band-pass filter as $\omega = 0 \rightarrow \infty$. We will look at this more closely in Example 6.11. With V, $v(t)$ is given by

$$v(t) = \|H(j\omega)\|A\cos(\omega t + \angle H(j\omega) + \theta)$$

If we consider the output of the circuit to be the voltage $v(t)$, the voltage across the resistor, then $H(j\omega)$, which is called the **transfer function**, describes how the circuit modifies the amplitude A and phase angle θ of the input $v_s(t)$ to obtain the amplitude $\|H(j\omega)\|A$ and phase angle $\angle H(j\omega) + \theta$ of the output.

If instead, the current $i(t)$ is the output of interest, then $Y(j\omega)$ in (6.51) is the transfer function from the input $v_s(t)$ to this output.

Example 6.11 _____

With the complex arithmetic capability of MATLAB, we can assess the performance of the circuit given in Fig. 6.11 for different values of circuit components. Let $R = 100\ \Omega$, $L = 11\ \text{mH}$, and $C = 0.01\ \mu\text{F}$. The following MATLAB script finds $H(j\omega)$, and plots the magnitude shown in Fig. 6.12.

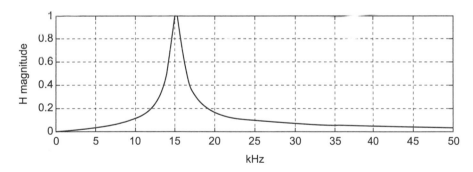

Figure 6.12   Magnitude frequency response of the RLC circuit.

```
clear all; clc;
R=100; % resistor value
L=11*10^-3; % inductor value
C= 0.01*10^-6; % capacitor value
f=linspace(0,5.0*10^4,1001); % frequency range
```

```
w=2.0*pi*f; % vector of frequencies
H_num=j*w*R*C; % transfer function numerator
H_den=1-L*C*w.*w+j*w*R*C; % transfer function denominator
H=H_num./H_den; % element by element complex divide
plot(f/1000,abs(H)); % plot band-pass filter magnitude
grid on
xlabel('KHz')
ylabel('H magnitude')
```

Program 6.7   Find and plot the magnitude frequency response.

From Fig. 6.12 we conclude that when the input sinusoid has a frequency given by 15K Hz, then the output sinusoid will have an amplitude larger than the amplitude of an output for any other input sinusoid frequency. If the frequency of any sinusoidal input is close to 15K Hz, then the output amplitude is close to the input amplitude, while if an input sinusoid has a frequency significantly below or above 15K Hz, then the output sinusoid will have a much smaller amplitude than the input. This is the behavior of a **band-pass filter**. We can control where the magnitude frequency response shown in Fig. 6.12 peaks by adjusting, for example, the value of the capacitor.

The frequency selective behavior of circuits like the one in Fig. 6.11 is widely applied in communication systems. In fact, we can say that if it were not possible to achieve such behavior, then the simultaneous activity of all AM radio, FM radio, cell-phone, wireless Internet, etc., communications would not be possible.

Example 6.12 _____

Let us apply the complex impedance concept to the circuit given in Fig. 6.13, and find the response (output) $v(t)$ to the input $v_s(t)$.

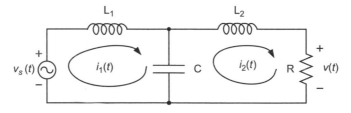

Figure 6.13   Circuit of a third-order Butterworth low-pass filter.

Let the component values be: $L_1 = 3/2$ H, $L_2 = 1/2$ H, $C = 4/3$ F, and $R = 1$ Ω. Suppose the input is the sum of two sinusoids given by

$$v_s(t) = 2 \cos\left(2\pi f_1 t - \frac{\pi}{4}\right) + 3 \cos\left(2\pi f_2 t - \frac{\pi}{6}\right)$$

where $f_1 = 0.2$ Hz and $f_2 = 2$ Hz. The input is shown in Figure 6.14. This input was plotted with the following script.

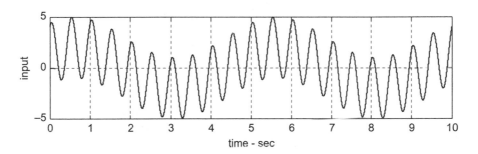

Figure 6.14   Voltage input.

```
clear all; clc;
% frequencies of the two components of vs(t)
f1=0.2; w1=2*pi*f1; f2=2.0; w2=2*pi*f2;
T_total=2/f1; % total plot time, which is two cycles of the f1=0.2 Hz sinusoid
N=5001; % specify number of time points
t=linspace(0,T_total,N); % time points
vs=2*cos(w1*t-pi/4)+3*cos(w2*t-pi/6);   % the input to the circuit
plot(t,vs)
grid on
xlabel('time - sec')
ylabel('input')
```

Program 6.8   Find and plot the input.

Now, replace each circuit component by its impedance and each sinusoidal current and sinusoidal voltage by its phasor resulting in the frequency-domain circuit shown in Fig. 6.15. Since the given circuit is described by a set of linear equations, we can find the steady-state response to $v_s(t)$ by finding the response to each sinusoidal component in $v_s(t)$ and then sum these responses. This is the same as considering $v_s(t)$ to be the series connection of two voltage sources, each producing one of the sinusoids in $v_s(t)$, and then applying the superposition principle.

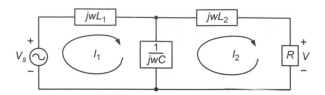

Figure 6.15   Frequency-domain circuit.

In the frequency domain (using impedances), the Kirchhoff voltage law (KVL) equations for the circuit meshes are given by

$$-V_s + j\omega L_1 I_1 + \frac{1}{j\omega C}(I_1 - I_2) = 0 \tag{6.53}$$

$$\frac{1}{j\omega C}(I_2 - I_1) + j\omega L_2 I_2 + R I_2 = 0 \tag{6.54}$$

which are two equations in the two unknowns I_1 and I_2. We must solve (6.53) and (6.54) with $\omega = \omega_1$ to find the phasor I_2, and then the contribution to $v(t)$ due to the first component in $v_s(t)$ can be found, and then again with $\omega = \omega_2$ to find the contribution to $v(t)$ due to the second component in $v_s(t)$. This is done by the following MATLAB script, and the resulting output is shown in Fig. 6.16, where you see that the high frequency component of $v_s(t)$ has been significantly attenuated (filtered) by the circuit.

```
% AC analysis of a third order Butterworth low-pass filter
clear all; clc
L1=3/2; L2=1/2; C=4/3; R=1; % specify circuit component values
f1=0.2; f2=2.0; % frequencies of sinusoidal components of the input signal
w=2*pi*[f1 f2];
Vs=[2*exp(-j*pi/4) 3*exp(-j*pi/6)]; % phasors of the two input components
T_total=2/f1; N=1001; % plot output over 2 cyles of the first input component
t=linspace(0,T_total,N);
v=zeros(1,N); % preallocate the output voltage vector to zero
for k=1:2 % analyze the circuit for each input component frequency
    % find impedances
    ZL1=j*w(k)*L1;
    ZL2=j*w(k)*L2;
    ZC=1/(j*w(k)*C);
    Z=[ZL1+ZC  -ZC; ... % set up the impedance matrix Z
        -ZC  ZC+ZL2+R];
```

```
    V=[Vs(k)   0]'; % mesh voltage vector
    Y=inv(Z); % invert the impedance matrix
    I=Y*V; % solve for the mesh current phasor vector
    % I(2) is the phasor of the current in mesh 2 for the frequency w(k)
    V(k)=R*I(2); % phasor of output voltage component
    v=v+abs(V(k))*cos(w(k)*t + angle(V(k))); % contribution to output for w(k)
end
plot(t,v)
grid on
xlabel('t - sec')
ylabel('output voltage')
```

Program 6.9   MATLAB program for AC circuit analysis.

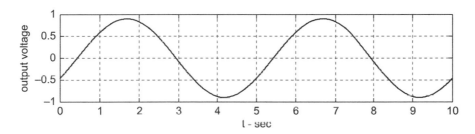

Figure 6.16   Output of the low-pass filter.

This plot shows that the circuit has almost entirely removed (filtered) the higher frequency sinusoidal contribution to $v_s(t)$, even though the amplitude of the higher frequency sinusoid is larger than the amplitude of the lower frequency sinusoid. It is said that the circuit is a **low-pass filter**, meaning that sinusoids at low frequencies get through the circuit almost unaltered, while sinusoids at high frequencies are attenuated (almost removed). This and other kinds of filtering are widely applied activities in signal processing and communication systems.

To find the transfer function $H(j\omega)$ from the input to the output $v(t)$, use (6.54) to express I_1 in terms of I_2, and substitute this into (6.53) resulting in

$$V = RI_2 = \frac{R}{R - \omega^2 L_1 RC + j\omega(L_1 + L_2) - j\omega^3 L_1 L_2 C} V_s = H(j\omega)V_s \qquad (6.55)$$

Notice that $H(j0) = 1$ and $\lim_{\omega \to \infty} H(j\omega) = 0$. This means that input sinusoidal signals with very high frequencies are attenuated by the circuit, while input sinusoidal signals with very low frequencies pass through the circuit. For example, see the difference between the input in Fig. 6.14 and the output in Fig. 6.16. This will be studied further, after some plotting tools have been presented in Chapter 9. A parameter that describes the frequency selective behavior of a low-pass filter is the **bandwidth** (BW), which is the highest frequency for which the magnitude squared of the transfer function is greater than one half its maximum value. The low-pass filter in this example has BW = 1 rad/sec.

6.9   Operational Amplifier

The **operational amplifier** (Op-Amp) is a very versatile device. It is a fundamental building block in many circuits designed for analog signal processing. The symbol of the Op-Amp, a triangle, is shown in Fig. 6.17. The left side of Fig. 6.17 shows how an Op-Amp receives electric power, which ranges from $-V_{Ref}$ to $+V_{Ref}$ with respect to the common ground (reference) terminal. An Op-Amp has two inputs, V_a and V_b, connected to the Op-Amp input terminals labeled with the plus and minus signs, respectively.

Basically, an Op-Amp is a circuit comprised of many transistors that is designed to have a few, but very important properties, which are:

1) The currents I_a and I_b into the plus and minus terminals of the Op-Amp are nearly zero, which means that the resistances between these terminals and the reference terminal are very large.
2) The output voltage V_o is given by $V_o = A (V_a - V_b)$, where A is very large, $A > 10^6$, and $-V_{Ref} \le V_o \le +V_{Ref}$.
3) The resistance between the output terminal and the reference terminal is nearly zero, which means that the output voltage is nearly an ideal voltage source.

It is said that according to property (1), an Op-Amp has a very **high input impedance**, according to property (2) an Op-Amp has a very **high gain**, and according to property (3) an Op-Amp has a very **low output impedance**. The output voltage V_o cannot be greater than $+V_{Ref}$, and it cannot be less than $-V_{Ref}$. If the input voltages V_a and V_b cause V_o to reach either $+V_{Ref}$ or $-V_{Ref}$, then it is said that the Op-Amp is **saturated**. In a schematic, for convenience, the power connections are usually not shown, as in the drawing on the right of Fig. 6.17.

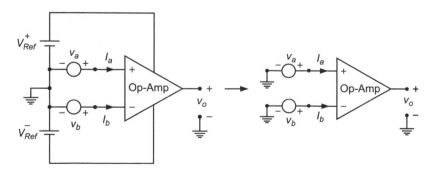

Figure 6.17   Operational amplifier circuit and simplified version.

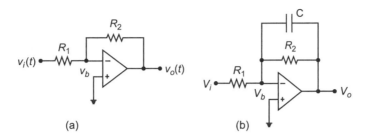

Figure 6.18 (a) Inverting amplifier and (b) low-pass filter.

The operation of a circuit made with Op-Amps depends on the kinds and configuration of additional components connected to the Op-Amps. For example, consider the two circuits shown in Fig. 6.18, where $v_a = 0$ and the input voltage $v_i(t)$ is due to some voltage source.

To find the relationship between $v_o(t)$, the output voltage, and $v_i(t)$, the input voltage, of the circuit in Fig. 6.18(a), apply KCL to the minus terminal of the Op-Amp. This gives

$$\frac{v_b - v_i(t)}{R_1} + \frac{(v_b - v_o(t))}{R_2} = 0 \tag{6.56}$$

where the current into the minus terminal of the Op-Amp is set to zero. The output voltage $v_o(t)$ of the Op-Amp is given by

$$v_o(t) = A(0 - v_b) \rightarrow v_b = -\frac{v_o(t)}{A} \tag{6.57}$$

Substituting (6.57) into (6.56) results in

$$\left(\frac{1}{R_1} + \frac{1}{R_2}\right)\left(\frac{v_o(t)}{A}\right) + \frac{v_i(t)}{R_1} + \frac{v_o(t)}{R_2} = 0$$

Since A is very large, the first term is negligible compared the second and third terms, and we get

$$v_0(t) = -\frac{R_2}{R_1} v_i(t) \tag{6.58}$$

This circuit is called an **inverting amplifier**. For example, to design an amplifier with a gain of −10, so that $v_o(t) = -10 v_i(t)$, use $R_2 = 100K\ \Omega$ and $R_1 = 10K\ \Omega$. Generally, Op-Amps are low-power devices, and resistor values are used such that currents through them are not more than a few mA.

Notice that the equations describing the circuit in Fig. 6.18(a) are linear equations. This means that if the input voltage is a sinusoid, then the output voltage is also a sinusoid with the same frequency as the input sinusoid.

Let us find the output voltage of the circuit shown in Fig. 6.18(b) when the input is a sinusoid given by $v_i(t) = K\cos(\omega t + \theta)$. Recall that for AC analysis of linear circuits we can replace voltage and current sources by their phasors and circuit components by their impedances, and then apply Kirchhoff's laws in terms of phasors and impedances. Applying KCL to the minus terminal of the Op-Amp in Fig. 6.18(b) gives

$$\frac{V_b - V_i}{R_1} + \frac{V_b - V_o}{R_2} + \frac{V_b - V_0}{1/j\omega C} = 0 \tag{6.59}$$

where $V_i = Ke^{j\theta}$, the phasor of the input. Based on (6.57) we have the phasor relationship $V_b = -V_o/A$, and therefore (6.59) becomes

$$\left(\frac{1}{R_1} + \frac{1}{R_2} + j\omega C \right) \left(\frac{V_o}{A} \right) + \frac{V_i}{R_1} + \left(\frac{1}{R_2} + j\omega C \right) V_o = 0$$

Compared to the second and third terms, the first term is negligible, and solving for V_0 results in

$$V_o = -\frac{R_2}{R_1(1 + j\omega R_2 C)} V_i, \quad H(j\omega) = -\frac{1}{R_1((1/R_2) + j\omega C)} \tag{6.60}$$

Equation (6.60) gives the transfer function $H(j\omega)$ of the circuit. We see that as the frequency of the input changes from $\omega = 0$ to $\omega \to \infty$, the transfer function magnitude decreases from R_2/R_1 to zero, respectively, which means that this circuit is a **low-pass filter**. With R_1 and R_2, we can design a low-pass filter to have a desired low frequency gain, and with C we can specify the BW of the filter.

If the resistor R_2 is removed, which is like $R_2 \to \infty$, then the transfer function becomes $H(j\omega) = -1/j\omega R_1 C$. Then, like (6.44), the circuit input-to-output relationship is given by

$$v_o(t) = \frac{-1}{R_1 C} \int v_i(t) dt \tag{6.61}$$

which means that the output of the circuit is proportional to the **integral** of the circuit input. This circuit is called an **integrator**.

Let us consider a few more useful circuits that were designed with Op-Amps. Consider the two circuits shown in Fig. 6.19.

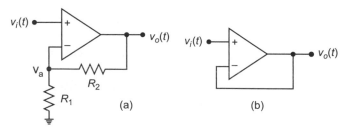

Figure 6.19   (a) Positive gain amplifier and (b) a buffer.

To find the relationship between $v_o(t)$, the output voltage, and $v_i(t)$, the input voltage, of the circuit in Fig. 6.19(a), apply KCL to the minus terminal of the Op-Amp. This gives

$$\frac{v_a}{R_1} + \frac{v_a - v_o(t)}{R_2} = 0 \tag{6.62}$$

For the Op-Amp we have

$$v_o(t) = -A(v_a - v_i(t)) \tag{6.63}$$

From (6.63) we get

$$v_a = v_i(t) + \frac{v_o(t)}{-A}$$

and since A is very large, $v_a \cong v_i(t)$. Substituting v_a into (6.62) gives

$$v_o(t) = \left(1 + \frac{R_2}{R_1}\right) v_i(t) \tag{6.64}$$

This circuit is a **noninverting amplifier**. For example, to amplify a signal $v_i(t)$ by a gain of 10, use $R_2 = 45\text{K }\Omega$ and $R_1 = 5\text{K }\Omega$ to get $v_o(t) = 10v_i(t)$. Note that depending on the maximum input level, the gain must be limited to avoid saturating the Op-Amp output.

Suppose we remove R_1 and replace R_2 by a short circuit. Then we get the circuit in Fig. 6.19(b). According to (6.64), we have $v_o(t) = v_i(t)$. The circuit is called a **buffer**. A buffer does not amplify an input signal. It does, however, have a very high input resistance and a very low output resistance. A buffer is often connected between a signal source, for example, some transducer such as a microphone, and a device that is intended to receive the output of the transducer. The buffer electronically isolates the signal source from whatever is intended to receive the signal source output.

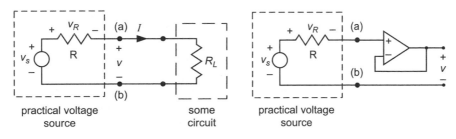

Figure 6.20 Buffer application.

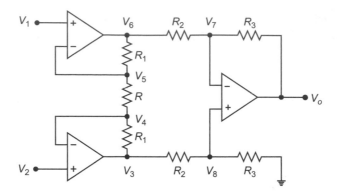

Figure 6.21   Schematic of an instrumentation amplifier.

Let us consider the practical problem of connecting a signal source to some circuit, as depicted in Fig. 6.20. On the left is a practical voltage source connected to some circuit, represented by the resistor R_L. Generally, a practical voltage source has an internal resistance. This is modeled by an ideal voltage source v_s in series with a resistor R. When a circuit is connected to the terminals (a) and (b) of the voltage source, a current I will cause a voltage drop V_R, and the voltage at the terminals will be $v = v_s - v_R = v_s - RI$. If R_L is small, then I may be large enough to make $v < v_s$. Therefore, the voltage at the terminals of the voltage source depends on the resistance of the circuit to which the voltage source is connected. It is said that the circuit **loads down** the voltage source.

To avoid the loading problem, a buffer is used, as shown on the right of Fig. 6.20, where the buffer output should be connected to the circuit. Since the input resistance of the buffer is very large, V_R will be small, and $v \cong v_s$.

Another very useful circuit is shown in Fig. 6.21. This circuit is called an **instrumentation amplifier**, and its output voltage is given by $V_o = K(V_2 - V_1)$, where the constant K is determined by the resistor values. This circuit has seven unknown node voltages, and it will be analyzed in Chapter 11 with the help of the MATLAB Symbolic Math Toolbox.

An instrumentation amplifier is widely used in audio, medical, and other electronic equipment where there is concern about environmental electrical noise.

6.10   Conclusion

In this chapter the complex number was defined in rectangular and polar forms. MATLAB also works with complex numbers given as scalars, vectors, and matrices. It was demonstrated that complex numbers play an important role in signal analysis to represent a periodic signal as a linear combination of complex conjugate exponential functions. The inductor and capacitor circuit components were introduced. For RLC circuit analysis, the concepts of a phasor and impedance of circuit components were introduced to convert integral–differential equations that describe the behavior of a circuit in the time domain into algebraic equations in the frequency domain. And, with MATLAB it was convenient to solve complex systems of linear equations. There is much more to find out. You should now know how to

- find the real and complex roots of a polynomial with MATLAB
- do arithmetic with complex numbers
- use many of the MATLAB built-in functions concerned with complex numbers
- apply Euler's identity for representing sinusoidal time functions
- use MATLAB for analyzing a periodic signal
- work with the concept of an impedance and apply MATLAB for AC circuit analysis
- investigate the frequency selective behavior of a circuit with MATLAB
- analyze some circuits designed with Op-Amps

Table 6.3 Built-in MATLAB functions introduced in this chapter

Function	Brief explanation
complex	Constructs a complex number
isreal	Checks to see if a complex number has a zero imaginary part
conj	Returns the complex conjugate of a complex number
real	Returns the real part of a complex number
imag	Returns the imaginary part of a complex number
roots	Finds the roots of a polynomial
poly	Finds the polynomial coefficients given its roots
angle	Finds the angle of a complex number
abs	Finds the magnitude of a real or complex number
subplot(m,n,p)	Breaks the Figure Window into an m-by-n matrix of $p = 1, \ldots$, mn small plots

Table 6.3 gives the MATLAB functions that were introduced in this chapter. Use the MATLAB help facility to learn more about these built-in functions, where you will also find out about many other related built-in functions.

Also, character strings were used to enhance plot output. In the next chapter built-in functions for character string construction, identification, and manipulation will be discussed.

Problems

Section 6.1

1) Manually, write each polynomial as a product of first-order factors: (a) $x^2 - 2x - 15$, (b) $-x^2 - x + 6$, and (c) $x^3 + 5x^2 + 8x + 6$ (hint: one of the roots is $x = -3$).
2) For each polynomial of Prob. (1) find the value(s) of x where the polynomial has a minimum or maximum. Give the minimum or maximum value of each polynomial.

Section 6.2

Where required in the following problems, use the following numbers:
$$c_1 = j, \ c_2 = 2+j3, \ c_3 = -3+j, \ c_4 = -4-j3, \ c_5 = -j3, \ c_6 = 2-j3, \ c_7 = 2e^{j\pi/4},$$
$$c_8 = 3e^{j\pi/2}, \ c_9 = 3e^{j5\pi/4}, \ c_{10} = 3e^{j7\pi/3}$$

3) Manually, obtain: (a) $c_1 + c_2 + c_3$, (b) $c_3 - c_4$, (c) $c_2 c_3$, (d) c_1/c_2, (e) c_2/c_3, and (f) c_5/c_6.

4) For each part of Prob. (3) give MATLAB statements to do the arithmetic.

5) Give a sketch of each of the complex numbers $c_1, \ldots, c_6$ in the complex plane.

6) (a) Find the real part of: $1/c_2$. Find the imaginary part of: (b) $c_3 c_4^*$, (c) c_5/c_6.

7) For each part of Prob. (6) give MATLAB statements to find the answer.

8) (a) Use the MATLAB function roots to find the roots of: $2x^6 + 3x^4 - x^3 + 4x - 5$.
   (b) Use the MATLAB function poly to find the polynomial with roots given by: $-j$, $+j$, 3, 2+j5, 2-j5.

9) Write a script to find the number of real roots of an Nth order polynomial. Use the function input to obtain N from the program user. Then, continue to use the function input to obtain the polynomial coefficients. Terminate the program if the user enters $N \leq 0$.

10) For any two complex numbers x_1 and x_2 prove that: $||x_1 + x_2|| \leq ||x_1|| + ||x_2||$. Hint: let $x_3 = x_1 + x_2$, and then x_1, x_2, and x_3 form a triangle in the complex plane, to which you can apply the triangle inequality.

Section 6.3

11) Manually, convert the complex numbers: $c_1, \ldots, c_6$ to polar form.

12) For each complex number of Prob. (11) give MATLAB statements to find the magnitude and angle.

13) Manually, convert the complex numbers: $c_7, \ldots, c_{10}$ to rectangular form.

14) Manually, obtain: (a) $c_7 c_8$, (b) $c_8 + c_9$, (c) c_{10}^*, and (d) c_7/c_9.

15) For each part of Prob. (14) give MATLAB statements to find the answer.

16) For any two complex numbers x_1 and x_2 prove that: (a) $||x_1/x_2|| = ||x_1||/||x_2||$, (b) $(x_1 x_2)^* = x_1^* x_2^*$, and (c) $\angle(x_1/x_2) = \angle x_1 - \angle x_2$.

17) Given is a polynomial function $f(x)$ of a complex variable x with real coefficients. Prove that $f^*(x) = f(x^*)$.

Section 6.4

18) Use Euler's identity to prove that $\cos(\alpha + \beta) = \cos(\alpha)\cos(\beta) - \sin(\alpha)\sin(\beta)$.

19) For the sinusoidal voltage $x(t) = -3\cos(20\pi t + (\pi/6))$ mV, find the amplitude, frequency, period, and phase angle. Give units with each answer.

20) Write a script to plot on one axes $x(t) = \cos(2\pi f t)$, where $f = 3, 6, 12$, and 18 Hz. Plot each signal over one period of the sinusoid with the lowest frequency. Use a sampling rate of $f_s = 300$ samples/sec. Include axes labels.

21) Find the phasor X of each sinusoid: (a) $x(t) = -3\cos(20\pi t + \pi/6)$, (b) $x(t) = 5\cos(120\pi t - \pi/4)$, (c) $x(t) = -7\sin(4\pi t - \pi/6)$, and (d) $x(t) = 2\sin(1000\pi t)$.

22) Use $\omega = 10\pi$ rad/sec, and find the sinusoidal time functions that have the phasors: (a) $X = 3e^{j\pi/3}$, (b) $X = e^{j\pi}$, (c) $X = -j$, and (d) $X = -3 + j4$.

23) With $\omega_0 = 4\pi$ rad/sec and $X = 4e^{j\pi/6}$, use (6.24) to write a script that plots $x(t)$ over two periods. Plot enough points to obtain a smooth looking plot. Include axes labels.

Section 6.5

24) One period of a rectangular wave $x(t)$ is given by

$$x(t) = \begin{cases} -2, -1 \le t < 1 \\ 4, \ 1 \le t < 3 \end{cases}$$

(a) Manually, give a sketch of $x(t)$ for $-5 \le t \le 7$. What is the fundamental frequency f_0?

(b) Find the trigonometric Fourier series coefficients a_0, a_k, and b_k, $k = 1, 2, \ldots$.

25) For the periodic function described in Prob. (24) find the complex exponential Fourier series coefficients.

26) For some square wave $x(t)$ with period $T_0 = 8$ secs, the complex exponential Fourier series coefficients are given by

$$X_k = (-2e^{jk\pi/4} + e^{-jk3\pi/4})\frac{\sin(k\pi/2)}{k\pi}, \ X_0 = -1/2$$

Write a script to plot two periods of the Fourier series of $x(t)$ given by (6.31). Obtain two plots, one with a total of 11 terms in the Fourier series and one with a total of 51 terms in the Fourier series. In each case, plot a total of 1001 points. Explain the difference between the two plots. To see the details, you may have to plot more than 1001 points.

27) For the square wave of Prob. 26, write a script to plot the magnitude spectrum versus frequency over the range $-50f_0 \le f \le 50f_0$, where f_0 is the fundamental frequency.

28) Write a script that applies Parseval's relation to find the average power of the signal described in Prob. 26. Use enough terms to make your result accurate to four significant digits.

Section 6.6

29) What does the impedance of (a) an inductor become as $\omega \to 0$ and as $\omega \to \infty$, (b) a capacitor become as $\omega \to 0$ and as $\omega \to \infty$. If the impedance goes to zero, then the component looks like a short circuit, and if the impedance goes to infinity, then the component looks like an open circuit.

30) In the circuit shown in Fig. P6.30, the voltage source is $v_s(t) = 5\sin(10,000t)$ volts, and $C = 0.25$ µF, $L = 10$ mH, and $R = 100$ Ω. Draw another circuit, where the voltage source, loop current, and output voltage are replaced by their phasors, and all components are replaced by their impedances.

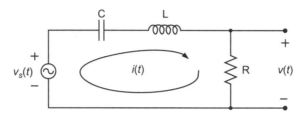

Figure P6.30   Series RLC circuit.

Section 6.7

31) (a) To the circuit shown in Fig. P6.30 apply KVL to obtain a loop equation in terms of phasors and impedances. Use component values given in Prob. 6.30. Then, write a MATLAB script to solve for the phasors of $i(t)$ and $v(t)$.
   (b) Continue the MATLAB script to plot $v_s(t)$ and $v(t)$ on the same axes over two periods.
   (c) Find the frequency of $v_s(t)$ that will cause $v(t)$ to have the largest amplitude.

32) (a) For the circuit shown in Fig. P6.32, find the transfer function $H(j\omega)$ from the input $v_s(t)$ to the output $v(t)$.

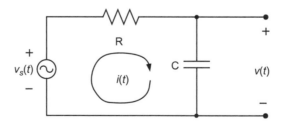

Figure P6.32   RC circuit.

   (b) Set $R = 442$ Ω and $C = 0.1$ µF, and write a MATLAB script to plot the magnitude squared frequency response for $0 \leq f \leq 10K$ Hz. What kind of a filter is this circuit? From the plot of the magnitude squared frequency response determine the bandwidth of the filter.

33) Let the input $v_s(t)$ of the circuit given in Fig. P6.32 be a square wave with period $T_o = 8$ msecs. Then, we can write

$$v_s(t) \cong \sum_{k=-K}^{K} X_k e^{jk\omega_o t}$$

where $\omega_o = 2\pi/T_o = 1000\pi/4$ rad/sec. Suppose the complex Fourier series coefficients are given by

$$X_k = (-2e^{jk\pi/4} + e^{-jk3\pi/4})\frac{\sin(k\pi/2)}{k\pi}, \quad X_0 = -1/2$$

Since the circuit is a linear circuit, we can employ superposition to find the output due to each sinusoid at the frequency $\omega = k\omega_o$ in $v_s(t)$, and then sum all the outputs due to each sinusoidal input to obtain $v(t)$. In Prob. 6.32 the transfer function $H(j\omega)$ was found. Use the component values given in Prob. 6.32. The phasor Y_k of each sinusoid in the output $v(t)$ is given by

$$Y_k = H(jk\omega_o)X_k, \quad Y_0 = H(j0)X_0$$

Summing the responses to each sinusoidal input gives

$$v(t) \cong \sum_{k=-K}^{K} Y_k e^{jk\omega_o t}$$

Write a MATLAB script to plot $v_s(t)$ and $v(t)$ over two periods. Notice that, while the input is discontinuous, the output of the low-pass filter is continuous. Execute the program for $K = 5$ and $K = 50$ to see the effect of the number of terms used in the Fourier series.

34) (a) Fig. P6.34 shows a circuit of a high-pass filter. Find the transfer function $H(j\omega)$ from the input $v_s(t)$ to the output $v(t)$.

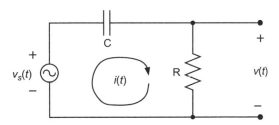

Figure P6.34   A high-pass filter.

(b) Using the same component values as in Prob. 6.32, write a MATLAB script to plot the magnitude squared frequency response for $0 \leq f \leq 10$K Hz. What is the bandwidth of this filter?

35) (a) Prove that if a periodic function is an even function, then the complex Fourier series coefficients are real. In this case, what is true about the trigonometric coefficient b_k?

(b) Prove that if a periodic function is an odd function, then the complex Fourier series coefficients are imaginary. In this case, what is true about the trigonometric coefficient a_k?

Section 6.8

36) The inverting amplifier given in Fig. 6.18(a) can be cascaded with another inverting amplifier, where the output of the first amplifier is connected to the input to the second amplifier. Overall, the result is a positive gain amplifier. Use two inverting amplifiers to design a positive gain amplifier with an overall gain of 100. Let each inverting amplifier have a gain of -10. Give the schematic.

37) The input voltage of a noninverting amplifier varies over the range $-100\,\mathrm{mV}$ to $+200\,\mathrm{mV}$. The power supply voltage is $V_{Ref} = 9$ volts. What is the most gain that can be used and avoid saturating the output?

38) Fig. P6.38 shows another way to use an Op-Amp. Following the derivation of (6.58), show that

$$v_o(t) = -\left(\frac{R_3}{R_1}v_1(t) + \frac{R_3}{R_2}v_2(t)\right)$$

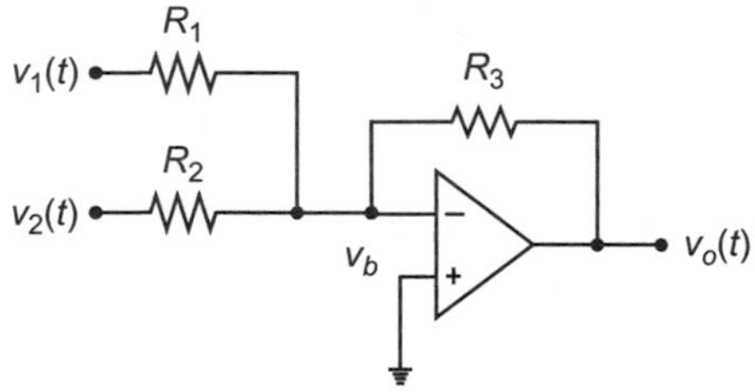

Figure P6.38   Summing amplifier.

CHAPTER 7

Character Data

While computing does include the processing of numerical data, more generally, computing is concerned with collecting, processing, and presentation of information. In a digital computer system, information is coded in binary format. The binary codes represent not only numerical and logical data, but also character data, which can have many kinds of meaning. For example, in a program file, all variable names and their values, key and reserved words, operations, data, and more are character strings. In this chapter you will learn how to

- create character strings
- manipulate character strings
- search character strings
- create and work with structure arrays
- create and work with cell arrays

7.1 Character Strings

A **character string** can be created in several different ways. The syntax of a direct way is

```
s = ' .... text .... '
```

where s, a valid MATLAB® variable name, is the character string, and the text is its value. With this syntax, the variable s is a one-row **array** that contains the ASCII (American Standard Code for Information Interchange) codes of the text characters. See Appendix B for a table of ASCII codes. The length of s is the number of characters in the text. This includes the ASCII codes for unprintable characters such as space, carriage return, tab, and others. To include a single quotation mark within the text, use single quotation marks twice. For example,

```
>> % the spaces before and after the minus sign are also characters
>> current = 'I2 - I1';
>> length(current)  % there are 7 elements in the character array current
ans =     7
>> % the elements of a character array are referenced like the elements of a matrix
>> current(4) % the fourth element in current is a minus sign
ans =      -
>> disp(current)  % same as disp('I2 - I1')
I2 - I1
>> comment = 'you''re doing very well'  %  '' counts as one character
comment =   you're doing very well
```

Table 7.1 gives built-in MATLAB functions concerned with creating character strings and character arrays. A **cell array** can be an array of character strings. However, a cell array, as we shall see later, provides a means to collect and organize data arrays of any size and any kind of data type.

Table 7.1   Built-in MATLAB functions concerned with character data

Built-in function	Brief description (cell arrays will be discussed in section 7.4)
s = [s1 s2...]	Concatenate character arrays into a new character array
C = {s1 s2...}	Create a cell array of character strings. Separate each row of the cell array with a semicolon (;). Notice the use of braces.
s = strcat(s1, s2,...)	Horizontally concatenate s1, s2, etc., which can be character arrays or cell arrays of strings
s = strvcat(s1, s2,...)	Vertically concatenate s1, s2, etc., which can be character arrays or cell arrays of strings
s = char(s1, s2,...)	Vertically concatenate character arrays s1, s2, etc., padding each input string with blank characters as needed such that each row contains the same number of characters
s = char(A)	Convert a numeric array A that contains positive integers of ASCII numeric codes into a character array
s = char(C)	Convert cell array C of character strings into a character array, and place each element of C into a row of the column character array s
B = double(A)	Convert a character array A or string s of ASCII codes into numeric ASCII codes
cellstr(s)	Create cell array of strings from character array s
blanks(n)	Create character string of n blanks
num2str(A)	Convert numeric matrix A into a character array
int2str(N)	Convert integer matrix N into a character array

Example 7.1 ————————————————

```
>> % concatenate three character strings
>> resistor_property = ['Ohm''s'  ' Law:'  ' v=Ri']
resistor_property = Ohm's Law: v=Ri
```

```
>> % or, use the function strcat to
>> % concatenate three character strings
>> resistor_property = strcat('Ohm''s', ' Law:', ' v=Ri')
resistor_property = Ohm's Law: v=Ri
>> length(resistor_property)
ans =        14
>> disp(['insert some' blanks(10) 'space'])   % display character string
insert some              space
>> x = [72 101 108 108 111]   % numeric vector of 5 ASCII codes
x =    72   101   108   108   111
>> y = [116 104 101 114 101 33]    % another numeric vector of 6 ASCII codes
y =   116   104   101   114   101    33
>> s = char(x,y)   % convert to character array and pad with blanks if necessary
s =
Hello
there!
>> size(s)   % s is a character array of 2 rows and 6 columns
ans -     2    6
>> z = double(s)   % convert characters to numeric values, ASCII codes
z =
     72   101   108   108   111    32
    116   104   101   114   101    33
>> % notice the addition of the ASCII code for a blank in the first row and
>> % sixth column of z
>> s1 = 'up'; s2 = 'down'; s3 = 'left'; s4 = 'right';
>> s = strvcat(s1,s2,s3,s4)   % vertically concatenate the four character strings
s =
up
down
left
right
>> size(s)    %  s is an array of characters, where each row has 5 elements
ans =    4    5
>> s(2,3)    % element of s in row 2 and column 3
ans =    w
>> % elements of s are characters
>> % braces, instead of brackets, are used to create a cell array
>> % create a cell array of character strings
>> direction = {'up', 'down'; 'left', 'right'}
direction =
    'up'      'down'
    'left'    'right'
>> %  in a cell array, the elements can have different sizes
>> direction{2,2}   % use braces to reference elements of a cell array
ans =     right
>> s = char(direction)   % convert cell array into a character array
```

```
s =
up
left
down
right
>> % each row of s has the same number of characters due to padding with blanks
>> C = cellstr(s)    % create cell array of strings from character array
C =
    'up'
    'left'
    'down'
    'right'
>> C{2}  %  elements of C are character strings
ans =      left
```

Sometimes it is useful to place numeric information in a plot title or some other character string. To do this, numerical data must be converted into character data. For example,

```
>> K = 100;    %  numeric value
>> % to include the value of K in a character string, in must be converted
>> % into a character string
>> % then, concatenate three character strings
>> title = ['Fourier series of x(t) using K = ', num2str(K),'  terms']
 title =   Fourier series of x(t) using K = 100  terms
```

Table 7.2 gives built-in MATLAB functions with which you can check for each kind of character data. See Table 5.6 for a list of the classes of data that MATLAB supports.

Table 7.2   Built-in functions concerned with checking character data

Built-in function	Brief description
ischar(s)	Returns 1, meaning true, if s is a character array and 0 otherwise
iscellstr(C)	Returns 1 if C is a cell array of character strings and 0 otherwise
isa(x,'classname')	Returns 1 if x is of a particular class of data. See Table 5.6 for class names.
isletter(s)	Returns a logical matrix with elements that are 1 for corresponding elements of s that are letters of the alphabetic and 0 otherwise
isscalar(s)	Returns 1 if s is a scalar and 0 otherwise
isspace(s)	Returns a logical matrix with elements that are 1 for corresponding elements of s that are unprintable characters, such as space, tab, etc., and 0 otherwise
isstrprop(s,'type')	Returns a logical matrix with elements that are 1 for corresponding elements of s that are of a particular type, such as alpha, digit, lower (lower case), upper (upper case), punct, (punctuation) etc., and 0 otherwise
isvector(v)	Returns 1 if v is a row vector, column vector or a scalar and 0 otherwise
validatestring(s, C, etc.)	Checks validity of text string s to see if s matches (case insensitive) a character string in cell array C and returns string in C. See MATLAB help facility for additional options

For example,

```
>> s = strvcat('The current equals 10 mA', 'The voltage equals 6.3 volts')
s =
The current equals 10 mA
The voltage equals 6.3 volts
>> where_letters = isstrprop(s,'alpha')    % find all alphabetic elements of s
where_letters =
  Columns 1 through 14
     1    1    1    0    1    1    1    1    1    1    1    0    1    1
     1    1    1    0    1    1    1    1    1    1    1    0    1    1
  Columns 15 through 28
     1    1    1    1    0    0    0    0    1    1    0    0    0    0
     1    1    1    1    0    0    0    0    0    1    1    1    1    1
```

With the logical matrix, where_letters, you can find all of the elements of s that are ASCII codes of alphabetic characters. Or, with where_digits = isstrprop(s,'digit') you can find all of the elements of s that are ASCII codes of digits.

7.2 Manipulate and Search Character Strings

Table 7.3 gives MATLAB built-in functions for manipulating character arrays, and Table 7.4 gives built-in functions for searching character arrays.

Table 7.3 MATLAB functions for manipulating character arrays

Built-in function	Brief description
deblank(s)	Strip trailing blanks from end of string
lower(S)	Convert character string to lower case
strjust(s,format)	Justify character array s, format = 'right','left' or 'center'
strrep(s1,s2,s3)	Replace occurrences of s2 in s1 with s3; s1, s2 and s3 can be cell arrays of the same dimension
strtrim(s), strtrim(C)	Remove leading and trailing unprintable characters from character string or cell array
upper(s)	Convert character string to upper case

Table 7.4 MATLAB functions for searching character arrays

Built-in function	Brief description
findstr(s1,s2)	Returns in a vector the starting indices of any occurrences of the shorter string in the longer string
regexp(s1,pat)	Returns the starting indices of any occurrences of a match of characters in s1 and a pattern of characters specified by pat

(Continues)

Table 7.4   (*Continued*)

Built-in function	Brief description
regexpi(s1,pat)	Case insensitive version of regexp
regexprep(s1,pat,s2)	Replaces all characters in s1 that match the pattern of characters specified by pat with the characters given by s2
sscanf(s,'format')	Reads data in the character string s according to the specified format; format syntax will be discussed in Chapter 8
strfind(s1,s2)	Returns in a vector the starting indices of all occurrences of the character string s2 in the character string s1; s2 must be shorter than s1
[a,b, ...] = strread(s)	Assigns to a,b, ... the numbers, delimited by blanks, in s; see the MATLAB help facility for more options
eval('expression')	Evaluate expression; see Chapter 2
strcmp(s1,s2)	Returns logical 1 if strings s1 and s2 are identical, and 0 otherwise
strcmpi(s1,s2)	Case-insensitive version of strcmp
strncmp(s1,s2,N)	Returns logical 1 if the first N characters of s1 and s2 are the same, and 0 otherwise
strncmpi(s1,s2,N)	Case-insensitive version of strncmp

Example 7.2

```
clear all; clc
% by including 's' in the input function argument list, MATLAB expects
% a character string input without requiring single quotes
your_name = input('What is your name? ','s');
initial = upper(your_name(1));  % use upper case for the first character
your_name = [initial,your_name(2:length(your_name))];
question = [your_name,', do you look forward to using MATLAB more? '];
answer = input(question,'s');
if strncmpi(answer,'yes',1) % compare the first characters
    disp('You will find MATLAB to be very useful.')
elseif strncmpi(answer,'no',1)
    disp('Read on anyway, you may change your opinion.')
else
    disp('A little more experience with MATLAB will help.')
    disp([your_name,', try using demo in the Command Window.'])
    disp('The audio/video tutorials are very informative.')
end
```

Program 7.1   Demonstration of text input.

Running this program gives, for example,

```
What is your name? wxyz
Wxyz, do you look forward to using MATLAB more? maybe
```

```
A little more experience with MATLAB will help.
Wxyz, try using demo in the Command Window.
The audio/video tutorials are very informative.
```

7.3   Structure Arrays

A scalar variable can have just one data type value, which could be numerical, logical, or a character. If a variable is an array, then all elements in the array must be the same data type. A **scalar structure** variable, which can have any valid MATLAB variable name, can have any number of variables associated with it, and each of the associated variables can be assigned any kind of data type. Each associated variable of the structure variable is called a **field** of the structure variable, and each field of the structure variable is given a name that can be any valid MATLAB variable name. The syntax for a scalar structure variable is

$$\texttt{structure\_variable\_name.field\_name\_1 = data\_1}$$
$$\vdots$$
$$\texttt{structure\_variable\_name.field\_name\_N = data\_N}$$

Here, a period is the delimiter between the structure variable name and the field name, the structure variable has *N* fields, and the data assigned to each field can be a matrix, vector, scalar, or array of any kind of data type, and even another structure variable.

Example 7.3

Create a scalar structure variable for the inventory of components in a circuit on a printed circuit board (PCB). Assume this is the third circuit on the PCB of some product. Prog. 7.2 creates a scalar structure variable named CKT_3, and associates with this variable nine fields to which are assigned character strings and numeric vectors of the values of circuit components. For this purpose, the nine fields are named:

- ckt_function, to describe what the circuit does
- schematic, to give a complete path name of the file, an enhanced meta file (emf), that contains a schematic of the circuit
- resistor_names, to list the notation used to identify all resistors
- resistor_values, to give the values, in K ohms, of all resistors
- capacitor_names, to list the notation used to identify all capacitors
- capacitor_values, to give the values, in μF, of all capacitors
- IC_names, to list the notation used to identify all integrated circuits (ICs)
- IC_parts, to give the part number of all ICs
- misc, to provide a description of any other circuit components

Additional fields can be included, for example, distributor SKU numbers, cost of each component, etc. Then, the program illustrates how to use various built-in MATLAB functions to report about the content of CKT_3. Notice that the elements of the structure variable and its fields are indexed in a manner similar to the way that we work with matrices. The program output follows.

```
% Program to create a structure variable and report about its content
clc; clear all
CKT_3.ckt_function = '4-bit adder';  % assign data to field ckt_function
CKT_3.schematic = 'C:\schematics\four_bit_adder.emf';  % file path name
CKT_3.resistor_names = char('R1','R2','R3','R4','R5','R6','R7','R8',...
      'R9','R10','R11','R12','R13');  % assign character strings
CKT_3.resistor_values = [10,10,10,10,10,10,10,10,0.333,0.333,...
      0.333,0.333,0.333];  % assign resistor_values in K Ohms
CKT_3.capacitor_names = char('C1','C2');  % assign character strings
CKT_3.capacitor_values = [0.01,100];  % assign capacitor values in uF
CKT_3.IC_names = char('IC1','IC2','IC3','IC4','IC5','IC6','IC7');
CKT_3.IC_parts = char('74LS08','74LS08','74LS32','74LS86','74LS86',...
      '74LS04','74LS04');  % assign part numbers as character strings
CKT_3.misc = char('5 red LED', '8-switch DIP');
disp 'Structure for inventory of PCB circuit #3'  % display a heading
disp(CKT_3)  % display the entire structure
disp 'IC parts field is'  % display a heading
disp(CKT_3.IC_parts)  %  display the IC parts field
disp(' ') % skip a line
% display the value of the first component in the resistor_names field
% get number of resistors and number of characters per name
[N_resistors N_chars] = size(CKT_3.resistor_names);
% concatenate 4 character strings and display the resulting string
disp([CKT_3.resistor_names(1,1:N_chars),' = ',...
    num2str(CKT_3.resistor_values(1)),'K Ohms'])
% concatenate 3 character strings and display the resulting string
disp(['The circuit has ',num2str(N_resistors),' resistors'])
[N_ICs M] = size(CKT_3.IC_names);  % get number of rows (ICs)
IC_names = char(CKT_3.IC_names);  % all rows will have the same length
IC_parts = char(CKT_3.IC_parts);  % all rows will have the same length
disp (' ')  % skip a line
disp 'List of IC parts'  % display heading
% concatenate 3 character arrays, each with N_ICs rows and
% display the resulting array, a table of IC parts
disp([IC_names,repmat(' = ',N_ICs,1),IC_parts])
```

Program 7.2 Program creates a scalar structure variable CKT_3 and reports about it.

```
Structure for inventory of PCB circuit #3
        ckt_function: '4-bit adder'
           schematic: 'C:\schematics\four_bit_adder.emf'
      resistor_names: [13x3 char]
     resistor_values: [1x13 double]
     capacitor_names: [2x2 char]
    capacitor_values: [0.0100 100]
            IC_names: [7x3 char]
            IC_parts: [7x6 char]
                misc: [2x12 char]
```

```
IC parts field is
74LS08
74LS08
74LS32
74LS86
74LS86
74LS04
74LS04
```

```
R1  = 10K Ohms
The circuit has 13 resistors
```

```
List of IC parts
IC1 = 74LS08
IC2 = 74LS08
IC3 = 74LS32
IC4 = 74LS86
IC5 = 74LS86
IC6 = 74LS04
IC7 = 74LS04
```

Take a look at the last statement in the program, where three character arrays are concatenated within the brackets to make one character array, a list of IC components. The elements of the first array, which consists of N_ICs rows, are each a row character array of the IC name. To use a space, equal sign, and space in the list, concatenation requires that the second character array also has N_ICs rows. The second array is formed with the built-in function **repmat**, which replicates the given three character string, ' = ', into a character array of dimension, N_ICs × 3 characters. Since the third character array, IC_parts, has N_ICs rows, the resulting character array has N_ICs rows, and after concatenation the result is one new character array with N_ICs rows, all having the same number of characters/row, that is displayed.

Example 7.3 demonstrated how to construct a scalar structure variable, with associated variables that can be any data type and any size. This concept can be extended to a structure array of structure variables to make one object out of scalar structures of any number of circuits.

Example 7.3 (continued) ───────────────────────────────

In Prog. 7.2, a scalar structure variable was defined. Suppose there are other circuits, such as circuit #1, circuit #2, and others, on the PCB, for which an inventory must be kept. For each circuit, we could create a scalar structure, like CKT_3. Or, redefine CKT_3 as PCB(3) to become

```
PCB(3).ckt_function = '4-bit adder';  % assign data to field ckt_function
PCB(3).schematic = 'C:\schematics\four_bit_adder.emf';  % file path name
PCB(3).resistor_names = char('R1','R2','R3','R4','R5','R6','R7','R8',...
    'R9','R10','R11','R12','R13');  % assign characters strings
PCB(3).resistor_values = [10,10,10,10,10,10,10,10,0.333,0.333,...
    0.333,0.333,0.333];  % assign resistor values
PCB(3).capacitor_names = char('C1','C2');  % assign character strings
PCB(3).capacitor_values = [0.01,100];  % assign capacitor values
PCB(3).IC_names = char('IC1','IC2','IC3','IC4','IC5','IC6','IC7');
PCB(3).IC_parts = char('74LS08','74LS08','74LS32','74LS86','74LS86',...
    '74LS04','74LS04');  % assign part numbers as character strings
PCB(3).misc = char('5 red LED', '8-switch DIP');
```

This defines the third element of a one-dimensional **structure array** named PCB. Each element of PCB is called a **record**. In the remainder of Prog. 7.2, CKT_3 must be replaced by PCB(3). Then, the contents of records PCB(1) and PCB(2) give the inventories of circuits #1 and #2, and so on. For example,

```
% inventory of PCB circuit #1
PCB(1).ckt_function = '1 KHz oscillator'; % assign data to ckt_function
PCB(1).schematic = 'C:\schematics\oscillator.emf';  % file path name
PCB(1).resistor_names = char('R1','R2'); % assign character strings
PCB(1).resistor_values = [8.3,6.8]; % assign resistor values in K Ohms
PCB(1).capacitor_names = 'C1';  % assign a character string
PCB(1).capacitor_values = 0.01;  % assign a capacitor values in uF
PCB(1).IC_names = 'IC1';
PCB(1).IC_parts = 'LM555'; % assign part number as a character string
PCB(1).misc = [];  % leave empty
% inventory of PCB circuit #2
PCB(2).ckt_function = 'non-inverting amplifier'; % assign data
PCB(2).schematic = 'C:\schematics\amplifier.emf';  % file path name
PCB(2).resistor_names = char('R1','R2'); % assign character strings
```

```
PCB(2).resistor_values = [10,50]; % assign resistor values in K Ohms
PCB(2).capacitor_names = []; % leave empty
PCB(2).capacitor_values = []; % leave empty
PCB(2).IC_names = 'IC1';
PCB(2).IC_parts = 'LM741';
PCB(2).misc = []; % leave empty
```

PCB(1) and PCB(2) must have the same fields as PCB(3). However, the fields of PCB(1) or PCB(2) do not have to have the same lengths or be the same data types. This depends on how the fields will be used. Depending on how the structure array is referenced, you can access a variety of data. For example, to display all IC parts, use

```
PCB.IC_parts    %  display all IC parts in each record

ans =   LM555
ans =   LM741
ans =
74LS08
74LS08
74LS32
74LS86
74LS86
74LS04
74LS04
```

By including indices, any data can be accessed. For example,

```
PCB(3).resistor_values(1:2)
ans =       10        10
```

Notice the similarity to the way elements of a matrix are referenced. However, you must be careful, since fields of different records can have different lengths. To see an overall description of the array structure, use

```
PCB    %  display overall description
PCB =
1x3 struct array with fields:
    ckt_function
    schematic
    resistor_names
    resistor_values
    capacitor_names
    capacitor_values
```

```
IC_names
IC_parts
misc
```

Here, the records of the structure array PCB are organized in a row.

There are several MATLAB built-in functions that work with structure arrays. These are listed in Table 7.5. The function **struct** provides a convenient way to create a scalar structure variable. Its syntax is given by

```
S = struct('field_1',values_1,'field_2',values_2, …)
```

where S is the name of the structure, field_1 is the first field name, values_1 is the data to be assigned to field_1, and this continues for as many fields as needed.

Table 7.5   MATLAB functions concerned with structure arrays

Built-in function	Brief description
struct	Create a scalar structure; see Example 7.3
isstruct(S)	Returns logical 1 if S is a structure, and 0 otherwise
setfield(S,'field',v)	Set the contents of the specified field to the value v
getfield(S,'field')	Returns the contents of the specified field
C = fieldnames(S)	Returns a cell array C of strings containing the structure S field names
S2 = orderfields(S1)	Order fields in structure array S1 into a structure array S2 with field names in ASCII dictionary order
isfield(S,'field')	Returns logic 1 if field is the name of a field in the structure array S, and 0 otherwise
rmfield(S,'field')	Remove the specified field from the N × M structure array S; the size of S is preserved
[S.fields] = deal(v)	Set all the fields with the named fields in the structure array S to the value v
[a,b,c, …] = deal(S.field)	Copy the contents of the structure array field with the name field to the variables a,b,c,…
struct2cell	See Table 7.6
cell2struct	See Table 7.6

Example 7.3 (continued) ⎯⎯⎯⎯⎯⎯⎯⎯⎯⎯⎯⎯⎯⎯⎯⎯⎯⎯⎯

Use the function **struct** to specify the inventory of circuit #1, which is given below.

```
clc; clear all
CKT_1 = struct('ckt_function','1 KHz oscillator',...
```

```
'schematic','C:\schematics\oscillator.emf',...
'resistor_names',char('R1','R2'),'resistor_values',[8.3,6.8],...
'capacitor_names','C1','capacitor_values',0.01,...
'IC_names','IC1','IC_parts','LM555',...
'misc',[]);
```

Assuming that CKT_2 and CKT_3 have been defined in the same way, we can obtain PCB as in Example 7.3 by concatenating these three scalar structure variables with

```
PCB = [CKT_1, CKT_2, CKT_3];
```

and then we get, for example,

```
PCB(1)
ans =
        ckt_function: '1 KHz oscillator'
           schematic: 'C:\schematics\oscillator.emf'
      resistor_names: [2x2 char]
     resistor_values: [8.3000  68.0000]
     capacitor_names: 'C1'
    capacitor_values: 0.0100
            IC_names: 'IC1'
            IC_parts: 'LM555'
                misc: []
PCB    %  display overall description
PCB =
1x3 struct array with fields:
    ckt_function
    schematic
    resistor_names
    resistor_values
    capacitor_names
    capacitor_values
    IC_names
    IC_parts
    misc
```

It is not hard to think of situations where a structure array can be used to record and access information, literally like using a file cabinet and paper folders, for example, a name/ address/phone book, courses with titles, instructors, scores and grades taken each semester, video/music library, etc. Once a structure array has been started, it is easy to expand it by adding more elements and report on information it contains.

MATLAB also has built-in functions that return structure arrays, one of which is the built-in function **dir** that returns a structure array of file names in a specified directory (folder). A syntax option for the built-in function dir is given by

```
D = dir('directory_name')  % can include a file specification
```

where D is an $N \times 1$ structure array of N records with the fields:

```
name, file name
date, date last modified
bytes, size in bytes
isdir, equals 1 if directory name is a directory, and 0 otherwise
datenum, modification date as a MATLAB serial date number
```

More functions concerned with directories will be discussed in the next chapter.

Example 7.4 _____

Suppose you have a folder (directory) of WAV files of songs, and you want to play a song. Each song is identified by its title, and maybe you do not remember exactly each title. It would be convenient to provide only enough about the title to find and play it.

In addition to the built-in functions that were described previously, Prog. 7.3 uses the following built-in functions.

- **wavread** returns the stereo samples of a WAV file and the sampling rate
- **wavplay** plays stereo samples at a specified sampling rate

In this application, the function play_WAV is located in the *Current Folder*, and the songs are located in the directory: C:\Music\. The names of the song files are: *.wav, where the asterisk is a wildcard. With this file specification, only files with the suffix .wav will be retrieved. For example,

```
>> song_titles = dir('C:\Music\*.wav')   % get structure array of WAV files
song_titles =
11x1 struct array with fields:
    name
    date
    bytes
    isdir
    datenum
```

```
>> % use char to create a character array of titles
>> titles = char(song_titles.name)
titles =
Aretha Franklin_Son Of A Preacher Man.wav
Pandit Jasraj.wav
Coltrane_Out of This World.wav
Dirty Vegas.wav
Fly Away.wav
Hot Buttered Soul.wav
One More Road to Cross.wav
Shaman_Drums.wav
Sweet Home Alabama.wav
Thriller.wav
Werewolves of London.wav
```

```
function play_WAV(Song_Word,seconds)
% Function to find and play a WAV file of a song
% For example: play_WAV('Fly',20)
% where Song_Word, a character string, is any part of the song name that
% is sufficient to find the song WAV file, and
% seconds, in seconds, is the length of time you want to play the song
song_word = lower(Song_Word); % change input string to lower case
%
song_directory = strcat('C:\','Music','\*.wav'); % build a path name
song_titles = dir(song_directory); % get structure array of song WAV files
%
N_titles = length(song_titles); % get number of records (WAV files)
%
if N_titles > 0 % check if no WAV files in the specified song directory
    i_song = 0; % initialize index pointing to a WAV file
    for i=1:N_titles % search all titles for song_word
        Title = song_titles(i).name; % get a title from a record
        title = lower(Title); % make search case insensitive
        % check if title has song_word part
        found_word = ~(isempty(findstr(song_word,title)));
        % if song WAV file found, set index
        if found_word
            i_song = i;
            break % found song, jump out of for loop
        end
    end
    if i_song == 0 % check if song was found
```

```
        disp('No song title match')
        % return from function play_WAV without playing a song
        return
    end
    % read the song WAV file
    [Song,fs] = wavread(strcat('C:\Music\',Title)); % get song
    T = 1/fs;  % sample time increment
    % Here, song consists of two columns, the left and right channels
    %
    % get the duration, in seconds, of the song
    [N_samples N_channels] = size(Song); % get number of samples
    duration = N_samples*T;  % seconds
    % limit play time to specified number of seconds
    if duration > seconds
        duration = seconds;
    end
    N_samples = floor(duration/T);  % get number of samples to play
    % get song samples
    song(1:N_samples,:) = Song(1:N_samples,:); % copy both channels
    disp(Title); % display tile of song
    wavplay(song,fs); % play song
else % get here if folder has no WAV files
    disp('There are no WAV files in this directory');
end
```

Program 7.3 Function to find and play a WAV file.

To play a song, use, for example,

```
>> play_WAV('Fly',120)  % specify any part of a title
Fly Away.wav
>> %  now, the audio device is playing the song
```

Another built-in MATLAB function that returns a structure is the function **solve** that symbolically solves systems of equations. Its syntax is given by

```
solution = solve('f1(var1,..., varN) = g1(var1, ..., varM)', ...
                 'f2(var1,..., varN) = g2(var1, ..., varM)', ...
                 'f3(var1,..., varN) = g3(var1, ..., varM)', ...
                                .
                 'fK(object1,...,objectN) = gK(var1, ..., varM)', ...
                 'solution_var1, ..., solution_varK')
```

where the expressions f and g can each involve any number of variables and a solution of any number of variables can be found in terms of the other variables.

Although symbolic math will be presented in Chapter 11, let us consider the following example.

Example 7.5 ───

In the circuit given in Fig. 7.1, solve for the voltage V_o in terms of the inputs V_1 and V_2.

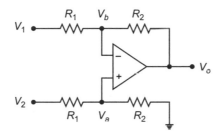

Figure 7.1   A one Op-Amp differential amplifier.

The KCL equations at nodes a and b are given by

$$\frac{V_2 - V_a}{R_1} = \frac{V_a}{R_2}$$

$$\frac{V_1 - V_b}{R_1} = \frac{V_b - V_o}{R_2}$$

and the Op-Amp gain equation is

$$V_o = A(V_a - V_b)$$

The following MATLAB statement finds V_a, V_b and V_o.

```
>>  voltages = solve('(V2 - Va)/R1 = Va/R2','(V1 - Vb)/R1 = (Vb - Vo)/R2', ...
                 'Vo =A*(Va - Vb)','Va,Vb,Vo')
```

The argument of the function solve contains three equations of five voltages and the three voltages V_a, V_b, and V_o to be found. The result is a structure given by

```
voltages =
    Va: [1x1 sym]
    Vb: [1x1 sym]
    Vo: [1x1 sym]
```

```
>> Vo = voltages.Vo   % the value of voltages.Vo is a character string
Vo = -(A*R2*V1 - A*R2*V2)/(R1 + R2 + A*R1)
```

The structure voltages has three fields that are symbolic (sym) variables, called objects. The third field name is V_o. If we let $A \rightarrow \infty$, then we get

$$V_o = \frac{R_2}{R_1}(V_2 - V_1)$$

which is a symbolic relationship between the inputs and the output. MATLAB also has a built-in function that finds the limit as $A \rightarrow \infty$, which will be discussed in Chapter 11.

7.4   Cell Arrays

A **cell array** is a collection of cells (sort of like containers) that can contain objects none of which must be the same size or data type as any other object contained by any other cell of the cell array. Each cell is an element of the cell array, which is referenced using indices in the same way as an element of a matrix is referenced. However, to specify or access the content of a cell in a cell array, braces must be used instead of parentheses. If you do use parentheses, then you are referencing a cell, not its content. For example, to define a 2×2 cell array, braces must be used, and the syntax is given by

```
C = {object_1    object_2; object_3    object_4}
```

where C is the cell array, and the row and column delimiters are the same as for a matrix. The braces mean that the contents of the cells of the cell array are being specified. An object in a cell of a cell array can even be another cell array.

Example 7.6 ⎯⎯⎯⎯⎯⎯⎯⎯⎯⎯⎯⎯⎯⎯⎯⎯⎯⎯⎯⎯⎯⎯⎯⎯⎯⎯⎯⎯⎯⎯

Recall that in Example 7.3, a three-record structure array PCB was defined. It can be placed in an element of a cell array, while other elements of the cell array can contain different kinds of objects. For example, the following statements were appended to the program that created PCB, and the results follow.

```
C = {'Inventory of PCB' 23711005; '12/14/2012' PCB};   % define a cell array
% the object in the cell located at (1,2) is meant to be a serial number
disp(C)   % display the cells of C
celldisp(C)   % display the contents of the cells of C
C11 = C(1,1)   % referencing an element to get cell attributes
c11 = C{1,1}   % get the content of the cell at location (1,1)

    'Inventory of PCB'     [   23711005]
    '12/14/2012'               [1x3 struct]

C{1,1} =     Inventory of PCB
```

```
C{2,1} =    12/14/2012
C{1,2} =     23711005
C{2,2} =
1x3 struct array with fields:
    function
    schematic
    resistor_names
    resistor_values
    capacitor_names
    capacitor_values
    IC_names
    IC_parts
    misc

C11 -      'Inventory of PCB'
c11 =      Inventory of PCB
```

Notice the difference between using parentheses and braces when referencing an element of the cell array. To access the content of PCB, use, for example,

```
% get resistor names in first record in structure array C(2,2)
C{2,2}(1).resistor_names
ans =
R1
R2
```

In this example we see that a cell array is a means to bundle a variety of kinds of information into one object.

Another way to create a cell array is with the built-in function **cell**, which has the syntax given by

   C = cell(N_rows,N_columns)

where the argument of the function cell is the desired dimension, and C is a cell array of empty cells. For example, in Example 7.6, we could have used

```
C = cell(2,2);  % set up a 4 element cell array
C{2,2} = PCB  % place an object in the (2,2) element of C
 C =
    []            []
    []      [1x3 struct]
```

Notice that three elements of C are still empty.

Table 7.6   Some MATLAB functions concerned with cell arrays

Built-in function	Brief description
C = cell(N,M)	Set up a cell array of empty cells
C = {s1 s2 ... }	Create a cell array of character strings; separate each row of the cell array with a semicolon (;)
celldisp(C)	Display the content of the cells in a cell array
iscell(C)	Returns logic 1 if object is a cell array, and 0 otherwise
iscellstr(C)	Returns logic 1 if C is a cell array of strings, and 0 otherwise
C = cellstr(s)	Create a cell array C of strings from a character array s
C = num2cell(A)	Converts a numeric array A into a cell array C by placing each element of A into a separate cell in C
S = cell2struct(C,fields,M)	Convert an $N \times M$ cell array C into a structure array S having N records; fields can be a character or a cell array of strings of M field names
C = struct2cell(S)	Convert an $N \times M$ structure array S with P fields into a $P \times N \times M$ cell array C; if S is a scalar structure array, then C will be a $P \times 1$ cell array
cellplot(C)	Display graphical depiction of cell array
C = fieldnames(S)	Returns a cell array of strings containing the field names associated with the structure array S.
[a,b,c, ...] = deal(C{:})	Copies the contents of the cell array C to the separate variables a,b,c,...; see MATLAB help facility for more options
[C{:}] = deal(S.field)	Copies the values of the S structure array field with name field to the cell array C; see MATLAB help facility for more options
D = sort(C)	Sort the strings in a cell array C of strings into ASCII dictionary order of strings in a cell array D

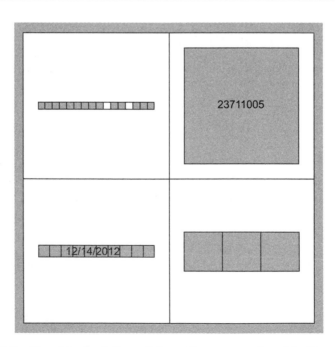

Figure 7.2   Graphic depiction of the cell array produced in Example 7.6.

Several built-in functions that work with cell arrays are given in Table 7.6. The **cellplot** function gives an interesting graphic that depicts the organization of objects in a cell array. For example, see Fig. 7.2, which was obtained with cellplot(C), where C is the cell array produced by the program given in Example 7.6.

7.5 Conclusion

With MATLAB, any combination of numerical, logical, and character data can be processed. Many built-in functions were given in Tables 7.1 through 7.6 to support processing character strings. You should now know how to

- create and combine character strings
- search and modify character strings
- convert character strings to numerical and logical data and back
- construct an organization of character strings and other data types
- bind together data of any type and size

Table 7.7 gives the additional built-in functions that were introduced in this chapter. Use the MATLAB help facility to learn more about these built-in functions, where you will also find out about many other related built-in functions.

Table 7.7 Built-In MATLAB functions introduced in this chapter

Built-in function	Brief description
rcpmat(A,N,M)	Replicates an object A into the elements of an N × M array
dir('path name')	Returns a structure array of file names in the directory pointed to by path name; see the MATLAB help facility for more options
wavread(music,fs)	Returns the left and right channels of sampled music in a WAV file and the sampling frequency, f_s
wavplay(music,fs)	Sends to the audio play device the samples contained in music and the sampling frequency, f_s
s = solve('f(x1,x2,x3,...) = g(x1,x2,x3,...)','x1,x2'.)	Solve a system of algebraic equations; solution given by s.x1 and s.x2

Some methods to input and output data have been used in this and previous chapters. In the next chapter a more detailed discussion about formatted input/output is given.

Problems

Section 7.1

1) Give MATLAB statements to define character data type variables as follows.
   (a) String named first, middle, and last of the name James Clerk Maxwell

(b) String named full by horizontally concatenating last, first, and middle with comma and a blank between last and first and a blank between first and middle. Give two ways.

(c) Array named Full by vertically concatenating last, first, and middle. Is there a difference between using the functions char and strvcat?

2) Using the results of Prob. P7.1, give MATLAB statements to obtain
   (a) last_n, first_n, and middle_n vectors of numeric ASCII codes. See Appendix B of ASCII codes to check results.
   (b) numeric array, Full_n, of ASCII codes. What is the size of Full_n?
   (c) What is the difference between Full_n and the result of num2str(Full_n)?
   (d) What is the difference between Full_n and the result of int2str(Full_n)?

3) Discuss the differences between x=pi, y=num2str(pi), and z=int2str(pi).

4) Give a MATLAB statement to define a cell vector named full_c of the variables last, first, and middle defined in Prob. P7.1. What is the size of full_c? Give full_c(1).

5) Give a MATLAB statement that defines a character string named Maxwells_equations that is the sentence: In 1865 Maxwell published, "A Dynamical Theory of the Electromagnetic Field.".

6) Given is a character vector S, the value of which is some sentence. For example, S = 'Now is the time to learn MATLAB.', and, S = 'The number pi = 3.1416 is an irrational number.'. Write a MATLAB program that creates a character array of all of the words in a sentence. Provide a copy of your program and several examples of using it.

Section 7.2

7) A character string Pi is given by: Pi = 'In long format, pi = 3.141592653589793.'. Give MATLAB statements that do or find the following.
   (a) The size of Pi.
   (b) The index in Pi where the word long starts.
   (c) Replace the word, "long," with, "double precision." Name the result Pi_too.
   (d) Determine if Pi and Pi_too are the same strings.
   (e) Determine if the first three characters in Pi and Pi_too are the same.

8) (a) Explain what the following MATLAB statements do.

```
numbers = char();
numbers = char(numbers,'three')
numbers = char(numbers,'one')
```

(b) Write a MATLAB script that
   1) starts with an empty character array named digits
   2) uses an input statement in a while-loop to obtain a digit name until the entry is not a digit name, when the while-loop is terminated

3) converts all upper case letters in an entry to lower case letters
4) appends an inputted digit name to digits only if the inputted digit name is not a duplicate of a previously entered digit name
5) continues after the while-loop to sort the elements of digits in ascending numerical value order.

Provide a copy of your program and examples of entering at least three digit names. Be sure that your examples demonstrate that the program meets all requirements.

9) Write a MATLAB function script, call it check_variable, that determines whether or not an input character string is a valid MATLAB variable name. Use an input statement within a while-loop of a program to obtain a character string. If the input is just a carriage return (enter), then terminate the while-loop. Otherwise, apply your function to the inputted character string. Let the function return logic 0 or 1. Provide a copy of your program, function, and several examples of using them.

Section 7.3

10) Given below is the table of a few electrical scientists.

Table P7.10   Electrical scientists

Name	Lived	Research field
Charles-Augustin de Coulomb	1736–1806	electrostatic force
Andre Marie Ampere	1775–1836	electrodynamics
Michael Faraday	1791–1867	electricity and magnetism
James Clerk Maxwell	1831–1879	electromagnetic field theory
Nikola Tesla	1856–1943	electromagnetic technology

Give MATLAB statements to
(a) create a scalar structure named electrical_scientists with field names: name, lived, and research_field
(b) display the structure
(c) display the second field
(d) get the contents of the field name

11) Write a MATLAB script that
(a) starts with using the function struct to initialize a scalar structure named resistors with field names: resistor_name and resistor_value, where the two fields are empty.
(b) continues to do the same for the structure capacitors and fields capacitor_names and capacitor_values.

(c) continues by using input statements within a while-loop to obtain resistor and capacitor information to be placed in the respective fields. Use questions with yes/no answers to find out about entering resistors and capacitors.

(d) continues by displaying tables of resistor and capacitor names and values.

(e) finishes by creating a structure array named components with elements resistors and capacitors.

Provide a copy of your program and a demonstration of using the program.

12) With the MATLAB statement: my_m_files = dir('*.m'), you can obtain a structure that contains among other fields a field of all m-file names in your *Current Folder*. Give MATLAB statements to do or find

(a) size of my_m_files

(b) field names of my_m_files

(c) first object in my_m_files

(d) assign to the variable mfile3 the third m-file name

(e) date when the second m-file was last modified

Section 7.4

13) Given are the following structures and character array.

```
names = struct('last',char('Coulomb','Ampere','Faraday', ...
        'Maxwell','Tesla'), ...
        'first',char('Charles','Andre','Michael','James','Nikola'))
lived = struct('born',[1736;1775;1791;1831;1856], ...
        'died',[1806;1836;1867;1879;1943])
research = char('electrostatic force','electrodynamics', ...
        'electricity and magnetism','electromagnetic field theory', ...
        'electromagnetic technology')
```

Give MATLAB statements to do or find

(a) assign the field names of the structure names to a cell array

(b) create a 1x3 cell array named ES, with empty cells

(c) place the structure names into the first element of ES

(d) get ES(1) and ES{1}. What is the difference between ES(1) and ES{1}?

(e) assign the first names from ES{1} to a cell array named first_names

(f) get the first name of Faraday from ES

(g) place into the second and third elements of ES the structure lived and the character array research

(h) display the content of the cells of ES

14) Given is the following program.

```
1    % Program to sort the names in a field of a structure array contained in a
2    % cell of a cell array
3    clear all, clc
4    names = struct('last',char('Coulomb','Ampere','Faraday', ...
     'Maxwell','Tesla'), ...
5      'first',char('Charles','Andre','Michael','James','Nikola'));
6    lived = struct('born',[1736;1775;1791;1831;1856], ...
7      'died',[1806;1836;1867;1879;1943]);
8    research = char('electrostatic force','electrodynamics', ...
9      'electricity and magnetism','electromagnetic field theory', ...
10     'electromagnetic technology');
11   ES = {names, lived, research};
12   Names{1} = ES{1}.last;
13   Last_Names = char(Names{1});
14   [n_names n_chars] = size(Last_Names);
15   for k=1:n_names
16      Last{k,1} = Last_Names(k,:);
17   end
```

(a) Explain the activity of lines (12) and (13).
(b) What is Last after the for-loop has completed?
(c) Give MATLAB statements that can be appended to the program to sort the elements of Last to obtain and display Last_sorted.

CHAPTER 8

Input/Output

To make a program interactive, it must be possible for the program to receive from the program user input data or information about where data can be found. And, the program must be able to output data and information in a form that is useful to the program user and possibly to another program. Graphical output will be discussed in Chapter 9. In this chapter, you will learn how to

- input data and information about where data can be found
- input information to guide program execution
- output structured and formatted data and information
- provide data and information to another program

8.1 Output

By building a character array A, the built-in function **disp**(A) is useful to output (display) a combination of numeric and character data. The result is the same as omitting a semicolon at the end of an assignment statement, except that the name of the array is not displayed.

8.1.1 Text Output

To provide access to data by programs other than m-files, it is useful to create a text file of the data. The built-in function **fprintf** can be used to do this. The function fprintf applies a format to the elements in each column of an array *A* and any additional array arguments. Its syntax is given by

```
fprintf(format, A1, A2, … )
```

where **format** is a character string that describes the output fields. The format can contain combinations of

- text to be printed
- percent sign followed by a conversion character
- operators that describe the output width, precision, and other options
- escape characters

For example,

```
>> fprintf('The number pi is: %19.16f, using %2u fractional digits.\n',pi,16)
The number pi is:  3.1415926535897931, using 16 fractional digits.
```

The format in this invocation of fprintf starts with text that is printed. The location of the first percent sign within the format string is the location where the first argument, pi, is placed in the output. Following the first percent sign, 19.16f means that pi should be printed using a field width of no less than 19 characters, 16 fractional digits, and fixed point notation. After the conversion character, f, there is additional text until the next percent sign, where the second argument is placed in the output. Following the second percent sign, 2u means that the second argument should be printed using a field width of no less than two characters and base ten unsigned integer notation. Then, after additional text there are the escape characters, \n, which mean to continue on a new line. You can assign the format character string to a variable, and use it in fprintf. For example

```
>> pi_format = 'The number pi is: %19.16f, using %2u fractional digits.\n';
>> fprintf(pi_format,pi,16)
```

You can also create a cell array of formats, where each cell of the cell array contains a format character string. For example,

```
>> N_formats = 5;  % numbers of formats to be used
>> formats = cell(1,N_formats)
formats =    []   []   []   []   []
>> formats{1} = 'pi is: %19.16f, using %2u fractional digits.\n'
formats =    [1x46 char]   []   []   []   []
>> fprintf(formats{1},pi,16)
```

Some of the **escape characters** are given in Table 8.1.

Table 8.1   List of some escape characters

Symbol	Brief description
''	Single quotation mark
% %	Percent character
\\	Backslash
\a	Alarm
\b	Backspace
\n	New line
\r	Carriage return
\t	Horizontal tab

Between the **percent sign** and the **conversion character** there can be other information as shown below. Table 8.2 gives some of the conversion characters.

```
(%)(identifier) (flags) (field width) (.precision)(conversion character(subtype) )
```

The **identifier**, with notation n$, specifies which of the arguments should be printed. For example,

```
>> fprintf('The number pi is: %2$ 19.16f, using %1$ 2u fractional digits.\n',16,pi)
The number pi is:  3.1415926535897931, using 16 fractional digits.
>>
```

Table 8.2   List of some conversion characters

Class	Conversion	Brief description
integer, signed	d or i	Base ten values
	ld or li	Subtype, 64-bit base ten values
	hd or hi	Subtype, 16-bit base ten values
integer, unsigned	u	Base ten value
	lu	Subtype, 64-bit base ten value
	hu	Subtype, 16-bit base ten value
floating point number	f	Fixed point notation
	e	Exponential notation
	E	Uppercase version of e
	g	The more compact of e or f
	G	Uppercase version of g
	bu	Subtype, double precision
	tu	Subtype, single precision
characters	c	Single character
	s	String of characters

The **flags** that can be inserted are listed in Table 8.3.

Table 8.3   List of flags

Flag	Action	Example
minus sign	Left justify	%−21.16f
plus sign	Print sign	%+19.16f
space	Insert space before value	% 19.16f
zero	Pad with zeros	%021.16f
pound sign	For conversion characters f, e and E print decimal point even if precision is zero; for g or G do not remove trailing zeros or decimal point	%#19.0f

The **field width** is the minimum number of characters to be printed. For conversion characters f, e, and E, **precision** is the number of digits to the right of the decimal point to print, and for conversion characters g and G, **precision** is the number of significant digits to print.

In a second syntax option you can use fprintf in an assignment statement, in which case it returns a count of the number of bytes it writes. For example,

```
>> count = fprintf('The number pi is: %19.16f, using %2u fractional digits.\n',pi,16)
The number pi is:  3.1415926535897931, using 16 fractional digits.
count =   67
```

A third option has the syntax given by

fprintf(file_ID, format, A1, A2, …)

where **file_ID** can be 1 for output to the screen (*Command Window*), 2 if the output is intended to be an error message, or a file identifier (greater than 2) obtained with the built-in function **fopen** for output to or input from a file (see Table 8.4).

Table 8.4   List of built-in functions concerned with data output

Function	Brief description
disp(A)	Display the array A on the screen
type m-file type file_name dbtype file_name	Displays the contents of the specified file in the Command Window; the function dbtype lists the m-file with line numbers
f = fullfile('disk_name', 'folder1_name', 'folder2_name', … ,'file_name')	Returns in f a character string of the path name to the file file_name; use **doc** fullfile for more options

(Continues)

Table 8.4   (Continued)

Function	Brief description
file_ID = fopen('file_name') file_ID = fopen('file_name','permission') [file_ID, message] = fopen('filename', ...)	Opens the file file_name and returns an integer file identifier file_ID; returns file_ID = −1 if file_name could not be opened. Permission possibilities are given in Table 8.5. Use doc fopen for more options.
fprintf	Writes formatted data; see this section
S = sprintf	Same as fprintf without the file_ID, except that results are returned to the character string S
dlmwrite	See this section and Example 8.1
fclose(file_ID)	Closes an opened file, where file_ID is an integer file identifier obtained from fopen; fclose('all') closes all open files
more on (off)	More on enables paging the output to the screen; output continues by pressing: return for one line at a time or the space bar for one page at a time. Type q to quit more.
fwrite(file_ID, A, 'precision')	Writes the elements of array A to a binary file in column order and translates the values of A according to the form and size described by the precision, where precision can be double, single, char, int64, uint64, etc. Use doc fwrite for more options.
save('file_name') save('file_name','var_1','var_2', ...) save('file_name','-struct','struct_name') save('file_name','-append') save('filename', ..., format)	Store all variables from the current Workspace in a MATLAB formatted binary file (mat-file) called file_name; save('file_name','var_1','var_2', ...) stores only the specified variables. Also saves in the specified format: '-mat' ('my_file.mat') or '-ascii' ('my_file.txt').
f_check = exist('A')	Returns: 0 if A does not exist, 1 if A is a variable in the workspace, 2 if A is an m-file on the MATLAB search path or if A is a full path name to a file, 5 if A is a built-in MATLAB function. See help for more options.
open('file_name') output = open('file_name')	Opens files such as m-files in editor window, fig-files in figure window, mat-files, p-files, etc; if opening a mat-file, then output is a structure that contains the variables in that file. Use doc open for more options.

While you can specify for input or output the complete path name of a file, when you use only a file name, MATLAB® uses the *Current Folder*. Therefore, specify a *Current Folder* that you want MATLAB to use.

Table 8.5   Permissions for fopen function

Permission	Action
r	Open file for reading (default)
w	Open or create new file for writing; discard existing content, if any
a	Open or create new file for writing; append data to the end of the file
r+	Open file for reading and writing
w+	Open or create new file for reading and writing; discard existing content, if any
a+	Open or create new file for reading and writing; append data to the end of the file

The **dlmwrite** function can also be used to output data to a text file. It combines opening a file and writing delimited data, and its syntax is given by

```
dlmwrite('textfile.txt',M,'delimiter', ...

     'datadelimiter','precision','dataprecision')
```

where textfile is a file name, M is a rectangular matrix of data, delimiter is a key word, datadelimiter is, for example, a space, a comma, a tab (\t), and more, precision is a key word and dataprecision is, for example, %19.16f. Use doc dlmwrite to see many more options.

Example 8.1

Using the functions dlmwrite, fprintf, and save, let us create text files of data with Prog. 8.1. The resulting files follow.

```
% Program to write a matrix of data to text files
clear all; clc;
f = 10; w = 2*pi*f; T0 = 1/f; % frequency and period
N = 9; t = linspace(0,T0,N); % obtain N samples over one cycle
s = sin(w*t); c = cos(w*t); % samples
SC_t = [t;s;c]; SC_tT = SC_t'; % collect data in rows and columns
dlmwrite('example1_data.txt',SC_tT,'delimiter',' ','precision','%10.6f')
% each output line comes from a row of SC_tT
type example1_data.txt
% create and open a text file with write permission
ID2 = fopen('example2_data.txt','w');
fprintf(ID2,' time    sine   cosine\n'); % include a heading
fprintf(ID2,'%6.4f %8.4f %8.4f\n',SC_t); % format the data
% each output line comes from a column of SC_t
```

```
type example2_data.txt
%
save('example3_data.txt','-ascii','SC_tT') % save ASCII coded data
type example3_data.txt
fclose('all');
```

Program 8.1 Demonstration of using the built-in functions dlmwrite, fprintf, and save.

dlmwrite			ID2 = 3			Save		
			time	sine	cosine			
0.000000	0.000000	1.000000	0.0000	0.0000	1.0000	0.0000000e+000	0.0000000e+000	1.0000000e+000
0.012500	0.707107	0.707107	0.0125	0.7071	0.7071	1.2500000e−002	7.0710678e−001	7.0710678e−001
0.025000	1.000000	0.000000	0.0250	1.0000	0.0000	2.5000000e−002	1.0000000e+000	6.1232340e−017
0.037500	0.707107	−0.707107	0.0375	0.7071	−0.7071	3.7500000e−002	7.0710678e−001	−7.0710678e−001
0.050000	0.000000	−1.000000	0.0500	0.0000	−1.0000	5.0000000e−002	1.2246468e−016	−1.0000000e+000
0.062500	−0.707107	−0.707107	0.0625	−0.7071	−0.7071	6.2500000e−002	−7.0710678e−001	−7.0710678e−001
0.075000	−1.000000	0.000000	0.0750	−1.0000	0.0000	7.5000000e−002	−1.0000000e+000	7.0448140e−016
0.087500	−0.707107	0.707107	0.0875	−0.7071	0.7071	8.7500000e−002	−7.0710678e−001	7.0710678e−001
0.100000	−0.000000	1.000000	0.1000	−0.0000	1.0000	1.0000000e−001	−2.4492936e−016	1.0000000e+000

With the function fprintf we have the most control over the format of the output file content. As a text file, data can be accessed by m-files and other programs that work with text files, for example, Microsoft Word® and WordPad®.

You can also save the Workspace from the desktop with the file menu sequence

   File → Save Workspace As ...

Then, the *Save to Mat-File Window* opens, and you can name a mat-file into which you want to save the Workspace.

Example 8.2

Recall that in Prog. 7.2 of Example 7.3, a scalar structure array, named CKT_3, was created that contains an inventory of components in a circuit. To save this information for later retrieval, the following MATLAB statements were appended to Prog. 7.2.

```
ckt_path = fullfile('C:','My_MATLAB_Programs','CKT_3.mat'); % path name
save(ckt_path,'-struct','CKT_3') % save structure CKT_3 into file CKT_3.mat
```

Here, the built-in function **fullfile** is used to create a complete path name. This could also have been accomplished with the statement

```
ckt_path = 'C:\My_MATLAB_Programs\CKT_3.mat';
```

Then, the built-in function **save** is used to create the file CKT_3.mat, and store in it the scalar structure array CKT_3. The second argument -struct, a qualifier, indicates to MATLAB that a structure array will be saved.

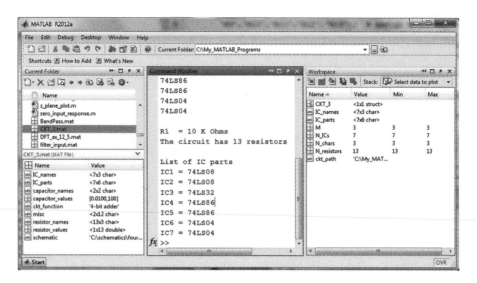

Figure 8.1   Desktop after Prog. 7.2 and the appended statements have executed.

The resulting MATLAB desktop is shown in Fig. 8.1. The *Workspace Window* shows all of the variables that were assigned values during execution of the program. Notice the complete path name that was produced by the function fullfile. The function save placed the mat-file in the folder My_MATLAB_Programs, which is the *Current Folder*. In the *Current Folder*, the file CKT_3.mat is highlighted, and its content description is shown in the window below the *Current Folder Window*. Since My_MATLAB_Programs is the current folder, the complete path name ckt_path could have been replaced by 'CKT_3.mat'.

Now, suppose we want to generate a report about the circuit. Prog. 8.2 does this, and the results follow.

```
% Program to generate a report about a circuit
clear all; clc
f_name = input('Enter a mat-file name: ','s'); % get file name string
mat_suffix = '.mat'; % used to append a suffix
s1 = lower(fliplr(f_name)); s2 = fliplr(mat_suffix); % put suffix in front
if ~strncmpi(s1,s2,4) % check for a match of the first 4 characters
   f_name = strcat(f_name,mat_suffix); % append suffix if missing
end
f_path = fullfile('C:','My_MATLAB_Programs',f_name); % create path to file
```

```
f_check = exist(f_path); % check if file exists
if f_check == 2 % function exist returns 2 if file exists
   ckt = open(f_path); % assign to ckt the structure in the file
   % report contents of the scalar structure ckt
   fprintf('Circuit function: %s\n',ckt.ckt_function) % print field
   fprintf('Path to schematic: %s\n',ckt.schematic)
   Rs=char(ckt.resistor_names);[Nr Mr]=size(Rs); % get resistor info
   Rvs=char(num2str(ckt.resistor_values')); % convert values to strings
   Cs=char(ckt.capacitor_names);[Nc Mc]=size(Cs); % get capacitor info
   Cvs=char(num2str(ckt.capacitor_values')); % convert values to strings
   ICn=char(ckt.IC_names);[NIC MIC]=size(ICn); % get IC info
   ICp=char(ckt.IC_parts); % get IC part numbers
   misc=char(ckt.misc);[Nm Mm]=size(misc); % get misc info
   N = [Nr,Nc,NIC,Nm]; N_rows = max(N)+1; ckt_comps=[]; % initialize report
   % pad different sized data with space and build table ckt_comps
   if Nr > 0 % check if there are any resistors
      resistors=[Rs,repmat(' = ',Nr,1),Rvs];[NR Mr]=size(resistors);
      resistors=[resistors;repmat(' ',N_rows-Nr,Mr)]; % add space
      ckt_comps=[ckt_comps,repmat(' ',N_rows,5),resistors]; % add resistors
   end
   if Nc > 0 % check if there are any capacitors
      capacitors=[Cs,repmat(' = ',Nc,1),Cvs];[NC Mc]=size(capacitors);
      capacitors=[capacitors;repmat(' ',N_rows-Nc,Mc)]; % add space
      ckt_comps=[ckt_comps,repmat(' ',N_rows,5),capacitors]; % add capacitors
   end
   if NIC > 0 % check if there are any integrated circuits
      ICs=[ICn,repmat(' = ',NIC,1),ICp];[NIC MIC]=size(ICs);
      ICs=[ICs;repmat(' ',N_rows-NIC,MIC)]; % add space
      ckt_comps=[ckt_comps,repmat(' ',N_rows,5),ICs]; % add ICs
   end
   if Nm > 0
      Misc=[misc;repmat(' ',N_rows-Nm,Mm)]; % add space
      ckt_comps=[ckt_comps,repmat(' ',N_rows,5),Misc]; % add misc parts
   end
   ckt_comps=ckt_comps'; [Nckt Mckt]=size(ckt_comps); % table size
   fprintf('\n Resistors in K Ohms and capacitors in uF\n\n'); % heading
   for k=1:Mckt
      fprintf('%s\n',ckt_comps(:,k)); % one column per output line
   end
else
   fprintf('The file could not be found.\n');
end
fclose('all');
```

Program 8.2   Program to report the inventory of a circuit given in a mat-file.

```
Enter a mat-file name: CKT_3
Circuit function: 4-bit adder
Path to schematic: C:\schematics\four_bit_adder.emf
```

Resistors in K Ohms and capacitors in μF

R1 =	10	C1 = 0.01	IC1 = 74LS08	5 red LED
R2 =	10	C2 = 100	IC2 = 74LS08	8-switch DIP
R3 =	10		IC3 = 74LS32	
R4 =	10		IC4 = 74LS86	
R5 =	10		IC5 = 74LS86	
R6 =	10		IC6 = 74LS04	
R7 =	10		IC7 = 74LS04	
R8 =	10			
R9 =	0.333			
R10 =	0.333			
R11 =	0.333			
R12 =	0.333			
R13 =	0.333			

8.1.2   Binary Output

The built-in function **fwrite** writes a bit stream of binary data to a file. By default it writes values from an array in column order as 8-bit unsigned integers (uint8). Its syntax is given by

```
fwrite(file_ID, A)  % use default precision
fwrite(file_ID, A, 'precision')  % precision options given in Table 8.6
```

where file_ID can be 1 (output to screen), 2 (output intended as an error message) or an integer file identifier obtained from fopen, A is a numeric or character array of data and where some precision options are given in Table 8.6.

Table 8.6   Some fwrite precisions

Precision	Brief description
uint	Unsigned integer, 4 bytes
uintn	$n = 8$(1 byte), 16, 32, 64(8 bytes)
ubitn	Unsigned integer, $n = 1, \ldots, 64$
int	Signed integer, 4 bytes
intn	Signed integer, $n = 8$(1 byte), 16, 32, 64(8 bytes)
bitn	Signed integer, $n = 1, \ldots, 64$
single	Floating point, 4 bytes
double	Floating point, 8 bytes
char*1	Character, 1 byte

The following programs demonstrate how to write different kinds of data to binary files. Associated with a file is a file position indicator. When a file is opened, the file position indicator is automatically set to zero, the beginning of the file (**bof**), and after writing to a file, the file position indicator points to the last file location where data was written, which may be the end of the file (**eof**). To read a file from the beginning, the file position indicator must point to the beginning of the file. Therefore, after a file write and before a file read, use the function **frewind** to set the file position indicator to the bof.

Example 8.3

```
% Demonstration of writing a character string to a binary file
clc; clear all; fclose('all');
Pi = sprintf('Pi equals: %10.7f\n',pi); % a character string
N_codes = length(Pi); % Pi is a row character vector of ASCII codes
ID1 = fopen('Binary_data_1.bin','w+') % open file for read and write
fwrite(ID1,Pi); % using default precision, uint8
frewind(ID1); % set file pointer to bof
S = fread(ID1,[1,N_codes]) % get a row vector using default precision
fclose(ID1);
char(S) % convert binary data to text

ID1 =    3
S =
 Columns 1 through 22
80 105 32 101 113 117 97 108 115 58 32 32 51 46 49 52 49 53 57 50 55 10
ans =  Pi equals: 3.1415927

% Demonstration of writing signed integers to a binary file
clc; clear all; fclose('all');
A = [-347 239;-701 3]; % each element of A requires no more than 11 bits
ID1 = fopen('Binary_data_2.bin','w+'); % open file for read and write
fwrite(ID1,A,'bit11'); % signed integer 11-bit precision
frewind(ID1); % set file pointer to beginning of the file
B = fread(ID1,[2,2],'bit11') % using same precision as for fwrite
fclose(ID1);

B =
 -347  239
 -701    3

% Demonstration of writing floating point numbers to a binary file
clc; clear all; fclose('all');
A = [pi sqrt(2);exp(1) acos(-1)];
ID1 = fopen('Binary_data_3.bin','w+'); % open file for read and write
position_indicator = ftell(ID1) % returns the file position indicator
fwrite(ID1,A,'double'); % signed double precision
frewind(ID1); % set file pointer to beginning of the file
```

```
format long
B = fread(ID1,[2,2],'double') % using same precision as for fwrite
position_indicator = ftell(ID1) % returns the file position indicator
fclose(ID1);

position_indicator =   0
B =
  3.141592653589793  1.414213562373095
  2.718281828459046  3.141592653589793
position_indicator =  32
% notice that the position indicator counts bytes
```

8.2 Input

The **input** function provides a means for a programmer to prompt a program user for input. Its syntax is given by

X = input(' .... prompt text ....')

where x is any valid MATLAB variable name. The prompt text can contain any number of the escape characters \n, which cause continuation on a new line for each occurrence of the escape characters, and then, a multi-line prompt text can be used. When the input statement is executed, MATLAB prints the prompt text in the *Command Window* and stops program execution, waiting for the program user to respond in the *Command Window*. The response can be any valid MATLAB expression followed by depressing the Enter Key. Then, MATLAB evaluates the expression, assigns the result to x and continues program execution. If nothing is entered and the Enter Key is depressed, then x is assigned an empty array. An input option has the syntax given by

X = input(' .... prompt text ....', 's')

where the input, without using single quotes, is treated as a character string that is assigned to x.

Another way to obtain direct input from a program user is with the built-in function **menu**, which has the syntax given by

choice = menu('menu_title','option1','option2', ... , 'optionN')

For example,

```
>> analysis_type = menu('Select the kind of circuit analysis method', ...
     'elemental','loop','nodal')
```

After this statement has executed, the selection menu shown in Fig. 8.2 is displayed, and program execution is stopped until one of the buttons is clicked. If an option button is clicked, then analysis_type is assigned an integer, 1, 2, or 3 corresponding to option1, option2, or option3. If instead, the menu window is closed, then menu returns 0. After any button is clicked the menu window closes.

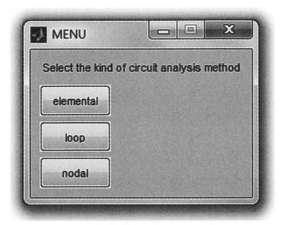

Figure 8.2   Selection menu.

If the loop (option2) button is clicked, then we get

```
analysis_type =   2
```

With this user input, if-elseif-else or switch-case methods can be used to control which script segment is executed.

Table 8.7 gives several built-in MATLAB functions concerned with accessing the content of a file.

Table 8.7   Some functions concerned with accessing data in a file

Function	Brief description
M = load('filename') load ('filename','var1', ..., 'varN')	Load data from a mat-file into the Workspace; mat-file can contain a structure or ASCII data. Or, load designated variables. Use doc load for more options.
S = fileread('filename')	Returns content of a file as a string vector
M = dlmread('filename') M = dlmread('filename','delimitier',R1,C1) M = dlmread('filename','delimiter',R)	Reads numeric data from the ASCII delimited file filename; R1 and C1 specify the upper left corner (row and column) where the data read starts. R = R1 C1 R2 C2] also specifies the lower right corner (row and column) where the data read ends.
f_path = which('filename')	Returns the path name of filename
A = fread(file_ID,size,'precision')	Reads binary data from file_ID; the number of elements is specified by size, where size can be N to read N elements into a column vector A, inf to read to the end of the file and [N,M] to read elements to fill an N × M matrix, in column order; some precision possibilities are given in Table 8.6. See Example 8.3.

(Continues)

Table 8.7 (*Continued*)

Function	Brief description
A = fscanf(file_ID,'format',size)	Same as fread, but converts data according to the specified format string
frewind(file_ID)	Sets the file position indicator to the beginning of the file
L = feof(file_ID)	Returns logic 1 if a read from the end of the file with file_ID has occurred
S = ferror(file_ID)	Returns an error message if an error occurred during the most recent file_ID input/output operation
S = fgetl(file_ID)	Returns as a string the next line of the file with file_ID; a −1 if an end-of-file occurred. The line terminator is not included.
S = fgets(file_ID)	Same as fgetl, but includes the line terminator
status = fseek(file_ID,offset,origin)	Sets the file position indicator to the byte with the specified offset relative to origin; offset, an integer, can be positive or negative and origin can be −1 for beginning of file, 0 for current file position and 1 for end of file. Status is 0 if successful and −1 if not successful
position = ftell(file_ID)	Returns the location of the file position indicator

Example 8.4

The files created by Prog. 8.1 will be used to demonstrate some of the functions given in Table 8.7. See Example 8.3 for demonstrations of using the **fread** function.

With the fgetl function you can read a file one line at a time. For example,

```
clear all; clc; close all
ID1 = fopen('example2_data.txt');
while 1
   txt_line = fgetl(ID1);   % get a line from file ID1
   if ~ischar(txt_line), break; end
   disp(txt_line) % display line from file ID1
   % could use instead:  if feof(ID1), break; end
end
fclose(ID1);
```

The output produced by this program is the same as the center columns of the table given in Example 8.1.

With the load function, the content of a file can be placed in the Workspace. For example,

```
>> % using command line format
>> load example3_data -ascii  % mat-file contains ascii coded data
```

With the fscanf function, you can read a file according to a size specification. For example, in Prog. 8.1, the dlmwrite function was used to store in a text file each row of a matrix using one space as a delimiter and ten spaces for each number. Therefore, each number occupies ten character spaces. Let us retrieve the fourth row of the matrix stored in example1_data.txt with the following script.

```
clear all; clc; close all
ID1 = fopen ('example1_data.txt');
N_chars = 10;  % number of characters used to store each number
status = fseek(ID1,9*(N_chars+1)+1,'bof')  % locate position indicator
for k =1:3
   row_two(k,1:N_chars) = fscanf(ID1,'%c',N_chars);  % get each number
   fseek(ID1,1,0);  % skip one character (the delimiter)
end
fclose(ID1);
row_two % this is a character array

status =    0
row_two =
  0.037500
  0.707107
 -0.707107
```

The output produced by this script is the fourth row of the left columns in the table given in Example 8.1.

We can also search a file by first retrieving it into a character string with the **fileread** function. For example,

```
>> S = fileread ('example1_data.txt')
S =
  0.000000   0.000000   1.000000
  0.012500   0.707107   0.707107
  0.025000   1.000000   0.000000
  0.037500   0.707107  -0.707107
  0.050000   0.000000  -1.000000
  0.062500  -0.707107  -0.707107
  0.075000  -1.000000   0.000000
```

```
 0.087500  -0.707107   0.707107
 0.100000  -0.000000   1.000000

>> S(35:44)   % first number in the second row
ans =   0.012500
```

The **dlmread** function is useful to read numeric data from an ASCII file produced by an m-file and other scripts.

Example 8.5

A data acquisition (**DAQ**) system was used to simultaneously sample the outputs of four electronic stethoscopes positioned on a person's chest in the vicinity of the heart. A recording of the heart sound is called a **phonocardiogram**. The DAQ software was set to sample each of the four input channels at the rate $f_s = 5,000$ samples/sec for 10 secs. Then, a text file, heart_sound.txt, was produced that contains 50,000 samples of each of the four phonocardiograms. The first few lines of this text file are shown below.

Simultaneous data acquisition of four stethoscope outputs

Channel 1	Channel 2	Channel 3	Channel 4
0.021488016532000	0.011297458343000	0.009650987007000	0.019350983292000
0.012722182546000	0.008381021024000	0.010532377762000	0.014182648164000
0.005507278486000	0.001055447980000	0.008542691504000	0.011926686962000
0.003813026290000	−0.007777469734000	0.005921087722000	0.006991390552000
0.003179968774000	−0.014918743956000	0.000258228366000	0.001760220377000
0.002405418634000	−0.015743206567000	−0.005977382491000	0.004876960718000
−0.00164907690200	−0.008207093478000	−0.008384890664000	0.001980448066000
−0.000740962604000	−0.001375831844000	−0.003824837180000	0.001073305203000
−0.001673472182000	0.001246459673000	0.000276527136000	−0.000185835933000
−0.005308368902000	−0.001171394409000	0.003352550373000	−0.001687166522000

Prog. 8.3 was used to import the data into MATLAB.

```
% Program to read a text file of digitized phonocardiograms
clc; clear all; close all
PCG_file_name = input('Enter a data file name: ','s');
PCG_file_name_txt = [PCG_file_name,'.txt']; % form complete file name
DAQ_output = dlmread(PCG_file_name_txt,'',0,0); % get data
%
[N_samples N_channels] = size(DAQ_output); % get array size
fs = 5000; T = 1/fs; % set sampling frequency and sample time increment
n = 0:N_samples-1; t = n*T; % sample time points
```

```
while 1
  channel = input('Enter a channel number: ');
  if channel < 1 || channel > N_channels
    disp('Terminated Program.')
    break;
  end
  y = DAQ_output(:,channel); % get data from one channel
  plot(t,y') % plot data
  grid on
  xlabel('Time - seconds'); ylabel('Phonocardiogram');
  title(['Data from Channel ',num2str(channel)]) % show channel number
end
```

Program 8.3   Import data with the dlmread function.

```
Enter a data file name: heart_sound
Enter a channel number: 4
```

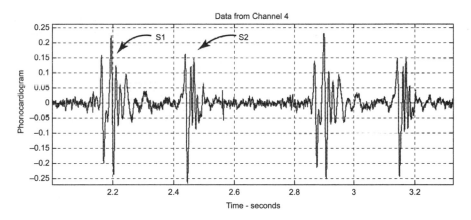

Figure 8.3   Phonocardiogram over two cardiac cycles.

```
Enter a channel number: 0
Terminated Program.
```

Program 8.3 plotted the data from 0 to 10 seconds. Fig. 8.3 was obtained by using the **zoom** feature in the MATLAB Figure Window. Fig. 8.3 shows the heart sound over two cardiac cycles. The sound called **S1**, the first heart sound, is a combination of the sounds caused by the closure of the **Mitral** and **Tricuspid** valves, over almost the same time range. The second heart sound, called **S2**, is caused by the closure of the **Pulmonary** and **Aortic** valves, where normally the Aortic valve begins to close first. To a medical practitioner, this plot can provide useful information about the state of cardiac function. A heart dysfunction can cause additional sounds, called **murmurs**, to appear in a phonocardiogram. The location of a murmur within the time duration of a cardiac cycle can help a medical practitioner to identify the kind of heart dysfunction that caused the murmur.

As often occurs, the information of interest about a physical process is not easy to obtain, and the data that is obtained includes various kinds of distortions. Fig. 8.3 illustrates this. There is a considerable amount of noise. With a little additional MATLAB program-ming, the heart beat rate can be determined with very good confidence using an inexpensive instrument and a noninvasive method. However, it would be particularly useful if, for example, the sound made by the aortic valve could be extracted from the S2 sound.

A program can only work on the content of a file after the data contained in the file has been transferred to the Workspace. MATLAB has a GUI that will enable you to import data from any kind of a file recognized by MATLAB and place the data in the Workspace. To use this GUI, go to the MATLAB desktop and click the menu sequence

File → Import Data

which opens the *Import Data Window* that shows all of the data files in the *Current Folder*. Click on the file of interest to highlight it, and then click the open button. Then, the file content is displayed in the *Import Wizard Window*, where you can elect to transfer the file content to the Workspace. Once the data is in the Workspace, it can be accessed from the *Command Window* or a script, like any defined MATLAB variable.

Another way to import data is with the built-in function **importdata**, with which you can load the content of a variety of kinds of data files into the Workspace. Use the MATLAB help facility (doc importdata) for more information.

8.3   File Management

MATLAB includes several built-in functions that are useful for managing files and folders. We have already discussed how to create a variety of files. Table 8.8 gives built-in functions for creating a folder, moving files from a folder to another folder and other activities.

Table 8.8   Functions concerned with file and folder management

Function	Brief description
isdir(fullfile('C:','My_MATLAB_ Programs')) isdir('Music')	Returns logic 1 if given name is a folder
S = dir('folder_name')	Returns an Mx1 structure array with fields: name, date, bytes, isdir and datenum; see Example 7.4
rmdir('folder_name') rmdir('folder_name','s')	Remove folder_name from the current folder when folder_name is empty; and use the 's' option to remove folder_name and all of its contents

(Continues)

Table 8.8 (*Continued*)

Function	Brief description
mkdir('folder_name')	Create folder_name within the current folder; use a complete pathname to create folder_name in another folder
cd('new_folder') current_folder = cd('new_folder')	Changes the current folder to the new_folder; and returns the current_folder name
movefile('source') movefile('source','destination')	Move the file or folder named source to the current folder; move file or folder named source to file or folder named destination
delete filename delete('fileName1','filename2', ...) delete(h)	Command syntax to delete filename; function argument syntax to delete files; delete graphic object with handle h; use wildcard * to delete all files with the same suffix
copyfile('source','destination')	Copy source file or folder to destination file or folder
F = ls	Returns an m × n character array of the names of files and folders in the current folder, where there are m names
zip('zip_file_name', 'name1', ..., 'nameN')	Creates a zip file with the name zip_file_name from the list of files and folders specified by strings name1 through nameN or cell array of strings that specify the files or folders; and * may be used as a wild card for all files having the same suffix
unzip('zip_file_name') unzip('zip_file_name','output_folder')	Unzip a zipped file named zip_file_name into the current folder; unzip into the output folder

Example 8.6

```
>> % save storage space and time to email files by compressing them
>> % text file heart_sound.txt uses 4.52 Mb of storage
>> zip('heart_sound.zip','heart_sound.txt')
>> % heart_sound.zip uses 1.53 Mb of storage
>> % unzip heart_sound.zip and place output in folder Music
>> unzip('heart_sound.zip','Music')
>>
>> isdir('Music') % check if Music is a folder in the current folder
ans =    1
>> % use a complete path name to check elsewhere
>>
>> old_folder = cd('Music')
old_folder =  C:\My_MATLAB_Programs
>> % now, Music is the current folder
>> cd(old_folder) % change back to C:\My_MATLAB_Programs
>>
>> mkdir('circuits') % create a folder named circuits within the current folder
>> % move circuit_report.m to sub-folder circuits
>> movefile('circuit_report.m','circuits')
```

```
>> % use wildcard * to move all mat-files starting with CKT
>> movefile('CKT*.mat','circuits')
>>
>> cd('circuits')  % make circuits the current folder
>> circuit_files = ls  % get a character array of all files in current folder
circuit_files =
.

..
CKT_1.mat
CKT_2.mat
CKT_3.mat
circuit_report.m
```

8.4   Sound

MATLAB has a few built-in functions with which you can send data to the audio output device of your computer. These functions are given in Table 8.9.

Table 8.9   Functions concerned with sound

Function	Brief description
beep on beep off	Turns the beep sound on and off
sound(audio_data,fs)	Sends audio_data to the speaker at sample rate fs; for a single channel, audio_data is an $m \times 1$ column vector, where m is the number of audio samples; audio_data can be an $m \times 2$ matrix, where the first column corresponds to the left channel and the second column corresponds to the right channel. Data outside the range $[-1,+1]$ will be clipped.
soundsc	Same as sound; and also automatically scales the data to the range $[-1,+1]$
wavplay	Play WAV files; see Example 7.4
[Y,fs,nbits] = wavread('file_name')	Returns in the two columns of Y the left and right sound channels of a wav file; f_s is the sampling frequency and nbits is the number of bits per sample. There is also an option about the data in Y.
wavwrite(Y,fs,nbits,'file_name')	Create a wav file
record = audiorecorder(fs,nBits, nChannels) recordblocking getaudiodata	Set up, record and store audio input; see Example 8.8
Y = wavrecord(N,fs)	Records N samples of an audio signal, sampled at a rate of f_s samples per second

Example 8.7 ————————————————————

Interesting effects are possible by assigning a handle to a plot. This example demonstrates using the **delete** and **sound** built-in functions. Prog. 8.4 simulates the movement of the second hand of a clock, including a tick sound. To achieve real-time operation, the built-in function **pause** is used.

```
% Animation of the second hand of a clock
clear all; clc; close all
fs = 44100; T = 1/fs; % set sampling frequency
f = 1500; w1 = 2*pi*f; w2 = 2*pi*(1.5*f); % set frequencies for tick sound
N = 4000; n = 0:N-1; t = n*T; % let tick last for about 0.1 sec
envelope = 0.5*(exp(-50*t)-exp(-100*t)); % tick sound envelope
tic_sound = envelope.^(sin(w1*t) + sin(w2*t)); % tick sound
angle_increment = 2*pi/60; % second hand angle increment
sec_angle = pi/2:-angle_increment:-3*pi/2; % once around
K = length(sec_angle);
xt = cos(sec_angle); % coordinates for circular movement
yt = sin(sec_angle); % coordinates for circular movement
Limit = 1.5; axis([-Limit Limit -Limit Limit]); %set x-axis and y-axis limits
hold on % all plots on the same figure
% draw a dial
plot(xt,yt,'Color','red','LineWidth',2) % plot outside circle
plot(0.15*xt,0.15*yt) % plot center hub
for k=1:K % plot tick marks
  plot([0.15*xt(k),xt(k)],[0.15*yt(k),yt(k)])
end
xt = 0.9*xt; yt = 0.9*yt; % adjust second hand length
loop_time = (63.435027-60)/60; % time to execute the following for loop once
% loop execution time, which is system dependent, was found with tic and toc
wait = 1 - loop_time; % time adjustment to account for loop execution time
% tic % used to start stop watch
for k=1:K % animate second hand
   sec_hand = plot([0,xt(k)],[0,yt(k)],'LineWidth',2.5); % assign a handle
   sound(tic_sound,fs) % send tick sound to audio device
   pause(wait) % wait for about 1 second
   delete(sec_hand) % delete previous second hand
end
% toc % used to find for loop execution time (63.435027 seconds) with wait=1
plot([0,xt(K)],[0,yt(K)],'LineWidth',2.5) % show second hand stopped
```

Program 8.4   Animation of the second hand of a clock.

————————————————————————————————————

In Example 8.7, the plot function was used with options that were not used before this example. One case is the statement

```
plot(xt,yt,'Color','red','LineWidth',2) % plot outside circle
```

which includes the arguments, "Color," "red," "LineWidth," 2. The plot of yt versus xt, which in this case makes a circle, is an object that can have many properties, for example, the color of the line and the width of the line. You can set these properties by specifying their values after naming them. The property names, "Color" and "LineWidth," are field names in an array structure that defines the object, where the desired value is given immediately following the field name. A line object can have many other properties to which you can assign values this way. This will be discussed in greater detail in the next chapter.

The other case is the statement

```
sec_hand = plot([0,xt(k)],[0,yt(k)],'LineWidth',2.5); % assign a handle
```

which also includes a property specification. Moreover, the plot function was used like in an assignment statement. However, the name, sec_hand, is not called a variable name. Instead, it is called a **handle**. The handle is a name, which can be any valid MATLAB name, of the plot object, the second hand. Its handle gives us a way to refer to the plot object, find out more information about it and set values, other than default values, to its properties. In the for loop, using its handle, the second hand is deleted to plot a second hand in another position. Handle graphics will also be discussed in greater detail in the next chapter.

Example 8.8 ———————————————————————————

This example demonstrates using built-in functions for recording audio input from a microphone. You can play back the input, plot it, and store it in a binary file for further processing. For example, for speech recognition, it is useful to find how the frequencies and amplitudes of the sinusoidal time functions that together make up the sound, change over the time duration of the sound. Program output follows Prog. 8.5, and Fig. 8.4 shows an example recording.

```
% Program to record microphone input
clear all; clc; close all
fs = 44100; N_bits = 16; N_chan = 1;
% specify sampling freq, number of bits per sample and 1 or 2 channels
mic_record = audiorecorder(fs,N_bits,N_chan); % set recording parameters
while 1
    disp('Enter zero record time to terminate recording program')
    rec_time = input('Enter record time in seconds: ');
    if rec_time <= 0
      disp('Program terminated.')
```

```
      break
   end
   disp('Press enter key to start recording')
   pause
   recordblocking(mic_record,rec_time); % recording
   disp('End of recording')
   answer = input('Do you want to hear the recording, Yes or No? ','s');
   if strcmpi(answer,'yes')
      play(mic_record) % play MATLAB formatted recording
   end
   Mic_record = getaudiodata(mic_record); % store data in an array
   answer = input('Do you want to save the recording, Yes or No? ','s');
   if strcmpi(answer,'yes')
      file_name = input('Enter a file name: ','s'); % get file name
      File_name = [file_name,'.bin']; % add suffix
      ID1 = fopen(File_name,'w+'); % open binary file for write
      fwrite(ID1,Mic_record,'double'); % store binary data
      fclose(ID1);
      % to read the file, use, for example,
      % ID1 = fopen(File_name);
      % x = fread(ID1,'double');
      % fclose(ID1)
   end
   answer = input('Do you want recording plotted, Yes or No? ','s');
   if strcmpi(answer,'yes')
      N_samples = floor(rec_time*fs); T = 1/fs;
      time_pts = [0:N_samples-1]*T; % time in seconds
      plot(time_pts,Mic_record)
      grid on
      xlabel('Time - seconds')
      ylabel('hello')
      title('Microphone Recording')
   end
end
```

Program 8.5 Program for recording audio input from a microphone.

```
Enter zero record time to terminate recording program
Enter record time in seconds: 2
Press enter key to start recording
End of recording
Do you want to hear the recording, Yes or No? yes
Do you want to save the recording, Yes or No? yes
Enter a file name: hello
```

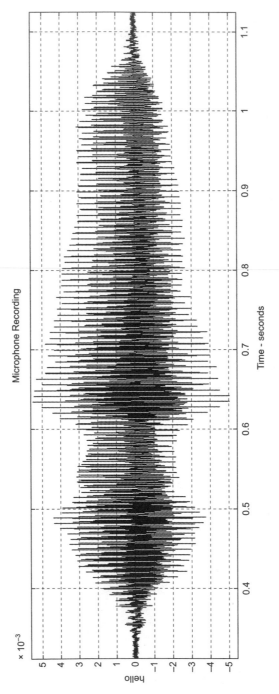

Figure 8.4 Recording of microphone audio input.

```
Do you want recording plotted, Yes or No? yes
Enter zero record time to terminate recording program
Enter record time in seconds: 0
Program terminated.
```

The zoom and label features of the MATLAB *Figure Window* were used to edit this plot.

8.5 Conclusion

With MATLAB, you can write interactive scripts that receive input from the program user and files produced by other programs, and output computing results to the program user and files. You should now know how to

- create files for output of text data
- create files for output of binary data
- display formatted output
- save and load the Workspace
- prompt a program user for input
- import text and binary data
- locate file positions
- manage files and directories
- write and read a structure array
- compress and decompress files
- use the MATLAB sound functions

As we have seen many times already, another important part of interactive computing is graphics, which is the topic of the next chapter.

Problems

In many of the following problems you will be asked to give MATLAB statements or write MATLAB scripts. You must also provide copies of statement results, scripts, and their outputs. Include any required or explanatory discussion.

Section 8.1

1) Suppose the following MATLAB statements have been executed.

```
birth_year = input('In what year were you born? ')
birth_date = input('What is your birth date, (mm/dd/yyyy)? ','s')
```

(a) Using the fprintf function, give a MATLAB statement that outputs a complete sentence starting with, "I was born in". Include the year, and be sure to include, \n.

(b) Using the num2str function, create an appropriate vector to use the disp function for outputting the same sentence as in part (a).

(c) Repeat part (a), but start the sentence with, "I was born on". Include the date.

2) Suppose the following MATLAB statement has been executed

```
Eulers_number = exp(1);
```

(a) Using the fprintf function, give a MATLAB statement that outputs a complete sentence starting with "The number e is: ". Use fixed point notation and include seven fractional digits in a field width of 10.

(b) Repeat part (a) using exponential notation.

3) In a statement that uses the fprintf function, give an example of using (a) \t, (b) %#, (c) %-15.5f, (d) %0, (e) several %n$, where n is an integer.

4) Give a MATLAB statement to

(a) type an m-file in your current folder to include line numbers

(b) obtain the complete path name of an m-file in your current folder

(c) open a new file test.txt for writing, and assign to it a file ID

(d) check if the function eig is a built-in function

5) Write a MATLAB script that creates an 11×11 array mult_tab of the multiplication table. The first row, starting in the second column, must be the ten digits 9-0, and the first column, starting in the second row, must be the ten digits 0-9. Place the character X in the top left corner (first row, first column). Describe what happens if you include the statement, more(1), just before the script segment that displays the table.

6) Write a MATLAB script that uses the dlmwrite function to write to a file, named squares.txt, a table of two columns for x and x^2, where $x = 0:0.1:1$.

7) Repeat Prob. P8.6, but use instead the fprintf function.

8) Repeat Prob. P8.6, but use instead the save function.

9) Write a MATLAB script that creates a structure, named my_info, with fields: last_name, first_name, birth_date, and address. Place the structure in a file, named MyInfo. Then, open the file, and use the fprintf function to display the structure content.

10) Write a MATLAB script that assigns to the variable Name a character string that is your name and assigns to the variable Age your age. Find the length of each variable. Using the fopen function, obtain a file ID to write into a file named name.bin. With the fwrite function write Name and Age into name.bin.

Section 8.2

11) Repeat Prob. 8.10, and append statements to locate the file position indicator at your age, and retrieve your age from the file.

12) Using the function input, give a MATLAB statement to enable a program user to enter a

(a) matrix and assign it to the variable X
(b) cell array of strings and assign it to the variable C
(c) character vector and assign it to the field of a structure Ckt.resistor_names
(d) numeric vector and assign it to the field of a structure Ckt.resistor_values

13) Assume that you have executed a program located in your current folder, and that one of the variables listed in the Workspace is named X.
    (a) Using the function format, give a MATLAB statement to save your Workspace in the file my_prog_data.mat. Now, assume that the Workspace has been cleared. Using the function format, give a MATLAB statement to load X into the Workspace.
    (b) Repeat part (a) using the command line format.
14) Write a MATLAB script that starts by opening one of your m-files and assigns a file ID. Then, read and display the file one line at a time until the end of file is reached.
15) Write a MATLAB script that starts by reading an m-file in your current folder into a string vector. Then, find the first occurrence of an assignment statement, and display that statement. *Hint*: Convert vector to class double, look up the ASCII codes for linefeed and carriage return and use the find function.
16) Assume the following program has executed.

```
% Program to write a matrix of data to a text file
clear all; clc;
f = 10; w = 2*pi*f; T0 = 1/f; % frequency and period
N = 9; t = linspace(0,T0,N); % obtain N samples over one cycle
s = sin(w*t); c = cos(w*t); % samples
TSC = [t',s',c']; % collect data in columns
dlmwrite('P8_16_data.txt',TSC,'delimiter',' ','precision','%10.6f')
% each output line comes from a row of TSC
```

    (a) Continue the script by using the function dlmread to read the file and display a table of the data.
    (b) Use the Import Data GUI to look at the data. Describe the display.
17) Use the MATLAB help facility, and find out how to use the function **importdata**. Import the data from the file created by the script given in Prob. P8.16. Describe what happens.

Section 8.3

18) Write a MATLAB script that starts by using the function input to obtain the name for a new folder, and create the new folder with the inputted name within your current folder. Then, obtain a structure array named my_m_files that includes in a field the names of all of your m-files in the current folder. Copy only those files that have a date within the past

month from the current folder to the new folder. *Suggestion*: look up built-in functions date and datevec to obtain the present date. Finally, make the new folder the current folder. Does the new folder name appear in the MATLAB desktop toolbar?

19) Give a MATLAB statement that creates a zip file of all of the m-files in your current folder. Then, give another MATLAB statement that unzips the zip file into another folder.

Section 8.4

20) The following MATLAB script uses the built-in function **chirp** to create samples of a sweep frequency signal.

```
% Program to generate and sound a chirp signal
% specify fs, sampling frequency
fs = 44100; T = 1/fs; total_time = 1; N = floor(total_time/T);
n = 0:N-1; time = n*T; % time points
fstart = 30; fend = 4000; % frequency range in Hz
x = chirp(time, fstart, total_time, fend); % get chirp signal samples
y = fliplr(x); z = [x,y]; % flip chirp signal and concatenate
% change time increment between samples by changing sampling frequency
fz = 0.1*fs;
sound(z,fz); % sound z at sampling rate fz
plot(time(1:3000),x(1:3000)); grid on % plot some points
```

(a) Enter this program into your computer and run it. Describe the sound. What does the fliplr function do to generate the second half of the sound?

(b) Change fz to 0.25*fs. How has the sound changed? The sound function uses the samples at time increments given by 1/fz secs. How has the time increment changed from the time increment used in part (a)?

(c) Change fz to 1.25*fs, and repeat part (b).

21) The following MATLAB script is a shortened version of Prog. 8.5.

(a) If you have a microphone that is plugged into the audio input of your computer, then enter the script and run it. Record for 2 secs, and speak one word into the microphone. Describe what happened.

```
% Program to record microphone input
clear all; clc; close all
fs = 44100; N_bits = 16; N_chan = 1;
% specify sampling freq, number of bits per sample and 1 or 2 channels
mic_record = audiorecorder(fs,N_bits,N_chan); % recording parameters
disp('Enter zero record time to terminate recording program')
```

```
rec_time = input('Enter record time in seconds: ');
if rec_time <= 0
  disp('Program terminated.')
else
  disp('Press enter key to start recording')
  pause
  recordblocking(mic_record,rec_time); % recording
  disp('End of recording')
  play(mic_record); % play MATLAB formatted recording
  Mic_record = getaudiodata(mic_record); % store data in a vector
  N_samples = length(Mic_record);
  T = 1/fs; % time increment
  time_pts = [0:N_samples-1]*T; % time in seconds
  plot(time_pts,Mic_record)
  grid on; xlabel('time - seconds'); ylabel('Microphone Recording')
end
```

(b) The vector Mic_record contains the samples of the audio signal, which can be used as input to the function sound. For example,

```
fx = 0.5*fs;
sound(Mic_record,fx)
```

Append these statements to the else part of the script, and try different values for fx. Describe what happened for fx = 0.5*fs and fx = 1.5*fs. You may have to change the play statement into a comment statement to avoid interference.

CHAPTER 9

Graphics

An outstanding feature of MATLAB® is its facility for visualizing data. The built-in function **plot** has already been used many times. It automatically scales the axes and locates grid lines, which make it convenient to use the plot function. Labels for the axes and a plot title can be added easily. The functions **bar** and **stem** to obtain bar charts and stem plots were also easy to use. The plots obtained with these and other two-dimensional (2-D) plotting functions can be edited extensively to create custom figures. Also, MATLAB has a variety of 3-D plotting functions. In this chapter, you will learn how to

- use additional 2-D plotting functions
- place multiple plots within a figure
- display the frequency response of an analog filter
- use 3-D plotting functions
- do translation and rotation of objects in 3-D space
- create animations
- customize graphics

9.1 Figure

A MATLAB graphic is a collection of graphic objects. Every graphic is created by first creating a figure graphic object. A figure object is a window that appears on the screen into which you can place other graphic objects using various MATLAB functions. The built-in function **figure** creates a new figure object using default property values, which you can alter. A newly created figure automatically becomes the current figure, which appears above all other figures on the screen until another figure object is either created or activated. The syntax options for the figure function are given by

```
figure
figure('PropertyName',PropertyValue, ...)
h = figure(...)
figure(g)
```

When the function figure is invoked without an argument, MATLAB creates a new figure object using default property values, assigns to the figure a number, which is called the **handle** of the figure, and makes the new figure object the current figure. The handle of a figure object is an integer. Its value is the next higher integer value after integer values assigned to previously created figures. For example, if figure 1 is the only figure that already exists, then invoking figure will create figure 2, and make it the current figure. Then, to make figure 1 the current figure, invoke the function figure with a handle operand, as in figure(1).

The second syntax allows you to create a new figure object and override default property values. First, the property name is given, and then comes its new value. For example, one property is color, the color of the figure background. The background color is described by a three-element row vector that specifies the intensity of red, green and blue (RGB) each with values ranging from 0 to 1. Thus, a color vector of [1 0 0] makes the background color red. The default background has a color vector given by [0.8 0.8 0.8], which makes the background light gray. To create a figure object with, for example, yellow as its background use

```
figure('Color',[1  1  0])
```

To associate a name with a magenta background figure, use the statement, for example,

```
figure('Color',[0  1  1],'Name','i-v plot')
```

In addition to color and name, a figure object has many properties. See Example 9.1, and use **doc figure** for detailed explanations.

With the third syntax, which looks like an assignment statement, you can associate a valid MATLAB variable name with the created figure object. The name is also called the **handle** of the figure object. The handle name is equivalent to the handle number assigned to the figure object. If a graphic object has a named handle, then it can be referred to by using its handle name. If a previously created figure has a handle name, then in the argument of the fourth syntax you can use its handle name to make it the current figure. By associating a graphic object with a handle name, you do not have to keep track of its handle number. The fourth syntax allows you to either create a new figure object by using an integer for the argument or make a previously created figure object the current figure by using its handle name or number.

Table 9.1 gives a list of built-in functions concerned with figure graphic objects.

Table 9.1 Built-in functions concerned with figure objects

Function	Brief description
clf clf(h)	Clear current figure of all graphic objects; clear the specified figure with handle h of all graphic objects
close close(h) close all	Close the current figure window; close the specified figure window with handle h; close all open figure windows
delete(h)	Delete the specified graphic object with handle h
gcf	Get current figure handle
gco	Get current object handle
get(h) F = get(h) get(h,'PropertyName') get(h,'Default')	Get all graphic object property values; return a structure F, where each field name is the name of a property and each field contains the value of the property; get value of specified property and get all default property values of a graphic object
propedit(h)	Open property edit window and edit graphic object properties
saveas(h,'FileName.ext') saveas(gcf,'FileName.ext')	Save figure with handle h to a file; ext can be, for example, bmp (bit map), emf (enhanced metafile), fig (MATLAB figure), jpg (JPEG image), pdf (portable document format)
set(h,'PropertyName1', PropertyValue1, ...)	Set named graphic object properties to specified values

Example 9.1

Let us exercise some of the built-in functions given in Table 9.1.

```
>> clc;  % clear command window
>> % create a new figure object (becomes the current figure) and assign a handle
>> diode_characteristic = figure(1) % the handle name is diode_characteristic
diode_characteristic =     1
>> gcf   % (get current figure) returns the current figure handle number
ans =    1
>> set(diode_characteristic,'Name','i-v plot')  % give figure 1 the name i-v plot
>> % this could also have been accomplished with: set(gcf,'Name','i-v plot')
>> figure(2)  % create a new figure; becomes the current figure
>> gcf  % get current figure handle
ans =     2
>> get(diode_characteristic)  % returns all figure 1 properties and their values
>> % the properties of figure 1 and their current values are given in Table 9.2
>> % delete the figure 1 object; figure 1 no longer exists
>> delete(diode_characteristic)
>> % you can also use, delete(1), to delete the figure 1 object
```

```
>> % get value of the color property and assign it to fig_color
>> fig_color = get(2,'Color')
fig_color =
      0.8000      0.8000      0.8000
>> % since figure 2 is the current figure, it can also be deleted with
>> delete(gcf)
```

Table 9.2   Properties and values of figure 1

Property name and current value	Property name and current value
Alphamap = [(1 by 64) double array]	ResizeFcn =
CloseRequestFcn = closereq	SelectionType = normal
Color = [(1 by 3) double array]	ToolBar = auto
ColorMap = [(64 by 3) double array]	Units = pixels
CurrentAxes = []	WindowButtonDownFcn =
CurrentCharacter =	WindowButtonMotionFcn =
CurrentObject = []	WindowButtonUpFcn =
CurrentPoint = [0 0]	WindowKeyPressFcn =
DockControls = on	WindowKeyReleaseFcn =
FileName =	WindowScrollWheelFcn =
IntegerHandle = on	WindowStyle = normal
InvertHardcopy = on	WVisual = [(1 by 32) char array]
KeyPressFcn =	WVisualMode = auto
KeyReleaseFcn =	BeingDeleted = off
MenuBar = figure	ButtonDownFcn =
Name = i-v plot	Children = []
NextPlot = add	Clipping = on
NumberTitle = on	CreateFcn =
PaperUnits = inches	DeleteFcn =
PaperOrientation = portrait	BusyAction = queue
PaperPosition = [(1 by 4) double array]	HandleVisibility = on
PaperPositionMode = manual	HitTest = on
PaperSize = [8.5 11]	Interruptible = on
PaperType = usletter	Parent = [0]
Pointer = arrow	Selected = off
PointerShapeCData = [(16 by 16) double array]	SelectionHighlight = on
PointerShapeHotSpot = [1 1]	Tag =
Position = [(1 by 4) double array]	Type = figure
Renderer = None	UIContextMenu = []
RendererMode = auto	UserData = []
Resize = on	Visible = on

Notice that the property, Name, has the value "i-v plot". Since no graphic objects have been placed on figure 1, many properties have no values or are assigned an empty matrix, while other properties have default values. Remember, you can inquire about the value of any property with, for example, get(gcf,'Type'), which, if figure 1 is the current figure, returns figure.

Graphic objects, for example, axes, that have been placed on a figure are called the **children** of that figure, and the figure itself is called the **parent**. All figure graphic objects are children of figure(0), which is called the **root object**. You can see the properties of figure(0) (the root object) with the statement **get(0)**, and you can set some of these properties to values other than their default values with the function **set**. However, you cannot delete the root object. Notice that since no graphic objects were placed on figure 1 in Example 9.1, Table 9.2 shows that the property, Children, is assigned an empty matrix. As axes graphic objects are placed on figure 1, their handles are entered as elements of the vector assigned to the Children property. You can also see in Table 9.2 that the Parent property has the value [0], meaning that the root object is the parent of figure 1.

The *Figure Window*, into which you can place graphic objects, has normalized horizontal and vertical dimensions from zero to one, where the point (0,0) is the lower left corner and the point (1,1) is the upper right corner of the *Figure Window*. These dimensions can be used to place graphic objects at specific locations within a figure.

A few examples, with more details to be given later, of the graphic objects that you can place on a figure object are:

Axes

Axes objects are children of figure graphic objects and are parents of graphic objects produced by, for example, plot functions. Axes objects define a reference frame in a figure window for the display of objects that are generally defined by data. All functions that draw graphics, for example, **plot**, **surf**, **mesh**, and **bar**, create an axes object if one does not exist. More about this will be discussed later.

Line

Line objects are the basic graphic objects used to create most 2-D and some 3-D plots. Line objects become children of axes objects.

Rectangle

Rectangle objects are 2-D filled areas having a shape that can range from a rectangle to an ellipse.

Surface

Surface objects are 3-D representations of matrix data created by plotting the value of each matrix element as a height above the x–y plane.

Text

Text objects are character strings. Functions, for example, title, xlabel, ylabel, zlabel, text, and gtext create text objects, which you can place at a default or specified position on the current axes. The text location is specified with respect to the current axes limits. For example,

```
>> figure(1)  % create a figure graphic object
>> % place, "Hello", in the center of the current axes
>> text('Position',[0.5 0.5],'String','Hello')
```

To place the text, an axes graphic object is automatically created. The axes object is a child of the figure object, and the text object is a child of the axes object. See the following examples for more details about placing graphic objects on a figure.

9.2 Plots

MATLAB includes a wide variety of 2-D plotting functions, and much can be done to customize plots, including plot line types and widths, line colors, marker types, grid notation, text and legend insertion, fonts, math symbol insertion, and more. Special commands and a graphics GUI make it convenient to customize a plot.

9.2.1 2-D Plots

Table 9.3 gives a list of some of the graphic objects that can be placed on a figure graphic object and on an axes graphic object.

Table 9.3 Some graphic objects that can be placed on a figure and axes

Function	Brief description
area(x,y)	Same as plot(x,y), except that area under the plot line is filled with a color
axes('position','default') axes('position',[xmin,ymin,width, height]) h = axes(('position',[xmin,ymin, width,height])	Place axes with a default size in the current figure; place axes at a specified position, xmin = 0 → 1, ymin = 0 → 1, width = 0 → 1, height = 0 → 1; assign a handle to the axes
bar(Y) bar(X,Y) bar(X,Y,width)	Draws length(Y) bars; draws the columns of the M-by-N matrix Y as M groups of N vertical bars; width > 1, bars overlap, width = 1, bars touch and width < 1, leaves space between bars
barh	Horizontal bar graph

(Continues)

Table 9.3 (*Continued*)

Function	Brief description
comet(Y) comet(X,Y) comet(X,Y,p)	Displays an animated comet plot of the vector Y; displays an animated comet plot of vector Y vs. X; uses a comet of length p*length(Y), where default is p = 0.1
compass(U,V) compass(Z) = compass(real(Z), imag(Z)) compass(U,V,'linespec')	Draws a graph that displays the vectors with components (U,V) as arrows emanating from the origin; for linespec, see plot function
drawnow	Causes a figure window and its children to update; see Example 9.11
feather(U,V)	Plots vectors with components U and V as arrows emanating from equally spaced points along a horizontal axis
fill(X,Y,c)	Fills the 2-D polygon defined by vectors X and Y with the color specified by c, where c can be, for example, 'r', or a color vector
fplot(function,limits)	Plot function; limits = [xmin xmax] or limits = [xmin xmax ymin ymax]; for example, fplot(@(t) [sin(2*pi*t), sin (10*pi*t)], [0 2])
[X Y] = ginput [X Y] = ginput(N)	Use the mouse to get an unlimited number of points from the current axes coordinates, where each point is returned by left-clicking the mouse, until the enter key is depressed; get a total of N points
gtext('string')	Causes cross-hairs to appear on the current figure, which can be moved with the mouse, to place the character string at the cross-hair position by left-clicking the mouse
N = hist(Y,M) N = hist(Y,X) hist(...)	Bins elements of Y into M bins with bin count returned in N; bins elements of Y into bins with centers specified by X; no return produces bar chart instead
image(X) imagesc(X)	Display matrix X as an image, where each element of X specifies the color of a rectangular patch in the image; scale data to use the full colormap range; see Section 9.4
legend	place legend on current axes
line(X,Y) line(X,Y,Z)	Adds the line in vectors X and Y to the current axes; if X and Y are matrices, one line per column is added; creates lines in 3-D coordinates
loglog(...)	Same as PLOT(...), except logarithmic scales are used
pie(X) pie(X,slices) pie(... , labels)	Draws a pie plot of the data in the vector X; slices, with same length as X, specifies the slices that should be pulled out from the pie; labels, a cell array, contains labels for each pie slice
plot(...)	See text
plotedit on	Turn on tools for editing and annotating plots

(*Continues*)

Table 9.3 (*Continued*)

Function	Brief description
plotedit off	
plotmatrix(X,Y)	Scatter plots the columns of X against the columns of Y
plotyy(X1,Y1,X2,Y2)	Plots Y1 versus X1 with y-axis labeling on the left and plots Y2 versus X2 with y-axis labeling on the right
polar(theta,r) polar(theta,r,'linspec')	Draws a plot using polar coordinates of the angle THETA, in radians, versus the radius r
scatter(X,Y) scatter(X,Y,S,C) scatter3(X,Y,Z, …)	Displays colored circles at the locations specified by the vectors X and Y, where S, in points squared, specifies the area of each point and C specifies the color; 3-D scatter plot
semilogx	Same as plot(…), except a logarithmic (base 10) scale is used for the x-axis
semilogy	Same as plot(…), except a logarithmic (base 10) scale is used for the y-axis
sphere	A built-in function that plots a sphere
stairs	Similar to plot(…), except produces a piecewise constant plot
stem(x,y) stem(… ,'filled') stem3(x,y,z)	Plots a marker at the data sequence and a vertical line from the marker to the x-axis; produces a stem plot with filled markers; use help stem for additional options
subplot(m,n,p)	Breaks the current *Figure Window* into an m × n matrix of small axes, selects the p-th axes for the current plot; use help subplot for additional options
text(x,y,'string')	Places the character string at the location (x,y) on the current axes, where (x,y) are in units from the current axes
title('text') title('text','Property1',Property Value1, …) title(ax_h, …) h = title(…)	Adds text above the current axis; sets the values of the specified properties of the title; adds the title to the specified axes; returns a handle of the text object used as the title
xlabel('text') xlabel('text','Property1',Property Value1, …) xlabel(ax_h, …) h = xlabel(…)	Adds text beside the x-axis of the current axes; sets the values of the specified properties of the xlabel; adds the xlabel to the specified axes; returns a handle to the text object used as the label
ylabel	Same as xlabel(…), except that text is vertically oriented

To place a plot graphic object on a *Figure Window*, axes must first be placed on the figure. Some functions automatically create an axes graphic object, if one does not already exist. Table 9.4 gives some built-in functions concerned with axes graphic objects. You can place multiple axes on a *Figure Window*. See the following examples.

Table 9.4 Built-in functions concerned with axis graphic objects

Function	Brief description
axis([xmin xmax ymin ymax]) ax = axis axis action	Set limits on the axes Returns in ax, a row vector, that contains ax = axis scaling of the current plot; action can be: auto, sets axis scaling to its default automatic mode manual, freezes the scaling at the current limits tight, sets the axis limits to the range of the data equal, sets the aspect ratio so that equal tick mark increments on the x, y, and z axis are equal in size square, makes the current axis box square in size normal, restores the current axis box to full size vis3D, freezes aspect ratio properties to enable rotation of 3-D objects and overrides stretch-to-fill off, turns off all axis labeling, tick marks and background on, turns axis labeling, tick marks and background back on see doc axis for more options
box on	Add box around current axes
cla	Clear current axes
gca	Get handle of current axes
grid on; grid off grid(ax_h, 'on')	Adds major grid lines to the current axes; adds major grid lines to the named axes
rectangle('Position', [x y w h])	Place a rectangle with left lower corner at location (x,y) on the current axes; w and h are width and height of the rectangle; use doc rectangle for many options
X_lim=xlim xlim([xmin xmax]) xlim(ax_h, ...)	Get limits of current axis; set limits of current axis; set limits of axis with handle ax_h
Y_lim=ylim ylim([ymin ymax]) ylim(ax_h, ...)	Get limits of current axis; set limits of current axis; set limits of axis with handle ax_h
Z_lim=zlim zlim([zmin zmax]) zlim(ax_h, ...)	Get limits of current axis; set limits of current axis; set limits of axis with handle ax_h

Depending on what you want a plot to look like and communicate, the syntax for the **plot** function varies considerably. When the plot function is invoked, it automatically creates a figure graphic object and an axes graphic object. Some syntax options are:

```
plot(y)
```

The argument y can be an $M \times N$ matrix. Each column of y is plotted in the same figure versus the row index.

```
plot(X1,Y1,...,XN,YN)
```

This version plots the elements of the vector *Yi* versus the elements of the vector *Xi*, $i = 1, \ldots, N$, on the same axes.

```
plot(X1,Y1,LineSpec, ... , XN,YN,LineSpec)
```

With the character string LineSpec you can specify line properties. Table 9.5 gives possible line specifications. For example, the command

```
plot(t,v,'-pg') % LineSpec = '-pg'
```

produces a line plot of the elements of v versus the elements of t, where the line is green and dashed, with a 5-pointed star located at every data pair $(t(i),v(i))$, $i = 1, \ldots, \text{length}(v)$. If you specify a marker (a 5-pointed star in this case), but not a line style, then only the markers are plotted. Many of the plotting functions given in Table 9.3 also accept linespec specifications, as given in Table 9.5.

Table 9.5   Line specifications

Specifier	Line style	Specifier	Marker type	Specifier	Color
-	solid (default)	+	Plus sign	r	Red
--	dashed	o	Circle	g	Green
:	dotted	*	Asterisk	b	Blue
-.	dash-dot	.	Point	c	Cyan
		x	Cross	m	Magenta
		s	Square	y	Yellow
		d	Diamond	k	Black
		^	Upward triangle	w	White
		v	Downward triangle		
		>	Rightward triangle		
		<	Leftward triangle		
		p	5-Pointed star		
		h	6-Pointed star		

```
plot(X1,Y1,LineSpec,'Property_Name',property_value)
```

With this syntax you can also specify line properties, such as LineWidth. The width of lines is specified in **points**, where 1 point = 1/72 inch.

```
h = plot(X1,Y1,LineSpec,'PropertyName',PropertyValue)
```

This syntax assigns the plot to a handle.

```
plot(axes_handle,X1,Y1,LineSpec,'PropertyName',PropertyValue)
```

This syntax allows you to place a plot graphic object on an axes other than the current axes, which enables you to place plots on the axes of any previously created figures or another axes in the current figure.

Example 9.2 ────────────────────────────────

This example demonstrates some of the functions concerned with plotting, while we also look at the operation of a **digital-to-analog converter** (DAC).

As has been mentioned before, a music CD stores samples (at the rate $f_s = 44,100$ samples/sec) of music as a sequence of 16-bit binary numbers. When a CD is played, each channel produces a 16-bit binary number every $T = 1/f_s$ seconds $= 22.7$ μsecs. To hear the music, each number sequence must be used to produce a continuous time signal.

A DAC receives a 16-bit number input, and it produces a constant output voltage that is proportional to the 16-bit input until T seconds later, when it receives the next 16-bit number input. This is depicted in Fig. 9.1. The input is $x(nT)$, a discrete time signal, where $n = 0, 1, 2, \ldots$, which is called the **discrete time index**. At the start, when $n = 0$, the first 16-bit number $x(0T)$ is received, and then T seconds later, the second 16-bit number $x(1T)$ is received, and so on. The output $x(t)$ is a voltage in the range $V_{Ref}^- \leq x(t) \leq V_{Ref}^+$, and it is a piecewise constant signal.

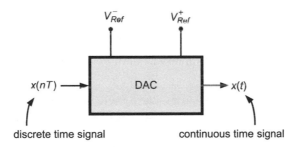

Figure 9.1   Block diagram of a digital to analog converter.

A music signal is a multi-frequency signal, typically consisting of a combination of many sinusoidal signals, where the amplitude and frequency f (20 Hz $< f <$ 20 KHz) of each sinusoid changes with time. Prog. 9.1 retrieves a stereo music signal and plots a few milliseconds of both channels of the music, as shown in Fig. 9.2.

```
% Program to plot a segment of music data
clc; clear all;
[song fs] = wavread('Thriller.wav'); T = 1/fs; % get music WAV file
[npts nch] = size(song); % get number of samples of each channel
while 1 % using a while loop to look at different music segments
    nstart = input('enter start index: '); % get index of start plot point
```

```
if nstart <= 0, break; end  % a way to exit the while loop
nend = input('enter end index: ');  % get index of end plot point
if nend <= nstart+1, break; end
left_channel_seg = song(nstart:nend,1);  % get left channel segment
right_channel_seg = song(nstart:nend,2);  % get right channel segment
n = 0:nend-nstart;  % plot index range
t = 1e3*n*T;  % plot time in milli-seconds
% use increased line width and red color and give plot a handle
left_ch = plot(t,left_channel_seg,'linewidth',1.5,'color','r')
hold on  % place additional graphic objects on the current axes
% use blue color and give plot a handle
right_ch = plot(t,right_channel_seg,'linewidth',1.5,'color','b')
% add a legend box
legend('left channel','right channel','location','southeast')
set(gcf,'color','w')  % set figure background to white
% set axis to increased line width and grey background
set(gca,'linewidth',2.0,'color',[.8 .8 .8])
xlabel('time - milliseconds')
ylabel('voltage')
title('segments of left and right channels of music')
grid on
end
```

Program 9.1   Program to plot segments of both channels of music.

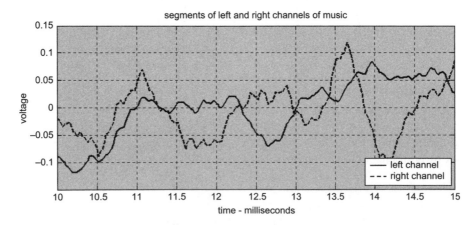

Figure 9.2   Plot of segments of the left and right channels of music.

Notice that the plot of each channel is assigned to a handle (to be used later), and that some line properties are set in the plot function argument. To place both channel plots on the same axes, while also specifying line width and color, the **hold** function is used. The syntax options of the hold function are given by

hold on

This retains the current plot and certain axes properties so that subsequent graphing commands add to the existing axes. If no current axes object exists before you call hold on, MATLAB creates a new axes object and retains the default properties. However, some axes properties change to accommodate additional graphic objects. For example, the limits of the axes increase when required by the data. Hold on sets the NextPlot property of the current figure and axes to add.

hold off

This resets axes properties to their default values before drawing new plots. Hold off is the default, and it sets the NextPlot property of the current axes to replace.

hold all

This holds the plot and the current line color and line style so that subsequent plotting commands do not reset these properties.

hold(axes_handle, ...)

This applies the hold to the axes identified by the handle axes_handle.

A **legend** also has several properties (Use **doc legend** to find more details about these properties.). In Prog. 9.1, the default legend location is changed to "southeast". The function **gcf** returns the current figure handle, and it is used to set the figure background color to white. The function **gca** returns the current axes handle, and it is used to set the line width of the axes border to 2.0 points and the background color to gray.

In addition to characters and numbers (with the int2str or num2str functions), you can also include in the **xlabel**, **ylabel**, and **title** character strings other characters such as Greek letters and relational operators by using **escape characters**. For example, if the x-axis in Fig. 9.2 is instead in microseconds, the xlabel statement could be: xlabel('time - \mu seconds'), where no space is necessary after the escape characters **\mu** to delimit them from the following characters, and the label will appear as: time − μseconds.

If you want to zoom in on a segment of the music signal, for example, from 10 to 15 milliseconds over the entered sample index range, then place the following statement after the left channel plot statement.

```
axis([10 15 -0.15 0.15]) % xmin = 10, xmax = 15, ymin = -0.15 and ymax = 0.15
```

To see all of the properties of the left channel plot graphic object, use

```
>> get(left_ch) % get all properties of the graphic object with the handle left_ch
```

These properties are shown in Table 9.6. The left and right channel line plots are children of the axes graphic object, which has the handle 173.0112. Some of these properties are informational, while others can be changed with the **set** function. For example, you can add data-point markers to the plot with the set function or by including the property name and value in the argument list of the plot function. However, in Fig. 9.2 there would be too many markers to see them as isolated objects.

Table 9.6   Properties of the left channel plot graphic object

DisplayName: 'left channel'	DeleteFcn: []
Annotation: [1x1 hg.Annotation]	BusyAction: 'queue'
Color: [1 0 0]	HandleVisibility: 'on'
LineStyle: '-'	HitTest: 'on'
LineWidth: 1.5000	Interruptible: 'on'
Marker: 'none'	Selected: 'off'
MarkerSize: 6	SelectionHighlight: 'on'
MarkerEdgeColor: 'auto'	Tag: ''
MarkerFaceColor: 'none'	Type: 'line'
XData: [1×1301 double]	UIContextMenu: []
YData: [1×1301 double]	UserData: []
ZData: [1×0 double]	Visible: 'on'
BeingDeleted: 'off'	Parent: 173.0112
ButtonDownFcn: []	XDataMode: 'manual'
Children: [0×1 double]	XDataSource: ''
Clipping: 'on'	YDataSource: ''
CreateFcn: []	ZDataSource: ''

The axes graphic object also has many properties, which can be obtained with

```
>> get(gca)  % get the properties of the current axes
```

Some of the properties are given in Table 9.7. To access any particular property, use, for example

```
>> get(gca,'FontSize')  % use the property name
ans =    10
```

You can assign all of the axes properties to a scalar structure, where each axes property name becomes a field name of the structure, with, for example,

```
>> axes_prop = get(gca);
```

To access any property, use, for example,

```
>> axes_prop.Box
ans =    on
```

Table 9.7 Some properties of the axes graphic object

AmbientLightColor = [1 1 1]	12
Box = on	12.5
Color = [0.8 0.8 0.8]	13
DrawMode = normal	13.5
FontName = Helvetica	14
FontSize = [10]	14.5
GridLineStyle = :	15
LineWidth = [2]	YLabel = [177.003]
NextPlot = add	YAxisLocation = left
TickLength = [0.01 0.025]	YLim = [−0.15 0.15]
TickDir = in	YScale = linear
Title = [179.011]	YTick = [(1 by 7) double array]
View = [0 90]	YTickLabel =
XLabel = [176.003]	−0.15
XAxisLocation = bottom	−0.1
XLim = [10 15]	−0.05
XScale = linear	0
XTick = [(1 by 11) double array]	0.05
XTickLabel =	0.1
10	0.15
10.5	Parent = [1]
11	Type = axes
11.5	Visible = on

With the set function you can change any axes property. For example, to change the *x*-axis tick locations, use the statement: set(gca,'XTick',[0 10 20 30]), or, for example, set(gca,'XTick',linspace(0,30,11)).

To illustrate the activity of a DAC, the **stairs** function is useful. Prog. 9.1 was modified to plot only the left channel, and the left channel plot statement was replaced by the following statement.

```
stairs(t,left_channel_seg,'LineWidth',1.5,'color','k')
```

The **stairs** function produces a piecewise constant continuous time plot, like the voltage output of a DAC. The toolbar of the *Figure Window* includes a zoom feature, and it was used to zoom in on a 2 millisecond time segment of the plot, as shown in Fig. 9.3.

In the *Command Window*, the following statement was executed.

```
>> gtext('piece-wise constant continuous time signal')
```

The **gtext** function, which you can also include in a script, causes MATLAB to place cross-hairs on the *Figure Window* that can be moved with the mouse to locate where the character string argument should be placed. Then, left-click the mouse, and the character string appears at the current cross-hair location, while the cross-hairs disappear.

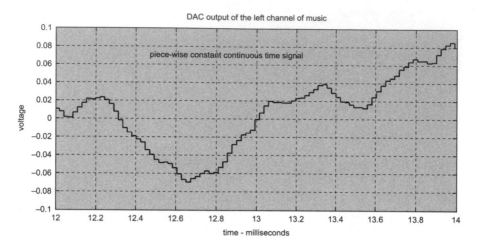

Figure 9.3   Illustration of the operation of a digital-to-analog converter.

Notice that the signal jumps from one constant value to another as when the next 16-bit binary number is received by the DAC. Discontinuities in the DAC output signal cause the signal to contain very high frequency sinusoidal components. These components are removed (almost entirely) with a low-pass filter, such as, for example, the circuit in Fig. 6.13, to produce an analog signal that is perceived to be substantially identical to the analog music signal that was sampled to make the CD.

As you may have noticed, each graphic object has a large number of properties, each with options that are too numerous to describe them all here. At first, this can be overwhelming. However, you should now be able to use **help** and **doc** to better understand the possibilities for customizing a graphic. Some trial and error may occur. Once you become accustomed to the intent of some of the graphic functions and their syntax options, the effort required to utilize the options becomes easier and worthwhile. Also, you should use **Product Help** to look up **object properties** and **functions** by category and alphabetical listing. For example, with

Product Help → Object Properties → Axes → Core Objects → Line

you can learn more about the properties of line graphic objects.

If you want a hardcopy of a graphic, then use **print** in command or function mode. The syntax options are given by

```
print  % send current figure to your current printer, same as print(gcf)
print('-fh') % send figure with handle h to your current printer
print -device -options % print device and options to control some characteristics
      function mode: print('-device','-options')
```

```
   print -device -options filename % specify a filename
            function mode: print('-device','-options','FileName')
```

There are many device options and options to control the output. For example,

```
print(h,'-dpsc','figure_2.ps') % save figure in color as a post script file
% save figure as a 24-bit bit map using 300 dpi resolution
print(h,'-dbmp','-r300','figure_1.bmp')
print(gcf,'-dmeta')   % send figure to clipboard in metafile format
% send figure to clipboard in bitmap format using screen resolution
print(gcf,'-dbitmap','-r0')
```

Use **doc print** to see the innumerable possibilities for printing and exporting graphic objects.

9.2.2 Multiple 2-D Plots

You can place many axes on a figure or place an axes on another axes to insert a plot within another plot.

Example 9.3 _____

Prog. 9.2 uses the built-in function **axes** to place four axes objects on a figure, where each axes object is assigned a handle for later reference. Fig. 9.4 is the program output.

```
clear all; clc; clf
fig_1 = figure(1); % create a figure graphic object and assign a handle
% figure(1) is now the current figure
ax_1 = axes('Position',[0.55,0.55,0.4,0.4]); % place axes in first quadrant
% position vector format is [left  bottom  width  height]
% ax_1 is now the current axes
angle = linspace(-pi,pi,61); % Xdata used for example plotting
y = sin(angle); % example Ydata
plot_1 = plot(angle,y) % assign this plot to a handle
set(ax_1,'XTick',-pi:pi/2:pi) % locate tick marks
set(ax_1,'XTickLabel',{'-pi','-pi/2','0','pi/2','pi'}) % annotate tick marks
% notice that a cell array is used here
grid on
xlabel('-\pi \leq \theta \leq \pi') % using stream modifiers
% for example, \theta appears as θ and \leq appears as ≤
ylabel('sin(\theta)')
title('Plot of sin(\theta)')
% locate text that points to the data point (-pi/4, sin(-pi/4))
text(-7*pi/32,sin(-pi/4),'\leftarrow sin(-\pi/4)',...
     'HorizontalAlignment','left')
```

```
% Change the line color to red and set the line width to 2 points
set(plot_1,'Color','red','LineWidth',2)
gtext('first axes') % place text in the current figure
% with the mouse you can specify the location of the text
ax_2 = axes('Position',[0.05,0.55,0.4,0.4]); % second quadrant
% ax_2 is now the current axes
grid on
gtext('second axes')
set(ax_1,'LineWidth',2) % demonstrates referral to a previously created axes
ax_3 = axes('Position',[0.1,0.1,0.3,0.3],'LineWidth',2); % third quadrant
plot(angle,y) % example plot in current axes
grid on
gtext('third axes')
ax_4 = axes('Position',[0.6,0.1,0.3,0.3]); % fourth quadrant
gtext('fourth axes')
```

Program 9.2   Program to demonstrate placing multiple axes on a figure.

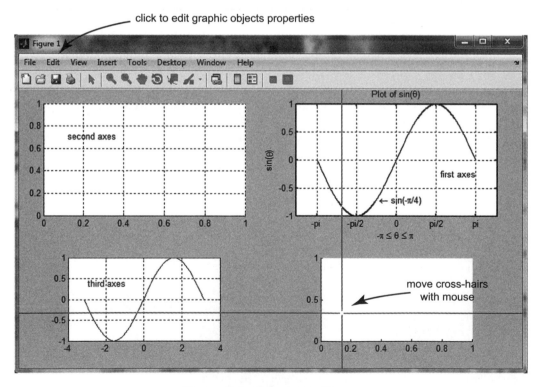

Figure 9.4   A four axes figure.

Notice that **stream modifiers** (also called escape characters) are used extensively in the character strings of text objects on the first axes to include Greek letters and mathematical symbols. To customize a plot, you can use stream modifiers in the character sequence of a string property in a text graphic object and in the string arguments of functions such as title, xlabel, and ylabel. To see an extensive table of stream modifiers use **doc text**, and then click, "Text Properties". Also, see the, "Annotating Graphics", subsection of the, "Graphics" section in the MATLAB Help Product menu.

The **subplot** function provides another way to place multiple axes on a figure. Its syntax is given by

```
subplot(m,n,p)
```

which creates m rows by n columns of axes in the current figure, where $p = 1, \ldots, mn$ counts the axes starting on the left of and along the top row, then the next row and so on. If the pth axes already exists, then subplot(m,n,p) makes the pth axes the current axes. Use help subplot for more syntax options.

Example 9.4 _____

For convenience, the transfer functions from the input $v_s(t)$ to the output $v(t)$ of the circuits in Fig. 6.11 (a band-pass filter) and Fig. 6.13 (a low-pass filter) are repeated here.

$$H_{BP}(j\omega) = \frac{j\omega RC}{1 - LC\omega^2 + j\omega RC} \tag{9.1}$$

$$H_{LP}(j\omega) = \frac{R}{R - \omega^2 L_1 RC + j\omega(L_1 + L_2) - j\omega^3 L_1 L_2 C} \tag{9.2}$$

Let us study the frequency response (magnitude and phase angle versus frequency) of these filters from different points of view. Prog. 9.3 uses the function **plotyy** to plot on the same axes the magnitude and phase angle, which have different scales, of a transfer function. The left y-axis is used for the magnitude, and the right y-axis is used for the phase angle. The **subplot** function is used to place the frequency response plots of the band-pass filter and the low-pass filter on the same figure. The program output is shown in Fig. 9.5. You can use color from black to other colors to distinguish the left and right axes and one plot from another. You can insert legends by clicking the legend button in the toolbar of the *Figure Window*. You can drag the legend box to a preferred location with the mouse.

```
% Program to demonstrate using the subplot function
clear all; clc; close all
f = linspace(0,2e4,501);w = 2*pi*f;f_KHz = f/1000; % frequency range
R = 330;L = 3*11e-3;C = 0.03e-6  % circuit component values from Example 6.9
HBP_num  =  j*w*R*C;
HBP_den  =  1-L*C*w.*w+j*w*R*C;
HBP  =  HBP_num./HBP_den; % band-pass filter frequency response
HBP1_magnitude  =  abs(HBP); % transfer function magnitude
HBP1_angle  =  pi/2 - unwrap(angle(HBP_den)); % transfer function angle
% the unwrap function is used to undo phase jumps by pi/2 and pi
figure(1) % create a figure window
%
subplot(2,1,1) % a two row by one column array of subplots, first subplot
% plot magnitude and phase angle vs frequency
% assign handles to the left and right axes and to the two plots
% ax_BP, a 1x2 matrix, gets the left and right axes handles
[ax_BP h_left h_right] = plotyy(f_KHz,HBP1_magnitude,f_KHz,HBP1_angle);
set(h_left,'LineWidth',1.5,'Color','k') % set plot line width and color
set(h_right,'LineWidth',1.5,'Color','k')
% get handle of left axis label, Ylabel, and set character string
set(get(ax_BP(1),'Ylabel'),'String','magnitude','Color','k');
set(ax_BP(1),'YColor','k') % set left axis color
% get handle of right axis label, Ylabel, and set character string
set(get(ax_BP(2),'Ylabel'),'String','phase angle - radians','Color','k')
set(ax_BP(2),'YColor','k') % set right axis color
set(ax_BP(2),'YTick',-pi:pi/2:pi) % locate right axis tick marks
set(ax_BP(2),'YTickLabel',{'-pi';'-pi/2';'0';'pi/2';'pi'}) % annotate
title('BP Filter Magnitude and Phase Angle')
xlabel('frequency - K Hz')
gtext(strcat('C  =  ',num2str(0.03),' \mu F')) % use mouse to place text
gtext('Phase')
grid on; hold all
C = 0.005e-6; % plot magnitude for a different capacitor value
HBP_num  =  j*w*R*C;
HBP_den  =  1-L*C*w.*w+j*w*R*C;
HBP  =  HBP_num./HBP_den; % band-pass filter frequency response
HBP2_magnitude  =  abs(HBP); % transfer function magnitude
plot(f_KHz,HBP2_magnitude,'k','LineWidth',1.5)
gtext(strcat('C  =  ',num2str(0.005),' \mu F')) % use mouse to place text
%
subplot(2,1,2) % second subplot
R = 1;L1 = 3/2;L2 = 1/2;C = 4/3; % circuit component values from Example 6.10
fc = 5e3;wc = 2*pi*fc;
```

```
% change component values to increase bandwidth to wc
% this requires a knowledge of analog filter design
L1 = L1/wc;L2 = L2/wc;C = C/wc;
HLP_den = (R-w.*w*L1*R*C+j*w*(L1+L2)-j*w.^3*L1*L2*C);
HLP = R./HLP_den; % low-pass filter frequency response
HLP_magnitude = abs(HLP); HLP_angle = -unwrap(angle(HLP_den));
[ax_LP h_left h_right] = plotyy(f_KHz,HLP_magnitude,f_KHz,HLP_angle);
set(h_left,'LineWidth',1.5,'Color','k')
set(get(ax_LP(1),'Ylabel'),'String','magnitude','Color','k');
set(ax_LP(1),'YColor','k')
set(h_right,'LineWidth',1.5,'Color','k')
set(get(ax_LP(2),'Ylabel'),'string','phase angle - radians','Color','k')
set(ax_LP(2),'YColor','k')
set(ax_LP(2),'YTick',-3*pi/2:pi/2:0) % locate tick marks
set(ax_LP(2),'YTickLabel',{'-3pi/2';'-pi';'-pi/2';'0'}) % annotate
title('LP Filter Magnitude and Angle')
xlabel('frequency - K Hz')
grid on
gtext(strcat('Band Width = ',int2str(5),' KHz')) % use mouse to place test
gtext('Phase')
```

Program 9.3 **Program to plot the frequency response of a band-pass and a low-pass filter.**

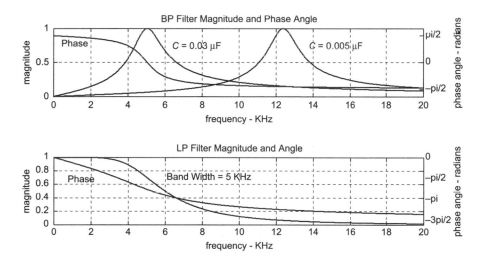

Figure 9.5 Frequency response of a band-pass and a low-pass filter.

Note that by changing C in the circuit of Fig. 6.11, we can select the frequency at which the BP filter is most responsive. The **pass-band** of a filter is defined to be the frequency range over which the magnitude frequency response is greater than $\|H(j\omega)\|_{max}/\sqrt{2}$. The pass-band of $H_{BP}(j\omega)$ ($C = 0.03\,\mu F$) extends from about 4.3K Hz to 6K Hz, and the pass-band of $H_{LP}(j\omega)$ extends from 0K Hz to 5K Hz. The frequency range of the pass-band is called the **bandwidth** (BW). The bandwidth of the band-pass filter is BW $= (6 - 4.3) = 2.7$K Hz, and the bandwidth of the low-pass filter is BW $= 5$K Hz.

Ideally, within the pass-band of a filter the magnitude frequency response should be near 1.0 (0 dB), and then input sinusoids having frequencies within the pass-band will not have their amplitudes diminished in the output. Also, outside of the pass-band of a filter, the magnitude frequency response should be near 0.0, and then input sinusoids having frequencies outside of the pass-band will not (almost not) contribute to the output. There are many methods to design such filters and a variety of electronic components with which to build them. The circuits we have analyzed are called **analog filters**, and they are only simple examples of the many possibilities.

Another perspective about the behavior of a filter can be obtained with a polar plot of its transfer function. Prog. 9.4, which was appended to Prog. 9.3, uses the function **polar** to obtain the polar plots shown in Fig. 9.6.

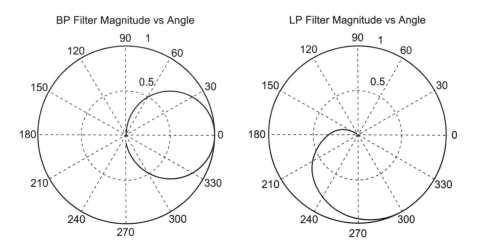

Figure 9.6   Polar plots of the frequency response of a band-pass and a low-pass filter.

```
% Program to demonstrate using a polar plotting function
figure(2) % start a new figure
subplot(1,2,1) % first subplot
polar(HBP1_angle,HBP1_magnitude,'k') % plot magnitude vs angle
title('BP Filter Magnitude vs Angle')
subplot(1,2,2) % second subplot
```

```
polar(HLP_angle,HLP_magnitude,'k')
title('LP Filter Magnitude vs Angle')
```

Program 9.4   Program to obtain polar plots.

The polar plot of $H_{BP}(j\omega)$ starts at the origin, where $\omega = 0$, $\|H_{BP}(j\omega)\| = 0$, and $\angle H_{BP}(j\omega) = \pi/2$. Then, as ω increases, $\|H_{BP}(j\omega)\|$ increases until $\|H_{BP}(j\omega)\| = 1$, where $\angle H_{BP}(j\omega) = 0$. As ω continues to increase, $\|H_{BP}(j\omega)\|$ decreases and $\angle H_{BP}(j\omega)$ continues to decrease until $\angle H_{BP}(j\omega) \to -\pi/2$. The polar plot of $H_{LP}(j\omega)$ starts at $\|H_{LP}(j\omega)\| = 1$ and $\angle H_{LP}(j\omega) = 0$, where $\omega = 0$. Then, as ω increases, $\|H_{LP}(j\omega)\|$ decreases until $\|H_{LP}(j\omega)\| \to 0$, where $\angle H_{LP}(j\omega) = -3\pi/2$.

Example 9.5 demonstrates how to use the function **ginput** to acquire the location within the coordinates of the current axes of cross-hairs that you can position with the mouse. The syntax for the ginput function is given by

   [X Y] = ginput(N)

When MATLAB executes this function, it places cross-hairs on the *Figure Window*, which can be moved around with the mouse. When you left-click the mouse, the coordinates of the cross-hairs are stored in the vectors X and Y until a total of N points have been acquired. X and Y will be scalars, if $N = 1$. If instead, the statement is: $[X \ Y] =$ ginput, then you can acquire points indefinitely until the enter key is depressed.

Given a set of coordinates (X,Y), $X(k)$, $k = 1, \ldots, N$ and $Y(k)$, $k = 1, \ldots, N$, the function **line** can be used to draw a line between the points $(X(1),Y(1))$ and $(X(2),Y(2))$, and then between $(X(2),Y(2))$ and $(X(3),Y(3))$ and so on, connecting all the points in the set of coordinates. The syntax for this function is given by

   line(X,Y,'Property1_Name',property1_value, ...)

With the ginput and line functions you can write a script to create a graphic object with a mouse. If X and Y are N by M matrices, then the function draws M lines using corresponding columns of X and Y for each line.

Example 9.5 ————————————————————————————

This example demonstrates the time-domain behavior of a serial adder. Prog. 9.5 uses the functions **ginput** and **line** to enable the program user to enter two 7-bit numbers, $a = a_6a_5a_4a_3a_2a_1a_0$ and $b = b_6b_5b_4b_3b_2b_1b_0$, with the mouse. Then, seven serially connected full-adders are used to find the 8-bit sum, $s = s_7s_6s_5s_4s_3s_2s_1s_0$, of the numbers. The output of the adder is displayed in time sequence starting with the least significant bit (LSB) s_0 of the sum to demonstrate that an adder designed this way cannot find, for example, the sum bit s_1 until after the carry bit c_1 has been generated by the full-adder that produced s_0.

Therefore, the most significant bit (MSB) s_7 of the sum cannot be known until logic signals have propagated serially through all full-adders. Faster adder logic circuits use carry-look-ahead generators to avoid this long propagation delay.

```
% Use 7 full-adders connected serially to add two 7-bit numbers
% Bits are entered with the ginput function and the mouse
% The carry and sum bits are found serially and plotted in time sequence
clear all; clc, close all
N_bits = 8; % use 8 bits to account for the sum most significant bit (MSB)
X_labels = {'bits','bits','time - \mu seconds','time - \mu seconds'};
Y_labels = {'a','b','carry','sum'}; % using cell arrays for labels
y = zeros(2,N_bits); % preallocate space
figure(1)
for k = 1:4 % set up four axes
    ax_h(k) = subplot(4,1,k); % get a handle for each axes
    axis([0 N_bits 0 1.05]) % axes limits
    set(gca,'XTick',0:1:N_bits) % place x-axis tick marks
    set(gca,'XTickLabel',N_bits:-1:0) % place x-axis tick mark labels
    set(gca,'YTick',[0 1])
    set(gca,'YTickLabel',{'0' '1'}) % label could be [0 1]
    grid on
    xlabel(X_labels(k)); ylabel(Y_labels(k)) % label axes
end
for k = 1:2 % get two 7-bit binary numbers
    axes(ax_h(k))
    y(k,1) = 0; % set MSB to zero
    % draw MSB with the line function
    line([0 1],[0 0],'LineWidth',1.5,'Color','r')
    for n = 2:N_bits % get 7 bits of each number
        [x,y(k,n)] = ginput(1); y(k,n) = round(y(k,n)); % use mouse input
        if n>1 && y(k,n) ~= y(k,n-1) % draw 0 to 1 or 1 to 0 transition
            line([n-1 n-1],[0 1],'LineWidth',1.5,'Color','r')
        end
        % draw line at bit level
        line([n-1 n],[y(k,n) y(k,n)],'LineWidth',1.5,'Color','r')
    end
end
y=fliplr(y); % make y(k,1) the LSB and y(k,N_bits) the MSB
s = zeros(1,N_bits);C = zeros(1,N_bits); % preallocate space
for n = 1:N_bits-1 % implement 7 full adders
    X = xor(y(1,n),y(2,n));
    s(n) = xor(C(n),X); % sum bit
    C(n+1) = (y(1,n) && y(2,n)) || (C(n) && X); % carry to next column
end
```

```
s(N_bits) = C(N_bits); % MSB of sum equals the carry out bit
s = fliplr(s); C = fliplr(C); % reverse again
for n = N_bits:-1:1  % start plot with LSB
    axes(ax_h(3)); % carry plot
    line([n n-1],[C(n) C(n)],'LineWidth',1.5,'Color','b') % draw bit
    pause(1) % pause to simulate serial activity of adder
    if n < N_bits && C(n) ~= C(n+1) % if carry bit changed, draw transition
        line([n n],[0 1],'LineWidth',1.5,'Color','b')
    end
    axes(ax_h(4)); % sum plot
    line([n n-1],[s(n) s(n)],'LineWidth',1.5,'Color','b') % draw bit
    if n < N_bits && s(n) ~= s(n+1) % if sum bit changed, draw transition
        line([n n],[0 1],'LineWidth',1.5,'Color','b')
    end
end
end
```

Program 9.5   Program to simulate the activity of a serial adder.

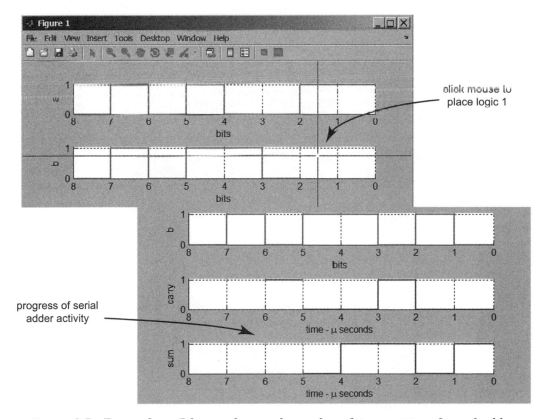

Figure 9.7   Entry of two 7-bit numbers and snapshot of time activity of serial adder.

Fig. 9.7 shows two screen captures. On the left, the bit b_1 is about to be entered. On the right, a screen shot was taken while the carry and sum bits were being produced in time sequence. Such diagrams are called **timing diagrams**.

Within Prog. 9.5 the function **pause** was used to slow down the display of the adder output. Notice that sum bit s_5 cannot become valid until carry bit c_5 becomes valid.

Example 9.6

In this example, the function **compass** will be used to plot the $2n$ roots of

$$x^{2n} = \lambda \tag{9.3}$$

where $\lambda = 1$, $j, -1$, and $-j$.
  Let us start with $\lambda = 1$, and get

$$x^{2n} = 1 = e^{j(k2\pi)}, \quad k = 0, 1, \ldots$$

and therefore, the roots x_k are given by

$$x_k = e^{j(k2\pi)/2n} = e^{j(k\pi)/n}, \quad k = 0, 1, \ldots, 2n - 1 \tag{9.4}$$

Notice that the magnitude of each root is $\|x_k\| = 1$. Also, for $k = 2n$ we get the same root as for $k = 0$, and therefore, there are a total of $2n$ roots given by (9.4).
  For $\lambda = j$, we get

$$x^{2n} = j = e^{j\pi/2} = e^{j(\pi/2 + k2\pi)}, \quad k = 0, 1, \ldots$$
$$x_k = e^{j(\pi/2 + k2\pi)/2n} = e^{j((\pi/4 + k\pi)/n)}, \quad k = 0, 1, \ldots, 2n - 1 \tag{9.5}$$

For $\lambda = -1$, we get

$$x^{2n} = -1 = e^{j\pi} = e^{j(\pi + k2\pi)}, \quad k = 0, 1, \ldots$$
$$x_k = e^{j(\pi + k2\pi)/2n} = e^{j((\pi/2 + k\pi)/n)}, \quad k = 0, 1, \ldots, 2n - 1 \tag{9.6}$$

And, for $\lambda = -j$, we get

$$x^{2n} = -j = e^{j3\pi/2} = e^{j(3\pi/2 + k2\pi)}, \quad k = 0, 1, \ldots$$
$$x_k = e^{j(3\pi/2 + k2\pi)/2n} = e^{j((3\pi/4 + k\pi)/n)}, \quad k = 0, 1, \ldots, 2n - 1 \tag{9.7}$$

Prog. 9.6 uses these formulas for the roots, and plots them with the function **compass**. As a check of (9.3), the function poly is used for the case $\lambda = -j$.

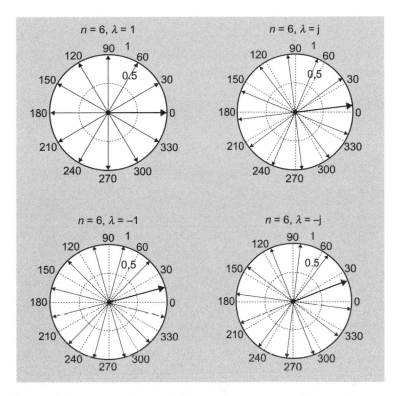

Figure 9.8   Demonstration of using the compass plotting function.

```
% Program to find and plot the roots of x^(2n) = lambda,where
% lambda = 1, j, -1 and -j
clear all; clc; close all
n = 6; % there will be 12 roots
k = 0:2*n-1; % index for roots
offset = [0 pi/4 pi/2 3*pi/4]; % rotation for each lambda
lambda = {' 1 ' ' j ' ' - 1 ' ' - j '}; % cell array of 4 values of lambda
figure(1)
for m = 1:4 % 4 values of lambda
    xk = exp(j*(offset(m)+k*pi)/n); % 2n roots
    subplot(2,2,m);
    compass(xk) % plot 2n roots, same as compass(real(xk),imag(xk))
    Title = strcat('n = ',int2str(n),', \lambda = ',lambda(m));
    title(Title)
end
poly(xk) % check for the case lambda = -j
```

Program 9.6   Program to find and plot the 2*n* roots of $x^{2n} = \lambda$, $\lambda = 1, j, -1$ and $-j$.

The following coefficients are for the case $\lambda = -j$.

```
ans = Columns 1 through 4  1.0000 0.0000 + 0.0000i 0.0000 - 0.0000i 0.0000 + 0.0000i
      Columns 5 through 8  -0.0000 - 0.0000i 0.0000 + 0.0000i -0.0000 + 0.0000i -0.0000 -
         0.0000i
      Columns 9 through 12   0.0000 - 0.0000i  0.0000 + 0.0000i  -0.0000 + 0.0000i
         0.0000 - 0.0000i
      Column 13                    0.0000 + 1.0000i
```

Fig. 9.8 shows the $2n = 12$ roots for each value of λ, where the wider arrow gives the root for $k = 0$.

Notice that when λ is real, the roots are symmetrical about the vertical axis.

9.3   Edit GUI

The functions **set** and **gca** and **handle** are useful to specify figure properties within a program and customize a graphic. A particularly convenient way to enhance and customize a graphic is with the tools that you can access from the *Figure Window*.

To illustrate the possibilities, let us work with an example figure that was produced with Prog. 9.7. The output of this program is shown in Fig. 9.9. There are three plots, which were obtained with the **plot**, **scatter,** and **stairs** plotting functions. This program does not include any property specifications other than default plot properties. Recall that the plot function automatically creates figure and axes graphic objects.

```
% Example figure
clear all; clc; close all
f = 1e3; w = 2*pi*f; T0 = 1/f; % frequency and period
N = 1000; T = T0/N; fs = 1/T; % number of samples, time increment, sampling rate
n = 0:2*N; t = n*T; tmsec = 1e3*t; % time over two cycles, time in milliseconds
x = sin(w*t); % signal sampled at the rate fs
plot(tmsec,x); % plot signal
N1 = 10; T1 = T0/N1; fs1 = 1/T1; % much lower sampling rate
n1 = 0:2*N1; t1 = n1*T1; t1msec = 1e3*t1;
x1 = sin(w*t1); % signal sampled at the rate fs1
hold all
scatter(t1msec,x1) % sampled points
stairs(t1msec,x1) % create a continuous time signal
```

Program 9.7   Program to produce plots of a signal, sampled signal, and reconstructed signal.

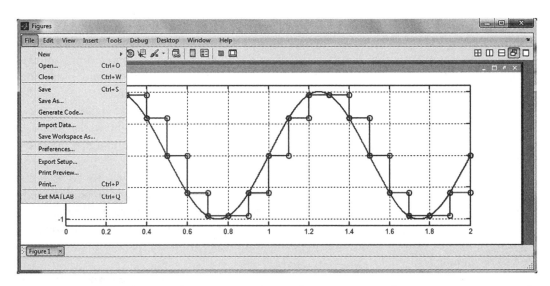

Figure 9.9 Plot of a signal, signal samples, and a reconstructed continuous time signal.

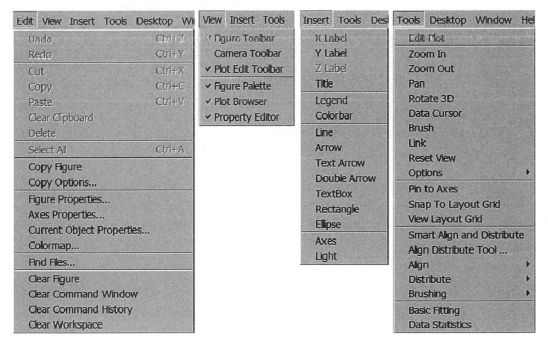

Figure 9.10 Edit, View, Insert, and Tools Menus in the Figure Window.

In Prog. 9.7, the function $x(t) = \sin(\omega t)$, $\omega = 2\pi(1000)$ rad/sec, is sampled at a high sampling rate, $f_s = 10^6$ samples/sec, to obtain a plot that looks like a continuous time signal. Then, $x(t)$ is sampled at the rate $f_s = 10^4$ samples/sec to obtain the sample points plotted with the function **scatter**. The **stairs** function is then used to reconstruct a piecewise constant continuous time signal. The timescale in this figure is in milliseconds.

Also shown in Fig. 9.9 is the File Menu, where you can print the figure or save it to print, copy and paste or edit later. If you want to print it, then preview the figure to see what a printed version will look like, and possibly change the size, margins, location, and more.

Fig. 9.10 shows four menus that you can access from the *Figure Window*. To paste the figure into some document, select the Copy Figure option in the Edit Menu. First, let us edit the figure. Selecting Edit Plot in the Tools Menu opens the window shown in Fig. 9.11, where more properties, shown in the window on the left (called the *Inspector Window*), can be accessed by clicking the More Properties button. Also, from the View Menu other windows are included.

In Fig. 9.11, the axes are selected by clicking on the axes name in the *Browser Window*. Then, a corresponding *Property Window* opens. You can see that the X-axis and Y-axis grid boxes have already been checked, and that an X-axis label has already been entered.

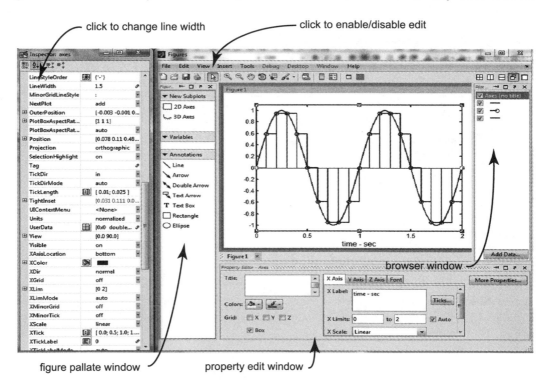

Figure 9.11   Axes properties and figure edit windows.

To enter a *Y*-axis label, click the *Y*-axis tab, and then type, for example, $x(t) = \sin(\backslash\text{omegat})$, which will set the *Y*-axis label to: $x(t) = \sin(\omega t)$. You can: enter a title, change the axes line and background colors, change the tick mark locations and tick labels by clicking the Ticks button, change the linear scale to, for example, a log scale, select a font and font size, enter a figure name, and much more. All axes properties are shown in the window on the left, where, for example, the line width can be changed.

From the *Figure Palette Window* and the tool bars you can insert: text with arrows, text boxes, a legend, and more. Use the zoom and pan features to display a portion of a plot. You can also remove and add back any plot by clicking the check boxes in the *Browser Window*. In Fig. 9.12, we will look at changing some plot properties.

Fig. 9.12 shows the result of the axes edit. Notice that the axes name in the *Browser Window* is the same as the title. To edit a plot, select it by clicking on its symbol or name in the *Browser Window*. You can give a plot a name by typing it into the Display Name box shown in Fig. 9.12. A graphic object can also be selected by clicking on it. Properties of the line (line width and color were changed to 1.5 and black, respectively) and scatter (marker color was changed to red) plots have already been changed. In Fig. 9.12, edit of the stairs plot, which is highlighted in the *Browser Window*, is in progress. The line width will be changed to 1.5.

Fig. 9.13 shows the final result, where a **legend** (click icon on the toolbar) has been inserted, and the **data cursor** (click icon on the tool bar) was used to show the coordinates of a point. Also, the axes background color was changed to off-white for better contrast against white paper.

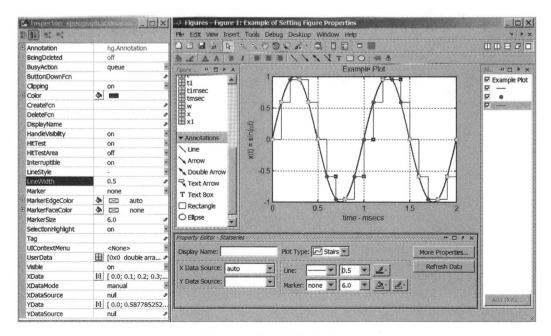

Figure 9.12   Edit of plots.

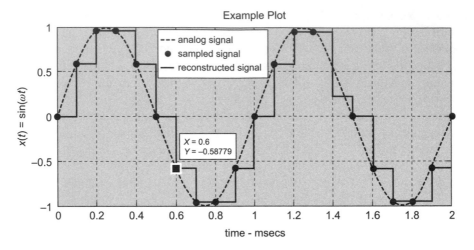

Figure 9.13   Example of customizing a figure.

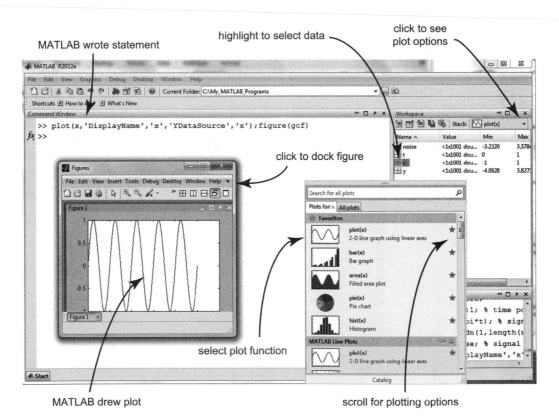

Figure 9.14   Creating a figure from the Workspace Window.

If you are not certain about how you want to display data, then start with the data in the *Workspace Window*. Let us work with data generated by the following script. After this script has executed, the *Command Window* will be empty, and the *Workspace Window* will show the vectors defined by the script.

```
clear all; clc;
t = 0:0.001:1; % time points
x = sin(10*pi*t); % signal
noise = randn(1,length(x)); % additive Gaussian noise
y = x + noise; % signal plus noise
```

In the *Workspace Window*, highlight a data vector, for example, *x*. Then, click the plotting function menu, and scroll to find the kind of plot you want to create. Click the function plot, and MATLAB will draw a plot, and generate a MATLAB plotting statement in the *Command Window*. This activity is shown in Fig. 9.14.

Now, you can edit the plot using all of the tools shown in Figs. 9.10–9.12. Or, you can also edit the plot with commands from the *Command Window*. First, the plot statement produced by MATLAB is reused and edited to plot *x* versus *t*. Some commands are shown in Fig. 9.15. As each statement is entered, the plot changes.

From the *Workspace Window* you can also create a figure with multiple line plots versus a designated variable. For example, let us create a figure that contains both *x* and *y*

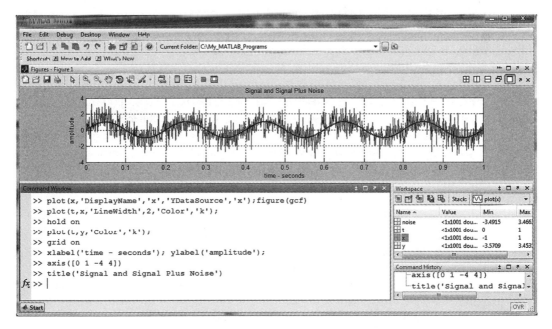

Figure 9.15   MATLAB commands used to edit a figure.

versus t. To do this, first highlight the t vector in the *Workspace Window*. While depressing the **Control Key** of the keyboard, use the mouse to highlight one or more (x and y data vectors in this case) other data vectors. Then, open the plot options menu in the *Workspace Window*, click on the, "all plots", tab (see Fig. 9.14.), scroll down to the, "plot as multiple series versus first input", option and click it. MATLAB will then produce a figure like the one shown in Fig. 9.15, except for the labels, which you can then edit in as before. Notice all of the other possible plot types that can be generated from the *Workspace Window*.

9.4   Color Map

An important property of a graphic object is the color of its parts, not only to make the graphic object more interesting, but also to better communicate the meaning of its parts and make the object easier to understand.

The function **colormap** is concerned with the colors used by filled parts of a figure graphic object. You cannot apply color maps to line plots. The syntax options are given by

```
colormap(map)
colormap('default')
cmap = colormap
colormap(ax,...) % uses the figure corresponding to axes ax
```

For example, if the statement: $C = $ colormap is executed after a figure object has been created, then C is assigned a 64×3 matrix, where each row $C(k,:)$, $k = 1, \ldots, 64$, is a color vector of RGB intensity weights, each ranging from 0 to 1. For the default colormap, the color vectors range from $C(1,:) = [0\ 0\ 0.5625]$ to $C(32,:) = [0.5000\ 1.0000\ 0.5000]$, and to $C(64,:) = [0.5000\ 0\ 0]$. A row index of C corresponds to a particular color.

Generally, a color map C is an $m \times 3$ matrix of color vectors. By defining a color map matrix C, you can design your own color scheme with the statement: colormap(C). Or, you can use the default color scheme with: colormap('default'). MATLAB includes a folder of color maps described by:

- *autumn* varies smoothly from red, through orange, to yellow.
- *bone* is a grayscale color map with a higher value for the blue component.
- *colorcube* contains as many regularly spaced colors in RGB color space as possible, while attempting to provide more steps of gray, pure red, pure green, and pure blue.
- *cool* consists of colors that vary smoothly from shades of cyan and to shades of magenta.
- *copper* varies smoothly from black to bright copper.
- *flag* consists of the colors: red, white, blue, and black.

- *gray* returns a linear grayscale color map.
- *hot* varies smoothly from black through shades of red, orange, and yellow, to white.
- *hsv* varies the hue component of the hue-saturation-value color model.
- *jet* ranges from blue to red, and passes through the colors: cyan, yellow, and orange.
- *lines* produces a color map of colors specified by the axes ColorOrder property.
- *pink* contains pastel shades of pink.
- *prism* repeats the six colors red, orange, yellow, green, blue, and violet.
- *spring* consists of colors that are shades of magenta and yellow.
- *summer* consists of colors that are shades of green and yellow.
- *white* is an all white monochrome color map.
- *winter* consists of colors that are shades of blue and green.

You can apply any of these standard color maps with the statement, for example, colormap (cool(128)), which applies a color matrix of 128 color vectors of the cool type. To see the colors of a color map, use the function **colorbar**. For example, the following statements produce the plots shown in Fig. 9.16.

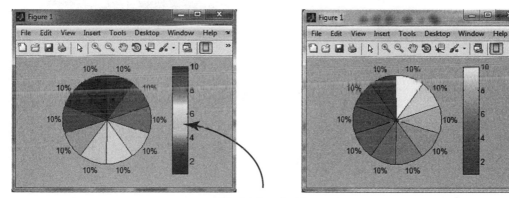

right click to adjust hue

Figure 9.16   Pie chart with default colormap (left) and summer colormap (right).

```
>> x = rand(1,1e6);  % get random numbers
>> n_bins = 10; bin_count = hist(x,n_bins);  % count points in n_bins bins
>> pie(bin_count);  colorbar;  % plot pie chart and display default colorbar
>> colormap('summer');  % change colormap
```

Although you can select the color of a graphic object with the function **colormap**, MATLAB automatically uses the default color map. Another way to select a color map is with

   Edit → Colormap

in the *Figure Window*. This will open the *Colormap Editor Window*, where you can specify one of the many standard color maps and adjust color hue.

Sometimes color can show a data or operation property better than any other way. For example, the following statements produce the plots shown in Fig. 9.17. The left image shows that the elements in the matrix *x* are randomly distributed numbers and neighboring elements seem to be unrelated, and the right image shows that neighboring elements in the matrix inverse of *x* are related.

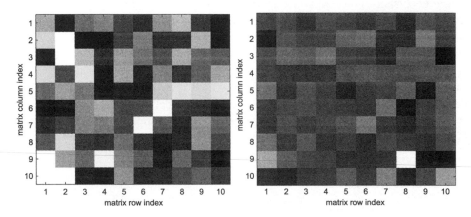

Figure 9.17   Images of a random matrix (left) and its matrix inverse (right).

```
>> clear all; clc; rng('default');  % reset random number generator
>> figure(1)
>> N = 10; x = rand(N); % N by N random matrix
>> imagesc(x); axis square; create image of colored points
>> xlabel('matrix row index'); ylabel('matrix column index');
>> figure(2)
>> y =  inv(x);
>> imagesc(y); axis square;  % image of random matrix inverse
>> xlabel('matrix row index'); ylabel('matrix column index');
```

9.5   3-D Plots

With the 3-D plotting functions provided by MATLAB you can produce line objects, solid objects, contour maps, and more, using a variety of coloration methods. You can control the viewpoint, rotate, zoom out or in on objects and place multiple objects on a figure.

9.5.1   3-D Line Plots

The **plot3** function plots lines in a three-dimensional (3-D) space with coordinates along the *x*, *y* and *z* axes. Its syntax options are given by

```
plot3(X1,Y1,Z1, ..., XN,YN,ZN)
plot3(X1,Y1,Z1,LineSpec, ... , XN,YN,ZN,LineSpec)
plot3(...,'PropertyName',PropertyValue,...)
h = plot3(...)
```

where, Xk, Yk and Zk, $k = 1, \ldots, N$, are vectors or matrices, and all other arguments are used in the same way as with the two-dimensional plot functions.

Example 9.7

Let us plot again the frequency response of the band-pass filter that was studied in Example 9.4, where the transfer function is given in (9.1). The following statement was appended to Prog. 9.3 to save the frequency response calculations.

```
save('BandPass.mat','HBP','w')
```

The frequency response shown in Fig. 9.5 consists of two plots, the magnitude and the phase angle versus frequency. Fig. 9.6 shows the magnitude and phase angle, but not explicitly the frequency. Prog. 9.8 uses the function **plot3** to plot the frequency response, where the real and imaginary parts of $H_{BP}(j\omega)$ are plotted along the x-axis and y-axis, respectively, versus frequency along the z-axis. The functions **xlim, ylim,** and **zlim** are used to get the limits of each axis. The function **view** is used to rotate the figure to see the plot from a preferred viewpoint. You can also rotate the figure by clicking on the rotate icon in the *Figure Window* toolbar. Then, drag the figure with the mouse to a preferred orientation. The result is shown in Fig. 9.18.

```
% plot real and imaginary parts of the frequency response versus frequency
clear all; clc; close all
% get band-pass filter frequency response calculations from Example 9.4
load('BandPass.mat'); % get data: HBP and w
f = w/(2*pi*1000); % convert to KHz
X = real(HBP); % plot real part along x-axis
Y = imag(HBP); % plot imaginary part along y-axis
plot3(X,Y,f,'linewidth',1.5) % plot frequency along z-axis
hold on
origin = zeros(2,1); % used to make axes lines stand out
plot3(xlim',origin,origin,'k','Linewidth',2) % x-axis
plot3(origin,ylim',origin,'k','Linewidth',2) % y-axis
plot3(origin,origin,zlim','k','Linewidth',2) % z-axis
[max_value index_max] = max(abs(HBP)); % locate maximum magnitude point
% plot a line with length equal to maximum magnitude of HBP
plot3([0 X(index_max)]',[0 Y(index_max)]',[f(index_max) f(index_max)]',...
    'LineWidth',2,'Color','r','Marker','o','MarkerEdgeColor','g')
```

```
grid on;
% view at 60 degrees azimuth (counter clockwise rotation about the z-axis
% starting at the x-axis) and 15 degrees elevation (up from the x-y plane)
view(60,15)
xlabel('x-axis, real part')
ylabel('y-axis, imaginary part')
zlabel('z-axis, f - KHz')
title('Frequency Response of Band-Pass Filter')
```

Program 9.8   Program to plot the frequency response of a band-pass filter.

The frequency response $H_{BP}(j\omega)$ starts at the origin, where $\omega = 0$, and moves into the first quadrant of the x–y plane, where the phase angle is positive, until it reaches the maximum value of its magnitude, which is the length of the red line. Then, as $\omega \to \infty$, $x \to 0$ and $y \to 0$.

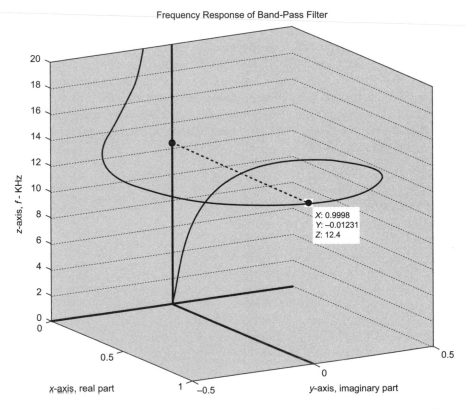

Figure 9.18   Snapshot of tracing the frequency response of a band-pass filter.

Notice that editing plots with built-in functions and tools within the *Figure Window* is done in the same way as editing 2-D plots. In fact, MATLAB actually creates 2-D plots as 3-D plots using **view(0, 90)**.

9.5.2   3-D Surface Plots

A single-valued surface in a 3-D Cartesian space is defined by

$$z = f(x, y) \tag{9.8}$$

where z, the height, depends on x and y in the plane of points (x, y). Several MATLAB functions concerned with surface plotting are given in Table 9.8.

A basic surface plotting function is the function **mesh** with syntax options given by

```
mesh(X,Y,Z)
mesh(Z)
mesh(...,C)
mesh(...,'PropertyName',PropertyValue,...)
mesh(axes_handle,...)
meshc(...)
meshz(...)
h = mesh(...)
```

Table 9.8 Some Built-in functions concerned with plotting

Function	Brief description
hidden on hidden off	Sets hidden line removal on for meshes in the current axes; sets hidden line removal off so you can see through meshes in the current axes
mesh(Z) mesh(X,Y,Z)	Creates a wire mesh from the elements of the N by M matrix Z, where x = 1:N and y = 1:M; Z specifies the surface height and color, which is proportional to the surface height
meshc(...)	Combination of mesh/contour plot
[X Y] = meshgrid(x,y)	Generate X and Y arrays for 3-D plots from the elements of the vectors x and y; the rows of X are copies of x and the columns of Y are copies of y
meshz	3-D mesh with a curtain
pan	Pan view of figure interactively
surf(X,Y,Z,C) surf(X,Y,Z)	Plots a colored parametric surface; the color scaling is determined by C, where the elements of C are used as indices into the current color map; surf(X,Y,Z) uses C = Z, making color proportional to surface height
surface(X,Y,Z,C)	Adds the surface into the current axes
surfc(X,Y,Z,C, ...)	Combination of surf/contour plot
surfl(X,Y,Z,C, ...)	Shaded surface with lighting
surfnorm(X, Y, Z)	Plots a surface and its normals from the surface with components (X,Y,Z)

where X, Y, and Z are each N by M matrices. The elements $Z(n,m)$, $n = 1$, ..., N and $m = 1$, ..., M of Z are each determined in (9.8) evaluated for the corresponding elements $X(n, m)$ of X and $Y(n,m)$ of Y. The function mesh creates a wire mesh connecting the elements of Z.

For example, to plot a wire mesh as x and y in (9.8) vary over the ranges $[-1,2]$ and $[0,3]$, respectively, where x and y are incremented by 0.5 and 1.0, respectively, we must find z for the (x,y) points: $(-1, 0)$, $(-1, 1)$, $(-1, 2)$, $(-1, 3)$, $(-0.5, 0)$, $(-0.5, 1)$, $(-0.5, 2)$, $(-0.5, 3)$, ... , $(2, 0)$, $(2, 1)$, $(2, 2)$, $(2, 3)$, for a total of 28 points. For all of these points, the matrices X and Y are given by

$$X = \begin{bmatrix} -1 & -0.5 & 0 & 0.5 & 1 & 1.5 & 2 \\ -1 & -0.5 & 0 & 0.5 & 1 & 1.5 & 2 \\ -1 & -0.5 & 0 & 0.5 & 1 & 1.5 & 2 \\ -1 & -0.5 & 0 & 0.5 & 1 & 1.5 & 2 \end{bmatrix}, \quad Y = \begin{bmatrix} 0 & 0 & 0 & 0 & 0 & 0 & 0 \\ 1 & 1 & 1 & 1 & 1 & 1 & 1 \\ 2 & 2 & 2 & 2 & 2 & 2 & 2 \\ 3 & 3 & 3 & 3 & 3 & 3 & 3 \end{bmatrix}$$

where the corresponding elements of X and Y are the points. The function **meshgrid** is useful to set up X and Y with the statements

```
>> x = [-1:0.5:2]; y = [0:1:3];
>> [X  Y] = meshgrid(x,y);
```

For information about the other mesh syntax options, use **doc mesh**.

Example 9.8

Let us see how to represent with a mesh plot the behavior of the function $G(s)$, a ratio of two polynomials given by

$$G(s) = \frac{(s - z_1) \ldots (s - z_m)}{(s - p_1) \ldots (s - p_n)} = \frac{\prod_{k=1}^{m}(s - z_k)}{\prod_{k=1}^{n}(s - p_k)} = \frac{P(s)}{Q(s)}, \quad m \le n \tag{9.9}$$

where s is a complex variable, $P(s)$ is an mth order polynomial with real coefficients and $Q(s)$ is an nth order polynomial with real coefficients. Write the complex variable s as $s = \sigma + j\omega$, where σ denotes the real part of s and ω denotes the imaginary part of s. Generally, for a complex value of s, $G(s)$ will be a complex number with magnitude $\|G(s)\|$ and angle $\angle G(s)$.

Since $G(s = z_k) = 0$, the z_k, $k = 1,\ldots$, m , which are the roots of $P(s)$, are called the **zeros** of $G(s)$. Furthermore, since $G(s = p_k) = \infty$, the p_k, $k = 1,\ldots$, n, which are the roots of $Q(s)$, are called the **poles** of $G(s)$.

Let us associate σ with the x-axis, ω with the y-axis, and $\|G(s)\|$ with the z-axis. Let us call the x–y plane, the **s-plane** instead. Here, the points (σ, ω) in the σ plane are the real and imaginary parts of the complex variable s. Therefore, we can write $z = \|G(\sigma, \omega)\|$, with which we will obtain a wire mesh plot using the function mesh.

Let us try the zeros of $G(s)$ given by $z_1 = -0.2$, $z_2 = -0.3$, and $z_3 = -0.4$, and the poles of $G(s)$ given by $p_{1,2} = -0.05 \pm j0.4$, $p_{3,4} = -0.05 \pm j0.5$, and $p_{5,6} = -0.05 \pm j0.6$. With these poles and zeros, $P(s)$ and $Q(s)$ are given by

$$P(s) = (s + 0.2)(s + 0.3)(s + 0.4) = s^3 + 0.9s^2 + 0.26s + 0.024$$
$$Q(s) = (s + 0.05 - j0.4)(s + 0.05 + j0.4)(s + 0.05 - j0.5)(s + 0.05 + j0.5)$$
$$(s + 0.05 - j0.6)(s + 0.05 + j0.6)$$
$$= s^6 + 0.3s^5 + 0.8075s^4 + 0.1565s^3 + 0.1992s^2 + 0.0191s + 0.0149$$

It is informative to see the poles and zeros located in the s-plane, as shown in Fig. 9.19, which is an equivalent description of $G(s)$.

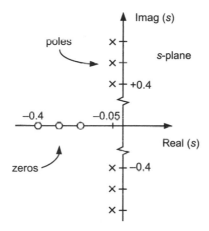

Figure 9.19   Pole-zero plot of G(s).

Prog. 9.9 produces the mesh plot shown in Fig 9.20. Notice that the peaks of $z = \|G(\sigma, \omega)\|$ occur above the poles of $G(s)$, and the troughs of $z = \|G(\sigma, \omega)\|$ occur above the zeros of $G(s)$.

```
% Program to produce a mesh plot of the magnitude of G(s)
clear all; clc; close all
Zeros = [-0.2 -0.3 -0.4]; % zeros of G(s)
P = poly(Zeros); N_P = length(P); % polynomial coefficients
p1 = -0.05+j*0.4; p3 = -0.05+j*0.5; p5 = -0.05+j*0.6
Poles = [p1 conj(p1) p3 conj(p3) p5 conj(p5)]; % poles of G(s)
Q = poly(Poles); N_Q = length(Q); % polynomial coefficients
```

```
sigma  =  [-0.3:0.02:0.2]; N_s = length(sigma); % real part of s
omega  = [-1:0.01:1]; N_o = length(omega); % imaginary part of s
[Sigma Omega]  =  meshgrid(sigma,omega); % s-plane grid
S  =  Sigma + j*Omega; % all values of s
Num_G  =  zeros(N_o,N_s); % initialize G(s) numerator for all s-plane values
Den_G  =  zeros(N_o,N_s); % initialize G(s) denominator for all s-plane values
S_pow  =  ones(N_o,N_s); % s^0 at all s-plane values
for k = 1:N_Q % evaluate P(s) and Q(s) for all s-plane values
    if k <= N_P
        Num_G  =  Num_G + P(N_P-(k-1))*S_pow;
    end
    Den_G  =  Den_G + Q(N_Q-(k-1))*S_pow;
    S_pow  =  S_pow.*S; % get next power of s at all s-plane values
end
Mag_G  =  abs(Num_G./Den_G); % magnitude of G(s) at all s-plane values
max_Mag_G  =  max(max(Mag_G)); % get maximum value
Mag_G  =  Mag_G/max_Mag_G;  % normalize G(s) magnitude
Mag_G_dB  =  20*log(Mag_G); % convert to deci-Bells
% find indices of all elements in Mag_G_dB less than -200 dB
[row_G,col_G]  =  find(Mag_G_dB < -200);
for k = 1:length(row_G) % clamp elements less than -200 dB to -200 dB
    Mag_G_dB(row_G(k),col_G(k))  =  -200;
end
mesh(Sigma,Omega,Mag_G_dB) % plot surface
xlabel('\sigma');ylabel('\omega');zlabel('magnitude of G(s) - dB')
[s_col]  =  find(sigma == 0); % find elements in sigma that are zero
mag_G_dB  =  Mag_G_dB(:,s_col); % get elements of Mag_G_dB where s = jw
z_row  =  zeros(1,N_o); % sigma = 0
hold on
% plot black line in surface where sigma = 0
plot3(z_row,omega,mag_G_dB,'LineWidth',1.5,'Color','k')
```

Program 9.9   Program to produce a mesh plot of $G(s)$.

The black line in Fig. 9.20 follows the mesh plot for $\sigma = 0$, where $G(s)$ becomes $G(j\omega)$ given by

$$G(j\omega) = \frac{(j\omega - z_1)(j\omega - z_2)(j\omega - z_3)}{(j\omega - p_1)\ldots(j\omega - p_6)} = \frac{P(j\omega)}{Q(j\omega)} \tag{9.10}$$

This is a function similar to the transfer functions given in (9.1) and (9.2). In fact, it is possible to design a circuit that has a transfer function given in (9.10), and the black line in Fig. 9.20 is the magnitude frequency response of the circuit. From the shape of the magnitude frequency response for $\omega \geq 0$ we see that $G(j\omega)$ is the transfer function of a band-pass filter, which peaks around $\omega = 0.5$ rad/sec.

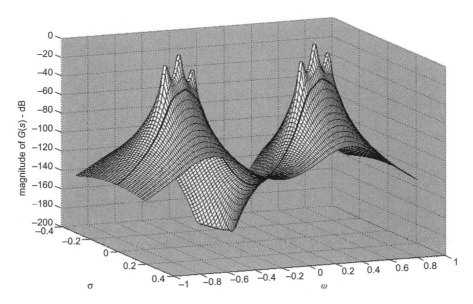

Figure 9.20   Mesh plot of the magnitude of G(s).

Another basic plotting function is the function **surf** with syntax options given by

```
surf(X,Y,Z)
surf(Z)
surf(...,C)
surf(...,'PropertyName',PropertyValue,...)
surf(axes_handle,...)
surfc(...)
h = surf(...)
```

The function surf creates a surface of connected colored patches, and uses inputs in the same way as the function mesh. With the matrix C, having the same dimension as the matrices X, Y, and Z, you can specify the patch colors. With the **colormap** function, you can create your own color scheme, or use one of the standard color schemes provided by MATLAB. Use **doc surf** for details about this.

Associated with the surf function is the function **shading**, which has the arguments: **faceted** (the default), **flat** (no grid lines), and **interp** (smooth color transitions). You can also control the opaqueness of the surface by using the function **set** to assign a value (0 to 1) to the surface graphic object property **FaceAlpha**.

Example 9.9

While the function meshgrid is convenient to set up a region in the x–y plane over which z-axis values are obtained, you can locate points in the x–y plane to form a region that has

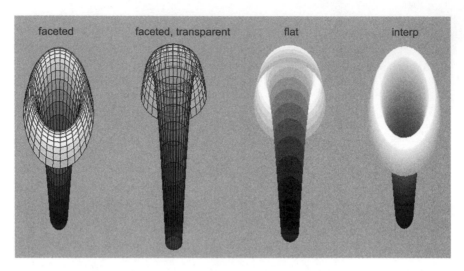

Figure 9.21   Illustration of shading.

any shape. In this example a circular region is formed to illustrate shading a 3-D graphic object. Prog. 9.10 produces the objects shown in Fig. 9.21.

```
% Program to produce a surface plot
clear all; clc; close all
N_r = 21; radius = linspace(0.25,1.0,N_r); % specify radial distance
N_theta = 30; d_theta = 2*pi/(N_theta-1); % specify angular increments
n = 0:N_theta-1; theta = n*d_theta; % angles
X = zeros(N_r,N_theta); Y = X; Z = X; % preallocate space
c_theta = cos(theta); s_theta = sin(theta); % x and y values
z = (exp(-radius)-exp(-4*radius)).*sin((pi/2)*radius); % variation with radius
for m = 1:N_r % set up x-y plane grid
    X(m,:) = radius(m)*c_theta;
    Y(m,:) = radius(m)*s_theta;
end
for n = 1:N_theta % z-axis value at every grid point
    Z(:,n) = z';
end
figure
subplot(1,4,1); % use 4 subplots in a row
surf(X,Y,Z); title('faceted'); % faceted
grid off; axis off
subplot(1,4,2);
h = surf(X,Y,Z); set(h,'FaceAlpha',0.5); title('faceted, transparent')
grid off; axis off
subplot(1,4,3);
```

```
surf(X,Y,Z); shading flat; title('flat'); % flat
grid off; axis off
subplot(1,4,4);
surf(X,Y,Z); shading interp; title('interp'); % interp
grid off; axis off
```

Program 9.10   Illustration of shading options.

The positioning and rotation were done in the *Figure Window*, using the pan and rotate icons.

9.5.3   3-D Rotation

When you use the pan, zoom, and rotate operations in the *Figure Window*, MATLAB does a translation and rotation. Recall the 2-D rotation matrix R that was obtained in Chapter 3, which rotated a vector defined by the coordinates (x, y) in the x–y plane. From a 3-D perspective, a rotation in the x–y plane is a rotation by an angle θ about the z-axis.

Let p denote a point with coordinates (x, y, z) in 3-D space. We can consider p to be a vector in a Cartesian space, where $p = [x \ \ y \ \ z]'$. Let ϕ, ψ, and θ denote rotations about the x, y, and z axes, respectively. Like in 2-D space, in 3-D space the vector p is multiplied by a rotation matrix R, which can be written as: $R = R_x R_y R_z$, where R_x, R_y, and R_z are rotation matrices about the x, y, and z axes, respectively. The rotation matrices are given by

$$R_x(\phi) = \begin{bmatrix} 1 & 0 & 0 \\ 0 & c(\phi) & -s(\phi) \\ 0 & s(\phi) & c(\phi) \end{bmatrix}, \ R_y(\psi) = \begin{bmatrix} c(\psi) & 0 & -s(\psi) \\ 0 & 1 & 0 \\ s(\psi) & 0 & c(\psi) \end{bmatrix},$$

$$R_z(\theta) = \begin{bmatrix} c(\theta) & -s(\theta) & 0 \\ s(\theta) & c(\theta) & 0 \\ 0 & 0 & 1 \end{bmatrix}$$

(9.11)

where c denotes the cosine function and s denotes the sine function. A rotation of the vector p results in the vector P given by

$$P = R p$$

(9.12)

To also translate the vector p, we add to (9.12) a vector $T = [T_x \ T_y \ T_z]'$, and the transformation becomes

$$P = R p + T$$

(9.13)

where T_x, T_y, and T_z are translations along the x, y, and z axes, respectively.

Example 9.10 _____

Let us rotate and translate the object shown in Fig. 9.22, which was produced by Prog. 9.11. The eight vertices of the object are specified by the vectors $p_1, \ldots, p_8$. These vectors are used to specify the end points of each line in the object. For reference after a rotation, the function **fill3** is used to fill one plane with the color red.

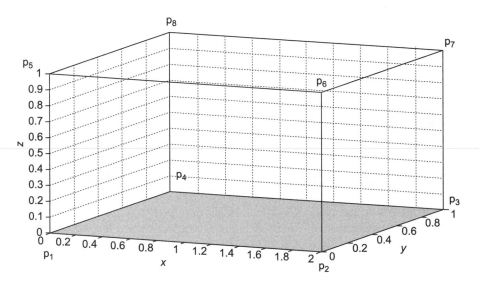

Figure 9.22   An object used to illustrate rotation in 3-D space.

```
% Program to draw and rotate an object in 3-D space
clear all; clc; close all
p = [0 0 0;2 0 0;2 1 0;0 1 0;0 0 1;2 0 1;2 1 1;0 1 1]; % vectors
% specify line end points with vector indices
lines = [1 2;2 3;3 4;4 1;1 5;2 6;3 7;4 8;5 6;6 7;7 8;8 5];
[N_lines, N_vectors] = size(lines); % get number of lines
figure(1); hold on
for k = 1:N_lines
    x = [p(lines(k,1),1) p(lines(k,2),1)]; % get end points along x-axis
    y = [p(lines(k,1),2) p(lines(k,2),2)]; % get end points along y-axis
    z = [p(lines(k,1),3) p(lines(k,2),3)]; % get end points along z-axis
    plot3(x,y,z,'b','LineWidth',2) % plot lines in 3-D space
end
x = p(1:4,1);y = p(1:4,2);z = p(1:4,3); % get vertex points of a plane
% color one plane in object as a reference plane after rotation
```

```
fill3(x,y,z,'r') % fill plane in 3-D space with a color
% saveas(gcf,'block','fig')% saves figure in file block.fig
grid on
xlabel('x'); ylabel('y'); zlabel('z')
figure(2)
% specify four rotations in degrees
x_angle = [30 0 0 45];y_angle = [0 30 0 45];z_angle = [0 0 30 45]; % angles
titles = {'x-axis rotation','y-axis rotation','z-axis rotation',...
    '3-D rotation and translation'}; % cell array of four titles
for K = 1:4 % do four rotations
    R = rotate_3D(x_angle(K),y_angle(K),z_angle(K)); % rotation matrix
    P = (R*p')'; % do rotation of all vectors
    if K == 4, P = P+1; end % include translation in 4th rotation
    subplot(2,2,K); hold on
    for k = 1:N_lines % plot lines defined by rotated vectors
        X = [P(lines(k,1),1) P(lines(k,2),1)];
        Y = [P(lines(k,1),2) P(lines(k,2),2)];
        Z = [P(lines(k,1),3) P(lines(k,2),3)];
        plot3(X,Y,Z,'b','LineWidth',2)
    end
    % control the axes limits
    axis tight % limit axes to data limits
    x_lim = xlim; % get x-axis limits
    x_min = floor(10*x_lim(1))/10; x_max = ceil(10*x_lim(2))/10; % 0.1
    set(gca,'XTick',x_min:0.5:x_max) % set x-axis tick marks
    y_lim = ylim;
    y_min = floor(10*y_lim(1))/10; y_max = ceil(10*y_lim(2))/10;
    set(gca,'YTick',y_min:0.5:y_max)
    z_lim = zlim;
    z_min = floor(10*z_lim(1))/10; z_max = ceil(10*z_lim(2))/10;
    set(gca,'ZTick',0.0:0.5:z_max)
    grid on
    title(titles(K)); xlabel('x');ylabel('y');zlabel('z')
    % color same plane of rotated object
    X = P(1:4,1);Y = P(1:4,2);Z = P(1:4,3);
    fill3(X,Y,Z,'r') % fill plane in 3-D space with a color
end
```

Program 9.11   Program to rotate an object in 3-D space.

Prog. 9.11 uses the function, rotate_3D, given by Prog. 9.12.

```
function R = rotate_3D(x_angle,y_angle,z_angle)
% function to find the rotation matrix
% convert angles in degrees to angles in radians
```

```
xr = 2*pi*x_angle/360; yr = 2*pi*y_angle/360; zr = 2*pi*z_angle/360;
sx = sin(xr); cx = cos(xr);
sy = sin(yr); cy = cos(yr);
sz = sin(zr); cz = cos(zr);
% assign the x, y, and z rotation matrices
Rx = eye(3);Ry = eye(3);Rz = eye(3); % initialize with 3 by 3 identity matrix
Rx(2,2) = cx;Rx(2,3) = -sx;Rx(3,2) = sx;Rx(3,3) = cx; % x-axis rotation
Ry(1,1) = cy;Ry(1,3) = -sy;Ry(3,1) = sy;Ry(3,3) = cy; % y-axis rotation
Rz(1,1) = cz;Rz(1,2) = -sz;Rz(2,1) = sz;Rz(2,2) = cz; % z-axis rotation
R = Rx*Ry*Rz; % rotation in 3-D space
end
```

Program 9.12   Function to compute a 3-D rotation matrix.

Prog. 9.11 produces the rotations shown in Fig. 9.23. The rotate icon in the *Figure Window* was used to give a clear view of each rotation.

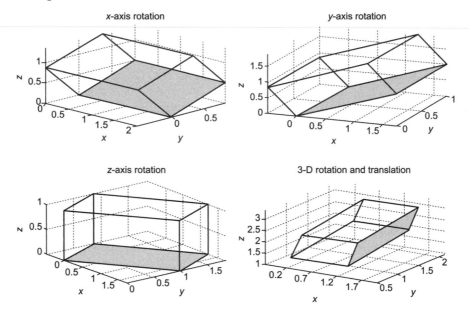

Figure 9.23   Four rotations of the object shown in Fig. 9.22.

9.6   Movies

There are several ways to create an animation. Basically, these methods have in common the following steps: (0) create a graphic, (1) display the graphic, (2) create a modified or new graphic, and (3) go back to step (1). A variation of this method is to first create all of the graphics, which requires much storage space, and then display one graphic after another.

Recall that axes graphic objects are children of figure graphic objects and line graphic objects are children of axes graphic objects. Each graphic object has a set of properties. When the plot function is invoked, it erases axes tick marks, labels, line plots, legends, and more to accommodate the new data, if these graphic objects exist, before it places a new line plot. This will not occur if previously the hold function was invoked, in which case, the new line graphic object is added to previously placed line plots. However, in an animation we want some previously placed graphic objects to be erased, for example, a line graphic object, while keeping other graphic objects in place. A way to do this is to update only those properties that we want to change from one graphic to another. Among the properties of a line graphic object are the **XData** and **YData** properties, which hold the line x–y data, respectively. The function **drawnow** updates the current figure graphic object and all of its children. If a figure graphic object exists and some properties have been changed, then invoking drawnow will update only the graphic objects with changed properties without replacing (erasing and redrawing) the unchanged graphic objects. This way, a graphic can be revised more quickly than invoking, for example, the plot function to revise a line graphic object.

Another property of a line graphic object is the **erasemode** property. It has the argument options: **normal**, **none**, **xor,** and **background**. This property controls the method MATLAB uses to erase and redraw objects and their children. In the normal (default) mode MATLAB erases and redraws objects based on an analysis to ensure that all objects are redrawn exactly. With the none option MATLAB does not erase objects. The xor option causes MATLAB to add the new line graphic object. Then, the new graphic will contain both the new line object and the previous line object, while the previous graphic only contains the previous line object. An exclusive or (xor) of corresponding pixels of the line objects in the new and previous graphic will then clear (erase) the previous line object in the new graphic. With the background option, the part of the previous graphic that is different from the new graphic disappears by redrawing it using the axes background color. For animation, where, depending on computer speed and the complexity of the object to be erased, erase time can be critical, and therefore, the xor option is used, because it is faster than the normal option.

Example 9.11

Let us work with the band-pass filter of Example 9.4, and investigate its time-domain behavior. The frequency response is given in Fig. 9.5, where the peak frequency response occurs for an input sinusoidal signal that has a frequency of about 5K Hz. To see the band-pass characteristic in the time-domain, we will apply an input sinusoidal signal, and sweep its frequency over a frequency range, while keeping a constant input amplitude. As the frequency of the input is changing, we will observe the input and output signals. To compute the amplitude and phase angle of the output, the transfer function given in (9.1) will be used. Prog. 9.13 produces an animation of the input and output as the input frequency is swept over the frequency range [500, 10,000] Hz.

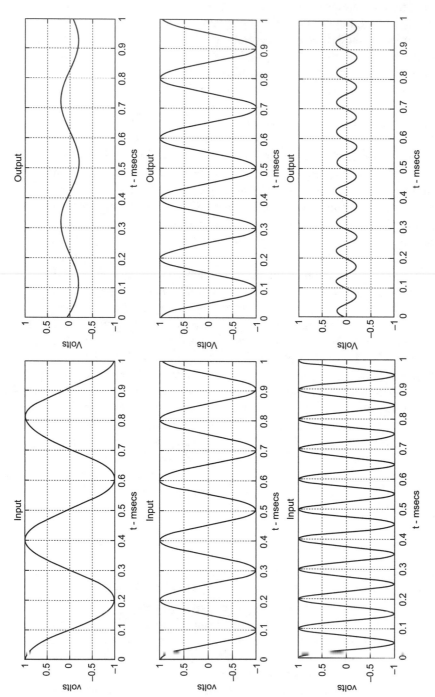

Figure 9.24   Three snapshots when $f \simeq 2{,}500$ *Hz,* $f \simeq 5{,}000$ *Hz, and* $f \simeq 10{,}000$ *Hz.*

```
% Program to animate the response to a swept frequency input
clear all; clc; close all
f_first = 500; f_last = 1.5e4; N_f = 501; % frequency range
f = linspace(f_first,f_last,N_f); w = 2*pi*f; % frequencies
T0 = 1/f_first; % period of sinusoid at frequency f_first
fs = 176400; T = 1/fs; N = ceil((T0/2)/T); % sampling frequency
% using a high sampling frequency to make plots look smooth
n = 0:N-1; t = n*T; t_msecs = 1e3*t; t_total = 1e3*T0/2; % time points
figure(1)  % create a figure graphic object
% create first input and output plots and assign handles
v_in = cos(w(1)*t); % input at frequency f_first, with a 1 volt amplitude
subplot(1,2,1); h_in = plot(t_msecs,v_in); % plot input
axis([0 t_total -1 1]); grid on; % input axes limits
xlabel('t - msecs'); ylabel('volts'); title('Input');
set(h_in,'EraseMode','xor') % method to erase previous line plot
R = 330;L = 3*11e-3;C = 3*0.01e-6; % band-pass filter circuit from Example 6.9
% band-pass filter frequency response
H = (1j*w(1)*R*C)/(1-L*C*w(1)^2+1j*w(1)*R*C);
v_out = abs(H)*cos(w(1)*t+angle(H)); % output at frequency f_first
subplot(1,2,2); h_out = plot(t_msecs,v_out); % plot output
axis([0 t_total -1 1]); grid on; % output axes limits
xlabel('t - msecs'); ylabel('volts'); title('Output');
set(h_out,'EraseMode','xor') % method to erase previous line plot
for k = 2:N_f  % sweep the frequency and plot
        pause(0.02); % slow down sweep
        v_in = cos(w(k)*t);  % input has amplitude equal to 1 volt
        subplot(1,2,1); set(h_in,'XData',t_msecs,'YData',v_in)
        % XData and YData are the data properties of the plot graphic object
        H = (1j*w(k)*R*C)/(1-L*C*w(k)^2+1j*w(k)*R*C);
        v_out = abs(H)*cos(w(k)*t+angle(H)); % output at frequency w(k)
        subplot(1,2,2); set(h_out,'XData',t_msecs,'YData',v_out)
        drawnow % update figure(1)
end
```

Program 9.13   Animation to show a sweep frequency response.

During a sweep of the input frequency, three snapshots of the animation are shown in Fig. 9.24. When the frequency of the input is much less than 5K Hz and much greater than 5K Hz, we see that the output amplitude is much less than the input amplitude. Since $\|H(j\,2\pi(5000))\| \simeq 1$, the output amplitude, when the frequency of the input is 5K Hz, is the same as the input amplitude.

A simple way to show an object in rotational motion is to change the viewing angles.

Example 9.12 ─────────────────────────────

Let us work with the object shown in Fig. 9.22. To retrieve the object, the following statement was inserted just after invoking the **fill3** function in Prog. 9.11.

```
saveas(gcf,'block','fig') % save figure in file block.fig
```

Prog. 9.14 causes the rectangular block to seem to rotate by incrementing the azimuth and elevation angles of the point of view with the function **view**.

```
% Program to rotate an object
clear all; clc; close all
open('block.fig'); % get object
% remove tick labels
set(gca,'XTickLabel',' ','YTickLabel',' ','ZTickLabel',' ')
set(gca,'TickLength',[0 0]) % remove tick marks
for k = 1:120
    view(k,k) % incrementing the azimuth and elevation viewing angles
    pause(0.05) % slow down rotation
end
```

Program 9.14   Example of using the function view to rotate an object.

The **camera** series of functions given in Table 9.9 provide more options to specify the viewing location. The following statements were appended to Prog. 9.14.

Table 9.9   Some built-in functions concerned with camera viewpoints

Function	Brief description
camdolly(dx,dy,dz)	Moves the camera position and camera target of the current axes by the amounts specified in dx, dy, and dz
camlookat	Views the objects that are children of the current axes
camorbit(dh,dv)	Rotates the camera position around the camera target by the amounts specified in dh, horizontal rotation, and dv, vertical rotation (both in degrees)
campan(dh,dv)	Pans (rotates) the camera target of the current axes around the camera position by the amounts specified in dh and dv (both in degrees)
camroll(da)	Rolls the camera of the current axes da degrees clockwise around the line which passes through the camera position and camera target
ova = camva camva(va)	Gets the camera view angle of the current axes; sets the camera view angle
camzoom(zf)	Zooms the camera of the current axes in (zf > 1) or out (0 < zf < 1)

```
camzoom(0.75) % object appears to be smaller
steps = 100; % specify number of translations
dstep = 2/steps; % specify translation increments
for k = 1:steps
    dx = dstep*sin(k*2*pi/steps); % circular translation
    dy = dstep*cos(k*2*pi/steps);
    camdolly(dx,dy,0); hold on
    pause(0.05) % slow down translation
end
camlookat % display the children of the current axes
```

After the object has completed its rotational movement, the camera zooms out and then the object moves in a circular pattern once.

A movie is a sequence of **frames**, where each frame contains a figure that is different from the figure in the previous frame. In MATLAB, each frame is stored as a column vector of a matrix *M* or any valid MATLAB variable name. Some built-in functions for creating and showing a movie are given in Table 9.10.

Table 9.10 Some built-in functions concerned with movies

Function	Brief description
[X,map] = frame2im(F)	Returns the indexed image X and associated ColorMap map from the single movie frame F
M(k) = getframe M(k) = getframe(h)	Returns a movie frame, a snapshot of the current axes; gets a frame from object h, where h is a handle of a figure or axes
F = im2frame(X)	Converts the indexed image X into a movie frame F using the current ColorMap
movie(M) movie(M,N) movie(M,N,fps) movie(h, ...)	Play the movie stored in the matrix M; plays the movie N times, where if N is a vector, the first element is the number of times to play the movie and the remaining elements are a list of the frames to play in the movie; plays the movie at fps frames per second; plays the movie in object h, where h is a handle of a figure or axes
M = moviein(N)	Preallocates space for a movie having N frames
image(C)	Displays matrix C as an image, where each element of C specifies the color of a rectilinear patch in the image; C can be a matrix of dimension $M \times N$ or $M \times N \times 3$, and can contain double, uint8, or uint16 data. When C is a 2-dimensional $M \times N$ matrix, the elements of C are used as indices into the current ColorMap to determine the color. When C is a 3-dimensional $M \times N \times 3$ matrix, the elements in C(:,:,1) are interpreted as red

(Continues)

Table 9.10 (Continued)

Function	Brief description
	intensities, in C(:,:,2) as green intensities, and in C(:,:,3) as blue intensities. Use **doc image** for details.
[X,map] = imread ('FileName')	Reads the indexed image in FileName into X and its associated colormap into map, where ColorMap values in the image file are automatically rescaled into the range [0,1]; file extensions can be, for example, jpg, gif and more. Use **doc imread** for details
imwrite(X,'FileName','fmt')	Writes the image X to the file specified by FileName in the format specified by fmt. Use **doc imwrite** for the many possibilities

Example 9.13

Let us make a movie that shows all of the standard color maps provided by MATLAB. Prog. 9.15 first creates a movie, and then shows it.

```
% Program to illustrate some standard color maps with a movie
clc; clear all; close all
% use a cell array to name the 17 standard color maps
C = {'autumn','bone','colorcube','cool','copper','flag','gray','hot',...
    'hsv','jet','lines','pink','prism','spring','summer','white','winter'};
N_maps = 17; N_titles = 2; N_ends = 1;
N_frames = N_maps + N_titles + N_ends;
M = moviein(N_frames); % preallocate space for N_frames frames
h = figure(1); % assign a handle to figure(1)
text(0.5,0.7,'Color Maps','fontsize',18,...
    'HorizontalAlignment','center',...
    'BackgroundColor',[.7 .9 .7]); % locate and specify a movie title
text(0.5,0.5,'Produced and Directed by','fontsize',12,...
    'HorizontalAlignment','center',...
    'BackgroundColor',[.7 .3 .7]); % credits
text(0.5,0.3,'Roland Priemer','fontsize',14,...
    'HorizontalAlignment','center',...
    'BackgroundColor',[.9 .3 .7]);
grid off; axis off
M(1) = getframe(h); % first movie frame, title frame
M(2) = M(1); % repeat first frame
clf; % clear figure(1)
```

```
for k = 1:N_maps
    x = 0.25+0.2*sin(2*pi*(k-1)/17); % specify axes location
    y = 0.20+0.2*cos(2*pi*(k-1)/17);
    axes('position',[x,y,0.5,0.5]); % position and size axes
    colormap(char(C(k))); % specify a standard color map
    sphere; % using the function sphere to draw a sphere
    axis equal; % correct aspect ratio to make sphere look circular
    title(C(k),'fontsize',14); % make color map name the title
    grid off; axis off
    M(N_titles+k) = getframe(h); % add frames to movie
    clf
end
text(0.5,0.5,'The End','fontsize',24,...
    'HorizontalAlignment','center',...
    'BackgroundColor',[.9 .1 .5]);
grid off; axis off
M(N_frames) = getframe(h); % last movie frame, end frame
clf; % prepare to show movie
dt_frame = 2; fps = 1/dt_frame; % show movie at 1 frame every 2 seconds
movie(h,M,1,fps);
pause(3); close(h); % wait and then close figure(1)
save('Color_Maps','M'); % save movie to a mat file
% for example, to show the movie, use: load Color_Maps, and: movie(M,1,0.5)
```

Program 9.15 Demonstration of creating a movie.

Any of the examples and animations in Figs. 9.6, 7, 8, 18, 20, 21, 22 and 24 can be made into a movie. Furthermore, you can also insert bmp, jpeg, and other image files into frames of a movie. Needless to say, creating a movie can be a lengthy process. However, through graphics and animation much information can be efficiently communicated.

Example 9.14

In Example 9.13, the entire figure content is replaced in the next frame. Usually, we only want to modify some part of a figure to create the next movie frame. Let us again work with the frequency response of Example 9.7, and make a movie of it. Prog. 9.16 retains all of the figure content in each movie frame except the magnitude of the frequency response as it changes with frequency. Fig. 9.25 shows three snapshots of the frequency response movie.

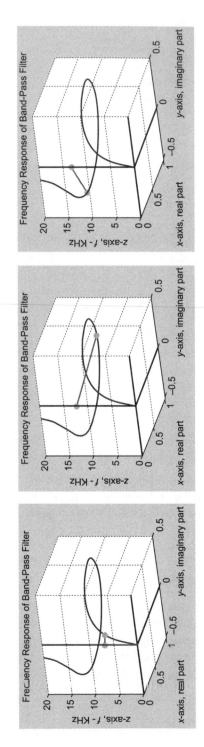

Figure 9.25   Three snapshots of the band-pass filter frequency response movie.

```
% Movie of the frequency response
clear all; clc; close all
% get band-pass filter frequency response calculations from Example 9.4
load('BandPass.mat'); % get data: HBP and w
f = w/(2*pi*1000); % convert to KHz
X = real(HBP); % plot real part along x-axis
Y = imag(HBP); % plot imaginary part along y-axis
N_w = length(w); FRM = moviein(N_w); % preallocate space for movie frames
h = figure(1); % create figure with a handle
plot3(X,Y,f,'linewidth',1.5); % frequency along z-axis
hold on; % add to figure
grid on;
xlabel('x-axis, real part')
ylabel('y-axis, imaginary part')
zlabel('z-axis, f - KHz')
title('Frequency Response of Band-Pass Filter')
origin = zeros(2,1); % make axes lines stand out
plot3(xlim',origin,origin,'k','Linewidth',2);
plot3(origin,ylim',origin,'k','Linewidth',2);
plot3(origin,origin,zlim','k','Linewidth',2);
% view at 60 degrees rotation about z-axis and up by 15 degrees
view(60,15)
% assign handle to magnitude line and set erasemode to xor
h1 = plot3([0 X(1)],[0 Y(1)],[0 0],...
    'lineWidth',2,'Color','r','Marker','o','MarkerEdgeColor','g',...
    'erasemode','xor');
FRM(1) = getframe(h); % get first frame of frequency response movie (FRM)
for k = 2:N_w % get remaining frames, changing the magnitude
    xd = [0 X(k)]; yd = [0 Y(k)]; zd = [f(k) f(k)];
    set(h1,'XData',xd,'YData',yd,'ZData',zd);
    FRM(k) = getframe(h); % update figure and get a movie frame
end
save('freq_response_movie','FRM')
clear all; close all; % clear and close everything
load('freq_response_movie','FRM') % get movie
h = figure(1); % create a figure in which to show movie
movie(h,FRM)
```

Program 9.16   Demonstration of modifying a part of a figure in each movie frame.

9.7 Conclusion

In this chapter we have explored many ways to visualize data with 2-D and 3-D graphics. It seems that there are endless creative possibilities for using MATLAB to generate graphic objects, especially those kinds of objects that help to understand technical problems and their solutions. You should now know how to

- create a figure graphic object and access and specify its properties
- create an axes graphic object and access and specify its properties
- use many of the 2-D plot functions and specify properties
- incorporate a color map
- place multiple 2-D and 3-D plots on a figure
- investigate the frequency response of a circuit
- make an animation
- rotate and translate objects in 3-D space
- create line plots in 2-D and 3-D space
- use various methods (command line, toolbars, program) to edit a graphic
- add a variety of annotations to a graphic
- create interactive programs
- make a MATLAB movie
- customize a graphic to meet your technical needs
- print and export graphics

All of the built-in MATLAB functions introduced in this chapter are briefly described in tables. Use the help, doc, and product help facilities for more details. Also, the demos in product help are very informative.

Problems

Section 9.1

1) Assume that the following MATLAB statements have been executed.

```
>> fig_2 = figure('Name','Input');
>> fig_1 = figure('Color',[0.2  0.5  0.8],'Name','Output');
```

Give MATLAB statements to
(a) assign the color vector of figure 1 to the variable c_1
(b) get the value of fig_2
(c) delete figure 2
(d) make the figure with the handle fig_2 the current figure
(e) create a structure, name it Fig_1, of the properties of figure 1
(f) close all figure windows

2) Assume that the MATLAB statements given in Prob. P9.1 have been executed. Then, the following statements are executed.

```
>> t = 0:0.01:1; plot(t,sin(4*pi*t)); grid on
>> propedit(2)
```

(a) In which figure window is the plot graphic object placed?

(b) Which figure window goes into the figure edit mode?

Give MATLAB statements to

(c) set the background color of the current figure window to red

(d) save the plot graphic object as an enhanced metafile named Signal

(e) place the text, "signal", in the bottom half and middle of the plot.

Section 9.2

3) Explain the outcome of executing the following MATLAB statements.

```
>> t = 0:0.001:2;  x1 = sin(2*pi*t);  x2 = -2*x1;
>> Y = [x1',x2'];
>> area(t,Y,'Facecolor',[0 0 0.9]); grid on;
```

Hint: Look at the outcome of executing bar(t,Y); grid on.

4) Assume that the following MATLAB statement has been executed.

```
>> t = 0:0.001:2;  x = exp(-2*t).*sin(10*pi*t);
```

Explain the difference between the outcomes of executing the following two MATLAB statements.

```
>> comet(t,x,0.01)
>> comet(t,x,0.99)
```

5) A MATLAB script starts as follows.

```
>> clear all; close all; clc;
>> axis([-2 2 -2 2]);   % automatically creates a figure object
```

Using the function fill in a for loop, continue the script to draw a checkerboard pattern of 1 by 1 squares alternating in blue and red colors. Set the figure background color to light green. Print the result.

6) Write a MATLAB script that sets up the same axis as in Prob. P9.5, uses the function ginput to obtain the coordinates of the corners of a five-sided polygon, then fills the polygon with a very light shade of blue, and then enables the user to place the text, "5-sided polygon", somewhere on the polygon. Print the resulting figure.

7) Write a MATLAB script that simulates an experiment that produces one of three possible outcomes, call them apple, orange, and banana that must occur with probabilities of 0.20, 0.35, and 0.45, respectively. To do an experiment, obtain y given by: $y = $ ceil

($a*x$), where x=rand and a=1e5. Then, if $1 \leq y \leq 20{,}000$, the experiment outcome is apple, or if $20{,}001 \leq y \leq 55{,}000$, the experiment outcome is orange or if $55{,}001 \leq y \leq 100{,}000$, the experiment outcome is banana. Do the experiment N times, and count the number of times each possible experiment outcome occurs. Finally, your script must produce a pie chart with the pie slice areas determined by the count of each possible experiment outcome. Use a cell array to label the pie slices apple, orange and banana. Run your program for $N = 1e6$ and $N = 1e7$, and print each pie chart. If you do the experiment three times, what is the probability that the three outcomes will each be orange, orange, and apple?

8) Write a MATLAB function script, named capacitor, that draws the circuit symbol of a capacitor. The function inputs are a two-element vector that gives the capacitor center location within a figure object and a character string that is either "vertical" or "horizontal" for the desired orientation of the capacitor.

9) Fig. P9.9 shows a voltage source $v(t) = 150 \sin(120\pi t + \pi/4)$ connected serially to an ideal diode and a resistor with $R = 3{,}000\ \Omega$.

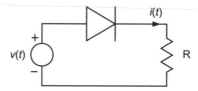

Figure P9.9   Serially connected voltage source, ideal diode, and resistor.

Write a MATLAB script that uses the subplot function to show four cycles of the voltage produced by the voltage source in the top plot and the current through the resistor in the bottom plot. Include grids, titles and x-axis, and y-axis labels, with the current scaled in mA.

10) The transfer function of a circuit is given by: $H(j\omega) = 1/((j\omega)^3 + 2(j\omega)^2 + 2(j\omega) + 1)$. Write a MATLAB script to find the magnitude and angle of $H(j\omega)$ for 101 values of ω over the range: $0 \leq \omega \leq 5$ radians/sec.

(a) Use the function polar to obtain a polar plot. Include in your script a statement that uses the print function to obtain a hardcopy of your polar plot. What kind of a filter has this kind of a frequency response?

(b) Continue your script, and use the function plotyy to plot the magnitude of $H(j\omega)$ with the left vertical axis and the phase angle of $H(j\omega)$ with the right vertical axis.

11) For the transfer function given in Prob. P9.10, write a MATLAB script that uses the function semilogy to obtain a plot of the magnitude of $H(j\omega)$ in dB versus frequency. Include axes labels and a title. Use a line specification to obtain a blue dashed line. Then, save the plot into a jpeg file named MagFreqResp.

12) Write a MATLAB script that places four axes on a figure, each with a handle, say axes1, axes2, axes3, and axes4. Make the dimension of each axes 0.3 by 0.3. Data is obtained with: $t=0.0{:}delta\_t{:}1.0$, $x=\sin(2*pi*t)$ and $y=x+0.1*\text{rand}(1,\text{length}(t))$, where delta_t$=0.05$. The data y is processed to obtain the data u, where $u(1)=y(1)/4$, $u(2)=(y(2)+2*y(1))/4$ and $u(n)=(y(n)+2*y(n\text{-}1)+y(n\text{-}2))/4$, for $n=3, \ldots,$ length(t). Like y, the data u is processed to obtain the data v. Plot the data x, y, u and v on axes1, axes2, axes3, and axes4, respectively, where each plot has a different color, and assign each plot to a handle, say xData, yData, uData, and vData. On each axes locate t tick marks at $t=0.0{:}0.25{:}1.0$, and annotate the tick marks. Set the axes border line widths to 2. Print the figure of four plots.

The data y is a noisy version of the data x. Discuss how the data u is different from the data y, and similarly compare the data v to the data u. Which data, u or v is more like the original data x?

13) Data is given by: $t = 0.0{:}0.05{:}1.0$ and $x = \sin(2*pi*t)+1.5*\sin(4*pi*t+pi/4)$ volts. Write a MATLAB script that produces a stem plot with red circular markers at the data points. Then, use the function stairs to obtain a piecewise constant plot, and finally use the function plot to obtain a line plot. Use the subplot function to obtain these three plots in one column with three rows. Include a title, x-axis label, and a t-axis label.

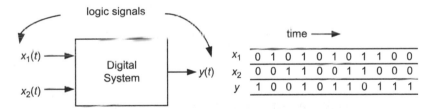

Figure P9.14   Digital system with two inputs and one output.

14) A digital system has two input logic signals and one output logic signal as shown in Fig. P9.14. Also shown in the figure is a table that gives inputs, starting on the left, and the output as they change every μsec. Write a MATLAB script that uses the function line to produce timing diagrams of the two inputs and output as given in the table. Use the subplot function to obtain three plots in one column and three rows.

Section 9.3

15) Fig. 9.9 shows a *Figure Window* that was created with Prog. 9.7. Fig. 9.9 also shows the File Menu. Enter the given MATLAB script into the *Edit Window*, and execute the script.

(a) Explain what can be accomplished with the Generate Code option in the File Menu of the *Figure Window*. Provide a copy of the script obtained from using this option.

(b) Pull down the Edit Menu, and select the appropriate menu option to introduce x and y grid lines.

(c) From the Edit Menu select the appropriate menu option to change the circular markers of the scatter plot to square markers.

(d) Enter x- and y-axis labels and a title. Print the resulting figure.

16) Execute the following statements in the *Command Window*.

```
>> clear all; clc
>> t = 0:0.01:10; x = exp(-t);
```

Highlight x in the Workspace. Pull down the plot options list, and select the semilogy plot. In the *Figure Window*, click the data cursor icon in the tool bar, and place a data marker at the smallest y-axis value of the plot. Give and explain the value given by Y of the data marker. Why is the plot a straight line with a negative slope?

Section 9.4

17) The built-in function **sphere** automatically creates a figure and an axes object. Then, it uses a default colormap to draw a sphere on the axes.

(a) In the *Command Window*, enter the following MATLAB statements.

```
>> sphere
>> colorbar
```

Explain what happened, and from the File Menu in the *Figure Window* print the figure.

(b) Execute the following statements.

```
>> binary_map = [0 0 0 0 1 1 1 1;0 0 1 1 0 0 1 1;0 1 0 1 0 1 0 1]'
>> colormap(binary_map)
```

For each binary number in the color map give the color.

(c) Give MATLAB statements and execute them to create a color map with no green or blue and red that varies from 0.1 to 1.0 in increments of 0.1. Explain what happened. In the *Figure Window*, click the rotate icon, and select a viewpoint from which to print the resulting sphere.

Section 9.5

18) The following MATLAB statements are the beginning of a script.

```
clear all; clc; close all
figure; axis([-5  5  -5  5  -5  5]);
hold on; grid on
origin  =  zeros(1,2);
plot3(xlim,origin,origin,'k','Linewidth',2);
plot3(origin,ylim,origin,'k','Linewidth',2);
plot3(origin,origin,zlim,'k','Linewidth',2);
```

(a) What are the limits of the x-axis?
(b) What does the function zlim return?
(c) What does this script do?
(d) Continue the script to plot a dashed red line from the point $(x, y, z) = (-4, -3, -5)$ to the point $(x, y, z) = (4, 3, 4)$. Include x, y, and z axis labels. Print the figure.

19) Write a MATLAB script that draws a red filled triangle in a 3-D space with corners given by $(-2, -3, 4)$, $(-2, 3, -4)$, and $(2, -3, -4)$. Include x, y, and z axis labels.
   (a) In the *Figure Window*, click the data cursor icon in the tool bar to place a marker at one of the triangle corners. Print the figure.
   (b) In the *Figure Window*, click on the rotate icon in the tool bar to obtain an x–y plane projection of the triangle. Print the figure.

20) Write a MATLAB script that draws in a 3-D space the frequency response of a circuit with transfer function $H(j\omega)$ given in Prob. P9.10, where the x and y axes give the real and imaginary parts, respectively, of $H(j\omega)$, and the z-axis gives the frequency for $-5 \leq \omega \leq 5$ rad/sec. Include x, y, and z axis labels. Plot enough points to see a smooth curve. Does the plot show that $H^*(j\omega) = H(-j\omega)$? Use the data cursor icon in the *Figure Window* to mark the curve where $\omega = 1$ rad/sec, and give the value of $\|H(j1)\|^2$. Provide a program listing and a copy of the figure.

21) A MATLAB script begins with

```
clear all; clc;
x  =  -1:0.01:1; y  =  -1:0.01:1;
[X Y]  =  meshgrid(x,y);
```

(a) Give the dimensions of X and Y.
(b) Use the function imagesc to obtain imagesc(X) and imagesc(Y). Explain the difference between the results.
(c) The variable z is given by $z = f(x,y) = x^2 y^3$. Continue the MATLAB script to evaluate z over the grid obtained with the function meshgrid, and use the function mesh to plot $z = f(x,y)$. Specify a MATLAB colormap other than the default colormap. Include a grid and x, y, and z axis labels. Provide a program listing and a copy of the figure.
(d) Explain the differences between the results from using mesh, meshc, and meshz.

22) A MATLAB script begins in the same way as in Prob. P9.21.
    (a) Repeat part(c) of Prob. P9.21, but use the MATLAB statement

```
h  =  surf(X,Y,Z);
```

to plot $z = f(x,y)$. Specify interp for shading. Print the figure.
    (b) Execute your script again with the MATLAB statements

```
axis([-0.5 0.5 -0.5 0.5 -1 1])
h  =  surf(X,Y,Z);
```

and

```
axis([-0.5 0.5 -0.5 0.5 -1 1])
h  =  surface(X,Y,Z);
```

Explain the difference between the resulting figures. Print both figures.

23) A ratio of two polynomials is given by

$$G(s) = \frac{s^2 + s/2 + 257/16}{(s + 1/2)(s^2 + s/2 + 17/16)} = \frac{(s - z_1)(s - z_2)}{(s - p_1)(s - p_2)(s - p_3)}$$

where s is a complex variable denoted by $s = \sigma + j\omega$.
    (a) What are the poles, p_1, p_2, and p_3 and zeros z_1 and z_2 of $G(s)$?
    (b) Write a MATLAB script that uses the surf function to plot the dB values of $\|G(\sigma, \omega)\|$, for $-1.5 \le \sigma \le 1.0$ and $-10 \le \omega \le 10$. Use enough points to obtain a smooth surface. Use grid points for σ and ω that do not exactly occur at any pole of $G(s)$. Also, normalize $\|G(\sigma, \omega)\|$ by its maximum value over your grid to make the maximum normalized value 0 dB. See Prog. 9.9.
    (c) If $G(j\omega)$, which means $\sigma = 0$, gives the frequency response of some circuit, then what kind of filtering activity does the circuit perform?

24) In a 3-D space, a 1×2 plane is located in the first quadrant of the x–y plane with a corner at the origin.
    (a) Write a MATLAB script to draw and fill with blue the plane. Use the subplot function to place the initial plane in the first position of a 2×2 arrangement of four plots. Use axis tight to specify the limits of the axes. Continue the script to rotate the plane through rotations about the x, y, and z axes by 30, 45, and 60 degrees, respectively, where each rotation starts with the initial plane. Show each rotation in the remaining subplot positions. Include axis labels and a title in each subplot. In the *Figure Window* click the rotate icon and rotate each subplot to clearly show that the desired rotation of the initial plane has been achieved. Print the figure.

(b) Continue the MATLAB script to combine the three rotations of part (a) into a single rotation matrix. Plot the plane, include axis labels, and print the figure.

25) The following MATLAB statements calculate the current through a p–n junction diode given the diode voltage v and the temperature t_C in degrees Centigrade.

```
I_sat = 1e-12; % saturation current

k = 1.3806503*10e-23;  % Boltzmann constant

q = 1.602176646*10e-19; % magnitude of the charge of an electron

t_K = t_C + 273.15; % temperature in degrees Kelvin

V_T = k*t_K/q; % thermal voltage

DiodeCurrent = I_sat*(exp(v/V_T) -1);
```

Write a MATLAB script that includes these statements to the plot in a 3-D space the diode current as it varies with voltage and temperature. The voltage should vary as v = 0.0:0.01:0.8, and the temperature should vary as t_C = −40:5:85, which is the standard industrial grade temperature range for electronic devices. Include axis labels and a title. Use the function view to see the plot from a viewpoint that clearly shows how current varies with voltage and temperature. Print the figure.

Section 9.6

26) Prog. 8.4 is a MATLAB script that displays the animation of the second hand of a clock. Modify this program to include the animation of a minute hand. This will require placing the for loop that displays the second hand into another for loop that displays the minute hand. Use a shorter red line for the minute hand. For testing, use a short pause to speed up clock activity.

27) Write a MATLAB script to animate rotation of a four cornered plane that starts by drawing a green filled plane centered at the origin of a 3-D space. A plane is defined by two vectors, call them p_1 and p_2. To center the plane, let the other two corners be defined by $p_3 = -p_1$ and $p_4 = -p_2$. The following MATLAB script should help you get started.

```
% Program to draw and rotate a plane in 3-D space

clear all; clc; clf

p = [3 -2 3;-1 2 -3;-3 2 -3;1 -2 3]; % DEFINE YOUR OWN VECTORS

lines = [1 2;2 3;3 4;4 1]; % specify line end points with vector indices

figure(1); hold on

for k = 1:4
```

```
x = [p(lines(k,1),1)  p(lines(k,2),1)]; % get end points along x-axis
y = [p(lines(k,1),2)  p(lines(k,2),2)]; % get end points along y-axis
z = [p(lines(k,1),3)  p(lines(k,2),3)]; % get end points along z-axis
plot3(x,y,z) % plot lines in 3-D space
end
x = p(1:4,1);y = p(1:4,2);z = p(1:4,3); % get vertices of the plane
fill3(x,y,z,'g') % fill plane in 3-D space with a color
axis([-5  5  -5  5  -5  5]); grid off; axis off
```

Continue the script.
  (a) Use the view function in a loop with pauses to simulate a smooth 3-D rotation.
      Print two snapshots of the green plane.
  (b) Repeat part (a), but use instead an appropriate camera function.
28) Write a MATLAB script that achieves the animation produced by Prob. P9.27, but uses instead the function movie by first creating a sequence of frames.
29) The transfer function of a circuit is given by

$$H(j\omega) = \frac{(j\omega)^2 + j\omega/2 + 257/16}{(j\omega + 1/2)((j\omega)^2 + j\omega/2 + 17/16)}$$

Let the input to the circuit be $x(t) = \cos(\omega t)$. The sinusoidal output of the circuit is given by $y(t) = \|H(j\omega)\| \cos(\omega t + \angle H(j\omega))$. Write a MATLAB script that creates a movie that shows in each frame a figure consisting of two subplots. The top subplot shows the input $x(t)$, and the bottom subplot shows the output $y(t)$. For each frame, let $0 \le t \le T_0$, and use enough points to see a smooth looking sinusoid. From frame to frame, let ω vary as $\omega = 0:0.2:10.0$. See how your movie shows the frequency response of the circuit with $T_0 = 20$ secs. What kind of filtering activity does the circuit perform?

CHAPTER 10

Debugging

When a program does not work, or seems to work, but does not produce results as intended, it is said to have bugs. A bug is a program error. Basically, there are two kinds of bugs: syntax errors and run-time errors. Syntax errors are, for example, incorrect or inconsistent spelling of reserved words, key words and variables, punctuation errors, unbalanced parentheses, and others. A run-time error may be an error in the logic of a program that implements some algorithm, and causes the program to produce incorrect results. Another kind of run-time error occurs when, for example, a variable value becomes **Inf** or **NaN**, which can prevent further acceptable computation. In this chapter, you will learn how to use MATLAB® facilities that can make the process of eliminating bugs more efficient, which is called **debugging**, including

- detection and correction of syntax errors
- use of built-in functions to locate and report errors and suggest improvements
- interruption of program execution to trace run-time and algorithm errors

10.1 Syntax Error Debugging

MATLAB includes several features that help to avoid syntax errors. Some of these features are enabled by selecting editor/debugger preferences with

$$\textbf{File} \rightarrow \textbf{Preferences} \rightarrow \textbf{Editor/Debugger} \rightarrow \begin{cases} \textbf{display} \\ \textbf{language} \\ \textbf{code folding} \end{cases}$$

Use **display** to, for example, enable highlighting the current line, **language** to, for example, select syntax colors, and **code folding** to, for example, enable coalescing statements within various code blocks. By folding code blocks it becomes easier to gain a perspective about the flow of a script.

A particularly helpful MATLAB feature is the code analyzer program **M-Lint** that runs automatically in the background as you write a script in the *Editor Window*. This is demonstrated in Fig. 2.16, where the vertical bar on the right side of the *Editor Window* shows not only syntax warnings and errors, but also how to fix some errors. Also, brief documentation is automatically provided as you write code.

Example 10.1 _____

Given a nonnegative real number x, design an algorithm that finds $y_0 = \sqrt{x}$, the positive square root of x. Let us apply the method of steepest descent, and minimize the cost function $f(y) = (y^2 - x)^2$, where $y > 0$ and $f(y_0) < f(y)$ for all positive $y \neq y_0$. This cost function comes from the given problem, where for a given x, we have $y_0^2 - x = 0$. At first, we do not know y_0, and must search for that $y = y_0$ that makes $y^2 - x = 0$. For any y, $y^2 - x$ can be positive or negative. However, $(y^2 - x)^2$ is always nonnegative, and when $y = y_0$ we have $(y^2 - x)^2 = 0$. According to the method of steepest descent (see (4.9)), where $y_{new} = y_{old} - \mu f'(y_{old})$.) we need $df/dy = 2(y^2 - x)(2y)$. The overall algorithm is implemented in Prog. 10.1, and a function and subfunction are given in Prog. 10.2. Some results are shown after the programs.

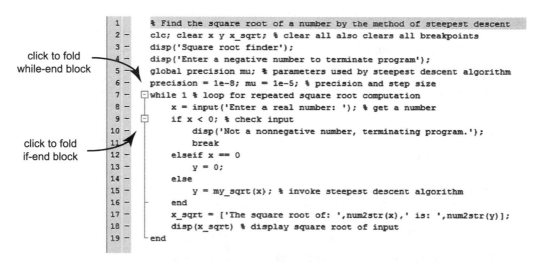

Program 10.1   Script of file sqrt_finder.m with code folding enabled, but not utilized.

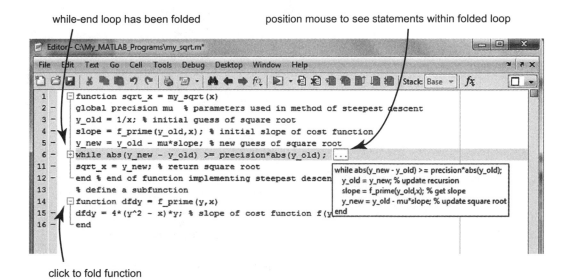

Program 10.2   Function m-file that implements method of steepest descent.

```
Square root finder
Enter a negative number to terminate program
Enter a real number: 4
The square root of: 4 is: 1.9999
Enter a real number: 100
The square root of: 100 is: 10
Enter a real number: 0.09
The square root of: 0.09 is: 0.30042
Enter a real number: -1
Not a nonnegative number, terminating program.
```

Suppose that Prog. 10.1 has been partially written as shown in Fig. 10.1. Notice that as you enter the argument(s) of almost any built-in function, the input function in this case, a window will pop up that shows you the argument syntax options. If you need more help, then click More Help, which causes another window to pop up that contains the help produced by using the doc facility, **doc input** in this case. Also, the right side of Fig. 10.1 automatically shows the execution status (green, orange, or red) of the code, with details about lines that can be improved.

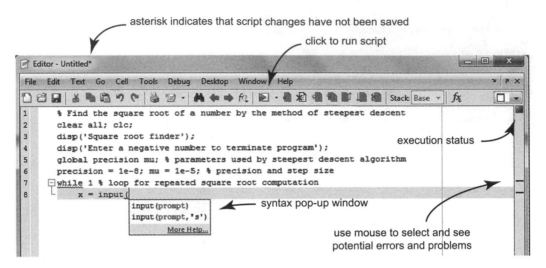

Figure 10.1   Editor Window showing potential problems detected by M-Lint.

With the built-in function **checkcode** (replacement of the function **mlint**) you can apply the code analyzer to an m-file and obtain M-Lint messages about syntax errors and possible program improvements. For m-files within the *Current Folder*, a syntax option is given by

```
checkcode('file_name')
```

For example, several syntax errors were introduced (but not shown) in Prog. 10.1. The result of using the function checkcode is shown below.

```
>> checkcode('sqrt_finder')  % equivalent to: checkcode sqrt_finder
L 8 (C 15-53): A quoted string is unterminated.
L 12 (C 14): Parse error at ' = ': usage might be invalid MATLAB syntax.
L 17 (C 54-67): A quoted string is unterminated.
L 20 (C 0): Program might end prematurely (or an earlier error confused M-Lint).
```

An alternative to the function checkcode is the built-in function **mlintrpt**, with which you can save the M-lint messages in a file. A syntax option is

```
mlintrpt('file_name')
```

For example, the statement: mlintrpt('sqrt_finder') generates the report shown in Fig. 10.2. Click on Learn More in the Code Analyzer Report to find much information about the meaning of the report.

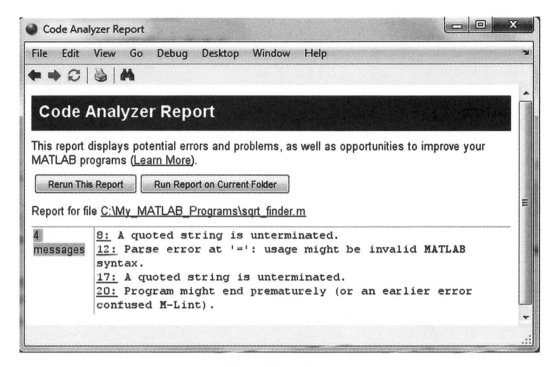

Figure 10.2 Code analyzer report.

To obtain a code analyzer report about all of the m-files in the *Current Folder* use

Action Button → **Reports** → **Code Analyzer Report**

This generates the reports shown in Fig. 10.3.

Not all syntax errors can be detected by the code analyzer. For example, a misspelled function name may not change a statement into an invalid MATLAB statement, but may cause errors in the program output. This depends on whether or not the misspelled function name is in the search path, and if it is in the search path, then it may still be a syntax error that will not be detected until run time.

10.2 Run-Time Error Debugging

Run-time error debugging can be very challenging. For some run-time errors, MATLAB stops program execution and displays a message about the kind of error that occurred and the file name and line number where the error was detected. It can be difficult to locate and correct the cause of such errors. Sometimes, a variable value becomes infinite (inf) or not a number (nan), which can produce incorrect results and may not cause an error message to be displayed. There may not be a unique correspondence between a script error and the

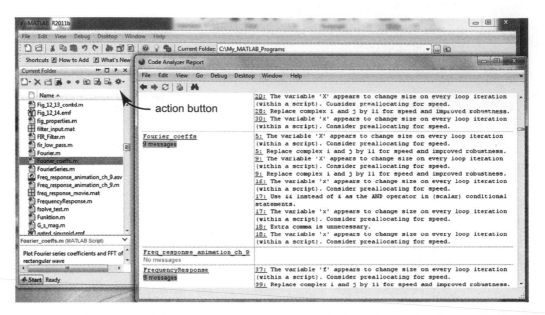

Figure 10.3   A part of the M-Lint message reports of all m-files in the Current Folder.

meaning of an error message. Without experience, some error messages may not be easily understood.

The source of a run-time error can be input data, syntax (a misspelled variable name, for example), script logic, or insufficient accounting for all problematic outcomes. When a MATLAB error message is clear or program output clearly points to an error in the script, there is no debugging problem. However, for complicated scripts, the efficiency of the debugging process depends on the skills of the programmer and the debugging facilities provided by the programming environment. MATLAB includes a variety of provisions for debugging.

An easy way to see the value of a variable from within the *Editor Window* is to position the mouse over the variable, as shown in Fig. 10.4. This causes a window to pop up that gives the class of a variable, the value of a scalar, and the size of a matrix. This way you can

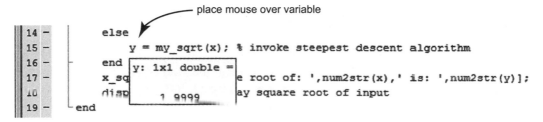

Figure 10.4   Example of finding properties of a variable with the mouse.

quickly check to see if some variables have values in expected ranges. Or, you can remove the semicolon at the end of lines that assign values to variables to stop suppressing their display when you run the program again.

Another way to check variables is to activate the *Command Window*, and enter the names of variables to display their values. Consider using functions such as disp, who, whos, size, isempty, isinf, isnan, iskeyword, all, any, find, and others to find out more about the condition of variables. Or, you can double click on the variable name in the *Workspace Window* to open the *Variable Editor Window*.

To trace statement execution in a file, use the function **echo**. Somewhere within a script or function, insert the statement: echo on, to display in the *Command Window* statements as they are executed. To stop echoing statements, use: echo off. You can specify the function file in which statements should be echoed with

```
echo 'file_name' on
```

Example 10.2

The beginning of the script of sqrt_finder.m, which is given in Prog. 10.1, is modified to

```
% Find the square root of a number by the method of steepest descent
clear all; clc
echo 'my_sqrt' on % echo statements in the function file my_sqrt.m
disp('Square root finder');
      :
```

Executing the file sqrt_finder.m produces the following output in the *Command Window*.

```
Square root finder
Enter a negative number to terminate program
Enter a real number: 1
global precision mu  % parameters used in method of steepest descent
y_old = 1/x; % initial guess of square root
slope = f_prime(y_old,x); % initial slope of cost function
dfdy = 4*(y^2 - x)*y; % slope of cost function f(y)
y_new = y_old - mu*slope; % new guess of square root
while abs(y_new - y_old) >= precision*abs(y_old); % check for convergence
end
sqrt_x = y_new; % return square root
end % end of function implementing steepest descent
The square root of: 1 is: 1
```

```
Enter a real number: -1
Not a nonnegative number, terminating program.
```

Here, we can check the sequence of statements that are executed in the function and its subfunction. The test case of $x = 1$ was used to ensure that the while-loop in my_sqrt.m is executed only once.

To investigate the state of variables before an error condition is reached, insert into the script wherever you want to stop program execution, the statement: **keyboard**. The function **keyboard** stops execution of an m-file and gives control to the keyboard. The MATLAB environment is said to be in the **keyboard mode**. This special status is indicated in the *Command Window* by the K before the prompt, K>>. In the *Command Window* you can examine or change variables, and all valid MATLAB statements and functions can be used.

Example 10.3

Suppose that the statement: keyboard, is inserted just after line 5 in the function script given in Prog. 10.2. After saving this function script as the m-file my_sqrt.m, the m-file sqrt_finder.m, which is given in Prog. 10.1, is opened and executed by clicking on the run button in the *Editor Window* toolbar. Fig. 10.5 shows the resulting keyboard mode. Notice that the *Workspace Window* shows the function my_sqrt.m work space, which is separate from the sqrt_finder.m work space, and that the value of a variable can be changed. To exit the keyboard mode, the function **return** must be used. In this case, execution of the file my_sqrt.m continues.

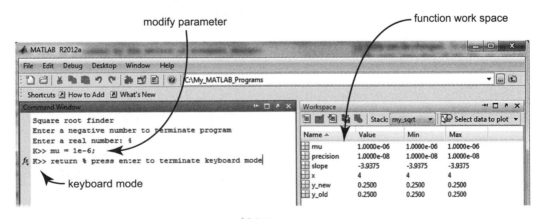

Figure 10.5 MATLAB keyboard mode.

10.2.1 Error and Warning Messages

MATLAB issues error messages and terminates program execution when a program includes code that it cannot execute. For example, an algorithm requires the inner product of two column vectors a and b, and a MATLAB program contains the statement: c = a*b, which produces the following result

```
>> c = a * b
```

Error using   *
Inner matrix dimensions must agree.

A programmer may not have known the dimensions of a and b when the script was written. After debugging with size(a) and size(b), the MATLAB statement is changed to: c = a' * b.

With the built-in function **error**, you can also include in a script MATLAB statements that issue error messages in response to computing conditions that violate algorithm requirements. A syntax option of the built-in function error is given by

error('message', var1, var2, ...)

where the message character string can include conversion specifiers, such as %d for a decimal number. See section 8.1.1 for format possibilities.

Example 10.4 _____

Let us reconsider Prog. 10.2 by understanding the given problem a little better. To find the square root of, for example, $x = 4$, the function $f(y)$ to be minimized is plotted in Fig. 10.6, which shows two minima. The behavior of the algorithm depends on three parameters: initial guess, step size, and precision.

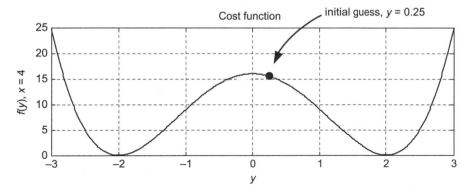

Figure 10.6   Cost function f(y) for x = 4.

Given x, the initial guess is obtained with $y = 1/x$, because if $x > 1$, then the initial guess will be less than x, and if $x < 1$, then the initial guess will be greater than x, which is expected of the square root of x. There may be good rationale for using another way to specify an initial guess, for example $y = x$. For $x = 4$, the initial guess is $y = 0.25$, where the slope of the cost function $f(y)$ is negative, and the next guess of the square root of x will be greater than $y = 0.25$. If the step size is too large, then the next guess could be much larger than $y_0 = 2$, and the cost function $f(y > y_0)$ may be larger than $f(y = 0.25)$. We want the cost function to decrease to zero, where $y = y_0$. However, when the next guess y is much larger than $y_0 = 2$, the slope of $f(y)$ is positive and big, which, with a large step size, may cause the following guess y to become negative. This understanding of what can go wrong in this process suggests that with each pass through the while loop to find the next guess of the square root of x, the cost function must not increase and the next guess must be positive. These constraints apply, regardless of the value of x. If the step size is too large, then the algorithm cannot converge. If the step size is too small, then the algorithm may take a long time to converge.

To make this application of the method of steepest descent more robust, it is modified to include checking the sign of the next guess of the square root of x and the change in the cost function through every iteration of the algorithm. In Prog. 10.3 the constraints are checked, and if a constraint is violated, an error message is issued and program execution is terminated. It would be useful to instead introduce a remedy for this problem, by, for example, automatically increasing and decreasing the step size.

```
function sqrt_x = my_sqrt(x)
global precision mu   % parameters used in method of steepest descent
y_old = 1/x; % initial guess of square root
slope = f_prime(y_old,x); % initial slope of cost function
y_new = y_old - mu*slope; % new guess of square root
if y_new < = 0 % only positive values allowed
    error('new root: %f is not positive',y_new)
elseif f_of_y(y_new,x) > f_of_y(y_old,x) % cost function must not increase
    error('new cost: %f is greater than old cost: %f',y_new,y_old)
end
while abs(y_new - y_old) > = precision*abs(y_old); % check for convergence
    y_old = y_new; % update recursion
    slope = f_prime(y_old,x); % get slope
    y_new = y_old - mu*slope; % update square root
    if y_new < = 0
        error('new root: %f is not positive',y_new)
    elseif f_of_y(y_new,x) > f_of_y(y_old,x)
        error('new cost: %f is greater than old cost: %f',y_new,y_old)
    end
end
```

```
sqrt_x = y_new; % return square root
end % end of function implementing steepest descent
% define a sub function
function dfdy = f_prime(y,x)
dfdy = 4*(y^2 - x)*y; % slope of cost function f(y)
end
% define a sub function
function cost_func = f_of_y(y,x) % cost function f(y)
cost_func = (y^2 - x)^2;
end
```

Program 10.3   Modification of Prog. 10.2 to include issuing error messages.

With the intention of speeding up convergence, set the step size to mu = 1.e-2 in line 6 of Prog. 10.1, which gives the following results.

```
Square root finder
Enter a negative number to terminate program
Enter a real number: 4
The square root of: 4 is: 2
Enter a real number: 100

Error using my_sqrt   (line 18)
new cost: 21.454051 is greater than old cost: 6.168742

Error in sqrt_finder   (line 15)
        y = my_sqrt(x); % invoke steepest descent algorithm

Square root finder
Enter a negative number to terminate program
Enter a real number: 0.81
The square root of: 0.81 is: 0.9
Enter a real number: 0.01

Error using my_sqrt   (line 7)
new root: -39899.960000 is not positive

Error in sqrt_finder   (line 15)
        y = my_sqrt(x); % invoke steepest descent algorithm
```

Notice that the output given by MATLAB includes the line number where the error occurred and the sequence of invoked m-files, which is called a **stack**. You can see the stack change in the *Stack Window* on the *Editor Window* toolbar. This enables you to trace the sequence of invoked functions.

A lesson given in Example 10.4 is that an important part of the debugging process is to understand the method of solution of a problem, and include in the program that implements a solution method checks in anticipation of problematic conditions.

Sometimes a condition may occur that is not acceptable, but does not warrant termination of program execution. To be made aware of this condition the MATLAB function **warning** can be used. One of its syntax options is

```
warning('message', var1, var2, ...)
```

Like the function error, the message, which can include conversion specifiers, is displayed, but program execution continues.

There will be occasions when a script has become very complicated, and program output is not what was expected. A thorough trace of selected parts of script flow and computation is necessary. This can be done with breakpoints.

10.2.2   Breakpoints

A **breakpoint** is a place in a script or function where the executing program or function has paused executing code. There are three types of breakpoints that you can set in a MATLAB script, which are

- **standard breakpoint**, that pauses file execution at a specified line in a file
- **conditional breakpoint**, that pauses file execution at a specified line in a file only under specified conditions
- **error breakpoint**, that pauses file execution of any file when it causes a specified type of warning, error, NaN or infinite value

In MATLAB, each executable statement in a script can be set to be a **standard breakpoint** location. If a particular statement is set as a standard breakpoint location, then just before this statement is about to be executed

- file execution is paused
- the paused file is opened in the *Editor Window*, if it was not already open
- the *Workspace Window* shows the work space of the paused file
- MATLAB goes into the debug mode as indicated by the K before the prompt, K>>

In the *Command Window* you can examine or change variables, and all valid MATLAB statements and functions can be used. You can edit all variables in the *Workspace Window*.

You can only set valid standard and conditional breakpoints at executable lines in saved files that are in the current folder or in folders on the search path. An executable line is preceded by a dash (-) located just to the right of the line number. To set a standard breakpoint in the *Editor Window,* click the dash that is just to right of the line number where you want file execution to pause. If the script has no syntax errors that can prevent file execution and it has been saved, then the dash will change to a red dot, indicating that a valid breakpoint has been set. You can see the saved status of a script by the presence of an asterisk after its file name (see Fig. 10.1). If a script has syntax errors that prevent file execution (Check execution status as shown in Fig. 10.1.), then the dot color will be gray instead of red, and the breakpoint is called an **invalid breakpoint**.

Example 10.5

To see the condition of variables and trace the activity of the function file my_sqrt.m, as given in Prog. 10.3, open the file and set a breakpoint at line 3, which is shown in Fig. 10.7.

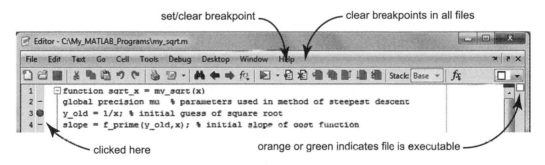

Figure 10.7 Set a valid breakpoint in my_sqrt.m.

Be sure to save script modifications before setting any breakpoints. To preserve breakpoint settings, you need not save the script again. Saving this function script will clear breakpoints. Instead, close the *Editor Window* shown in Fig. 10.7.

The function my_sqrt.m is invoked in the file sqrt_finder.m. Before running the file sqrt_finder.m, open it in the *Editor Window*, and change the statement: clear all, in line 2 (see Fig. 10.1) to: clear x y x_sqrt, because the statement: clear all, also clears all breakpoints. Save sqrt_finder.m, and to run it, for example, name it in the *Command Window*. Instead, let us set another breakpoint at line 15, as shown in Fig. 10.8, and run the file sqrt_finder.m by clicking on the run button in the toolbar of the *Editor Window*.

```
14 -        else
15 ●            y = my_sqrt(x); % invoke steepest descent algorithm
16 -        end
17 -        x_sqrt = ['The square root of: ',num2str(x),' is: ',num2str(y)];
18 -        disp(x_sqrt) % display square root of input
19 -    end
```

Figure 10.8 Breakpoint set in file sqrt_finder.m.

When the breakpoint at line 15 is reached, the *Command Window* and *Workspace Window* appear as shown in Fig. 10.9. MATLAB is in the debug mode, and sqrt_finder.m is paused as shown in Fig. 10.10.

Figure 10.9   Command window showing line 15.

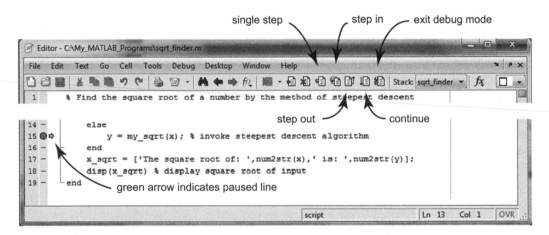

Figure 10.10   File sqrt_finder.m paused.

At line 15, you can single step to line 16 by clicking the **single step** button on the toolbar. During this step my_sqrt.m is executed, and since there is a breakpoint at line 3 of my_sqrt.m, my_sqrt.m is paused, as shown in Fig. 10.11. If instead you click the **step in** button, execution is continued in my_sqrt.m at line 2, the first line in the file. You can continue to single step through my_sqrt.m, and when another function is reached, you can step into it to single step through it. At any line you can click the **step out** button to step out

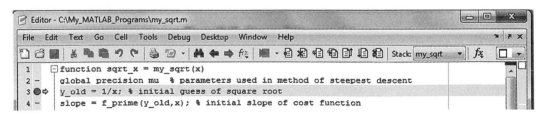

Figure 10.11   File my_sqrt.m paused.

of a function, or continue execution until the next breakpoint is reached by clicking the **continue** button.

If you want to execute several lines of code without single stepping through them, place the cursor anywhere on a line to which you want to execute code, and use

Debug → Go Until Cursor

It is particularly useful that when you step into a function, the *Workspace Window* shows the work space of the function, and you can see function variable values as they change from one line to the next line by single stepping through the function.

Other ways to set a breakpoint are (1) position the cursor in an executable line and then click the set/clear breakpoint button on the toolbar and (2) from the Debug menu, select Set/Clear Breakpoint.

The debugging possibilities demonstrated in Example 10.5 can also be achieved from the *Command Window* with debugging functions given in Table 10.1. For example,

```
K>> dbtype sqrt_finder
1     % Find the square root of a number by the method of steepest descent
2     clc; clear x y x_sqrt;  % clear selected variables
3     disp('Square root finder');
         .
         .
18        disp(x_sqrt) % display square root of input
19    end
K>> dbstatus
Breakpoint for my_sqrt is on line 3.
Breakpoint for sqrt_finder is on line 15.
K>> dbclear in sqrt_finder at 15   % clear breakpoint
K>> dbquit
```

Table 10.1 Debugging mode functions

Function	Brief description
dbstop	Set breakpoint
dbclear	Remove breakpoint
dbcont	Resume execution
dbdown	Change local work space context
dbstatus	List all breakpoints
dbstep	Execute one or more lines from present breakpoint
dbtype	List m-file with line numbers
dbup	Change local work space context
dbquit	Quit debug mode
dbmex	Enable mex-file debugging; use doc dbmex for details
dbstack	List who called whom

Another useful kind of breakpoint is the **conditional breakpoint**, which pauses execution of a script at a specified line only if a specified condition is satisfied. You can elect to set a standard or conditional breakpoint by right-clicking the dash next to a line number. In response to selecting a conditional breakpoint, a dialog box will open, where you can enter an expression that evaluates to a logical value. Or, you can change a standard breakpoint into a conditional breakpoint by right-clicking the standard breakpoint. After you have entered an expression into the dialog box, click OK in the dialog box, and the breakpoint dot color changes to yellow, indicating that it is a conditional breakpoint.

Example 10.6

To keep track of the number of iterations through the loop of my_sqrt.m, let us add a counter and conditional breakpoint as shown in Fig. 10.12.

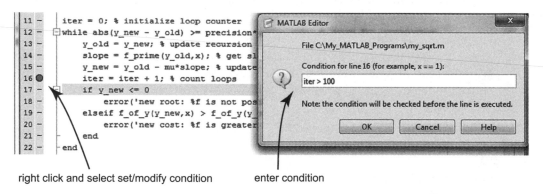

right click and select set/modify condition   enter condition

Figure 10.12   Setting a conditional breakpoint.

After 101 loop iterations, execution will pause at line 16. You can then look at the function work space to see the progress of the algorithm to gain more insight about its behavior.

MATLAB issues error/warning messages when an illegal operation is encountered in a script. For some illegal operations, for example, division by zero, it assigns the result inf and continues file execution. This may cause incorrect program output. For almost all other illegal operations file execution is stopped after the error message has been issued. Instead, it may be useful to enable **error breakpoints**. When MATLAB detects an error, it pauses file execution, enters the debug mode, opens the file in which the error occurred and positions, the pause indicator (green arrow) in the line after the line in which the error was detected. To enable a file to be opened in the debug mode, from the Desktop or any other menu bar use

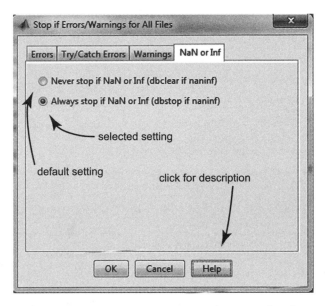

Figure 10.13   Dialog box to set error/warning breakpoints.

Debug → Open Files when Debugging

To enable debugging in response to MATLAB detected errors/warnings, you must enable error breakpoints with

Debug → Stop if Errors/Warnings

which opens the dialog box shown in Fig. 10.13, where in the Nan or Inf tab the Always Stop option was selected. Repeat the selected option in the Errors and Warnings tabs. The default settings are automatically reinstated when you end the MATLAB session. Click on help in each of the tabs for a description of the type of breakpoint. Use the help facility, and do a search for breakpoints to see extensive documentation.

Example 10.7 _____

To see the effect of a NaN or Inf error breakpoint as enabled with the dialog box shown in Fig. 10.13, suppose the statements

```
elseif x == 0
    y = 0;
```

are removed from the m-file sqrt_finder.m. Then, an input of $x = 0$ will cause an Inf error in my_sqrt.m, which results in the response shown in Fig. 10.14.

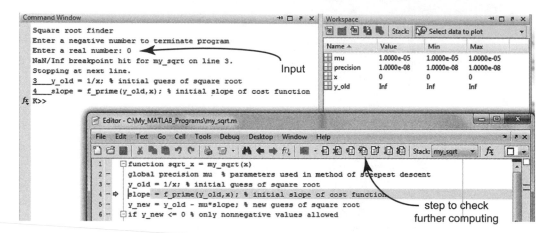

Figure 10.14   Handling of a NaN or Inf breakpoint.

After you have resolved the cause of errors you can disable breakpoints with

Debug $\rightarrow$ Clear Breakpoints in All Files

10.3   Conclusion

MATLAB provides numerous ways to track down bugs. Debugging success depends on your ingenuity, not only to find bugs, but also on the way you build scripts to make it easier to find bugs. For example, it is useful to

- understand the problem and method of solution as thoroughly as possible
- modularize scripts that implement the method of solution
- anticipate computing issues and exercise each module
- provide input test cases and expected results
- document scripts

You should now know how to

- fold program blocks
- utilize M-Lint to find, report, and fix syntax errors
- use the function keyboard to track computation
- apply the function echo to track code

- issue error and warning messages
- set up and work with standard and conditional breakpoints
- step through a script and function
- use debug functions
- set and reset error breakpoints

There is much more to be learned about the MATLAB debugging facilities. Since we are all prone to make programming mistakes, debugging experience will help to make debugging easier. Table 10.2 gives a list of the built-in functions that were introduced in the chapter.

Table 10.2   Built-In MATLAB functions introduced in this chapter

Function	Brief explanation
checkcode	Apply M-Lint code analyzer
mlint	Replaced by checkcode
mlintrpt	Apply M-Lint code analyzer to all files in a folder
echo	Display statements as they are executed
keyboard	Switch to keyboard mode
error	Display an error message and terminate program execution
warning	Display a warning message

Problems

Section 10.1

1) Describe the meaning of the three colors of the execution status button in the *Editor Window*.
2) Describe the difference between the two colors (red and orange) of the dashes in execution status column.
3) (a) Describe how to enable code folding.
   (b) Key in Progs. 10.1 and 10.2, and save them in your *Current Folder*. Enable code folding, and fold every possible structure. Provide copies of the programs before and after code folding. The folded while-loop in Prog. 10.2, shown in the highlighted window, is given below.

```
while abs(y_new - y_old) >= precision*abs(y_old); % convergence check
    y_old = y_new; % update recursion
    slope = f_prime(y_old,x); % get slope
    y_new = y_old - mu*slope; % update square root
end
```

These programs will be used in many of the following problems.

4) (a) Change line 12 of Prog. 10.1 to: elseif x = 0. What happened to the status execution button?

(b) Position the mouse on the red dash in the status execution column. What appears? Is this useful for fixing the syntax error? Fix the syntax error. What happened to the status execution button?

5) (a) Change line 9 of Prog. 10.1 to: if x > 0. Does this cause a syntax error or a run-time error? What happened in the execution status column? Fix line 9.

(b) Change line 6 of Prog. 10.2 to: while (y_new - y_old)> = precision*abs(y_old). Does this cause a syntax error or a run-time error?

6) Implement the changes described in Probs. P10.4 and P10.5. Use the function **check-code** to get a report. Does the report show all errors introduced in Probs. P10.4 and P10.5. Provide a copy of the report. Undo the errors.

Section 10.2

7) (a) Describe four different ways that you can see the value or properties of a variable after a program has executed and you suspect a run-time error.

(b) Insert: **echo on**, between lines 6 and 7 of Prog. 10.1, and insert: **echo off**, after the last statement of the program. Execute the program with an input of 2 and then −1. Describe what happens in the *Command Window* as the program executes. Then, remove the echo statements.

8) Edit Prog. 10.2 to enter the statement: **keyboard**, between lines 4 and 5. Save the program.

Open and execute Prog. 10.1 with an input of 2. When MATLAB goes into the keyboard mode, find and give the value of the variable slope. Does the variable slope appear in the function work space? Then, in the *Command Window*, enter the statement: **return**, to exit the keyboard mode. Remove the statement: keyboard, from your program.

9) To make Prog. 10.2 more robust (less prone to initiate undesirable activities) several checks were inserted into it, as shown in Prog. 10.3. The first error check occurs after line 5 in Prog. 10.2 (see Prog. 10.3). The kinds of error checks and their messages depend on an understanding of the problem, the solution method, and program implementation of the solution method.

(a) Describe the purpose of the first error check.

(b) In your Prog. 10.2, enter between lines 5 and 6 the statement: maxiter = 1e4; iter = 0;. Then, just before the while-loop end statement, enter the statement: iter = iter +1; % counting iterations. Follow this with an if-end structure to check iter equals maxiter, and a statement to issue a warning message if this is true. Execute Prog. 10.1 with an input of 2. Does your program issue a warning message? If not, reduce mu in Prog. 10.1, and try again.

(c) Modify your programs to make maxiter a global variable. Provide a copy of your programs.

10) (a) What is a standard breakpoint?
   (b) There are six buttons in the *Editor Window* toolbar that are concerned with breakpoints. The left most button is named: Set/clear breakpoint. Describe how this button is used to set a standard breakpoint.
   (c) From left to right, name and describe the activity of the remaining five buttons concerned with breakpoints.

11) Open your Prog. 10.1. Click the dash just to the right of the number 7 of line 7, which has the while statement. The dash should change to a red dot, which indicates that a regular breakpoint has been set.
   (a) Run the program. Notice the green arrow in line 7, which indicates that line 7 will be executed next. What kind of prompt is shown in the *Command Window*? This shows that MATLAB is in the debug mode.
   (b) Click the Step button. Notice that the green arrow has moved to line 8. Click the step button again, and in the *Command Window* enter the number 2. Click the step button repeatedly until the green arrow reaches line 15. Instead of clicking the Step button, click the Step in button. Describe what happened.
   (c) Instead of clicking the Step button, click the Step out button. Describe what happened.
   (d) Click the Continue button. Describe what happened. Is MATLAB still in the debug mode?
   (e) Run the program again. When the program pauses at line 7, click the Exit debug mode button. Describe what happened.

12) Repeat Prob. P10.11, parts (a) and (b), and open the *Stack Window*. Give its content. Explain what is meant by a stack.

13) What is an invalid breakpoint?

14) While in debug mode, give statements that you can enter in the *Command Window* to
   (a) type a copy with line numbers of your Prog. 10.1.
   (b) get a status of breakpoints
   (c) see the stack content
   (d) set a breakpoint at line 17

15) In line 5 of your Prog. 10.2, the variable y_new is assigned a value. Set a standard breakpoint on line 6, the while statement. Describe how this breakpoint can be changed to a conditional breakpoint that causes the program to pause at line 6 only if y_new is negative. After you have done this, what is the color of the breakpoint dot?

16) In the *Editor Window* menu bar, open the Debug menu.
   (a) Name and describe the purpose of every menu item.
   (b) Clear all breakpoints. In Prog. 10.1, place the cursor anywhere on line 7, and do

   Debug → Set/Clear Breakpoint

Run the program. After the program has paused at line 7, place the cursor anywhere on line 17, and do

Debug → Go Until Cursor

Describe what happened.

17) Describe how to enable an error breakpoint.

CHAPTER 11

Symbolic Math

In MATLAB®, all numeric computation involves the application of arithmetic operations and functions to array variables (also called objects) of the class double (double precision floating point), where even a scalar is treated like an array. With the functions in the MATLAB Symbolic Math Toolbox you can perform symbolic computations. For example, suppose a, b, c, and d are symbolic variables, which means that they do not have a numeric value. Then, the MATLAB statement d = a*b*c/b, which is a symbolic computation, produces: d = a*c. Symbolic objects can be symbolic variables, numbers, expressions, and matrices.

At the core of the toolbox is the MAPLE system of programs for symbolic computation. Symbolic MATLAB statements are passed to MAPLE, which performs the symbolic computations and returns the results to MATLAB.

In this chapter, you will learn how to symbolically:

- solve systems of algebraic equations
- perform matrix operations
- differentiate a function given in symbolic function form
- evaluate both definite and indefinite integrals of symbolic integrands
- find the Fourier series coefficients of a periodic signal

and do **variable precision arithmetic** (VPA).

11.1 Symbolic Objects and Expressions

To distinguish between symbolic computation and arithmetic computation, objects to be used for symbolic computation must be declared to be of the class symbolic. Table 11.1 gives several MATLAB functions concerned with the symbolic data type. Unless otherwise specified, all symbolic variables are complex symbolic objects, having a real part and an imaginary part.

Table 11.1   Functions concerned with symbolic variables

Function	Brief description
y = ceil(x)	Round symbolic x toward plus infinity
y = char(x)	Convert symbolic scalar or array to a string
y = conj(x)	Take complex conjugate of symbolic x
y = double(x)	Assigns to y a numeric value represented by the symbolic object x
y = fix(x)	Round symbolic x toward zero
y = floor(x)	Round symbolic x toward minus infinity
y = frac(x)	Get symbolic fractional part of x
y = imag(x)	Get symbolic imaginary part of x
y = intn(x)	Convert symbolic matrix(scalar) to signed n-bit integer matrix(scalar), where n = 8, 16, 32, or 64
y = real(x)	Get symbolic real part of x
y = round(x)	Round symbolic x toward nearest integer
d = size(X)	Returns the size of a symbolic array
y = sym x property	Declare a variable to be of the sym class, where property can be: real,
y = sym('x','property')	positive, a flag and clear; see the following text
syms x1 x2 ... property	Declare variables to be of the sym class
y = symvar(f)	See the following text
y = uintn(x)	Convert symbolic matrix(scalar) to unsigned n-bit integer matrix(sclar), where n = 8, 16, 32 or 64

The function **syms** is used to declare that a variable is a symbolic object. All symbolic object names must be valid MATLAB names. The syntax is given by

```
syms name1 name2 … property
```

where property can be **real** (objects are real), **positive** (objects are real and positive), **clear** (used to restore objects to complex), or omitted. For example:

```
>> syms a  b   % command format to declare that a and b are symbolic objects
>> % MATLAB enters the names in the Workspace Window, and
>> % note that MATLAB does not echo the result
>> syms('c','d')   % function format to declare that c and d are symbolic objects
>> % get a symbolic expression for the real part of the symbolic variable a
>> real(a)
ans = a/2 + conj(a)/2
>> imag(b)  % get a symbolic expression for the imaginary part of b
ans = -(b*i)/2 + (conj(b)*i)/2
>> e = real(c)  %  assign the real part of c to another symbolic object
e = c/2 + conj(c)/2
>> who  % show all objects in the Workspace Window
Your variables are:
a  b c   d  ans  e
```

The function **sym** is used to construct a symbolic object. There are several syntax options.

1) `y = sym(x)`

where x can be a character string, numeric scalar or matrix, or a function handle. If the input argument x is a string, the result is a symbolic number or variable. If the input argument is a numeric scalar or matrix, the result is a symbolic representation of the given number(s). For example:

```
>> clear all   % clear workspace
>> a = sym(b)   % argument is not numeric and not a character string
??? Undefined function or variable 'b'.
>> a = sym('b')   % argument is a character string, which is assigned to a
a =
b
>> whos
  Name       Size                 Bytes  Class     Attributes
    a        1x1                    112  sym
>> c = a/b
??? Undefined function or variable 'b'.
>> c = a/sym(b)
??? Undefined function or variable 'b'.
>> c = a/sym('b')   % symbolic character string divided by a symbolic character string
c =
1
d = a/'b'   % symbolic character string divided by a character string
d =
1
>> e = sym('2')   % argument is a character string, which is assigned to e
e =
2
>> f = sym(2)   % argument is numeric, which is assigned to f
f =
2
>> whos
  Name       Size                 Bytes  Class     Attributes
    a        1x1                    112  sym
    c        1x1                    112  sym
    d        1x1                    112  sym
    e        1x1                    112  sym
    f        1x1                    112  sym
>> g = e/2   %  same as e/'2'
g =
```

```
1
>> h = f/2  %  same as f/'2'
h =
1
```

Note that MATLAB does not indent a symbolic response, as it does a numeric response, for example:

```
>> n = 3  % numeric assignment
n =
    3
```

In this text, a numeric response will be shown as $n = 3$ to conserve space. Space will also be conserved for symbolic responses.

To construct a symbolic object that represents a numeric matrix, use, for example:

```
>> b = sym([-5.3  3])
b = [ -53/10, 3]
```

When a number with a fractional part is assigned to a symbolic object, MATLAB converts the number to the closest rational number. For example, computing with 1/3 is not an approximation, while computing with 0.3333333... is an approximation because a finite number of digits must be used. The sym function is used to create a variety of symbolic objects. For example:

```
>> c = sym(1/3) % note that this does not produce an arithmetic result
c = 1/3
>> d = sym('x')  %  define a symbolic object
d = x
>> y = c*d + 0.43  % define a symbolic expression
y = x/3 + 43/100
>> % form symbolic representation of the square root of -1, named e
>> e = sym(sqrt(-1))
e = i
>> f = 3*e - 2  % define a symbolic object that represents a complex number
f = 3*i - 2
>> f = e*f
f = -2*i -3
>> g = @(a)  a^2+3  % g is the handle of an anonymous function, and echo result
g = @(a)  a^2+3
>> h = sym(g) % create a symbolic version of the function g, named h
h = a^2 + 3
```

When you use numeric data in a symbolic expression, MATLAB attempts to preserve accuracy by retaining the details of the computation. Symbolic computation is done with rational arithmetic. For example:

```
>>  clear all
>>  syms a b
>>  c = a + b^2 + 2/3   %  2/3 is not replaced by 0.666666666  ....
c   =   b^2 + a + 2/3
>>  d   = c + 0.47  % this is computed with rational arithmetic
d   =   b^2 + a + 341/300
%  here, MATLAB converted 0.47 to a rational number and combined it with 2/3
```

With the function **symvar** you can find the variables in an expression that are of the sym class. For example:

```
>> symvar(d)
ans   =   [a, b]
```

Sometimes it is better to prevent floating point arithmetic computation within an expression by declaring a number to be of the sym class. For example, the arithmetic computation given by

```
>>  clear all
>>  a = sqrt(2), b = sqrt(3)  % causes floating point arithmetic computations
a   =   1.4142
b   =   1.7321
>> c = a*b
c   =   2.4495
```

may not be preferred because a, b, and c are approximations and the details of the computation are lost. Instead, use

```
>>  d  = sym(sqrt(2)), e  = sym(sqrt(3))
d   =   2^(1/2)
e   =   3^(1/2)
>>  f = d*e  % get a symbolic representation of d*e
f   =   2^(1/2)*3^(1/2)
```

Here, f is not an approximation. With the function **double** you can convert a symbolic number to a floating point number. For example:

```
>>  g = double(f)  % will produce the value of f using floating point arithmetic
g   =   2.4495
```

Remember that all floating point computation is done in double precision. The number of digits displayed is controlled by the format function. Since g is an approximation, in further computation f should be used instead, which will preserve accuracy.

2a) **y = sym('x','real')**

where real(y) = x and imag(y) = 0.

2b) **y = sym('x','positive')**

where real(y) is positive and imag(y) = 0.

2c) **y = sym('x','flag')**

where **flag** is one of r (rational), which is the default, d (decimal), e (estimate error), or f (floating point). For example:

```
>> clear all
>> % convert the number 1/3 to a symbolic object
>> r = sym(1/3), d = sym(1/3, 'd'), e = sym(1/3, 'e'), f = sym(1/3, 'f')
r = 1/3
d = 0.333333333333333
e = 1/3 - eps/12
f = 6004799503160661/18014398509481984
>> % f is the binary to decimal conversion of the floating point number N*2^E
>> % for example
>> log2(18014398509481984)  % this gives -E
ans = 54
```

2d) **y = sym('y','clear')**

where y is restored to be a complex symbolic object with flag = r (the default).

Notice that, for example, the statements syms x property and $x = \text{sym}('x','property')$ are equivalent. The function **syms** is useful to conveniently declare multiple variables of the sym class, while the function **sym** provides more options, but allows for declaring only one variable at a time.

3) **A = sym('A',[M N])** or **A = sym('A%d%d',[M N])**

which creates an M by N matrix of scalar symbolic objects. The names of the elements of A have the form Am_n or Amn. If A is a vector, then the names of the elements have the form An or Am. For example:

```
>> A = sym('A',[2 3])  % create a 2 by 3 symbolic matrix
A =
[A1_1,  A1_2,  A1_3]
```

```
[A2_1,   A2_2,   A2_3]
>> A(2,1)  % get the object in row 2 and column 1
ans   =   A2_1
```

You can work with symbolic matrices like you can work with numeric matrices. For example:

```
>> clear all
>>  A = sym('A%d%d',[2 2])  % create a 2 by 2 symbolic matrix
A   =
[ A11, A12]
[ A21, A22]
>> B = sym('B%d%d',[2 3])  % create a 2 by 3 symbolic matrix
B   =
[ B11, B12, B13]
[ B21, B22, B23]
>> C = A*B  % symbolic multiplication
C   =
[ A11*B11 + A12*B21, A11*B12 + A12*B22, A11*B13 + A12*B23]
[ A21*B11 + A22*B21, A21*B12 + A22*B22, A21*B13 + A22*B23]
>> D = B(2,:)  % retrieve some part of a symbolic matrix using colon notation
D   =
[ B21, B22, B23]
>> E = inv(A)  % find the symbolic inverse of A
E   =
[ A22/(A11*A22-A12*A21), -A12/(A11*A22-A12*A21)]
[ -A21/(A11*A22-A12*A21), A11/(A11*A22-A12*A21)]
>>  F = A*E  % check if inverse
F   =
[ (A11*A22)/(A11*A22-A12*A21) - (A12*A21)/(A11*A22-A12*A21), 0]
[ 0, (A11*A22)/(A11*A22-A12*A21) - (A12*A21)/(A11*A22-A12*A21)]
>>  simplify(F)  % use the function simplify to simplify the elements of F
ans   =
[1, 0]
[0, 1]
```

Example 11.1 _____

This example demonstrates the difference between floating point arithmetic and rational arithmetic. For example:

```
>> clear all; format long
>> rng('default')  % initialize random number generator
>> % create a 3 by 3 matrix with elements selected randomly from
>> %  among (0, 1/6, ... , 5/6)
```

```
>> A = (randi(6,3) - 1)/6
A =
    0.833333333333333    0.166666666666667    0.166666666666667
    0.333333333333333    0.333333333333333    0.500000000000000
                    0    0.500000000000000    0.666666666666667
>> B = inv(A)
B =
    0.857142857142858    0.857142857142856   -0.857142857142856
    6.857142857142851  -17.142857142857128   11.142857142857135
   -5.142857142857139   12.857142857142847   -6.857142857142851
>> C = A - inv(B)   % ideally the result should be zero
C =
    1.0e-015 *
   -0.333066907387547   -0.083266726846887   -0.138777878078145
   -0.610622663543836                    0                    0
   -0.789491928622333                    0                    0
>> D = sym(A)
D =
[ 5/6, 1/6, 1/6]
[ 1/3, 1/3, 1/2]
[   0, 1/2, 2/3]
>> E = inv(D)   % get the exact inverse of A
E =
[   6/7,    6/7,   -6/7]
[  48/7, -120/7,   78/7]
[ -36/7,   90/7,  -48/7]
>> F = D - inv(E)
F =
[ 0, 0, 0]
[ 0, 0, 0]
[ 0, 0, 0]
```

Note the difference between C and F. This shows that if B is used to solve $AX = Y$, the solution is an approximation, while if E is used to solve $DX = Y$, the solution is exact. This difference can become significant, depending on the dimension of A, its elements, and condition number.

Sometimes there is a need to use a function name or operator acronym for more than one purpose. In this case, the function name or operation acronym is said to be **overloaded**. For example, the acronym eq (or $==$) compares, by default, two objects of the class double (and other numeric classes), and gives a logical result. This acronym is overloaded, because it also refers to a valid operation, which is defined within the Symbolic Math

Table 11.2 Functions concerned with symbolic expressions

Function	Brief description
f = collect(f,x) f = collect(f)	Regards each element of the symbolic matrix f as a polynomial in x and rewrites f in terms of the powers of x; uses the default variable determined by symvar
c = eq(a,b)	The result is true if the elements of a and b are symbolically equal; and since eq does not expand or simplify the expressions before making the comparison, use c = simplify(a−b) == 0 for a mathematically equal test
f = expand(f)	Writes each element of a symbolic matrix f as a product of its factors; expands polynomials, trigonometric, exponential, and logarithmic functions
ezplot(f,[xmin,xmax]) ezplot(f,[xmin,xmax,ymin, ymax])	Plots the symbolic expression f(x) over xmin < x < xmax; plots f(x,y) = 0 over xmin < x < xmax and ymin < y < ymax; see doc sym/ezplot for more details
y = factor(f)	Factors a symbolic scalar polynomial or matrix of polynomials f; if f is an integer, then f is factored into a product of prime numbers
pretty(f)	Manipulates a symbolic expression f to look more like a mathematical expression
y = simple(f)	Tries several different algebraic simplifications and returns the shortest; use doc sym/simple for more details
y = simplify(f) y = simplify(f,N)	Returns a simplified version, if possible, of an expression; searches for a simplification in N steps, default value is N = 50
subs(f,{a,b...},{sym('x'),2,...})	Replaces symbolic variables in the expression f with values; see the following text
[g,r] = subexpr(f,r)	Rewrites f in terms of a common subexpression r
y = symvar(f)	Finds all symbolic objects in f, where f can be an expression or a matrix

Toolbox, among two objects of the sym class. When you get help with: help eq, the response concerns the default version of eq. To obtain help within the Symbolic Math Toolbox context, you must use **help sym/eq**. Many function names, for example, inv, are overloaded to reuse them in a context other than a numeric context.

Table 11.1 includes several functions that are usually applied to numeric data. For example, the functions **ceil**, **fix**, **floor**, and **round** are concerned with rounding a number. When the argument of the function is of the class double, then the usual rounding operation is applied. However, when the argument is a symbolic object, a symbolic rounding operation is utilized. For example:

```
>> clear all
>> a = sym(-1.6)  % assign a symbolic number to a
a   =   -8/5
>> % examples of symbolic rounding
>> b = [ceil(a), fix(a), floor(a), frac(a), round(a)]
b   =   [ -1, -1, -2, -3/5, -2]
```

The function **frac** returns the fractional part(s) of a symbolic scalar(matrix).

You can formulate symbolic expressions with symbolic scalars and matrices according to the conventional rules of arithmetic with numeric variables. Table 11.2 gives some functions concerned with symbolic expressions. For example:

```
>> clear all;  syms a b  % declare two symbolic objects
>> c = a + b  % assign an expression to c
c = a + b
>> d = c^2  % assign to d an expression in terms of another expression
d = (a + b)^2
>> d = expand(d)  %  multiply out
d = a^2 + 2*a*b + b^2
>> e = c*d
e = (a + b)*(a^2 + 2*a*b + b^2)
>> e = collect(e,a)  % reorganize e as a polynomial in powers of a
e = a^3 + (3*b)*a^2 + (3*b^2)*a + b^3
>> syms s
>> [f,s] = subexpr(e,s) % find a common subexpression and replace it with s
f = a^3 + 3*b*a^2 + 3*s*a + b*s
s = b^2
>> syms r
>> [g,r] = subexpr(f,r)  % do it again
g = a*r + 3*a*s + 3*b*r + b*s
r = a^2
>> pretty(e)  % make e look like a mathematical expression
     3     2        2           3
  a  + (3 b) a  + (3 b  ) a + b
 >> h = expand(sin(a+b))
h = cos(a)*sin(b) + cos(b)*sin(a)
```

Once a symbolic expression has been defined, we must be able to replace symbolic objects within the expression with other symbolic objects or numeric objects to evaluate the expression. This is done with the function **subs**, which has the syntax given by

new_expression = subs(old_expression, old, new)

where **old** is a string or cell array of strings of one or more of the objects in old_expression that are to be replaced, and **new** is a corresponding string or cell array of strings of the replacement objects. If all symbolic objects in an expression are replaced by numeric data, then the expression is evaluated. Use **doc sym/subs** for more syntax options. For example:

```
>> clear all;  syms w t b  % declare three symbolic objects
>> s = sin(w*t + b);  % assign an expression to s
>> d = s^2  % assign to d an expression in terms of another expression
```

```
d  =  sin(b + t*w)^2
>> d  =  subs(d,{w},{10*pi})   %  replace w with a numeric value
d  =  sin(b + 10*pi*t)^2
>> e  =  subs(d,{b,t},{pi/4, 0.25})   % evaluate d for b = pi/4 and t = 0.25
e  =   0.5000
A  =  sym('A%d%d',[2 2]);   % create a symbolic matrix
A = subs(A,{A},{[1  2;3  4]})   % replace the elements with numeric values
A  =
[ 1,  2]
[ 3,  4]
```

Table 11.3 gives some functions that are especially useful within the context of symbolic expressions.

Table 11.3   Functions of symbolic variables

Function	Brief description
f = poly2sym(C,x)	Converts polynomial coefficients in a vector C to a symbolic polynomial f of x
C = sym2poly(f)	Returns a vector C of polynomial coefficients of a symbolic polynomial f
C = mod(A,B)	Returns in a symbolic matrix C the result of A-N.*B, where N = floor(A./B)
y = limit(f,x,a) y = limit(f,x,u,'direction')	Returns the limit of f as x → a; from direction left or right
y = symsum(f,x) y = symsum(f,x,a,b)	Returns the indefinite summation of terms given by f with respect to x; returns the definite summation from a to b
y = taylor(f,n,a)	Returns the (n-1)th order Taylor series expansion about the point a of the expression f
C = coeffs(f,x)	Returns in the vector C the symbolic coefficient expressions in the symbolic expression f with respect to an object x
[num,den] = numden(f)	Converts each element of f to a rational form where the numerator and denominator are relatively prime polynomials with integer coefficients

Example 11.2

Recall that in Example 6.9, Parseval's relation was applied to prove that

$$\sum_{k=1}^{\infty} \frac{1}{k^2} = \frac{\pi^2}{6}$$

This result can also be found with the function **symsum** as follows:

```
>> syms k
>> S = symsum(1/k^2,1,inf)
S =   pi^2/6
```

A **geometric series** is given by

$$S = \sum_{n=0}^{N-1} r^n \tag{11.1}$$

This summation can be found with

```
>> syms r n N
>> S = symsum(r^n,n,0,N-1)
S = piecewise([r = 1, N], [r <> 1, (r^N - 1)/(r - 1)])
```

This result means that S is given by

$$S = \begin{cases} N, & r = 1 \\ \dfrac{1 - r^N}{1 - r}, & r \neq 1 \end{cases} \tag{11.2}$$

The Taylor series expansion of a function $f(x)$ about a point $x = x_0$ is given by

$$f(x) = \sum_{n=0}^{\infty} (x - x_0)^n \frac{f^{(n)}(x_0)}{n!} \tag{11.3}$$

where

$$f^{(n)}(x_0) = \left. \frac{d^n f(x)}{dx^n} \right|_{x=x_0}$$

Equation (11.3) is called a Maclaurin series if $x_0 = 0$. The function **taylor** can be used to find a Taylor series expansion of a given function. For example:

```
>> syms x
>> f = cos(x);
>> cx = taylor(f,20)  % returns the first 20 terms of the Maclaurin series
cx = - x^18/6402373705728000 + x^16/20922789888000 - x^14/87178291200 +
x^12/479001600 - x^10/3628800 + x^8/40320 - x^6/720 + x^4/24 - x^2/2 + 1
>> ezplot(cx); grid on  % plot the symbolic function cx for -2pi < x < 2pi
>> subs(cx,x,pi/4) % evaluate the series for pi/4
ans =    0.7071
>> subs(cx,x,pi)
ans =   -1.0000
>> format long  % display enough digits to see nonzero coefficients
>> sym2poly(cx) % gather coefficients into a vector
ans =
  Columns 1 through 6
-0.000000000000000     0   0.000000000000048   0  -0.000000000011471   0
  Columns 7 through 12
 0.000000002087676     0  -0.000000275573192   0   0.000024801587302   0
```

```
Columns 13 through 18
-0.001388888888889   0   0.041666666666667   0   -0.500000000000000   0
Column 19
1.000000000000000
```

The complex symbolic data type is well suited for AC circuit analysis, where we would like to find a transfer function of a circuit in terms of component symbols to better understand how component values affect responses to sinusoidal inputs. First, let us see how we can work symbolically with Euler's identity. For example:

```
>> clear all
>> j = sym(sqrt(-1))    %  declare j to be a symbolic object
j = i
>> syms w t real   % declare real symbolic objects
>> % define a symbolic complex exponential function
>> X = exp(j*w*t)
X = exp(t*w*i)
>> f = X + conj(X)  % see if Euler's identity occurs
f = 1/exp(t*w*i) + exp(t*w*i)
>> % use the function simplify to obtain a simplified version
>> % of the expression t
>> g = simplify(f)
g = 2*cos(t*w)
>>  % check for Euler's identity
>> h = simplify((X - conj(X))/(j*2))
h = sin(t*w)
```

This shows that we can work symbolically with Euler's identity, and use it to symbolically obtain a sinusoidal function given its symbolic phasor.

Some functions in the Symbolic Math Toolbox do not work with j and i interchangeably. Instead, while, for example, i^2 becomes -1, j^2 is retained as j^2, which means that terms containing a power of j are not simplified to participate in further simplification. We will replace j with i to take advantage of simplification possibilities.

Example 11.3 _____

Let us work again with the circuit given in Fig. 6.11, which is a band-pass filter. In Section 6.6 the concept of an impedance is introduced, which is useful to find the sinusoidal response of a circuit when the input is a sinusoidal signal. By replacing each circuit component by its impedance, we can transform the problem of solving integral-differential equations that result from applying Kirchhoff's laws to an algebraic problem to find the phasor of a desired response. The impedance of a resistor R is $Z_R = R$, the impedance of a capacitor C is $Z_C = 1/(j\omega C)$ and the impedance of an inductor L is $Z_L = j\omega L$. Let us employ the MATLAB Symbolic Math Toolbox to find the transfer function given by (6.51).

```
>> clear all
>> % i (sqrt(-1)) is of class double, which is assigned to a symbolic object
>> j = sym(i)
j = i
>> syms w t real; syms L C R positive; syms VS % phasor of the input
>> % by specifying a property, we avoid unnecessary complex computations
>> ZR = R, ZC = 1/(j*w*C), ZL = j*w*L % define each symbolic impedance
ZR = R
ZC = -i/(C*w)
ZL = L*w*i
>> % obtain the phasor of the response by voltage division
>> V = (VS/(ZR+ZC+ZL))*ZR
V = (R*VS)/(R + L*w*i - i/(C*w))
>> H = V/VS  % get the symbolic transfer function
H = R/(R + L*w*i - i/(C*w))
>> H = simplify(H)  % see if MATLAB can simplify H
H = (C*R*w*i)/(- C*L*w^2 + C*R*w*i + 1)
>>  % this gives the same transfer function given by (6.51)
```

In Example 6.10 the circuit component values were given to be $C = 0.01\ \mu F, L = 11\ mH$, and $R = 33\ \Omega$. Recall that the magnitude of H peaks at $\omega \approx 2\pi(15000)$. For example:

```
>> % evaluate the transfer function
>> H_value = subs(H,{R,C,L,w},{33,1e-08,0.011,2*pi*15175})
H_value = 1.0000 - 0.0007i
```

where H_value is of the class double. If the input is $v_s(t) = A\cos(2\pi(15175)t + \theta)$, then the output is $v(t) = \|H\|A\cos(2\pi(15175)t + \angle H + \theta) \cong A\cos(2\pi(15175)t + \theta)$, as predicted by the frequency response plot of $\|H(j\omega)\|$ versus ω, where $\|H(j2\pi(15175))\| \cong 1$ and $\angle H(j2\pi(15175)) \cong 0$. To obtain the function H($j\omega$), use, for example:

```
>> H_jw = subs(H,{R,C,L},{33,1e-08,0.011})  %  substitute only the component
      values

H_jw = (1558380939346983*w*i)/(4722366482869645213696*(-(8510837770086989*w^2)/
7737125245533626718119264+(1558380939346983*w*i)/4722366482869645213696 + 1))

>> H_jw = simplify(H_jw)  % see if MATLAB can simplify H_jw

H_jw = (25532513310260969472*w*i)/
(-8510837770086989*w^2+25532513310260969472*w*i + 7737125245533626718119264)
```

It may be tempting to define symbolic objects that are a combination of variables, for example:

```
>> jw = sym('jw')  % combine variables that often occur in combination
jw = jw
>> jw*jw
ans = jw^2
```

This does not permit further simplification, while, for example:

```
>> j = sym(i) ; syms w real
>> j*w*j*w  %  or  (j*w)^2
ans = -w^2
```

does simplify. While j^2 evaluated to -1, this does not occur in all Symbolic Math Toolbox functions.

11.2 Variable Precision Arithmetic

In floating point arithmetic, round off errors can occur with every division, multiplication, and addition (subtraction), and conversion from binary to decimal may cause an additional round off error. Floating point arithmetic provides an accuracy of about 16 digits that may get worse, as round off errors accumulate in extensive computations.

Symbolic computation is done with rational arithmetic, which means that for rational numbers round off errors cannot occur. For example:

```
>> clear all
>> a = sym(sqrt(pi)) % since pi is irrational, the result, a, cannot be exact
a = 3991211251234741/2251799813685248
>> % this is the closest rational number to pi^1/2 using two 16 digit integers
>> b = a^2  % given a, b is exact
b = 15929767251982786841597085337081/5070602400912917605986812821504
>> c = sym(.44444)  % just another example number
c = 11111/25000
>> d = b + c  % example of rational arithmetic
d = 568229555720141373200005826335844493/1584563250285286751870879006720000 0
>> %  d requires even more digits in the numerator and denominator
>> e = double(d)  % while d is exact, e is an approximation
e =    3.5860
```

As you can see, to preserve accuracy, the number of digits required in the numerator and denominator of each rational number is not limited, except by memory space. Therefore, symbolic computation can require increasing computing time and memory space.

Variable precision arithmetic (vpa) is a compromise between floating point arithmetic and rational arithmetic. The two functions **digits** and **vpa** are concerned with variable precision arithmetic. With the function digits you can specify the minimum number of significant digits to be used in variable precision arithmetic. The syntax is given by

```
digits(d)   % specify the minimum number d of digits to be used, 1 < d < 2^29 + 1
% get the number of digits in current use, default value is d = 32
d = digits
```

The syntax of the function **vpa** is given by

```
x = vpa(expression)   % evaluate the expression using the current value of d
x = vpa(expression, d)   % evaluate the expression using d significant digits
```

For example:

```
>> clear all
>> syms w real
>> H_jw = (25532513310260969472*w*i)/ ...
(-8510837770086989*w^2+25532513310260969472*w*i + 773712524553362671
81195264);
>> % evaluate the transfer function using a minimum of 40 digits
>> %  replace w with 2*pi*15175 rad/sec
>> H_value = vpa(subs(H_jw,w,2*pi*15175),40)
H_value = 0.9999994834171082036533562131808139383793 -
          0.0007187368258941159932309816760209741914878*i
>> H_mag = vpa(abs(H_value),40)
H_mag = 0.9999997417085207235256741927690262002331
```

Note that H_value from the function vpa is of class sym, which then becomes the input to the overloaded function abs to produce the 40-digit result for H_mag. With floating point arithmetic we get

```
>> format long
>> abs(double(H_value))
ans = 0.999999741708521
```

which is a rounded version of H_mag. This difference is not consequential in this application. However, for analysis and design procedures that involve an extensive amount of arithmetic, the possibility of round off error build-up should be investigated. See Example 3.21. By applying vpa over a range of d, you can find how the accuracy of a result depends on computation precision.

11.3 Algebra

All of the arithmetic work with matrices that is described in Tables 3.1 (basic operations), 3.2 (arithmetic), 3.3 (vectorization), 3.5 (manipulation), and 3.6 (construction) can also be done with symbolic matrix objects. Also, many high-level matrix operations are included in the Symbolic Math Toolbox. Some of these functions are given in Table 11.4.

Table 11.4 Some functions concerned with matrix algebra

Function	Brief description
[Q,R] = quorem(A,B)	Returns the element by element division of A by B and the remainder
det(A)	Compute determinant of a symbolic matrix
eig(A)	Compute symbolic eigenvalues of a matrix
expm(A)	Compute symbolic matrix exponential
inv(A)	Compute symbolic matrix inverse
poly(A)	Find characteristic polynomial of a symbolic matrix
rref(A)	Compute row-reduced echelon form of a symbolic matrix
svd(A)	Find singular value decomposition of a symbolic matrix

Before going into details, let us solve a nonlinear algebraic equation.

Example 11.4

Consider again the i-v characteristic of a diode, which is repeated here for convenience.

$$i = I_s(e^{v/V_T} - 1) \tag{11.4}$$

Let us solve for v in terms of i. The function **solve** will be used to do this. The solution is found with

```
>> syms i  Is  v  VT;
>> v = solve('i = Is*(exp(v/VT) - 1)', 'v')  % place equation in quotes
v =   VT*log(i/Is + 1)
```

The first argument of the function solve is the equation we want to solve, and the last argument declares that we want to solve for v, the dependent object. Here, the result is also assigned to v. To avoid manipulating complex objects, the declaration statement should have been: syms Is VT positive; syms i v real. To have i again equal $sqrt(-1)$, use clear i.

The function **solve** has a few syntax options, and it can be used to solve a single equation and a system of equations. For a single equation, syntax options are given by

```
y = solve(f(object))
y = solve(f(object1, object2, ..., objectN),'solution_object')
```

where the solution, which is assigned to y, of the equation $f = 0$ is found. The function symvar is used to find all symbolic objects in f. If there is more than one solution, y becomes a vector. For example:

```
>> clear all
>> syms a b c x
>> f = a*x^2 + b*x + c;   % solve the equation f(x) = 0
>> % the function symvar is automatically used to find
>> % symbolic objects in f
>> X = solve(f,'x')
 -(b + (b^2 - 4*a*c)^(1/2))/(2*a)
 -(b - (b^2 - 4*a*c)^(1/2))/(2*a)
>> %   X becomes a symbolic vector with 2 elements
>> X(1)
ans = -(b + (b^2 - 4*a*c)^(1/2))/(2*a)
>> % get a value for some specific coefficients
>> x1 = subs(X(1),{a,b,c},{1 1 1})
x1 = -0.5000 - 0.8660i
>> % get a more accurate value
>> x1 = vpa(subs(X(1),{a,b,c},{1 1 1}), 40)
```

When a function involves more than one symbolic object, you must specify the object to be found. For example:

```
>> X = solve(f,'x')   % or X = solve(f,x)
```

Another syntax option is required when an equation is given instead of a function expression. To solve an equation, the syntax becomes

```
y = solve('f(object) = g(object)')
y = solve('f(object1, ...,objectN) = g(object1, ...,objectM)',
    'solution_object')
```

where f and g must be expressions. The entire equation must be placed in single quotes. See Example 11.4, where the statement to find v could also have been

```
>> v = solve(i - Is*(exp(v/VT) - 1), 'v')  % this sets the expression to zero
```

And, for example:

```
>> % find where the quadratic and straight line intersect
>> X = solve('x^2 + x + 1 = 5*x +10')
X =
13^(1/2) + 2
2 - 13^(1/2)
```

In this case there is only one solution object that is assigned to X, and the two solutions are elements of X.

To solve several equations in as many unknowns, make each equation an argument of the function solve separated by commas. Either of two syntax options can be used. These are given by

```
[Y1, ...,YK] = solve('f1(object1,..., objectN)=g1(object1, ..., objectM)', ...
                     'f2(object1,..., objectN)=g2(object1, ..., objectM)', ...
                     'f3(object1,..., objectN)=g3(object1, ..., objectM)', ...

                          . . .

                          . . .

                          . . .

                     'fK(object1,...,objectN)=gK(object1, ..., objectM)', ...
                     'solution_object1, ..., solution_objectK')
```

where there are K equations in K unknowns. The alternative to this syntax is to replace the left side of the statement with a single object that is returned as a structure, where the field names are the solution object names, which are the solutions of the equations. See Example 11.5.

Example 11.5

Let us analyze the circuit given in Fig. 11.1 to find the output V_o in terms of the two inputs V_1 and V_2.

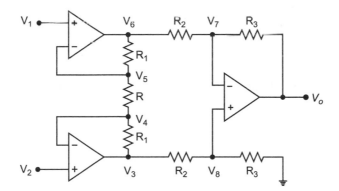

Figure 11.1 Instrumentation amplifier.

In this circuit there are seven unknown node voltages, which are V_3, V_4, V_5, V_6, V_7, V_8, and Vo. Applying KCL gives

```
(V4 - V3)/R1   +   (V4 - V5)/R   =   0
(V5 - V6)/R1   +   (V5 - V4)/R   =   0
        (V8 - V3)/R2   +   V8/R3   =   0
(V7 - V6)/R2   +   (V7 - Vo)/R3   =   0
```

where the currents into the operational amplifiers (op-amps) are set to zero. For each op-amp, which was introduced in Chapter 6, we have

```
V6  =  A(V1 - V5)
V3  =  A(V2 - V4)
Vo  =  A(V8 - V7)
```

where A, which is very large, is the gain of each op-amp. We have a total of seven equations in seven unknowns, which can be solved to find Vo with Prog. 11.1

```
clear all; clc
syms R R1 R2 R3 A positive  % declare symbolic circuit components
syms V1 V2 V3 V4 V5 V6 V7 V8 Vo real % declare symbolic voltages
% enter equations and specify the dependent variables
V = solve('(V4-V3)/R1 + (V4-V5)/R  = 0', ...
    '(V5-V6)/R1 + (V5-V4)/R  = 0', ...
    '(V8-V3)/R2 + V8/R3  = 0', ...
    '(V7-V6)/R2 + (V7-Vo)/R3  = 0', ...
    'V6  = A*(V1 - V5)  = 0', ...
    'V3  = A*(V2 - V4)', ...
    'Vo  = A*(V8 - V7)', ...
    'Vo,V3,V4,V5,V6,V7,V8');
% V is a scalar structure, with field names Vo, V3, V4, ..., V8
Vo  = V.Vo % get Vo from the structure V
Vo  = limit(Vo,A,inf)  % get Vo for large A
Vo  = simplify(Vo)  % see if MATLAB can simplify the expression for Vo
pretty(Vo)  % display a more conventional looking expression
```

Program 11.1   Program to analyze the circuit of a differential amplifier.

and the program output is given by

```
Vo  = -(A^2*R3*(R*V1 - R*V2 + 2*R1*V1 - 2*R1*V2))/((R + 2*R1 + A*R)*(R2 + R3 +
    A*R2))
Vo  = -(R3*(R*V1 - R*V2 + 2*R1*V1 - 2*R1*V2))/(R*R2)
Vo  = -(R3*(R + 2*R1)*(V1 - V2))/(R*R2)

      R3 (R + 2 R1) (V1 - V2)
  -   ──────────────────────
              R R2
```

The function pretty does not produce an expression form as we might prefer. A modified expression for V_o is given by

$$V_o = \frac{R_3}{R_2}\left(1 + 2\frac{R_1}{R}\right)(V_2 - V_1)$$

This expression shows that with the single resistor R we can control the gain of the differential amplifier. This analysis could also have been performed by setting up a symbolic matrix equation, followed by a symbolic matrix inversion.

A differential amplifier is useful, because if each input contains the same electrical noise, for example, 60 Hz noise, then this noise is canceled in the output. This common noise is called the **common mode**, and the amplifier does **common mode rejection**. A differential amplifier is used in medical and audio instrumentation and other electronic systems, where there is concern about environmental electrical noise.

To keep the problem relatively simple, idealized assumptions about each op-amp were used. A more in-depth knowledge about the electrical properties of op-amps will produce a more realistic input to output relationship. For example, the input to output relationship obtained here implies that the inputs can be signals at any frequency, where in fact, as the frequency of the inputs increases, say to 1 MHz, the op-amps start to exhibit frequency selective behavior, where their outputs begin to decrease, sort of like the action of a low-pass filter. Special additional circuitry is employed to compensate for such effects to make it possible to process signals at even higher frequencies. Such issues can make electronic circuit design challenging.

AC circuit analysis can be tedious, especially when we want a symbolic expression of a transfer function. However, with the Symbolic Math Toolbox, we must only specify the equations to solve.

Example 11.6

Consider again the circuit given in Example 6.10, where in terms of impedances the circuit is shown in Fig. 6.15, which is repeated in Fig. 11.2 for convenience. Applying KVL to the two meshes gives

$$-V_s + j\omega L_1 I_1 + \frac{1}{j\omega C}(I_1 - I_2) = 0$$

$$\frac{1}{j\omega C}(I_2 - I_1) + j\omega L_2 I_2 + R I_2 = 0$$

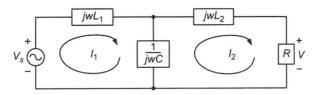

Figure 11.2   Frequency domain circuit.

Prog. 11.2 uses the function solve to solve these two equations in two unknowns to find the phasors of the two mesh currents. Then, the phasor V of the output and the transfer function from the input V_s to the output V are found. The program continues by substituting component values into the transfer function, and then the function solve is used again to find the bandwidth of the filter. Finally, the function **ezplot** is used to plot a symbolic function versus a specified symbolic object, the frequency.

```
clear all; clc
syms Vs I1 I2 V  % declare complex symbolic objects
syms R L1 L2 C positive  % declare positive component symbolic objects
syms w f real % declare real frequency symbolic objects
% solve loop equations for the current phasors
I = solve('i*w*L1*I1 + (I1 - I2)/(i*w*C) = Vs', ...
          '(I2 - I1)/(i*w*C) + i*w*L2*I2 + R*I2 = 0','I1','I2');
V = R*I.I2;  % get the phasor for the output voltage
H = V/Vs; % get the transfer function
H = simplify(H) % let MATLAB try to simplify the transfer function
pretty(H) % display an expression more like a mathematical expression
%  specify component values
R = sym(1e4);
L1 = sym(0.1492);L2 = sym(0.0497);
C = sym(0.001326e-6);
f = sym('f*(2*pi)'); % replace the frequency in rad/sec with frequency in Hz
%  substitute the component values into the transfer function
%  and replace the frequency
H = subs(H,{'R' 'L1' 'L2' 'C' 'w'},{R,L1,L2,C,f}) % H becomes a function of f
H2 = H*conj(H); % get the magnitude squared of the transfer function
H2 = expand(H2); % multiply out all terms
f = solve(H2-1/2,'f'); % solve for the bandwidth (in Hz) of the circuit
f = double(f); % convert bandwidth solution to floating point
BW = ['The bandwidth of the filter is: ',num2str(f(1)),' Hz'];
disp(BW);
ezplot(H2,[-5e4 5e4]);grid on  % plot the magnitude squared versus f
xlabel('frequency - Hz');ylabel('magnitude-squared');
title('Frequency Response');
```

Program 11.2   AC analysis of the low-pass filter given in Example 6.10.

The program output is given by

```
H =    -(R*i)/(L1*(- C*L2*w^3 + C*R*w^2*i + w) + L2*w - R*i)
```

$$
-\quad \frac{R\ i}{L1\ (-\ C\ L2\ w^3 + C\ R\ w^2\ i + w) + L2\ w - R\ i}
$$

```
H  =
-(10000*i)/(- (11886893975491556183333*pi^3*f^3)/1511157274518286468382720000000 +
   (2391729170119025389*pi^2*f^2*i)/3022314549036572936765544+(1989*pi*f)/5000
      -10000*i)
```
The bandwidth of the filter is 15999.9378 Hz

    The frequency response magnitude squared is shown in Fig. 11.3. The **bandwidth** (BW) of the circuit is the frequency where the magnitude squared, which is given by $H(j\omega)H^*(j\omega)$, has the value 1/2. This means that at $\omega = $ BW, the magnitude is down by a factor of $\sqrt{2}$ from its peak value. The magnitude squared is an even function of frequency, which makes $f(2)$ in Prog. 11.2 equal to $-f(1)$.

    Notice that the preferred symbol j for $\sqrt{-1}$ is not used in Prog. 11.2, because presently the function solve in the Symbolic Math Toolbox is not set up to do symbolic complex arithmetic with j. Instead, the conventional mathematics symbol i is used. It is significant that after component values were substituted into the symbolic transfer function, the expression for the transfer function is exact. From the program output we can write

$$H(j\omega) = \frac{R}{(R - \omega^2 L_1 RC) + j\omega((L_1 + L_2) - \omega^2 L_1 L_2 C)}$$

which is the same result found manually in Example 6.10.

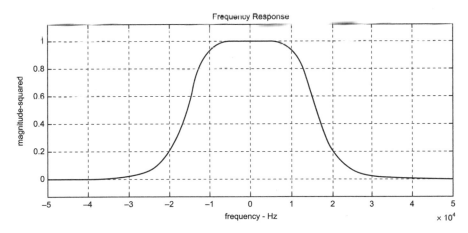

Figure 11.3 Magnitude squared frequency response of a low-pass filter.

11.4 Differentiation

The function **diff** is used to obtain the derivative of a symbolic expression. The syntax is given by

```
y  =  diff(f(object))
```

```
y = diff(f(object1, object2, ..., objectN,'deriv_object')
y = diff(f(object1, object2, ..., objectN,'deriv_object',n)
```

If the function expression f is a function of only one symbolic object, then the first syntax can be used. With the second syntax, you can specify the object with respect to which diff takes a derivative. With the third syntax you can specify the number of times diff should take a derivative. For example:

```
>> syms b c t w x real; syms a positive
>> f = exp(-a*t)   % decaying exponential for a > 0
f = 1/exp(a*t)
>> g = diff(f,t)
g = -a/exp(a*t)   % f has a negative slope
>> h = sin(w*t)*f  % exponentially decaying sinusoid
h = sin(t*w)/exp(a*t)
>> p = diff(h,t)
p = (w*cos(t*w))/exp(a*t) - (a*sin(t*w))/exp(a*t)
>> q = diff(1/t)
q = -1/t^2
>> r = diff(log(x))
r = 1/x
>> s = diff(sin(t)/t)   % derivative of the sinc function
s = cos(t)/t - sin(t)/t^2
>> diff(a*t^4 + b*t^3 + c*t^2,t,3)   % take third derivative with respect to t
ans = 6*b + 24*a*t
```

The function **limit** can be used to find the derivative of a function. Let us use the following definition of a derivative:

$$\frac{df(x)}{dx} = \lim_{h \to 0} \frac{f(x+h/2) - f(x-h/2)}{h} \tag{11.5}$$

This is called a **two-sided derivative**. For any particular point $x = x_0$, it gives the slope of the tangent at that point. Other possible derivative definitions can be

$$\frac{df(x)}{dx} = \lim_{h \to 0} \frac{f(x) - f(x-h)}{h} \quad \text{and} \quad \frac{df(x)}{dx} = \lim_{h \to 0} \frac{f(x+h) - f(x)}{h} \tag{11.6}$$

which are called **left-sided** and **right-sided** derivatives, respectively. A two-sided derivative of a function at a point $x = x_0$ exists (meaning defined and finite) if and only if the function is continuous in the neighborhood of $x = x_0$ and the left-sided and right-sided derivatives exist and are equal at $x = x_0$.

Let $f(x) = ax + b$, and we get

```
>> clear all
>> syms a x b h
```

```
>> f  =  a*x+b;   % define a function
>> f_plus  =  subs(f,x,x+h/2)    % function to the left of x
f_plus  =  b + a*(h/2 + x)
>> f_minus  =  subs(f,x,x-h/2)   % function to the right of x
f_minus  =  b - a*(h/2 - x)
>> limit((f_plus - f_minus)/h,h,0)   %  apply derivative definition
ans  =  a
```

Let $f(x) = \exp(-ax)$, and we get

```
>> f  =  exp(-a*x);
>> f_plus  =  subs(f,x,x+h/2)
f_plus  =  1/exp(a*(h/2 + x))
>> f_minus  =  subs(f,x,x-h/2)
f_minus  =  exp(a*(h/2 - x))
>> limit((f_plus - f_minus)/h,h,0)
ans  =  -a/exp(a*x)
```

Consider the function $u(x)$ defined by

$$u(x) = \frac{1}{2}(1 + x/|x|) = \begin{cases} 1, & x > 0 \\ 0, & x < 0 \end{cases} \tag{11.7}$$

This function is called the **unit step function**. If the argument x of $u(x)$ is positive, then the function value is 1, and if the argument of $u(x)$ is negative, then the function value is 0. The unit step function is commonly used to turn other functions on and off.

If the unit step function is shifted, as with $y(x) = u(x - x_0)$, then $y(x) = 0$ for $x < x_0$, and $y(x) = 1$ for $x > x_0$. The unit step function can be reversed with $y(x) = u(-x)$, and then $y(x) = 1$ for $x < 0$ and $y(x) = 0$ for $x > 0$.

Example 11.7 _____

Let us use the unit step function to define a function of time that is just one cycle of a sinusoidal function of time. Given is the sinusoidal function $x(t) = \sin(\omega_0 t)$. Its period is $T_0 = 2\pi/\omega_0$ seconds. Let the function $g(t)$ be defined by

$$g(t) = u(t) - u(t - T_0) = \begin{cases} 1, & 0 < t < T_0 \\ 0, & t < 0 \text{ and } t > T_0 \end{cases} \tag{11.8}$$

Therefore, $y(t) = x(t) g(t)$ is zero for $t < 0$ and $t > T_0$. The function $g(t)$ is called a **gate function**. For $0 < t < T_0$, $y(t) = x(t)$. Prog. 11.3 evaluates and plots, using ezplot, the symbolic function $y(t)$. The program output is shown in Fig. 11.4.

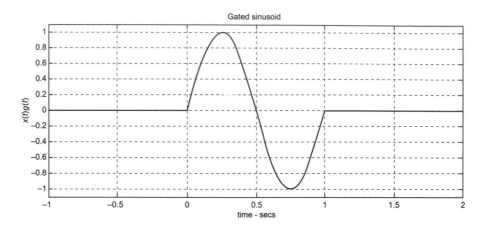

Figure. 11.4   A gated sinusoidal signal.

```
clear all; clc
syms w0 t T0 real  % declare symbolic objects
w0  =  sym(2*pi); % symbolic frequency
T0  =  sym(w0/(2*pi)) % symbolic period
u  =  1/2*(1+t./abs(t)) % unit step function
uT0  =  1/2*(1+(t-T0)./abs(t-T0)) % time shifted unit step function
g  = u - uT0  % gate function
x  =  sin(w0*t)
y  =  x.*g  % gated sinusoidal signal
ezplot(y,[-1 2])  % using ezplot to plot a symbolic function
xlabel('time - secs')
ylabel('x(t)g(t)')
title('Gated sinusoid')
grid on
```

Program 11.3   Demonstration of using the unit step function to gate a signal.

The unit step function $u(t)$ is discontinuous at $t = 0$. Let us apply the function **limit** to see what it produces as t goes to zero. For example:

```
>> syms t real
>> limit((1/2)*(1+t/abs(t)),t,0)
ans  =  NaN
>> %  NaN means not a number
```

```
>> limit((1/2)*(1+t/abs(t)),t,0,'left')   % find the limit from the left
ans =   0
>> limit((1/2)*(1+t/abs(t)),t,0,'right')   % find the limit from the right
ans =   1
>> % at t = 0, u(t) is not defined
```

Since the left and right limits are different, the derivative at $t = 0$ does not exist. If (11.5) is applied, we get

$$\frac{du(t)}{dt}\bigg|_{t=0} = \lim_{h \to 0} \frac{1-0}{h} \to \infty$$

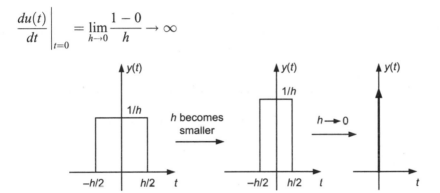

Figure 11.5   Limiting process to define the impulse function $\delta(t)$

To work with this dilemma, a special function is defined based on the function $y(t)$ described in Fig. 11.5, where

$$\int_{-\infty}^{+\infty} y(t)dt = 1, \ \forall h \tag{11.9}$$

and

$$\lim_{h \to 0} y(t) = \infty$$

When $h \to 0$, we have

$$y(t) = \delta(t) = \begin{cases} \infty, & t = 0 \\ 0, & t \neq 0 \end{cases} \tag{11.10}$$

and with (11.9) we get

$$\int_{-\infty}^{t} \delta(\tau)d\tau = \begin{cases} 0, & t < 0 \\ 1, & t > 0 \end{cases} = u(t) \tag{11.11}$$

The function $\delta(t)$ is called the **impulse function** (also called the **Dirac delta function**), and with (11.11) we have

$$\frac{du(t)}{dt} = \delta(t) \tag{11.12}$$

signifying that at $t = 0$ the slope of $u(t)$ is infinite. The function $\delta(t)$ is located at $t = 0$. It can be located anywhere on the timescale by shifting it, as with $\delta(t - t_0)$, which is an impulse function located at $t = t_0$.

The impulse function $\delta(t)$ occurs for zero time, and when it occurs it is infinite. Physical phenomena cannot behave this way. However, the integral of the impulse function is finite, and its properties are widely utilized in engineering.

Example 11.8

Consider the i-v relationship of a capacitor, shown in Fig. 6.10, where

$$i(t) = C\frac{dv(t)}{dt}$$

Suppose that the voltage across a capacitor changes suddenly. For example, this can happen when a switch is closed at some time, say $t = t_0$, to connect an electrical power source, which, for example, produces V volts, to a circuit. It is common practice to model this with

$$v(t) = Vu(t - t_0)$$

where $v(t) = 0$ before the switch is closed and $v(t) = V$ after the switch is closed. According to the i-v relationship of a capacitor, the current is given by

$$i(t) = CV\delta(t - t_0)$$

This means that at the time $t = t_0$ an infinite current will occur for zero time. Physically, the voltage cannot change instantaneously, and the current cannot become infinite for zero time. However, the voltage may change in a small amount of time, and the current, depending on the capabilities of the electrical power source, can be large for a small amount of time. Such a current surge could cause some device in the circuit to fail.

Now consider the i-v relationship of an inductor, shown in Fig. 6.10, where

$$v(t) = L\frac{di(t)}{dt}$$

If at some time $t = t_0$, the current changes suddenly, from $i(t) = I$ to $i(t) = 0$, then this can be modeled with

$$i(t) = Iu(-t + t_0) = \begin{cases} I, & t < t_0 \\ 0, & t > t_0 \end{cases}$$

According to the i-v relationship of an inductor, the voltage across the inductor is given by

$$v(t) = -LI\delta(-t + t_0)$$

This means that an infinite reverse voltage will occur for zero time. Physically, the current cannot change instantaneously, and the voltage cannot become infinite for zero time. However, the current may change in a very small amount of time, and the voltage can be large for a small amount of time. Such a voltage surge could cause some device in the circuit to fail.

Preventing current and voltage surges in electrical machines, appliances, medical instrumentation, digital circuits, and other electrical systems can make electronic circuit design challenging.

11.5   Integration

Given a function $f(x)$, let $F(x)$ denote the indefinite integral of $f(x)$ given by

$$F(x) = \int f(x)dx \tag{11.13}$$

Then, $f(x) = dF(x)/dx$, and $F(x)$ is also called the **antiderivative** of $f(x)$. The MATLAB function **int** can be used to try to find $F(x)$. The syntax is given by

```
F  =  int(f(object))
F  =  int(f(object1, object2,  ..., objectN),int_object)
```

The function int may not be able to find $F(x)$ of some $f(x)$ because

- $F(x)$ does not exist in closed form
- $F(x)$ is an unfamiliar function
- the int function program cannot find $F(x)$
- there is not enough computing time or memory space to find $F(x)$.

When this occurs, the function int returns, "int(f)". The syntax for a definite integral of $f(x)$ from $x = a$ to $x = b$ is given by

```
      F  =  int(f(object),a,b)
      F  =  int(f(object1, object2, ..., objectN),int_object,a,b)
```

For example:

```
>> syms x t w a b n
>> j = sym(i);
>> f = x^n;
>> f = subs(f,n,-1)
f = 1/x
>> F = int(f)
F = log(x)
>> f = x^n;  F = int(f,x)
F = piecewise([n = -1, log(x)], [n <> -1, x^(n + 1)/(n + 1)])
>> f = subs(f,n,2);  F = int(f)
F = x^3/3
>> f = sin(w*t);  F = int(f,t)
F = -cos(t*w)/w
f = exp(j*w*t);  F = int(f,t)
F = -(exp(t*w*i)*i)/w
>> % integral of cos(wt) using Euler's identity
>> f = (exp(j*w*t) + exp(-j*w*t))/2; F = int(f,t)
F = sin(t*w)/w
>> % definite integral of the sinc function
>> f = sin(t)/t; F = int(f,-inf,inf)
F = pi
>> % Gaussian probability density function
>> f = exp(-x^2/2)/sqrt(2*pi)
f = 2251799813685248/(5644425081792261*exp(x^2/2))
% definite integral of the Gaussian probability density function
>> F = int(f,-inf,inf)
F = (2251799813685248*2^(1/2)*pi^(1/2))/5644425081792261
>> double(F)  % probability of any outcome
ans = 1
>> a = sym(0);  b = sym(inf);
>> f = exp(-2*t);  int(f,a,b)
ans = 1/2
>> u = (1+t/abs(t))/2;  % unit step function
>> % unit step function shifted to the right by 1 second
>> uT1 = subs(u,'t','t-1')
uT1 = (t - 1)/(2*abs(t - 1)) + 1/2
```

```
>> % unit step function time shifted to the right by 3 seconds
>> uT3 = subs(u,t,t-3)
uT3 = (t - 3)/(2*abs(t - 3)) + 1/2
>> g = uT1 - uT3  %  gate function from t = 1 to t = 3
>> int(g,0,4)  % integral from t = 0 to t = 4 of the gate function
ans = 2
```

Example 11.9

Depending on the given periodic signal $x(t)$, manually finding the Fourier series coefficients with

$$X_k = \frac{1}{T_0} \int\limits_{t_0}^{t_0 + T_0} x(t)e^{-jk\omega_0 t}\, dt \qquad (11.14)$$

can be a challenging task. Let us use the function int to find the X_k.

First, we will find the X_k of a signal, where we can easily find them. Let $x(t)$ be given by

$$x(t) = 3\cos(2\omega_0 t + \pi/4) - 7\sin(5\omega_0 t - \pi/3)$$

for some fundamental frequency ω_0. This signal is periodic with period $T_0 = 2\pi/\omega_0$. Applying Euler's identity gives

$$
\begin{aligned}
x(t) &= 3\frac{e^{j(2\omega_0 t + \pi/4)} + e^{-j(2\omega_0 t + \pi/4)}}{2} - 7\frac{e^{j(5\omega_0 t - \pi/3)} - e^{-j(5\omega_0 t - \pi/3)}}{j2} \\
&= \frac{7}{j2}e^{j\pi/3}e^{-j5\omega_0 t} + \frac{3}{2}e^{-j\pi/4}e^{-j2\,\omega_0 t} + \frac{3}{2}e^{j\pi/4}e^{j5\omega_0 t} - \frac{7}{j2}e^{-j\pi/3}e^{j5\omega_0 t}
\end{aligned}
$$

and by inspection we have: $X_{-5} = 7e^{j\pi/3}/j2$, $X_{-2} = 3e^{-j\pi/4}/2$, $X_2 = X_{-2}^*$ and $X_5 = X_{-5}^*$. In rectangular form, the coefficients are given by

```
>> X_2 = conj(3*exp(-j*pi/4)/2)
X_2 =   1.0607 + 1.0607i
>> X_5 = conj(7*exp(j*pi/3)/(j*2))
X_5 =   3.0311 + 1.7500i
```

These coefficients are also found with Prog. 11.4, and the program output follows.

```
% Program to find the Fourier series coefficients of a periodic signal
clear all; clc
syms t k real; % declare symbolic objects
j = sym(i);
w0 = sym(10*pi); % declare some fundamental frequency
```

```
T0  =  sym(2*pi/w0);  % period
x  =  3*cos(2*w0*t + pi/4) - 7*sin(5*w0*t - pi/3); % specify signal
e  =  exp(-j*k*w0*t); % exponential part of X_k integral
Xk  =  int(x*e,t,0,T0)/T0; % integral for Fourier series coefficients
Xk  =  simple(Xk) % simplify as best as possible and display result
N  =  10; % get coefficients for k = -N to k = +N
K  =  -N:N; % vector of index values
X_k  =  zeros(1,2*N+1); % preallocate space
X  =  sym('X%d%d',[1 2*N+1]); % preallocate symbolic space
for n = 1:2*N+1 % loop to get values of Fourier series coefficients
    X(n)  =  limit(Xk,k,K(n)); % get value for k = K(n)
    X_k(n)  =  double(X(n)); % convert symbolic value to floating point
end
Coeffs  =  [K;real(X_k);imag(X_k)]; % set up output for display
disp('The Fourier series coefficients are:')
disp('k    real part  imaginary part')
fprintf('%i    %f    %f \n',Coeffs)
```

Program 11.4   Program to find Fourier series coefficients with (11.14).

Xk =   (7*(- 5 + 3^(1/2)*k*i)*(1/exp(2*pi*k*i) - 1))/(4*pi*(k - 5)*(k + 5)) + (3*2^(1/2)*(- 2 + k*i)*(1/exp(2*pi*k*i) - 1))/(4*pi*(k - 2)*(k + 2))

The Fourier series coefficients are as follows:

k	real part	imaginary part
-10	0.000000	0.000000
-9	0.000000	0.000000
-8	0.000000	0.000000
-7	0.000000	0.000000
-6	0.000000	0.000000
-5	3.031089	-1.750000
-4	0.000000	0.000000
-3	0.000000	0.000000
-2	1.060660	-1.060660
-1	0.000000	0.000000
0	0.000000	0.000000
1	0.000000	0.000000
2	1.060660	1.060660
3	0.000000	0.000000

```
 4          0.000000     0.000000
 5          3.031089     1.750000
 6          0.000000     0.000000
 7          0.000000     0.000000
 8          0.000000     0.000000
 9          0.000000     0.000000
10          0.000000     0.000000
```

Prog. 11.4 becomes more useful for periodic signals with Fourier series coefficients that are not easy to obtain manually with (11.14). Consider again the sawtooth signal shown in Fig. 6.5. One period of this signal is given by

$$x(t) = t + 1/4, \quad -1/4 \leq t < 3/4$$

Using this $x(t)$ in Prog. 11.4, where $T_0 = 1$, results in

```
Xk  =   ((1/exp((3*pi*k*i)/2))*(1 - exp(2*i*k*pi) + 2*pi*k*i))/
        (4*pi^2*k^2)
```

The Fourier series coefficients are as follows:

```
  k      real part      imaginary part
-10      0.000000        0.015915
 -9     -0.017684        0.000000
 -8      0.000000       -0.019894
 -7      0.022736        0.000000
 -6      0.000000        0.026526
 -5     -0.031831        0.000000
 -4      0.000000       -0.039789
 -3      0.053052        0.000000
 -2      0.000000        0.079577
 -1     -0.159155        0.000000
  0      0.500000        0.000000
  1     -0.159155        0.000000
  2      0.000000       -0.079577
  3      0.053052        0.000000
  4      0.000000        0.039789
  5     -0.031831        0.000000
  6      0.000000       -0.026526
  7      0.022736        0.000000
  8      0.000000        0.019894
```

```
 9     -0.017684     0.000000
10      0.000000    -0.015915
>>
```

The expression for X_k can be simplified further to become

$$X_k = \frac{j}{2\pi k} e^{-j3\pi k/2} = \frac{j}{2\pi k} e^{-j(2\pi - \pi/2)k} = \frac{j}{2\pi k} e^{-j2\pi k} e^{j\pi k/2} = \frac{j}{2\pi k} e^{j\pi k/2}$$

which agrees with the results given in Example 6.8.

11.6   Conclusion

The Symbolic Math Toolbox contains a wide variety of functions that can be applied to problems where we would like to retain parameter symbols to better see their effect in problem solutions. This is especially useful to see how a design procedure can be formulated. Now that you have completed this chapter, you should know how to symbolically

- declare the class of a scalar or matrix object
- create expressions
- work with rational arithmetic
- use many of the special symbolic functions, for example, simplify, limit, subs, and many more
- do variable precision arithmetic
- solve systems of algebraic equations
- do DC and AC circuit analysis
- differentiate
- plot a function
- integrate
- find the Fourier series coefficients of a periodic signal, and
- know how to work with the unit step and impulse functions.

In the next chapter, we will apply some of the material from this and previous chapters to more advanced problem areas, including continuous and discrete time signal analysis and continuous and discrete time system analysis.

Problems

Section 11.1

1) (a) Using the command format, give a MATLAB statement that declares the variables var1 and xdot to be symbolic objects.
   (b) Repeat part (a) using the function format.

2) For each of the following MATLAB statements, assume the workspace is clear. Give or explain the result. If there is an error, explain what is wrong. If possible, give a corrected statement.

(a) syms a, b, c

(b) syms ('a', 'b')

(c) syms a; b = real(a); c = conj(a)

(d) a = 2.36; syms a

(e) a = sym(3.1/0.43); b = ceil(a); c = char(a); d = double(a); e = double(c); size(e)

(f) a = sym(−3.13/5.77); b = fix(a); c = frac(a)

(g) a = sym(2.6^0.45)

(h) A = sym(A,3)

3) Repeat Prob. 11.2.

(a) r = sym(s)

(b) t = sym('x'); u = t/sym(v); w = t/x; y = t/'x'

(c) a = sym(2^8 - 1); a/2; b = a/2; c = a/'2'; d = char(a); e = d/2; e = d/'2'

(d) a = sqrt(3); b = sym(sqrt(3)); c = a/double(sqrt(b)); d = a/sqrt(b)

4) Give a MATLAB statement that creates a

(a) real symbolic variable x.

(b) 5 by 2 matrix A of scalar symbolic objects.

(c) For the result of part (b), give a MATLAB statement that assigns the second column of A to b.

5) What is the result of the following MATLAB statements?

(a) y = sym('x' + 1); char(y)

(b) syms w; x = sym(w + 1); y = sym(x^2); z = x^2

(c) syms a b; c = sym(a + 3*b^2); d = a + 3*b^2; e = d^2 - c; symvar(e)

(d) syms A a t; x = sym(A*exp(a*t)); w = A*exp(a*t); y = x^−2

(e) W = sym(exp(i*pi/3)); double(W^2) % in symbolic math, i is used for sqrt(−1)

6) Give MATLAB statements to create a symbolic function for

(a) $y = at^2 + 2t + 3$

(b) $y = e^{at}\sin(wt + b)$

(c) $y = (a + b)^3 + c(a + b)^2 + d(a + b) + e$

7) (a) Create symbolic functions for (1) $z = (at + b)$, (2) $w = e^{iz} + e^{-iz}$, and (3) $y = \sin(z) + \cos(z)$.

(b) Apply the function **collect** to obtain $f = z^3 + 2z + 3$ as a polynomial in t. Then, use the function **pretty**.

(c) In y, use the function **subs** to set $a = \pi$ and $b = \pi/3$, and then apply the function **ezplot** to plot y^2 over $0 \le t \le 2$.

(d) Get the function $g = w^3$, and apply the function **simplify** to g.

(e) Apply the function **expand** to g of part (d), and then apply the function simplify.

(f) Let $r = a*t$, and apply the function **subexpr** to the result of part (e).

8) Write a MATLAB script to create a symbolic chirp signal $x(t) = \sin(zt)$, where $z = at + b$. Use ezplot to display $x(t)$ over $0.1 \leq t \leq 2$ sec, where the frequency changes over $0.1 \leq f \leq 3$ Hz. Use the function subs to replace the symbolic objects a and b with values. Provide a copy of the program and the plot.

9) (a) Give MATLAB statements to define the symbolic function $y = 1/x$. Apply the function **limit** to find $y(x \to 0)$, $y(x \to 0)$ from the left, $y(x \to 0)$ from the right, $y(x \to -\infty)$ and $y(x \to \infty)$.

(b) Given is the vector of coefficients $a = [2\ -3\ 0\ 5]$. Using the function **poly2sym**, give MATLAB statements to obtain a symbolic polynomial function f of the variable x.

(c) Given are A = sym('A%d%d',[2 2]) and B = sym('B%d%d',[2 2]). Use the subs function to assign [14 7;2 43] to A and [2 3;4 5] to B. Explain each element in the result C = mod(A,B).

10) Manually, prove that the sum of a geometric series is given by (11.2).

11) Write a MATLAB script that creates a symbolic expression for $W = e^{-i2\pi/N}$. Continue the script to use the function **symsum** to find

$$h(n) = \sum_{k=0}^{N-1} W^{nk}$$

Clarify the piecewise result.

12) Write a MATLAB script that declares x, k, and N to be symbolic objects and uses a **for loop** to find $x(N)$ given by

$$x(N) = \sum_{k=1}^{N} \frac{1}{k^2}$$

for $N = 10$, 100, and 1000. Compare each x with $\pi^2/6$. Provide copies of your program and results.

13) Using the function symsum, write a MATLAB script that finds a symbolic expression for $H(z)$, where

$$H(z) = \sum_{n=0}^{\infty} nz^n$$

If z can be a complex number, then what restriction must be imposed on z for the series to converge?

14) Write a MATLAB script to find a symbolic expression for the Taylor series expansion of $y = f(x) = 1/(x^3 + 1)$ for

(a) $n = 9$ and $x_0 = 0$, and name the result w, and

(b) $n = 9$ and $x_0 = 1$, and name the result z.

(c) Compare y, w, and z for $x = 0 : 0.1 : 2$. Use the function **sym2poly** to obtain numeric coefficients for w and z. Provide plots of $y(x) - w(x)$ and $y(x) - z(x)$. Are the results of parts (a) and (b) useful for all x? Will the series approximation improve if n is increased, for example, $n = 51$?

Section 11.2

15) (a) Give MATLAB statements that (1) specify that 20 digits must be used, (2) construct the symbolic object $x = 5\cos(\omega t - \pi/8)$, where ω (use w) is declared to be a positive symbolic object and t is declared to be a real object, and (3) use the function **vpa** to evaluate x for $\omega = 4\pi$ rad/sec and $t = 1/16$ sec.

(b) Repeat part (a) for $x = (X/2)e^{j\omega t} + (X^*/2)e^{-j\omega t}$, where $X = 5e^{-j\pi/8}$.

16) Rational arithmetic can give more accurate results than floating point numeric arithmetic. To demonstrate this, find the inverse matrix of a matrix. With $\delta = 1e\text{-}8$, let A be given by

$$A = \begin{bmatrix} 1+\delta & 1-\delta & 1+\delta \\ 1 & 1+\delta & 1-\delta \\ 1 & 1 & 1+\delta \end{bmatrix}$$

This matrix is ill-conditioned. Write a MATLAB script that starts as follows:

```
clear all; clc
format long
d  =  1e-8;   % used for single precision error
A  =  [1+d   1-d   1+d;1   1+d   1-d;1   1   1+d]
AI  =  inv(A)
B  =  AI*A   %  ideally, B should be the identity matrix
```

(a) Continue the script by constructing a symbolic object of A, name it C. Then obtain CI, the inverse of C, and get $D = CI * C$. Compare B and D. What is the most number of digits used to express an element of CI? Use the function whos to find the memory space occupied by AI and CI.

(b) Rational arithmetic can take up much memory space. To see how memory requirements can be controlled, continue the script by constructing a symbolic object of A through the function **vpa** using 20 digits, name it E. Then, continue to use the function vpa to find EI, the inverse of E and $F = EI * E$. Compare B and F.

(c) Repeat part(b) using 30 digits.

Provide copies of your program, all matrices, and discussion concerned with results.

Section 11.3

17) Declare a matrix A to be A $=$ sym('A%d%d',[2 2]). Give statements to find
   (a) d $=$ determinant of A
   (b) B $=$ inverse matrix of A
   (c) p $=$ characteristic polynomial of A
   (d) L $=$ eigenvalues of A. Use the function subs to replace A in L with
       $A = [1\ -1;2\ 1]$.
   (e) $A./C$ and remainder, where C $=$ sym('C%d%d',[2 2])

18) Given is the op-amp circuit shown in Fig. P11.18. KCL gives

$$\frac{V_a}{R_1} + \frac{V_a - V_o}{R_2} = 0$$

The op-amp gain equation is

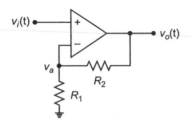

$$V_o = -A(V_a - V_i)$$

Figure P11.18   Non-inverting amplifier.

   (a) Write a MATLAB script that uses the function **solve** to solve the two equations for
       V_a and V_0. Write the KCL equation as $V_a/R_1 = (V_0 - V_a)/R_2$. Place the solution in
       a structure V.
   (b) Then, let $V_o = V.V_0$, and use the function **limit** to find V_0 as $A \to \infty$.

19) Write a MATLAB script that uses the function solve to solve the two equations in two
   unknowns given by $2a - 1/b = 0$ and $a - 3b = 1$, for a and b.

20) Given is the RLC circuit shown in Fig. P11.20.
   (a) The input $v_s(t)$ is a sinusoid with phasor V_s. Using impedances of the components,
       and phasors V_R and V_C for the node voltages, write two node equations.
   (b) Write a MATLAB script that uses the function solve to solve the two equations in
       two unknowns for the node voltage phasors V_R and V_C, and place the result in a
       structure V. Let $V_C = V.V_C$, and find the transfer function $H(j\omega)$.

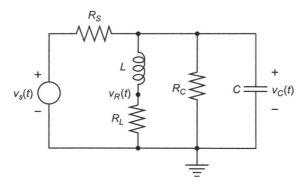

Figure P11.20 RLC circuit.

(c) Substitute in the component values $R_L = 10\ \Omega$, $L = 10\ \text{mH}$, $C = 0.1\ \mu\text{F}$, $R_C = 1\ \text{M}\Omega$, and $R_S = 500\ \Omega$ into $H(j\omega)$, and use the function ezplot to plot the transfer function magnitude squared over $-20\ \text{KHz} \le f \le 20\ \text{KHz}$. What kind of filtering activity does this circuit do?

Provide copies of your script, output (transfer function expressions), and the frequency response plot.

Section 11.4

21) (a) Write a MATLAB script that starts by defining the symbolic object $y = \tan(x)$. Use the function **diff** to obtain $dy\_dx = dy/dx$. Use ezplot to plot y and dy_dx.

(b) Continue the script to apply the two-sided derivate definition, and use the function **limit** to find dy_dx as $h \to 0$.

22) Over one period $T_0 = 4$, a period $x_p(t)$ of a periodic signal $x(t)$ is given by

$$x_p(t) = \begin{cases} 0, & t < 0 \\ 1, & 0 < t < 1 \\ -2, & 1 < t < 2 \\ 3, & 2 < t < 4 \\ 0, & t > 4 \end{cases}$$

(a) Over $-4 < t < 8$, give a sketch of $x(t)$.

(b) Using unit step functions, give an expression for $x_p(t)$.

(c) In terms of the impulse functions, give an expression for the derivative of $x_p(t)$.

23) (a) If the voltage across a capacitor changes significantly (going up or going down) in a very short amount of time, then describe what the current through the capacitor must be like in each case.

(b) If the current through an inductor changes significantly (going up or going down) in a very short amount of time, then describe what the voltage across the inductor must be like in each case.

24) Consider the definition of the unit step function given by (11.7). Declare x to be a real symbolic object. Use ezplot to plot $u(x)$. Use the function diff to obtain $d(x)$, the derivative of $u(x)$. Use the function subs to obtain values of $u(x)$ and $d(x)$ for $x = -0.1, 0$, and 0.1. Explain the results.

Section 11.5

25) Write a MATLAB script that uses the function int to find the Fourier series coefficients X_k of the periodic function $x(t)$ described in Prob. 11.22. Use the function simplify to see if the results can be simplified. Hint: Use the sum of three integral operations. Provide copies of your program and the Fourier series coefficients.

26) Consider the integral

$$Y(j\omega) = \int_{-\infty}^{\infty} y(t)e^{-j\omega t} dt$$

This is called the **Fourier transform** of the time function $y(t)$. It plays a very important and fundamental role in signal and system analysis. Write a MATLAB script that uses the function int to find the Fourier transform of

(a) $y(t) = e^{-at}u(t)$, $a > 0$. Hint: You can omit the unit step function by changing the lower limit to zero. Explain the result.

(b) $y(t) = u(t + T/2) - u(t - T/2)$. Use subs to replace T with $T = 2$, and use ezplot to plot $Y(j\omega)$ for $-4\pi \le \omega \le 4\pi$.

27) (a) Write a MATLAB script that produces $x(t)$, which is four cycles of a sinusoidal pulse train. Each cycle is one cycle of either $\sin(wt)$ or $\sin(wt+\pi/2)$. The sinusoid to be used in each cycle is determined by four bits $B = [b_0, b_1, b_2, b_3]$. Therefore, each cycle of $x(t)$ can be given by $\sin(wt + b_k\pi /2)$. The period T_0 of each cycle is given by $T_0 = 1/f$ sec. For $B = [1\ 0\ 0\ 1]$, create $x(t)$. Use $w = 2\pi f$, where $f = 1K$ Hz. Use ezplot to plot $x(t)$ over $0 \le t \le 4$ msec.

(b) Given $x(t)$, let $y(t) = x(t)\sin(wt)$, and use the function int to integrate $y(t)$ over one cycle at a time, and get

$$c_0 = \int_0^T y(t)dt, \quad c_1 = \int_T^{2T} y(t)dt, \dots, \quad c_3 = \int_{3T}^{4T} y(t)dt, \quad T = T_0$$

Given $C = [c_0, c_1, c_2, c_3]$, can you determine B?

28) Consider the integral

$$Y(s) = \int_{-\infty}^{\infty} y(t)e^{-st}dt$$

This is called the **Laplace transform** of the time function $y(t)$. It plays a very important and fundamental role in system analysis. Write a MATLAB script that uses the function **int** to find the Laplace transform of

(a) $y(t) = \cos(\omega t)u(t)$. Hint: You can omit the unit step function by changing the lower limit to zero.

(b) $y(t) = e^{-at}\cos(\omega t)u(t)$, where $a > 0$.

CHAPTER 12

Signals and Systems

In the previous chapters, many built-in functions were introduced that are concerned with fundamental mathematics used to solve practical problems. MATLAB® also includes many more built-in functions that are concerned with solving higher level mathematical problems that occur in a variety of fields of science and engineering. Some of these functions and the mathematical background to understand the meaning of results returned by these functions are presented in this chapter.

In this chapter, you will learn how to use MATLAB to:

- do spectral analysis of stationary and non-stationary signals
- find the transient and steady-state response of linear and time invariant continuous and discrete time systems
- obtain a state variable description of a system
- apply ordinary differential equation solvers
- find the frequency response of linear and time invariant systems
- operate discrete time systems

12.1   Signal Analysis

An important class of signals is periodic signals, for which

$$x(t) = x(t + T_0)$$

for some period T_0 and all t. In Chapter 6 we found that $x(t)$ can be written as a **complex exponential Fourier series** given by

$$x(t) = \sum_{k=-\infty}^{\infty} X_k e^{jk\omega_0 t} \tag{12.1}$$

where $\omega_0 = 2\pi/T_0$ rad/sec is called the **fundamental frequency**. The X_k, called the **complex Fourier series coefficients**, are found with

$$X_k = \frac{1}{T_0} \int_0^{T_0} x(t)\, e^{-jk\omega_0 t}\, dt \qquad (12.2)$$

for $k = -\infty, \ldots, -1, 0, 1, \ldots, \infty$. Equation (12.2) is called the **analysis equation.** With X_k, $x(t)$ can be reconstructed using (12.1), which is called the **reconstruction equation,** and it is said that the X_k are a **frequency domain description** of the signal. The idea of writing a periodic time function or signal in terms of a sum of complex exponential functions is a fundamental concept that is widely applied in engineering and physics. If $x(t)$ satisfies the **Dirichlet conditions**, which are that $x(t)$ must

1)   have a finite number of extrema in any given time interval
2)   have a finite number of discontinuities in any given time interval
3)   be absolutely integrable over a period

then we make no distinction between a given periodic signal $x(t)$ and its Fourier series representation. All practical (can be physically synthesized or occur naturally) periodic signals satisfy these conditions.

If $x(t)$ is discontinuous at a time point $t = t_d$, then for $t = t_d$ the Fourier series converges to the average value of $x(t)$ about $t = t_d$, which is $(x(t_d^-) + x(t_d^+))/2$. For example, see Fig. 6.6.

The X_k are in general complex numbers, which can be written in polar form to get

$$X_k = \|X_k\| e^{j\angle X_k}$$

Recall that $X_{-k} = X_k^*$, and then $\|X_{-k}\| = \|X_k\|$, which makes $\|X_k\|$ an even function of k, and $\angle X_{-k} = -\angle X_k$, which makes $\angle X_k$ is an odd function of k. Using Euler's identity, we found in Chapter 6 that (12.1) becomes

$$x(t) = \|X_0\| + 2\sum_{k=1}^{\infty} \|X_k\| \cos(k\omega_0 t + \angle X_k) \qquad (12.3)$$

This shows that $2\|X_k\|$ gives the amplitude and $\angle X_k$ gives the phase of the sinusoidal component of $x(t)$ at the frequency $\omega = k\omega_0$ rad/sec. To conveniently assess the nature of a signal $x(t)$, the magnitude and angle of X_k are plotted versus k. The number sequence X_k is also called the **spectrum** of $x(t)$. See Example 6.9.

Recall that in Chapter 6, it was found that the average power of a periodic signal is given by

$$P = \frac{1}{T_0} \int_0^{T_0} x^2(t)\,dt = \sum_{k=-\infty}^{\infty} X_k X_{-k} = \sum_{k=-\infty}^{\infty} \|X_k\|^2 \tag{12.4}$$

Therefore, $\|X_k\|^2$ gives the average power of each sinusoidal component of $x(t)$. A plot of $\|X_k\|^2$ versus k (frequency) shows the **power spectrum** of $x(t)$.

To be practical, we cannot use an infinite number of terms in (12.1) or (12.3) to reconstruct $x(t)$. We must truncate the series to use a finite number terms, and write

$$x(t) \cong \sum_{k=-M}^{+M} X_k e^{jk\omega_0 t} = \|X_0\| + 2\sum_{k=1}^{M} \|X_k\|\cos(k\omega_0 t + \angle X_k) \tag{12.5}$$

for some positive integer M. This makes the approximation of $x(t)$ a **bandlimited signal**, where the sinusoidal component with the highest frequency has the frequency given by $M\omega_0$ rad/sec, which is called the **bandwidth** (BW) of the approximation of $x(t)$.

Example 12.1

Find the complex Fourier series coefficients of

$$x(t) = 6\cos\left(\omega_a t + \pi/4\right) + 4\sin\left(\omega_b t - \pi/3\right)$$

If $x(t)$ is periodic, then the frequency of each sinusoidal component of $x(t)$ must be an integer multiple of a fundamental frequency ω_0, which requires that for some integers r_1 and r_2

$$\begin{aligned} \omega_a &= r_1\omega_0 \\ \omega_b &= r_2\omega_0 \end{aligned} \rightarrow \frac{\omega_a}{r_1} = \frac{\omega_b}{r_2} \rightarrow r_2 = \frac{\omega_b}{\omega_a}r_1$$

Case 1. If $\omega_a = 10\pi$ rad/sec and $\omega_b = 30\pi$ rad/sec, then $r_2 = 3r_1$. Therefore, let $r_1 = 1$ and $r_2 = 3$, and $\omega_0 = 10\pi$ rad/sec, which gives $T_0 = 2\pi/\omega_0 = 0.2$ sec. Select the integers r_1 and r_2 to give the smallest period T_0. This signal is bandlimited, with BW $= 3\,\omega_0$ rad/sec.

Case 2. If $\omega_a = 20\pi$ rad/sec and $\omega_b = 35\pi$ rad/sec, then $r_2 = 7r_1/4$. Therefore, let $r_1 = 4$ and $r_2 = 7$, and $\omega_0 = 5\pi$ rad/sec, which gives $T_0 = 2\pi/\omega_0 = 0.4$ sec. This signal is bandlimited, with BW $= 7\,\omega_0$ rad/sec.

Case 3. If $\omega_a = 10$ rad/sec and $\omega_b = 30\pi$ rad/sec, then $r_2 = 3\pi r_1$. Therefore, r_1 and r_2 cannot both be integers, which means that for these frequencies, $x(t)$ is not periodic.

For cases (1) and (2), the Fourier series coefficients can be found with (12.2). However, since $x(t)$ is given in terms of sinusoids, the coefficients can be identified by using Euler's identity, and the given signal becomes

$$x(t) = 6\frac{e^{j(\omega_a t+\pi/4)} + e^{-j(\omega_a t+\pi/4)}}{2} + 4\frac{e^{j(\omega_b t-\pi/3)} + e^{-j(\omega_b t-\pi/3)}}{j2}$$

$$= 3\,e^{j\pi/4}e^{j\omega_a t} + 3\,e^{-j\pi/4}e^{-j\omega_a t} + 2\,e^{-j\pi/3}e^{-j\pi/2}e^{j\omega_b t} + 2e^{j\pi}e^{j\pi/3}e^{-j\pi/2}e^{-j\omega_b t}$$

Case 1. Since $\omega_a = \omega_0$ and $\omega_b = 3\omega_0$, then in view of (12.1), we get $X_1 = 3e^{j\pi/4}$, $X_{-1} = 3e^{-j\pi/4}$, $X_3 = 2e^{-j5\pi/6}$, $X_{-3} = 2e^{j5\pi/6}$, and all other Fourier series coefficients are zero.

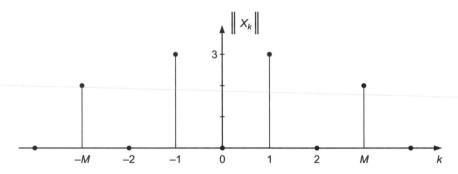

Figure 12.1   Magnitude spectrum of a bandlimited signal, $M = 3$ and $f_0 = 5$ Hz.

The magnitude spectrum of $x(t)$ is shown in Fig. 12.1, which contains **spectral lines** at $k = -3, -1, 1,$ and 3. From Fig. 12.1, we see that the amplitude of the sinusoid at the frequency $\omega = \omega_0$ ($k = -1$ and 1) is 6, and that $x(t)$ contains only one other sinusoidal component with frequency $\omega = 3\,\omega_0$ and amplitude equal to 4. For real-world signals we may not have an expression for $x(t)$. However, if we can find a signal's spectrum, then we can come to some conclusions about its behavior.

Case 2. Since $\omega_a = 4\,\omega_0$ and $\omega_b = 7\,\omega_0$, then in view of (12.1), we get $X_4 = 3e^{j\pi/4}$, $X_{-4} = 3e^{-j\pi/4}$, $X_7 = 2e^{-j5\pi/6}$, $X_{-7} = 2e^{j5\pi/6}$, and all other Fourier series coefficients are zero. The magnitude spectrum contains spectral lines at $k = -7, -4, 4,$ and 7.

To dissect real-world signals with (12.2) requires that we have an expression for $x(t)$, and generally this is not available. Real-world signals come from, for example, (1) a medical instrument that produces an electrocardiogram, (2) a gas sensor in the exhaust manifold of an engine, (3) a microphone in a cell phone that receives not only speech sounds but also background noise, and innumerable other situations where we study the spectrum of a signal to better understand the physical phenomenon from which the signal originated.

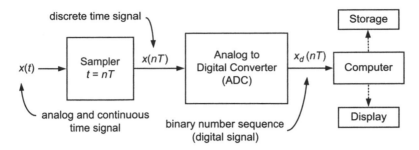

Figure 12.2   A data acquisition system.

To process real-world signals a **data acquisition system**, as depicted in Fig. 12.2, is used, where the discrete time signal $x_d(nT)$ is processed by digital means to gain insight about the nature of the analog signal $x(t)$. A discrete time signal is obtained by uniformly sampling a continuous time signal, which is expressed by

$$x(nT) = x(t)|_{t=nT} \tag{12.6}$$

where n , an integer, is called the **discrete time index** and T is the **sample time increment**. The sampling rate is called the **sampling frequency** $f_s = 1/T$ Hz (samples/sec). If T is known and fixed, then, for convenience, we write the discrete time signal as $x(n)$ instead of $x(nT)$.

This method of processing analog signals is utilized in innumerable applications of Electrical and Computer Engineering, because of the extensive kinds of computing that can be done, which is difficult, if not impossible, to achieve with analog circuitry. See Section 5.5 for a discussion about the components in Fig. 12.2. Here, we will not make a distinction between the discrete time signal $x(n)$ and its binary representation $x_d(n)$.

Processing of continuous time signals by digital means (**digital signal processing**, DSP) has become a viable processing mode over analog means for several reasons, some of which are digital signal processors, microcontrollers and microprocessors are inexpensive, programmable, reproducible, consume low power, have computing speeds suitable for signals with bandwidths beyond base band video (6 MHz), and can operate in extreme environments. Some broad application areas of DSP are automotive industry, consumer electronics, communication systems and medical systems. Since real-world signals are continuous in time, the technologist must properly interpret results of processing signals by digital means. This requires an understanding of the basic tools used for DSP. MATLAB provides an extensive set of tools for DSP.

12.1.1   Discrete Fourier Transform

Let us evaluate (12.2) numerically using Euler's method of integration (see Section 4.7.1). To do this, split one period of $x(t)$ into an integer N number of time segments, where

$$T = \frac{T_0}{N} \quad \rightarrow \quad T_0 = NT \tag{12.7}$$

and T is the time width of each segment. Then, $x(n), n = 0, 1, \ldots, N-1$, makes up N samples of $x(t)$ over one period. In fact, we can think of $x(n)$ for $-\infty < n < +\infty$ as a periodic discrete time signal with period N. Recall that Euler's integration method approximates the integral with a summation of rectangular areas, and we get

$$
\begin{aligned}
X_k &\cong \frac{1}{T_0} \sum_{n=0}^{N-1} x(n) e^{-jk\omega_0 nT} T \\
&= \frac{T}{T_0} \sum_{n=0}^{N-1} x(n) e^{-jk\frac{2\pi}{T_0}nT} = \frac{T}{NT} \sum_{n=0}^{N-1} x(n) e^{-jk\frac{2\pi}{NT}nT} = \frac{1}{N} \sum_{n=0}^{N-1} x(n) e^{-j\frac{2\pi}{N}kn}
\end{aligned}
\tag{12.8}
$$

where $x(n) e^{-jk\omega_0 nT}$ is the integrand of (12.2) evaluated at the beginning of each time segment and T is the segment width. As N is increased, making T smaller, the approximation given by (12.8) gets better. Let

$$
X(k) = \sum_{n=0}^{N-1} x(n) e^{-j\frac{2\pi}{N}kn}, k = 0, 1, \ldots, N-1
\tag{12.9}
$$

and then $X_k \cong \frac{1}{N} X(k)$. The algorithm given by (12.9) is called the **discrete Fourier transform** (DFT) of $x(n)$, $n = 0, 1, \ldots, N-1$. It is the **analysis equation** of the periodic discrete time signal $x(n)$. Equation (12.9) is also called an **N-point DFT**.

The DFT, $X(k)$, has an unexpected property. Let us evaluate (12.9) for $k = m$ and $k = m + rN$ for any positive or negative integers m and r. This gives

$$
\begin{aligned}
X(m+rN) &= \sum_{n=0}^{N-1} x(n) e^{-j\frac{2\pi}{N}(m+rN)n} = \sum_{n=0}^{N-1} x(n) e^{-j\frac{2\pi}{N}mn} e^{-j\frac{2\pi}{N}rNn} = \sum_{n=0}^{N-1} x(n) e^{-j\frac{2\pi}{N}mn} e^{-j2\pi rn} \\
&= \sum_{n=0}^{N-1} x(n) e^{-j\frac{2\pi}{N}mn} = X(m), e^{-j2\pi rn} = 1
\end{aligned}
$$

This shows that $X(k)$ is a periodic function of k, with period N. Therefore, it is conventional to evaluate (12.9) only for one period with $k = 0, 1, \ldots, N-1$. Recall that $\omega = k\omega_0$, or $f = k f_0 = k(1/T_0)$ is the frequency scale of $X(k)$. The time duration T_0, over which $x(t)$ is sampled, determines the **frequency resolution** $f_0 = 1/T_0$ Hz. Therefore, $X(k)$ versus frequency is periodic with period $f = N f_0 = N/T_0 = 1/T = f_s$. For any integers m and r, where $f = m f_0$, we have

$$
X(f) = X(f + r f_s)
\tag{12.10}
$$

Furthermore, since $X_k^* = X_{-k}$, $X^*(k) = X(-k)$, and since $X(k)$ is periodic, then

$$
X^*(k) = X(N-k), \quad k = 0, 1, \ldots
\tag{12.11}
$$

Example 12.2

Let us apply (12.9) to

$$x(t) = 6\cos(10\pi t + \pi/4) + 4\sin(30\pi t - \pi/3)$$

This is the signal given in case (1) of Example 12.1, where $T_0 = 0.2$ sec. Let us try $N = 20$, and then $T = T_0/N = 10$ msec, giving $f_s = 1/T = 100$ Hz (samples/sec). Prog. 12.1 uses (12.9) to find and plot the magnitude and angle of $X(k)$ for $-20 \le k \le 39$, a range of three periods of $X(k)$. Notice that in the program the summation over n for each k in (12.9) was obtained with a dot product of the two vectors x and argk.

```
% Application of the DFT
clear all; clc
wa  =  10*pi; wb  =  30*pi; T0  =  0.2; % specify signal frequencies and period
N  =  20; T  =  T0/N; % obtain N samples over one period of x(t)
n  =  0:N-1; t  =  n*T; % time points
x  =  6*cos(wa*t+pi/4)+4*sin(wb*t-pi/3); % sample signal for one period
Nk  =  -N:2*N-1; Nks  =  length(Nk); % DFT index range for three periods of X(k)
arg  =  exp(-j*2*pi*n'/N); % compute exponential terms over time only once
argk  =  ones(N,1); % initialize exponential terms to start at k = -N
for k = 1:Nks  % loop for all frequencies f from -fs to 2fs-f0
    Xk  =  x*argk; % sum of terms over time
    magX(k)  =  abs(Xk); % magnitude of DFT
    % get angle only if magnitude is greater than numerical noise
    if magX(k) > 1e-7
        angleX(k)  =  angle(Xk);
    else
        angleX(k)  =  0;
    end
    argk  =  argk.*arg; % update all frequency terms
end
subplot(2,1,1); stem(Nk,magX); grid on
xlabel('frequency index'); ylabel('magnitude')
title('Magnitude Spectrum of x(t)')
subplot(2,1,2); stem(Nk,angleX); grid on
xlabel('frequency index'); ylabel('angle - radians')
title('Phase Spectrum of x(t)')
```

Program 12.1   Program to compute the DFT of $x(n)$, $n = 0, 1, \ldots, N$-1.

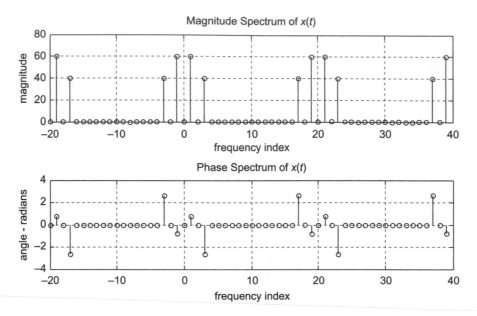

Figure 12.3   Magnitude and angle of the DFT output.

The DFT of $x(n)$ is shown in Fig. 12.3. Notice that $\|X(k)\|$ ($\angle X(k)$) is an even (odd) function of k. One period of $X(k)$ extends from $k = -10$ to $k = 9$, or $-f_s/2 \leq f < f_s/2$. If $\|X(k)\|$ is multiplied by $1/N$, then this period of $X(k)$ gives exactly the Fourier series coefficients, where the magnitudes are shown in Fig. 12.1. The next period of $X(k)$ shown in Fig. 12.3 extends from $k = 10$ to $k = 29$, which is a translation of the previous period of $X(k)$ centered about $k = 0$ (or $f = 0$) to the period centered about $k = N = 20$ (or $f = f_s$). Since it is conventional to compute $X(k)$ for $k = 0, 1, \ldots, N - 1 = 19$, which is also one period of $X(k)$, the first half ($k = 0, \ldots, 9$) of $X(k)$ gives $X(k)$ for positive k, and the second half ($k = 10, \ldots, 19$) gives $X(k)$ for negative k. Here, $X(-1) = X^*(1) = X(N - 1) = X(19)$ and $X(-3) = X^*(3) = X(N - 3) = X(17)$.

We can obtain the Fourier series coefficients of a periodic signal $x(t)$ exactly with the DFT, but only under special circumstances. The conditions are as follows:

1) $x(t)$ must be sampled over its period, or any integer multiple of its period, such that $T_0 = N T$.
2) $x(t)$ must be bandlimited, which means that BW $= M\omega_0$ must be finite.
3) referring to Fig. 12.3, for example, N must be large enough to satisfy $N - M > M$, which means that translations of $X_k, k = -M, \ldots, M$, to integer multiples of N must not overlap.

Then, one period of $X(k)/N$ gives exactly the Fourier series coefficients. Condition (3) becomes

$$N > 2M \quad \rightarrow \quad \frac{N}{T_0} > 2\frac{M}{T_0} \quad \rightarrow \quad \frac{1}{T} > 2M f_0 \quad \rightarrow \quad f_s > 2f_c \tag{12.12}$$

where $f_c = M f_0$ is the BW of $x(t)$. The condition, $f_s > 2f_c$, is called the **Sampling Theorem**, a very important result in signal and image processing by digital means. Given a sampling frequency f_s, we presume that a signal $x(t)$ consists of sinusoidal components with frequencies in the range $0 \leq f < f_s/2$.

To gain another perspective about the restriction, $f_c < f_s/2$, suppose $x(t) = \cos(2\pi f_0 t)$ is sampled at the rate $f_s = 8000$ Hz, where $f_0 = 1000$ Hz. This signal has spectral lines at $-f_0$ and $+f_0$ Hz. The signal and its samples are shown in Fig. 12.4(a). If we do not know the spectrum of $x(t)$, then this $f_s = 8000$ Hz implies that we expect $x(t)$ to consist of sinusoids with frequencies in the range $0 \leq f < 4000$ Hz. Now, let us sample $x(t) = \cos(2\pi(f_s - f_0)t) = \cos(14000\pi t)$ and $x(t) = \cos(2\pi(f_s + f_0)t) = \cos(18000\pi t)$ at the same sampling rate $f_s = 8000$ Hz. The results are shown in Fig. 12.4(b,c).

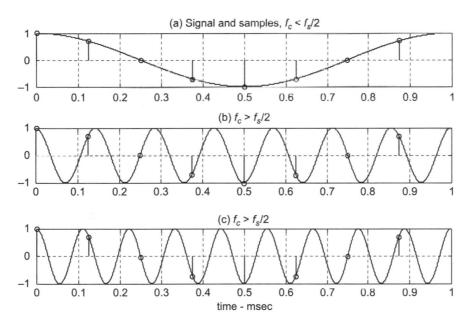

Figure 12.4   Data obtained from sinusoids with frequencies f_0, $f_s - f_0$, and $f_s + f_0$.

The plots in Fig. 12.4(b,c) show the same data obtained from one period of the sinusoid with frequency f_0. According to (12.10), the DFT will produce the same result for each set of data points. From the data alone, we cannot determine the frequency of the sinusoid that was sampled to obtain the data. To avoid this ambiguity, we must satisfy the sampling theorem, which will enable us to conclude that the data samples were obtained from the signal shown in Fig. 12.4(a), where $f_0 < f_s/2$. Prog. 12.2 was used to produce Fig. 12.4.

```
% Demonstration of signal frequency ambiguity introduced by sampling
clear all; clc
f0  =  1000; w0  =  2*pi*f0; T0  =  1/f0; % specify signal frequency and period
Np  =  1000; Tx  =  T0/Np; % obtain a smooth looking plot of x(t)
Nx  =  0:Np-1; tx  =  Nx*Tx; % time points
x  =  cos(w0*tx); % signal
subplot(3,1,1); plot(1e3*tx,x,'k'); grid on % time in milli seconds
N  =  8; T  =  T0/N; fs  =  1/T;  % sample time increment and sampling frequency
Ns  =  0:N-1; t  =  Ns*T; % sample time points
xs  =  cos(w0*t); % signal samples
hold on; stem(1e3*t,xs,'r') % plot samples versus time in milli seconds
fx  =  -f0+fs; wx  =  2*pi*fx;  % translating -f0 by fs
Np  =  7000; Tx  =  T0/Np; % obtain a smooth looking plot of x(t)
Nx  =  0:Np-1; tx  =  Nx*Tx; % time points
x  =  cos(wx*tx); % signal
subplot(3,1,2); plot(1e3*tx,x,'k'); grid on
xs  =  cos(wx*t); % signal samples
hold on; stem(1e3*t,xs,'r')
fx  =  +f0+fs; wx  =  2*pi*fx;  % translating +f0 by fs
Np  =  9000; Tx  =  T0/Np; % obtain a smooth looking plot of x(t)
Nx  =  0:Np-1; tx  =  Nx*Tx; % time points
x  =  cos(wx*tx); % signal
subplot(3,1,3); plot(1e3*tx,x,'k'); grid on
xs  =  cos(wx*t); % signal samples
hold on; stem(1e3*t,xs,'r')
```

Program 12.2   Program to demonstrate signal frequency ambiguity.

While Prog. 12.1 shows the details of implementing (12.9) to compute a DFT, the algorithm of (12.9) is not an efficient way to compute the DFT. The built-in MATLAB function **fft** (fast Fourier transform) computes a DFT, but much more efficiently than the algorithm described by (12.9). Its syntax is given by

$$\mathbf{X} = \mathbf{fft(x,N)}$$

where the vector x contains N samples of a signal $x(t)$, and the complex vector X contains one period of $X(k)$ for $k = 0, 1, \ldots, N - 1$. The DFT algorithm does not require that the N samples of $x(t)$ come from one period of a periodic signal or from a periodic signal.

The meaning of $X(k)$ must be properly interpreted based on the circumstances in which the N samples of $x(t)$ were obtained.

Example 12.3

Let us apply the FFT to analyze the signal given by

$$x(t) = 6 \cos(20\pi t + \pi/4) + 4 \sin(35\pi t - \pi/3)$$

This is the signal given in case (2) of Example 12.1, where $T_0 = 0.4$ sec, and its BW is $f_c = 17.5$ Hz. Let us try $N = 20$, and then $T = T_0/N = 20$ msec, giving $f_s = 1/T = 50$ Hz. With this f_s, we presume that $x(t)$ can have sinusoidal components with frequencies anywhere in the range $0 \leq f < 25$ Hz. Since $f_s > 2f_c = 35$ Hz, this f_s satisfies the sampling theorem. Prog. 12.3 obtains and plots the magnitude of $X(k)/N$, which is shown in Fig. 12.5.

```
% Application of the fast Fourier transform (FFT)
clear all; clc
wa  =  20*pi; wb  =  35*pi; T0  =  0.4; % specify signal frequencies and period
N   -  20; T  =  T0/N; % obtain N samples over one period of x(t)
Nn  =  0:N-1; t  =  Nn*T; % time points
x   =  6*cos(wa*t+pi/4)+4*sin(wb*t-pi/3); % sample signal for one period
X   -  fft(x,N)/N; % get DFT with FFT and convert to Fourier series coefficients
Xk_mag  =  abs(Xk); f  =  Nn/T0; % magnitude and frequency points
stem(f,Xk_mag); grid on
xlabel('frequency - Hz'); ylabel('magnitude')
title('Magnitude Spectrum of x(t)')
```

Program 12.3   Application of the fft algorithm.

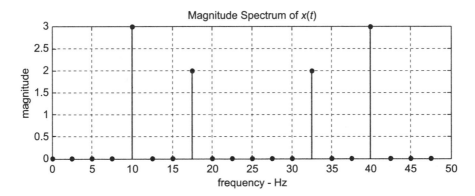

Figure 12.5   The magnitude spectrum versus frequency.

The magnitude spectrum over the frequency range $0 \leq f < f_s/2 = 25$ Hz is exactly the magnitude spectrum of $x(t)$. Given this magnitude spectrum, we conclude that $x(t)$ consists of two sinusoids with frequencies of 10 and 17.5 Hz. The spectral lines at $f = 40$ Hz and $f = 32.5$ Hz come from the spectral lines of $x(t)$ at $f = 40 - f_s = -10$ Hz and $f = 32.5 - f_s = -17.5$, respectively.

Now, let us see what can happen when we do not satisfy the sampling theorem. Let $N = 12$, and then $f_s = N/T_0 = 30$ Hz, which does not satisfy the sampling theorem. With this f_s, we presume that $x(t)$ can have sinusoidal components with frequencies anywhere in the range $0 \leq f < 15$ Hz. If the BW of $x(t)$ is not known, which is likely in a practical situation, then this choice for f_s implies that we believe that $x(t)$ contains sinusoids with frequencies in the range $0 \leq f < 15$. In Prog. 12.3, N was set to $N = 12$, and the resulting magnitude spectrum is shown in Fig. 12.6.

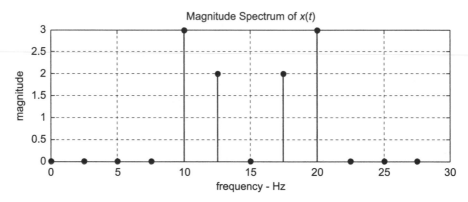

Figure. 12.6   A magnitude spectrum versus frequency that includes aliasing error.

The magnitude spectrum over the frequency range $0 \leq f < f_s/2 = 15$ shows two spectral lines. The spectral line at the frequency $f = 10$ Hz comes from a sinusoid in $x(t)$. However, the second spectral line occurs at $f = 12.5$ Hz, and $x(t)$ does not contain a sinusoid at this frequency. If we believe that $f_s = 30$ Hz does satisfy the sampling theorem, then we conclude that $x(t)$ does contain a sinusoid with $f = 12.5$ Hz. In fact, if we use the resulting X_k to reconstruct a continuous time function, then it would appear to consist of two sinusoids with frequencies 10 and 12.5 Hz. The sinusoid with the frequency 12.5 Hz is incorrect, and it is called an **alias**.

When the sampling theorem is not satisfied, **aliasing error** will occur.

Example 12.4

The signal shown in Fig. 12.7 is not bandlimited, and it also has discontinuities. Its period is $T_0 = 3$ sec, and $\omega_0 = 2\pi/3$ rad/sec.

The complex Fourier series coefficients of this signal are given by

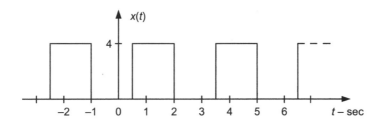

Figure 12.7 A periodic signal with discontinuities.

$$X_k = \frac{1}{3}\int_0^3 x(t)e^{-jk\frac{2\pi}{3}t}dt = \frac{1}{3}\int_{0.5}^{2.0} 4e^{-jk\frac{2\pi}{3}t}dt - \frac{4}{k\pi}e^{-jk\frac{5\pi}{6}}\sin(k\frac{\pi}{2}),\; k \neq 0 \qquad (12.13)$$

and $X_0 = 2$. Prog. 12.4 plots the magnitude spectrum of $x(t)$ using (12.13) and by taking the FFT of $N = 48$ samples of $x(t)$ over one period. With this N, $f_s = 16$ Hz. The result is shown in Fig. 12.8.

```
% Plot Fourier series coefficients and FFT of a rectangular wave
clear all; clc
ks  =  -24:47; Nks  =  length(ks); % frequency index range
for k = 1:24 % loop for negative frequency indeces
    X(k) = (4/(ks(k)*pi))*exp(-j*ks(k)*5*pi/6)*sin(ks(k)*pi/2);
end
X(25)  =  2; % X0
for k = 26:Nks % loop for positive frequency indeces
    X(k) = (4/(ks(k)*pi))*exp(-j*ks(k)*5*pi/6)*sin(ks(k)*pi/2);
end
X_mag  =  abs(X);
stem(ks,X_mag,'k'); grid on; axis([-24 48 0 2])
T0  =  3; N  =  48; T  =  T0/N; % parameters for sampling x(t)
for n = 1:N % loop to sample x(t)
    t  =  (n-1)*T;
```

```
    if t < 0.5, x(n) = 0;
    elseif t >= 0.5 & t < 2, x(n) = 4;
    else, x(n) = 0;
    end
end
X = fft(x,N); % get DFT with FFT algorithm
X_mag = abs(X)/N; % approximate Fourier series coefficients
ks = 0:47; % frequency index range for DFT
hold on
stem(ks,X_mag,'r')
```

Program 12.4   Program to compare a DFT computation to actual Fourier series coefficients.

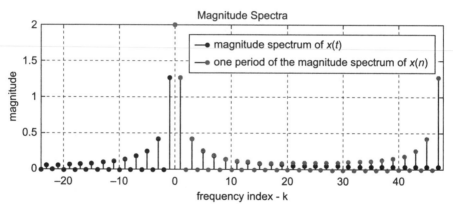

Figure 12.8   Magnitude spectrum of a signal that is not bandlimited.

The index range that corresponds to $0 \leq f < f_s/2$ is $0 \leq k < N/2 = 24$. The plot of $\|X(k)\|/N$ shows that while the DFT (FFT) incurs aliasing error in every spectral line, it becomes more significant as $k \to 24$. This aliasing error can be reduced by increasing N, which increases the sampling frequency.

A useful application of the DFT is to determine if a signal contains any sinusoidal components given data that includes a substantial amount of noise.

Example 12.5 _____

Prog. 12.3 was modified to become Prog. 12.5, which obtains noisy samples of the $x(t)$ given in Example 12.3. Fig. 12.9 shows $x(t)$ and samples of $x(t)$ plus noise over one period.

The magnitude spectrum, given in Fig. 12.10, was obtained using data obtained over one and two periods of $x(t)$.

```
% Application of the fast Fourier transform (FFT)
clear all; clc
wa  =  20*pi; wb  =  35*pi; T0  =  0.4; % specify signal frequencies and period
N  =  1000; T  =  T0/N; % obtain a smooth looking plot of x(t)
Np  =  0:N-1; t  =  Np*T; % time points
x  =  6*cos(wa*t+pi/4)+4*sin(wb*t-pi/3); % signal
figure(1); plot(t,x,'k'); grid on
N  =  20; T  =  T0/N; % sample time increment
Nn  =  0:N-1; t  =  Nn*T;
x  =  6*cos(wa*t+pi/4)+4*sin(wb*t-pi/3); % signal samples
rng('default'); % initialize random number generator to default seed
xpn  =  x+2*randn(1,N); % signal samples plus Gaussian noise
hold on; stem(t,xpn,'r')
X  =  fft(xpn,N); % get the DFT with the FFT
Xk_mag  =  abs(X)/N;
f0  =  1/T0; f  =  f0*Nn; % frequency resolution and frequency points
figure(2); stem(f,Xk_mag); grid on
xpn  =  [xpn,x+randn(1,N)]; % add noisy samples over a second period
X  =  fft(xpn,2*N); % get the DFT with the FFT
Xk_mag  =  abs(X)/(2*N);
f0  =  1/(2*T0); Nn  =  0:2*N-1; f  =  f0*Nn; % freq resolution and freq points
hold on; stem(f,Xk_mag,'r')
```

Program 12.5   Application of the DFT to noisy data.

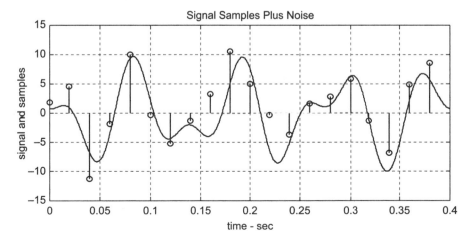

Figure 12.9   One period of signal samples plus noise, $f_s = 50$ HZ.

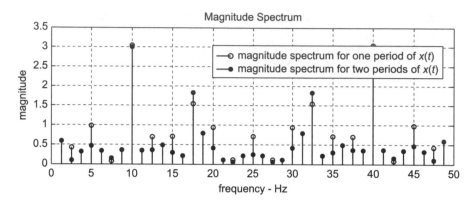

Figure 12.10   *Magnitude spectrum using data obtained over one and two periods of* $x(t)$.

If only the sampled data shown in Fig. 12.9 is available, then it would be difficult to conclude that the signal $x(t)$ contains two sinusoidal components at the frequencies 10 and 17.5 Hz, as Fig. 12.10 shows. Fig. 12.10 also shows that increasing the time duration over which samples of a signal are obtained improves the frequency resolution and decreases the degree to which additive noise degrades DFT results.

12.1.2   Inverse Discrete Fourier Transform

The number sequence $x(n)$, $n = 0, 1, \ldots, N-1$ that produced $X(k)$, $k = 0, 1, \ldots, N-1$ can be retrieved from $X(k)$ with the **inverse DFT** (IDFT). The IDFT of $X(k)$ is given by

$$x(n) = \frac{1}{N} \sum_{k=0}^{N-1} X(k)\, e^{j\frac{2\pi}{N}nk}, \quad n = 0, 1, \ldots, N-1 \tag{12.14}$$

It is the **reconstruction equation** of the discrete time periodic number sequence $x(n)$ that can be derived by sampling (12.1).

The MATLAB function **ifft** efficiently computes the IDFT. Its syntax is given by

```
x  =  ifft(X, N)
```

where X, which must satisfy (12.11), contains an N-point DFT and x contains the number sequence $x(n)$, $n = 0, 1, \ldots, N-1$.

Example 12.6

Consider again the signal $x(t) = 6\cos(10\pi t + \pi/4) + 4\sin(30\pi t - \pi/3)$, for which $\omega_0 = 10\pi$ rad/sec and $T_0 = 0.2$ sec. Let $N = 15$, making $f_s = 75$ Hz. Recall that the Fourier series coefficients are given by $X_1 = 3e^{j\pi/4}$, $X_{-1} = 3e^{-j\pi/4}$, $X_3 = 2e^{-j5\pi/6}$, $X_{-3} = 2e^{j5\pi/6}$, and all other Fourier series coefficients are zero. Prog. 12.6 reconstructs the number sequence $x(n)$, $n = 0, 1, \ldots, N-1$, shown in Fig. 12.11.

```
% Application of the inverse fast Fourier transform (IFFT)
clear all; clc
wa   =  10*pi; wb   =  30*pi; T0   =  0.2; % specify signal frequencies and period
N    =  1000; T     =  T0/N; % obtain a smooth looking plot of x(t)
Np   =  0:N-1; t    =  Np*T; % time points
x    =  6*cos(wa*t+pi/4)+4*sin(wb*t-pi/3); % signal
figure(1); plot(t,x,'k'); grid on
N    =  15; T     =  T0/N; % number of sample time points
X    =  complex(zeros(1,N),zeros(1,N)); % most of the elements of X are zero
% X(2) holds the k = 1 element and X(N) holds the k = N-1 element
X(2)  =  3*exp(j*pi/4); X(N)   =  conj(X(2)); % applying (12.12)
% X(4) holds the k = 3 element and X(N-2) holds the k = N-3 element
X(4)  =  2*exp(-j*5*pi/6); X(N-2)  =  conj(X(4)); % applying (12.12)
X    =  N*X; % scale to DFT result
x    =  ifft(X,N); % get the inverse DFT
Nn   =  0:N-1; t    =  Nn*T; % using t for x(n) to compare results to x(t)
hold on; stem(t,real(x),'r'); % plot the number sequence
```

Program 12.6   Demonstration of using the function ifft.

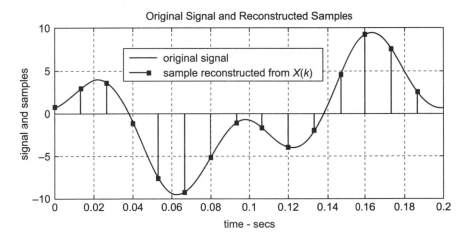

Figure 12.11   Reconstruction of samples of a signal with the IDFT.

Notice that any N can be used that satisfies the sampling theorem. We consider $x(n)$ (a representation in the **time domain**) and $X(k)$ (a representation in the **frequency domain**) to be equivalent information about $x(t)$.

MATLAB includes many built-in functions for signal analysis. Only a few of them have been used in examples. Table 12.1 gives a few more functions. You should use the **doc** help facility to see examples of the application of these and other functions useful for signal analysis.

Table 12.1   Some MATLAB functions for signal analysis

Function	Brief description
detrend	Removes the mean value or linear trend from a signal; useful before applying the fft algorithm
goertzel	Returns the DFT at a specified frequency
dct	Discrete cosine transform
idct	Inverse discrete cosine transform
hilbert	Returns the input signal with a 90 degree phase shift
cceps	Complex cepstral analysis
icceps	Inverse complex cepstrum
rceps	Returns the real cepstrum of a real data sequence

12.1.3   Windows

Now, let T_x denote the actual period of $x(t)$. In a practical situation, the actual period T_x of a periodic signal $x(t)$ may not be known. Using $x(t)$ over a time duration $T_0 \neq T_x$ can cause DFT results to be very different from results obtained with $T_0 = T_x$. If $T_0 \neq T_x$, the frequency $f_0 = 1/T_0$ Hz is not the correct fundamental frequency $1/T_x$ Hz, and the DFT will produce spectral lines at frequencies that are not integer multiples of $1/T_x$ Hz. Whatever T_0 may be, $x(t)$ over this time duration is considered to be one period $x_p(t)$ of a periodic function $\hat{x}(t)$. A period $T_0 \neq T_x$ is likely to introduce discontinuities at the period boundaries of $\hat{x}(t)$, regardless of the duration T_0. This means that even though $x(t)$ may be bandlimited, $\hat{x}(t)$ will not be bandlimited.

The **unit step function** (11.7) was introduced in Chapter 11, and its definition, repeated here for convenience, is given by

$$u(t - t_0) = \begin{cases} 1, & t > t_0 \\ 0, & t < t_0 \end{cases}$$

where t_0 is an arbitrary time point. Notice that the unit step function is one when its argument is positive, and it is zero when its argument is negative.

Let us model $x_p(t)$ with

$$x_p(t) = x(t)\,(u(t) - u(t - T_0)) = x(t)\,w_R(t)$$

where $x(t)$ is truncated by the gate function $w_R(t)$, called a **rectangular window**. If $T_0 = T_x$, then any time segment of $x(t)$ that has a duration of T_0 sec can be one period of $x(t)$. However, if $T_0 \neq T_x$, then $x_p(t)$ depends on where the window is positioned, and $\hat{x}(t)$ will not be unique.

Example 12.7

Consider the signal

$$x(t) = 1.5\ \cos(10\pi t + \pi/4) + \sin(30\pi t - \pi/3)$$

where $T_x = 0.2$ sec. The period determines the fundamental frequency $f_x = 1/T_x = 5$ Hz. This signal has spectral lines at $f = 5$ and 15 Hz. The following statements plot $x(t)$ over three periods as shown in Fig. 12.12.

```
>> clear all;
>> fs  =  44100; T  =  1/fs;  t  =  -0.1:T:0.5;
>> x  =  1.5*cos(10*pi*t+pi/4)+sin(30*pi*t-pi/3);
>> plot(t,x); grid on; axis([-0.1 0.5 -2.5 2.5]);
>>  xlabel('time - sec'); ylabel('x(t)')
```

Suppose we do not know the period of $x(t)$, and assume the period is some $T_0 \neq T_x$. Let $T_0 = 0.25$ sec, and then, the fundamental frequency of $\hat{x}(t)$ becomes $f_0 = 1/T_0 = 4$ Hz.

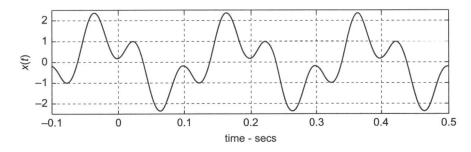

Figure 12.12   Plot of $x(t)$ over three periods.

The periodic function $\hat{x}(t)$ is shown in Fig. 12.13, where one period occurs for $0 \le t < 0.25$ sec. Notice that discontinuities occur at period boundaries, for example, at $t = 0.0$ and $t = 0.25$. This means that even though $x(t)$ is bandlimited (see Fig. 12.1.), $\hat{x}(t)$ is not bandlimited

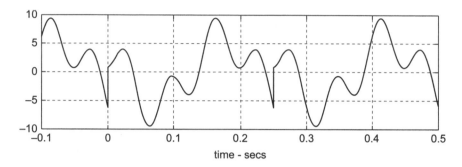

Figure 12.13   Periodic signal $\hat{x}(t)$ with $T_0 = 0.25\,sec.$

Let us try $N = 40$, and then $T = T_0/N = 6.25$ msec, giving $f_s = 1/T = 160$ Hz. Prog. 12.7 applies the FFT to $\hat{x}(t)$, and plots the magnitude and phase spectra shown in Fig. 12.14.

```
% Application of the DFT
clear all; clc
wa  =  10*pi; wb  =  30*pi; T0  =  0.25; % specify total sample time
N  =  40; T  =  T0/N; % obtain N samples of x(t)
Nn  =  0:N-1; Nk  =  Nn; t  =  Nn*T; f  =  Nk/T0; % time and frequency points
x  =  1.5*cos(wa*t+pi/4)+sin(wb*t-pi/3); % sample signal
%
X  =  fft(x,N); % use FFT to get DFT
magX  =  20*log10(abs(X)); % magnitude of DFT in dB
angleX  =  angle(X); % angle of DFT
subplot(2,1,1); stem(f,magX); axis([0 160 -40 40]); grid on
ylabel('magnitude - dB')
title('Estimate of Magnitude Spectrum')
subplot(2,1,2); stem(f,angleX); grid on
xlabel('frequency - Hz'); ylabel('angle - radians')
title('Estimate of Phase Spectrum')
```

Program 12.7   Application of the FFT algorithm.

Since $\hat{x}(t)$ is not bandlimited, Fig. 12.14 shows that there is substantial **aliasing error**. Also, the magnitude spectrum peaks occur at $f = 4$ and $f = 16$ Hz, indicating that $x(t)$ may

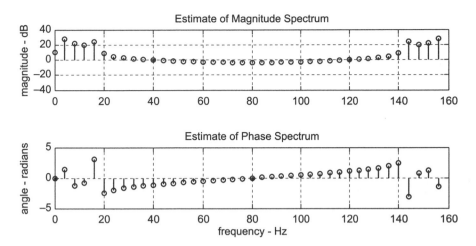

Figure 12.14   Estimate of the magnitude and phase spectra of x(t).

contain sinusoidal components at or close to these frequencies. However, there are also comparable spectral lines at $f = 8$ and $f = 12$ Hz, and $x(t)$ does not have sinusoidal components at these frequencies. The actual two spectral lines of $x(t)$ at $f = 5$ and $f = 15$ Hz in Fig. 12.1 have spread into neighboring frequencies, because $x(t)$ was truncated over a time range other than T_x. This is called **leakage error**, which can be reduced if $T_0 \to r\, T_x$ for a positive integer r.

Truncating $x(t)$ with a rectangular window introduces discontinuities at period boundaries of $\hat{x}(t)$. Such discontinuities can be reduced if not eliminated by using one of the window functions included in the MATLAB Signal Processing Toolbox, some of which are listed in Table 12.2.

Table 12.2   Some MATLAB window functions

Function	Brief description
bartlett	Triangular shaped window
blackman	Blackman window
chebwin	Chebyshev window
gausswin	Gaussian window
hamming	Raised cosine window
hann	Hann window
kaiser	Kaiser window
rectwin	Rectangular window

The **Hann window function** is given by

$$w_H(t) = \begin{cases} (1 - \cos(2\pi t / T_0))/2, & 0 \le t \le T_0 \\ 0, \text{ otherwise} \end{cases}$$

The syntax for the built-in function **hann** is

```
H  =  hann(N,'periodic')
```

The elements of the vector H are N samples of $w_H(t)$ for $t = nT$, $n = 0, 1, \ldots, N - 1$, where $T = T_0/N$. If instead of 'periodic', 'symmetric' is used, then $T = T_0/(N - 1)$, making the last element of H equal to $w_H(t = T_0)$.

Example 12.7 (continued) _____

With the Hann window, $x_p(t) = x(t)\, w_H(t)$ becomes the signal shown in Fig. 12.15, which was obtained by inserting into Prog. 12.7, just before the **fft** statement, the following statements:

```
figure(1)
plot(t,x,'r'); grid on; % plot x using rectangular window
xlabel('time - secs'); hold on
H  =  hann(N,'periodic')'; % get samples of the Hann window
plot(t,H,'b') % plot Hann window
x  =  x.*H; % windowed data
plot(t,x,'k'); hold off; % plot windowed data
figure(2)
```

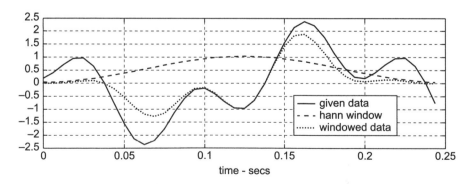

Figure 12.15   Given data (red), Hann window (blue), and windowed data.

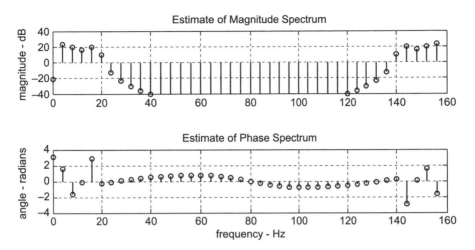

Figure 12.16 *Estimates of the magnitude and phase spectra of x(t) with the*
Hann window.

Comparing Figs. 12.14 and 12.16 shows that using a window other than a rectangular window can significantly reduce aliasing error. The leakage error has been reduced only slightly. However, if T_0 is increased, say to $T_0 = 3(0.25)$ sec, while keeping the sampling frequency at $f_s = 160$ Hz $(N = 120)$, then leakage error can be reduced significantly, as seen in Fig. 12.17. By increasing T_0 from $T_0 = 0.25$ sec to $T_0 = 0.75$ sec, the frequency resolution has been increased from a spectral line every $f_0 = 4$ Hz to every $f_0 = 4/3$ Hz, which shows two distinct spectral peaks in Fig. 12.17.

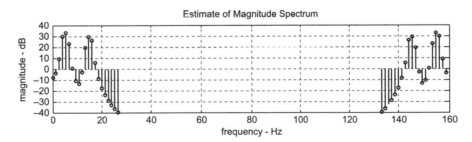

Figure 12.17 *Estimate of the magnitude spectrum of x(t) with the Hann window.*

12.1.4 *Non-Stationary Signals*

The signals that we have studied have a fundamental property in common. They are **stationary signals**. This means that their spectral content does not change with time. As we applied concepts of Fourier analysis to a signal, it was presumed that the signal is stationary. Stationary signals are distinguished from signals with a spectral content that changes with time, called **non-stationary signals**. A commonly used expression to model a non-stationary signal $x(t)$ is given by

$$x(t) = \sum_{k=1}^{K(t)} A_k(t) \cos(\omega_k(t)\, t + \theta_k(t)) \tag{12.15}$$

where the number of sinusoidal components and amplitude, frequency, and phase of each component can change with time.

Given are N samples of $x(t)$ that were obtained at the rate $f_s = 1/T$ Hz. To utilize the DFT for the spectral analysis of a non-stationary signal it is assumed that over each short time interval of a set of short time intervals the signal is substantially stationary. For some integer M, let each short time interval have a time duration given by $T_0 = MT$ sec. This segmentation is depicted in Fig. 12.18. The first time interval starts at $n = m_1$, usually $m_1 = 0$. Then, each following short time interval overlaps the previous time interval. For example, the second time interval starts at $n = m_2$, and it overlaps the first time interval. This continues until there is not enough data for another complete short time interval. Over N samples, the number of time intervals, denoted by N_0, depends on M and the amount of time interval overlap.

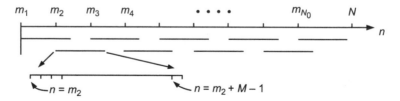

Figure 12.18   Segmentation of N samples into N_0 segments.

The signal over each time interval is windowed and sampled to obtain M samples of $x_p(t) = x(t)\, w(t)$, where $w(t)$ is some window function of duration T_0. After incorporating a window function into (12.9), the DFT becomes an algorithm called the **short time Fourier transform (STFT)** given by

$$X(k,m) = \sum_{n=0}^{M-1} w(n)\, x(n+m)\, e^{-j\frac{2\pi}{M}nk}, \quad k = 0,\, 1, \ldots,\, M-1 \tag{12.16}$$

where k is the frequency index such that $f = k f_0 = k/T_0$ Hz and m, where $m = m_i$, $i = 1, \ldots, N_0$, is the time index that specifies the beginning of a short time interval. For each m, (12.16) is an M-point DFT.

The STFT is a function of two variables. As a function of frequency it gives the same information about a periodic continuous time signal as does (12.9) applied to samples of $x_p(t)$. The time duration of the window function determines the frequency resolution. As a function of time, the STFT gives the spectral content of a signal as its spectral content changes from one time interval to the next time interval. If, for example, $m = 0, 1, 2, \ldots$, then an M-point DFT is computed at each discrete time point, which can require much computing time and provides the highest time resolution of $X(k, m)$. This also incurs the greatest amount of overlap between successive time intervals. Commonly used values of m are $m_i = (i - 1)(M/2), i = 1, 2, \ldots, N_0$, which means that there is a 50% data overlap from a time interval to the next time interval. An appropriate set of values of m depends on how rapidly the spectral content of a signal changes with time.

The built-in MATLAB function **spectrogram**, which is contained in the Signal Processing Toolbox, computes the STFT. A syntax option is given by

```
[S,Freq,Time,P] = spectrogram(x,window,noverlap,nfft,fs)
```

where x is a vector of N samples of $x(t)$, window is a vector of M samples of some window function, noverlap is the number of samples that successive intervals overlap, nfft is the number of windowed data samples used by the fft function that computes the DFT and fs is the sampling rate. For each m_i the elements of a column of S, which has N_0 columns, are given by (12.16) for $k = 0, 1, \ldots, (M + 2)/2$ if M is even and $k = 0, 1, \ldots, (M + 1)/2$ if M is odd, Freq is a column vector of frequencies k/T_0, Time is a row vector of the midpoints of the N_0 time intervals and P, which has the same dimension as S, has elements given by

$$P(k, m) = w_p\, S(k, m)\, S^*(k, m), \quad w_p = \frac{2}{f_s \sum_{n=1}^{M} w^2(n)}$$

For $k = 0$, the numerator of the normalizing factor w_p must be 1 instead of 2. Each column of P is the **power spectral density** of the windowed data over the corresponding time interval.

Example 12.8

Let us asses the spectral content of the signal shown in Fig. 12.19. Prog. 12.8 produced the magnitude spectrum, which is shown in Fig. 12.20, using all of the sampled data. Since $T_x = 0.5$, there is a spectral line at integer multiples of 2 Hz in Fig. 12.20.

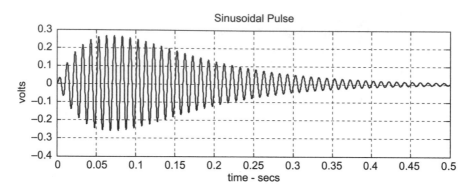

Figure 12.19   A sinusoidal (100 Hz) pulse.

```
% Spectral analysis of a sinusoidal pulse
clc; clear all
f0  =  100; w0  =  2*pi*f0; a  =  -10; b  =  -20;  % signal parameters
fs  =  8e3; T  =  1/fs; % sampling rate and time increment
Tx  =  0.5; N  =  floor(Tx/T); Tx  =  N*T; t  =  0:T:Tx-T; % sample times
x  =  (exp(a*t)-exp(b*t)).*sin(w0*t); % get signal samples
figure(1)
plot(t,x); grid on; xlabel('time - secs'); ylabel('volts');
title('Sinusoidal Pulse')
X  =  fft(x); mag_X_dB  =  20*log10(abs(X)/N); % apply fast Fourier transform
f0  =  1/Tx; k  =  0:N-1; f  =  k*f0; % prepare frequency scale
figure(2)
N_max  =  floor(N/16); f_max  =  f(N_max);
plot(f(1:N_max),mag_X_dB(1:N_max)); grid on; % plot portion of spectrum
xlabel('frequency - Hz'); ylabel('magnitude - dB')
title('Magnitude Spectrum')
% specify parameters for a spectrogram with high time and freq resolution
window_pts  =  1024; noverlap  =  window_pts - 128; nfft  =  window_pts;
window  =  bartlett(window_pts); % using the Bartlett window
[S,freq,time,P]  =  spectrogram(x,window,noverlap,nfft,fs);
f_index  =  find(freq < f_max); Freq  =  freq(f_index); % frequency axis range
p_dB  =  10*log10(P); P_dB  =  p_dB(f_index,:); % convert to dB
figure(3)
surf(time,Freq,P_dB); axis([0 0.5 0 f_max -120 -20]);
xlabel('time - sec'); ylabel('frequency - Hz');
```

Program 12.8   Program to plot the spectrogram of a sinusoidal pulse.

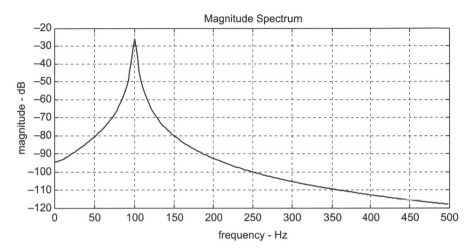

Figure 12.20   Magnitude spectrum of a sinusoidal pulse.

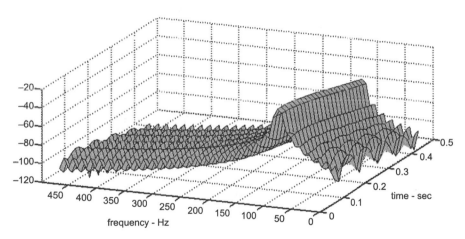

Figure 12.21   Spectrogram using the Bartlett window.

Prog. 12.8 also obtained the spectrogram shown in Fig. 12.21. Since the overlap from a time interval of $M = 1024$ samples to the next time interval is $100 (1024 - 128)/1024 = 87.5\%$, the spectrogram shows a power spectral density over 1024 samples every $128T = 16.0$ msec. The plot shows that the spectral content of the signal does not change with time, which means that the signal shown in Fig. 12.19 is substantially a stationary signal.

The spectral content of speech signals, which are non-stationary signals, is of particular interest for speech recognition, speaker identification and reduction of background noise. The ability to verbally interact with machines, particularly in the automotive industry, is an intensively studied problem.

Example 12.9

Prog. 12.9 finds and plots the spectrogram of the one word speech sound shown in Fig. 12.22. This sound was recorded with Prog. 8.5, using $f_s = 44100$ Hz.

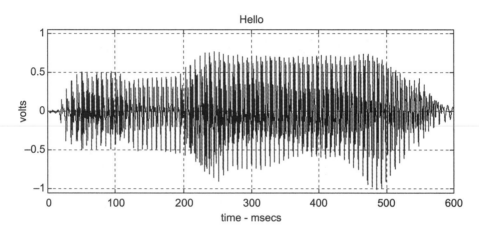

Figure 12.22   Recording of the one word speech sound, "hello".

```
% Program to graph the spectrogram of a one word speech signal
clear all; clc;
File_name = 'hello.bin'; % specify binary file name
ID1 = fopen(File_name); % open file
word = fread(ID1,'double'); fs = 44100; % get binary recording
x = word(44500:72500); % extract word sound part
n = 0:length(x)-1; t = 1000*n/fs; % time in milli seconds
figure(1);
plot(t,x); grid on; axis([0 600 -1 0.6])
xlabel('time - msecs'); ylabel('volts'); title('Hello')
figure(2) % new figure
% specify spectrogram parameters
window_length = 2048; noverlap = window_length - 64; nfft = window_length;
window = hann(window_length,'periodic'); % get Hann window values
x = sqrt(window_length)*x; % normalize sound
[S,Freq,Time,P] = spectrogram(x,window,noverlap,nfft,fs);
dB_P = 10*log10(P); % convert to dB
```

```
% for plotting, clamp dB_P at min_dB
min_dB  =  -100; dB_index  =  find(dB_P < min_dB); dB_P(dB_index)  =  min_dB;
% for plotting, limit maximum frequency
max_f  =  5000; f_index  =  find(Freq > max_f); max_f_index  =  f_index(1);
freq  =  Freq(1:max_f_index); max_f  =  Freq(max_f_index);
dB_p  =  dB_P(1:max_f_index,:);
surf(Time,freq,dB_p); % surface plot of spectrogram
axis([-0.05 0.65 0 max_f min_dB 10]); shading interp % axis limit
```

Program 12.9   Program to obtain the spectrogram of a signal.

Fig. 12.23 shows the spectrogram versus time and frequency. In this figure, color, which changes from blue to red with the z-axis, was changed to a gray scale in the *Figure Window* with

edit $\rightarrow$ colormaps $\rightarrow$ tools $\rightarrow$ standard colormaps $\rightarrow$ gray

Since the shift from one time interval to the next is 64 samples, Fig. 12.23 shows a power spectral density every $64T = 1.4512$ msec of 2048 samples of $x_p(t)$ with a frequency resolution of $f_0 = 1/(2048T) = 21.5$ Hz.

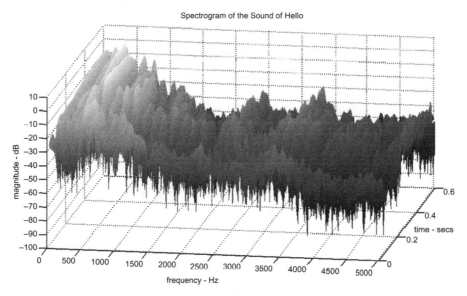

Figure 12.23   Spectrogram of the one word sound of hello.

Many kinds of signals, for example, music signals, biological signals, and video signals are processed by continuous and discrete time systems to modify their spectral content. In the next section we will investigate how such systems work.

12.2   Continuous Time Systems

A **linear and time invariant (LTI)** continuous time system (**CTS**) is modeled by

$$y^{(N)}(t) + a_{N-1}y^{(N-1)}(t) + \cdots + a_0 y^{(0)}(t) = b_M r^{(M)}(t) + \cdots + b_1 r^{(1)}(t) + b_0 r^{(0)}(t)$$

$$(12.17)$$

where $r(t)$ is called the **system input**, $y(t)$ is called the **system output** (also called **response**). This is an Nth order differential equation, where $r^{(m)}(t)$, $m = 0, 1, \ldots, M$, denotes the mth derivative of $r(t)$, $r^{(0)}(t) = r(t)$, $y^{(n)}(t), n = 0, 1, \ldots, N$, denotes the nth derivative of $y(t)$, and $y^{(0)}(t) = y(t)$. Unless stated otherwise, assume that $M \leq N$. It is also conventional to use the notation $\dot{y}(t)$ and $\ddot{y}(t)$ to mean $y^{(1)}(t)$ and $y^{(2)}(t)$, respectively. However, for higher order derivatives, the dot notation is not convenient.

While an input can be applied at any time, it is conventional to apply inputs starting at a reference time given by $t = 0$. Unless stated otherwise, assume that $r(t < 0) = 0$. Given an input $r(t)$, then all of the terms in the right side of (12.17) can be combined into one function of time $g(t)$, called the **forcing function**. If $g(t) = 0$, then (12.17) is called a **homogeneous equation**.

To find the complete response (solution of (12.17)) requires N **initial conditions** (ICs). Usually, these conditions are given by

$$y^{(N-1)}(0^-) = S_{N-1}, \ \ y^{(N-2)}(0^-) = S_{N-2}, \ldots, \ \ y^{(1)}(0^-) = S_1, \ \ y^{(0)}(0^-) = S_0 \quad (12.18)$$

for some constants $S_n, n = 0, 1, \ldots, N - 1$, where $t = 0^-$ is the time just prior to the time when the input is applied. The set of ICs is called the **state of the system** just prior to applying an input.

Equation (12.17) is a **time invariant model** of a system, because the coefficients, $a_n, n = 0, 1, \ldots, N - 1$, and $b_m, m = 0, 1, \ldots, M$, are constants (do not vary with time). This means that if $y(t)$ is the response to $r(t)$, then the response to $r(t - t_0)$ is $y(t - t_0)$, for any positive or negative time shift t_0. If any coefficient varies with time, then (12.17) is a **time varying model**.

Equation (12.17) is a **linear model** of a system, because if, for zero ICs, $y(t) = y_1(t)$ is the response (solution of (12.17)) when $r(t) = r_1(t)$ and if, for zero ICs, $y(t) = y_2(t)$ is the response when $r(t) = r_2(t)$, then, for zero ICs, $y(t) = c_1 y_1(t) + c_2 y_2(t)$ is the response when $r(t) = c_1 r_1(t) + c_2 r_2(t)$ for any constants c_1 and c_2. This is the **linearity property**.

Equation (12.17) is a **causal model** of a system, because the response $y(t)$ at any time $t = t_0$ only depends on the input $r(t)$ for $t \leq t_0$. If a system is **causal**, then the present output depends only on the present and past behavior of the input. All physical systems are causal systems. Consider, for example, a system model given by $\dot{y}(t) + a_0 y(t) = b_0 r(t + 1)$.

At any time t, the response $y(t)$ depends on the input one second into the future, and this system is said to be an **acausal** (or noncausal) system.

Example 12.10 _____

Here are some examples of causal LTI CTS. Recall the band-pass filter shown in Fig. 6.11. If (6.49) is differentiated once, then we get the filter input $v_s(t)$ to output $v(t)$ relationship given by

$$\frac{d^2v(t)}{dt^2} + \frac{R}{L}\frac{d\,v(t)}{dt} + \frac{1}{LC}v(t) = \frac{R}{L}\frac{d\,v_s(t)}{dt} \tag{12.19}$$

which is a second-order differential equation. With the application of an input voltage $v_s(t)$, the circuit operates according to the properties of its components and their interconnection to produce the voltage $v(t)$ across the resistor. If the circuit model given by (12.19) is accurate, then the solution of (12.19) anticipates the behavior of the physical circuit.

The low-pass filter shown in Fig. 6.13 has the input–output relationship given by

$$\frac{d^3v(t)}{dt^3} + \frac{R}{L_2}\frac{d^2v(t)}{dt^2} + \frac{1}{C}\left(\frac{1}{L_1}+\frac{1}{L_2}\right)\frac{dv(t)}{dt} + \frac{R}{L_1L_2C}v(t) = \frac{R}{L_1L_2C}v_s(t) \tag{12.20}$$

which is a third-order differential equation.

Let us find the input–output relationship of the circuit given in Fig. 12.24.

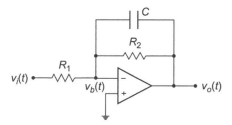

Figure 12.24   An active RC low-pass filter.

Summing the currents leaving the negative terminal of the op-amp gives

$$\frac{v_b(t) - v_i(t)}{R_1} + \frac{v_b(t) - v_o(t)}{R_2} + C\frac{d}{dt}(v_b(t) - v_o(t)) = 0 \tag{12.21}$$

Since $v_o(t) = A(0 - v_b(t))$, then, for very large A, $v_b(t) = -v_o(t)/A$ is negligible compared to $v_i(t)$ and $v_o(t)$, and (12.21) becomes

$$C\frac{dv_o(t)}{dt} + \frac{v_o(t)}{R_2} = -\frac{v_i(t)}{R_1} \qquad (12.22)$$

This is a first-order differential equation that models the behavior of the LTI analog circuit (CTS) shown in Fig. 12.24.

If a causal differential equation, time invariant or not, does not have the structure of (12.17), where each term in a sum of terms is a coefficient multiplied by one of $y^{(N)}(t)$, $y^{(N-1)}(t), \ldots, y^{(0)}(t)$ or one of $r^{(M)}(t), \ldots, r^{(1)}(t), r^{(0)}(t)$, then it cannot satisfy the linearity property, and it is called a **nonlinear differential equation**.

Example 12.11

The circuit shown in Fig. 12.25 converts an AC voltage, for example, $v_s(t) = A \cos(\omega t + \theta)$ into a DC voltage $v(t)$. The resistor is intended to represent a device that is powered by the AC to DC converter. Depending on the degree to which the capacitor discharges through the resistor while the diode is not conducting, $v(t)$ can be a nearly constant voltage. To describe the behavior of this circuit, apply KVL $(v_d(t) = v_s(t) - v(t))$, KCL $(i_d(t) = i_C(t) + i_R(t))$, and the diode characteristic to get

$$C\frac{dv}{dt} + \frac{v(t)}{R} = I_s(e^{(v_s(t)-v(t))/V_T} - 1) \qquad (12.23)$$

which is a first-order nonlinear and time invariant differential equation.

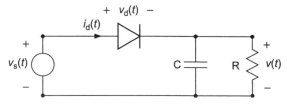

Figure 12.25   AC to DC converter.

The differential equations given in Examples 12.10 and 12.11 are **ordinary differential equations** (ODEs). An ODE is a differential equation that contains functions of only one independent variable, denoted by t in our discussion, and derivatives of these functions with respect to the one independent variable. Most generally, an ODE is written as

$$F(y^{(N)}(t),\ y^{(N-1)}(t),\dots,\ y^{(1)}(t),\ y^{(0)}(t),\ t) = 0$$

which is in **implicit form**. In **explicit form**, an ODE is written as

$$y^{(N)}(t) + F(y^{(N-1)}(t),\dots,\ y^{(1)}(t),\ y^{(0)}(t),\ t) = 0 \tag{12.24}$$

where $y^{(N)}(t)$ is written as a function of $g(t)$ and lower order derivatives of $y(t)$. The ODEs given in Examples 12.10 and 12.11 are in explicit form.

Ordinary differential equations are distinguished from equations that involve functions of more than one independent variable and their derivatives, called partial derivatives, with respect to anyone of the independent variables.

Given a system model, such as (12.17) or (12.24), we are interested to solve the equations of the model to gain insight into the behavior of the physical system, and thereby learn how to design a system to achieve desired system behavior.

12.3   Response of LTI Continuous Time Systems

The solution of (12.17) (complete response) can be written as

$$y(t) = y_{zi}(t) + y_{zs}(t)$$

where the function $y_{zi}(t)$, called the **zero-input response** (also called the **natural response)**, is the response when the input $r(t)$ is zero. It is the response due to the ICs given by (12.18), and it is the solution of

$$y_{zi}^{(N)}(t) + a_{N-1}y_{zi}^{(N-1)}(t) + \cdots + a_0\, y_{zi}^{(0)}(t) = 0 \tag{12.25}$$

which is a homogeneous equation. If the ICs are all zero, then $y_{zi}(t) = 0$. The function $y_{zs}(t)$, called the **zero-state response** (also called the **forced response**), is the response to the input $r(t)$ under zero ICs, and it is the solution of

$$y_{zs}^{(N)}(t) + a_{N-1}\, y_{zs}^{(N-1)}(t) + \cdots + a_0\, y_{zs}^{(0)}(t) = b_M r^{(M)}(t) + \cdots + b_1\, r^{(1)}(t) + b_0\, r^{(0)}(t) \tag{12.26}$$

If the input $r(t)$ is zero, then $y_{zs}(t) = 0$. The difference between $y(t)$ in (12.17) and $y_{zs}(t)$ in (12.26) is due to the ICs.

12.3.1   Zero-Input Response

We can gain an important insight into the behavior of LTI CTS by solving (12.25). The solution $y_{zi}(t)$ must be a function such that a linear combination of it and its derivatives equals zero for all t. Let us try $y_{zi}(t) = K e^{\lambda t}$ for some constants K and λ, because the derivative of an exponential function is the exponential function, and to achieve the equality of (12.25) for all t, every term in the sum of terms on the left side must be the same kind of function. Substituting this $y_{zi}(t)$ into (12.25) gives

$$K \lambda^N e^{\lambda t} + a_{N-1} K \lambda^{N-1} e^{\lambda t} + a_{N-2} K \lambda^{N-2} e^{\lambda t} + \cdots + a_1 K \lambda^1 e^{\lambda t} + a_0 K \lambda^0 e^{\lambda t} = 0 \qquad (12.27)$$

Since K cannot be zero, because then $y_{zi}(t) = 0$, and $e^{\lambda t} \neq 0$, (12.27) reduces to

$$Q(\lambda) = \lambda^N + a_{N-1} \lambda^{N-1} + a_{N-2} \lambda^{N-2} + \cdots + a_1 \lambda^1 + a_0 = 0 \qquad (12.28)$$

where $Q(\lambda)$ is called the **characteristic polynomial**. Equation (12.28) is called the **characteristic equation**, which has N roots, $\lambda_n, n = 1, 2, \ldots, N$. Notice that (12.28) can be found by inspection of (12.17). Let us assume that the roots are distinct. The λ_n can be real numbers or complex numbers that occur in complex conjugate pairs, because the coefficients $a_n, n = 0, 1, \ldots, N - 1$ are real numbers.

For each λ_n there is a solution of (12.25) given by $K_n e^{\lambda_n t}$, where we must still find K_n. Therefore, since (12.25) is a linear equation, then

$$y_{zi}(t) = \sum_{n=1}^{N} K_n e^{\lambda_n t} \qquad (12.29)$$

is also a solution of (12.25). To find the values of the $K_n, n = 1, 2, \ldots, N$ we can apply the N ICs at $t = 0^-$ to get

$$
\begin{aligned}
y^{(0)}(0^-) &\rightarrow K_1 + K_2 + \cdots + K_N = S_0 \\
y^{(1)}(0^-) &\rightarrow \lambda_1 K_1 + \lambda_2 K_2 + \cdots + \lambda_N K_N = S_1 \\
&\vdots \\
y^{(N-1)}(0^-) &\rightarrow \lambda_1^{N-1} K_1 + \lambda_2^{N-1} K_2 + \cdots + \lambda_N^{N-1} K_N = S_{N-1}
\end{aligned}
\qquad (12.30)
$$

which are N equations in the N unknowns $K_n, n = 1, 2, \ldots, N$ that can be written as $\Lambda K = S$, where

$$
\Lambda = \begin{bmatrix} 1 & 1 & \cdots & 1 \\ \lambda_1 & \lambda_2 & \cdots & \lambda_N \\ \vdots & \vdots & \vdots & \vdots \\ \lambda_1^{N-1} & \lambda_2^{N-1} & \cdots & \lambda_N^{N-1} \end{bmatrix}, \quad
K = \begin{bmatrix} K_1 \\ K_2 \\ \vdots \\ K_N \end{bmatrix}, \quad
S = \begin{bmatrix} S_0 \\ S_1 \\ \vdots \\ S_{N-1} \end{bmatrix} \qquad (12.31)
$$

Table 12.3   Characteristic modes

Distinct root of (12.28)	Contributions to response
$\lambda = 0$	K, a constant
$\lambda = \sigma$, a real number	$K\,e^{\sigma t}$, increasing, $\sigma > 0$, or decreasing, $\sigma < 0$ exponential function
$\lambda = \pm j\omega$	$2\|K\|\cos(\omega t + \angle K)$, a sinusoidal function
$\lambda = \sigma \pm j\omega$	$2\|K\|e^{\sigma t}\cos(\omega t + \angle K)$, an exponentially increasing, $\sigma > 0$, or exponentially decreasing, $\sigma < 0$, sinusoidal function

Then, $K = \Lambda^{-1}S$.

The time domain behavior of the response to ICs depends on the $\lambda_n, n = 1, 2, \ldots, N$, and each $e^{\lambda_n t}$ is called a **characteristic mode** of the system. If a natural frequency λ_n occurs with multiplicity k, then the corresponding k characteristic modes are $t^m e^{\lambda_n t}, m = 0, 1, \ldots, k - 1$. For all possible kinds of values of λ, Table 12.3 gives the kinds of characteristic modes that contribute to $y_{zi}(t)$.

For example, suppose two values of λ are $\lambda = \sigma \pm j\omega$, and, since $y_{zi}(t)$ is a real time function, the contribution to $y_{zi}(t)$ is

$$
\begin{aligned}
K\,e^{(\sigma+j\omega)t} + K^*e^{(\sigma-j\omega)t} &= \|K\|e^{j\angle K}\,e^{(\sigma+j\omega)t} + \|K\|e^{-j\angle K}e^{(\sigma-j\omega)t} \\
&= \|K\|e^{\sigma t}\big(e^{j(\omega t + \angle K)} + e^{-j(\omega t + \angle K)}\big) \\
&= 2\|K\|e^{\sigma t}\cos(\omega t + \angle K)
\end{aligned}
$$

Notice that λ is multiplied by t, and therefore, the unit of λ must be sec^{-1}. The roots of the characteristic equation are called the **natural frequencies** of the system, which are an intrinsic property of the system and do not depend on the input or ICs. The imaginary part of λ is the frequency of characteristic mode oscillation. Furthermore, $\lim_{t \to \infty} y_{zi}(t) = 0$, if and only if all of the natural frequencies have negative real parts.

Example 12.12

Find the zero-input response of the low-pass filter described by (12.20). Let the component values be $L_1 = 3/2\ H$, $L_2 = 1/2\ H$, $C = 4/3\ F$, and $R = 1\ \Omega$, and suppose the ICs are $v^{(2)}(0^-) = 0$, $v^{(1)}(0^-) = 1$, and $v(0^-) = 1$. Equation (12.20) becomes

$$v^{(3)}(t) + 2v^{(2)}(t) + 2v^{(1)}(t) + v(t) = v_s(t) \tag{12.32}$$

and we must solve

$$v_{zi}^{(3)}(t) + 2v_{zi}^{(2)}(t) + 2v_{zi}^{(1)}(t) + v_{zi}(t) = 0$$

Prog. 12.10 finds the natural frequencies and plots the zero-input response, which is shown in Fig. 12.26.

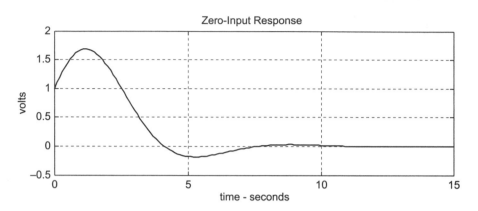

Figure 12.26   Zero-input response.

```
% Example of finding the zero-input response
clear all; clc
a  =  [1  2  2  1];  % specify coefficients of characteristic polynomial
disp('The natural frequencies are:')
lambda  =  roots(a)  % get natural frequencies
lambda  =  lambda.';  % transpose without conjugating
S  =  [1; 1; 0];  % specify initial conditions (the initial state)
Lambda  =  zeros(3,3); % preallocate space
Lambda(1,:)  =  ones(1,3); % first row
for n  =  2:3  % remaining rows of Lambda
    Lambda(n,:)  =  Lambda(n-1,:).*lambda;
end
K  =  inv(Lambda)*S; % get coefficients of exponential terms
t  =  0:0.05:15; % specify time points
v_zi  =  zeros(1,length(t)); % initialize sum
for n  =  1:3 % sum exponential terms
    v_zi  =  v_zi + K(n)*exp(lambda(n)*t);
end
plot(t,real(v_zi)); % imaginary parts can be very small nonzero numbers
grid on; xlabel('time - seconds'); ylabel('volts')
title('Zero-Input Response')
```

Program 12.10   Program to find the natural frequencies and the zero-input response.

The natural frequencies are as follows:

```
lambda  =
  -1.0000
  -0.5000 + 0.8660i
  -0.5000 - 0.8660i
```

Since the natural frequencies all have negative real parts, $v_{zi}(t)$ goes to zero as t increases.

12.3.2 Zero-State Response

Assume that $M = N$, and to find the solution $y_{zs}(t)$ of (12.26), let us first consider a simpler problem given by

$$w^{(N)}(t) + a_{N-1}w^{(N-1)}(t) + \cdots + a_0 w^{(0)}(t) = r(t) \tag{12.33}$$

where the ICs are zero and the forcing function is $g(t) = r(t)$. For a given $r(t)$, suppose we can solve (12.33) for $w(t)$. If both sides of (12.33) are multiplied by b_0 to obtain

$$b_0 w^{(N)}(t) + a_{N-1} b_0 w^{(N-1)}(t) + \cdots + a_0 b_0 w^{(0)}(t) - b_0 r(t) \tag{12.34}$$

then, if the forcing function is $g(t) = b_0 r(t)$, the solution of (12.34) is $b_0 w(t)$. If we differentiate both sides of (12.33) and multiply by b_1 to obtain

$$\left(b_1 \frac{dw}{dt}\right)^{(N)} + a_{N-1}\left(b_1 \frac{dw}{dt}\right)^{(N-1)} + \cdots + a_0\left(b_1 \frac{dw}{dt}\right)^{(0)} = b_1 \frac{dr}{dt} \tag{12.35}$$

then, if the forcing function is $g(t) = b_1 r^{(1)}(t)$, the solution of (12.35) is $b_1 w^{(1)}(t)$. Since (12.33) is a linear equation, then if the forcing function is

$$g(t) = b_N r^{(N)}(t) + \cdots + b_1 r^{(1)}(t) + b_0 r^{(0)}(t)$$

which is the right side of (12.26), the solution is

$$y_{zs}(t) = b_N w^{(N)}(t) + \cdots + b_1 w^{(1)}(t) + b_0 w^{(0)}(t) \tag{12.36}$$

Thus, if we can find $w(t)$, then with (12.36) we can find $y_{zs}(t)$. Soon, this method will be applied to find $y_{zs}(t)$ for a particular $r(t)$.

12.3.3   State Variables

The ODE solvers provided by MATLAB solve systems of first order differential equations. Any N^{th} order ODE can be converted into a system of N first order differential equations. If the ODE is given by (12.17), which describes an LTI system, then we can convert (12.17) into

$$\dot{x}(t) = A\,x(t) + B\,r(t) \tag{12.37}$$

$$y(t) = C\,x(t) + D\,r(t) \tag{12.38}$$

where $x(t) = [x_1(t)\ x_2(t)\cdots x_N(t)]'$ is an $N \times 1$ column vector, called the **state vector**, $\dot{x}(t)$ means $dx/dt = [\dot{x}_1(t)\ \dot{x}_2(t)\cdots \dot{x}_N(t)]'$, and the dimensions of the constant matrices A, B, C and D are $N \times N$, $N \times 1$, $1 \times N$, and 1×1, respectively. Equation (12.37) is called the **state equation**, and (12.38) is called the **output equation**. Together, (12.37) and (12.38) are called a **state variable description** of an LTI system.

To solve (12.37) for $x(t)$ and find $y(t)$ with (12.38), which is the solution of (12.17), requires the initial state $x(0^-)$, which we can find given the ICs of (12.18). Using (12.38) gives

$$y(t) = C\,x(t) + D\,r(t), \quad \rightarrow \quad y(0^-) = S_0 = C\,x(0^-) + D\,r(0^-), \quad \rightarrow \quad C\,x(0^-) = S_0$$

where $r(t < 0) = 0$ was used. Again using (12.38) gives

$$y^{(1)}(t) = C\,x^{(1)}(t) + D\,r^{(1)}(t) = C(A\,x(t) + B\,r(t)) + D\,r^{(1)}(t), \quad \rightarrow \quad C\,A\,x(0^-) = S_1$$

Continue with derivatives of $y(t)$ until we get

$$C\,A^{N-1}x(0^-) = S_{N-1}$$

Using matrix notation gives

$$\begin{bmatrix} C \\ C\,A \\ \vdots \\ C\,A^{N-1} \end{bmatrix} x(0^-) = S, \quad S = \begin{bmatrix} S_0 \\ S_1 \\ \vdots \\ S_{N-1} \end{bmatrix} \quad \rightarrow \quad O\,x(0^-) = S$$

where O, which is called the **observability matrix**, is an $N \times N$ matrix. If O is non-singular, then $x(0^-) = O^{-1}\,S$. With $S = 0$ or $x(0^-) = 0$ (a vector of zeros), we get $x(t) = x_{zs}(t)$ with (12.37) and $y(t) = y_{zs}(t)$ with (12.38).

The matrices A, B, C, and D must be found such that the state variable description gives the same input $r(t)$ to output $y(t)$ relationship as (12.17). There are an infinite number of possibilities. A commonly used conversion method starts with (12.33) by defining

$$
\begin{aligned}
x_N(t) &= w^{(0)}(t), & \rightarrow & & \dot{x}_N(t) &= x_{N-1}(t) \\
x_{N-1}(t) &= w^{(1)}(t), & \rightarrow & & \dot{x}_{N-1}(t) &= x_{N-2}(t) \\
x_{N-2}(t) &= w^{(2)}(t), & \rightarrow & & \dot{x}_{N-2}(t) &= x_{N-3}(t) \\
&\vdots & & & & \\
x_2(t) &= w^{(N-2)}(t), & \rightarrow & & \dot{x}_2(t) &= x_1(t) \\
x_1(t) &= w^{(N-1)}(t), & \rightarrow & & \dot{x}_1(t) &= w^{(N)}(t) = r(t) - (a_{N-1}\,w^{(N-1)}(t) + \cdots + a_0 w^{(0)}(t)) \\
& & & & &= r(t) - (a_{N-1}x_1(t) + \cdots + a_0 x_N(t))
\end{aligned}
$$

$$(12.39)$$

Using matrix notation, the equations of (12.39) become

$$
\dot{x}(t) =
\begin{bmatrix}
-a_{N-1} & -a_{N-2} & \cdots & \cdots & -a_1 & -a_0 \\
1 & 0 & \cdots & 0 & 0 & 0 \\
0 & 1 & 0 & 0 & 0 & 0 \\
\vdots & 0 & \ddots & \ddots & \vdots & \vdots \\
0 & \vdots & 0 & 1 & 0 & 0 \\
0 & 0 & \cdots & 0 & 1 & 0
\end{bmatrix}
x(t) +
\begin{bmatrix}
1 \\ 0 \\ 0 \\ \vdots \\ 0 \\ 0
\end{bmatrix}
r(t)
\qquad (12.40)
$$

which defines the matrices A and B in (12.37). Equation (12.40) is a set of coupled N first-order LTI differential equations.

Like (12.36), we can find $y_{zs}(t)$ with

$$
\begin{aligned}
y_{zs}(t) &= b_N w^{(N)}(t) + b_{N-1}w^{(N-1)} + \cdots + b_1 w^{(1)}(t) + b_0 w^{(0)}(t) \\
&= b_N w^{(N)}(t) + b_{N-1}x_1(t) + \cdots + b_1 x_{N-1}(t) + b_0 x_N(t) \\
&= b_N(r(t) - (a_{N-1}x_1(t) + \cdots + a_0 x_N(t))) + b_{N-1}x_1(t) + \cdots + b_1 x_{N-1}(t) + b_0 x_N(t)
\end{aligned}
$$

$$(12.41)$$

Gathering terms and using matrix notation, (12.41) becomes

$$
y_{zs}(t) = [\,b_{N-1} - b_N a_{N-1} \quad b_{N-2} - b_N a_{N-2} \quad \cdots \quad b_1 - b_N a_1 \quad b_0 - b_N a_0\,]x(t) + [b_N]r(t)
$$

$$(12.42)$$

which defines the matrices C and D in (12.38). If $M < N$, in which case $b_N = 0$, (12.42) becomes

$$y_{zs}(t) = [0 \quad \cdots \quad 0 \quad b_M \quad b_{M-1} \quad \cdots \quad b_0]x(t) + [0]r(t) \tag{12.43}$$

If we set $x(0^-) = O^{-1}S$, then (12.42) or (12.43) gives $y(t)$, the complete solution of (12.17).

Recall from (3.55) that the eigenvalues of a matrix, such as the matrix A in (12.37), are the roots of

$$Q(\lambda) = |(\lambda I - A)| = 0 \tag{12.44}$$

To be specific, consider the case $N = 3$, and let us find $Q(\lambda)$ using functions in the Symbolic Math Toolbox. The following statements give $Q(\lambda)$.

```
>> A  =  sym('A%d%d',[3 3])   % create a symbolic 3x3 matrix
A  =
[ A11, A12, A13]
[ A21, A22, A23]
[ A31, A32, A33]
>> syms a2 a1 a0  L  % L denotes lambda
>> A  =  [-a2 -a1 -a0; 1 0 0; 0 1 0]   % define A
A  =
[ -a2, -a1, -a0]
[   1,   0,   0]
[   0,   1,   0]
>> I  =  sym('I%d%d',[3 3])   %  create a symbolic identity matrix
>> I  =  [1 0 0;0 1 0;0 0 1];
>> Q  =  det(L*I - A)   % get characteristic polynomial
Q  =    L^3 + a2*L^2 + a1*L + a0
```

This $Q(\lambda)$ is the same polynomial as in (12.28), for the case $N = 3$. This is an example of the fact that the eigenvalues of the A matrix in (12.37) are the natural frequencies of the LTI CTS, and (12.44) is also the characteristic equation.

The MATLAB Control System Toolbox contains the function **tf2ss** that receives the coefficients of (12.17) and returns the matrices A, B, C, and D. Its syntax is given by

```
[A, B, C, D]  =  tf2ss(b,a)
```

For example, let $a = [a_N \cdots a_1 \ a_0] = [1 \ 2 \ 4 \ 6 \ 8]$ and $b = [b_N \cdots b_1 \ b_0] = [0 \ 0 \ 1 \ 3 \ 5]$, then we get

```
>> a  =  [1  2  4  6  8];  %  N  =  4 order LTI differential equation
>> b  =  [0  0  1  3  5];  %  M  =  2
```

```
>> % convert LTI differential equation to state variable form
>> [A, B, C, D] = tf2ss(b,a);
>> B_transpose  =  B';  %  will display B' to save space
>> %  the state variable description is
>> A,B_transpose,C,D
A  =
      -2     -4     -6     -8
       1      0      0      0
       0      1      0      0
       0      0      1      0
B_transpose  =        1      0      0      0
C  =       0      1      3      5
D  =       0
```

The built-in function **ss2tf** converts a state variable model of an LTI system into the model given by (12.17). Its syntax is

$$[\text{b} \quad \text{a}] \quad = \quad \text{ss2tf}(A,B,C,D)$$

Another useful function in the Control System Toolbox is the function **obsv**, which receives the matrices A and C, and returns the observability matrix O. Its syntax is given by

$$O \quad = \quad \text{obsv}(A,C)$$

For example, with the matrices A and C, as obtained above, we get

```
>> O  =  obsv(A,C)     % get observability matrix
O  =
       0      1      3      5
       1      3      5      0
       1      1     -6     -8
      -1    -10    -14     -8
>> det_O  =  det(O)     % check if the observability matrix is nonsingular
det_O  =     159
```

With the observability matrix we can convert the ICs given by (12.18) into the initial state required by (12.37).

The matrices A, B, C, and D can be combined into an object with the built-in function **ss**, which has the syntax given by

$$\text{LTI\_sys} \quad = \quad \text{ss}(A,B,C,D)$$

For example, with the matrices A, B, C, and D, as obtained above, we get

```
>> LTI_sys  =  ss(A,B,C,D);
>> % the matrices A, B, C and D are contained in the fields of LTI_sys
>> A = LTI_sys.A
A  =
     -2     -4     -6     -8
      1      0      0      0
      0      1      0      0
      0      0      1      0
```

The object LTI_sys can be used in another syntax of **obsv**, given by

$$O = obsv(\textbf{LTI\_sys})$$

which is equivalent to obsv(LTI_sys.A,LTI_sys.C). You should browse through the Control System Toolbox to see the many built-in functions concerned with control system analysis and design.

Analytical methods that find the solution of (12.17), an LTI CTS, are well developed for $r(t)$ that can be expressed in terms of functions like those given in Table 12.3, while other than for special cases, there are no analytical methods that find the solution of nonlinear differential equations. Numerical approximation methods are applied to solve nonlinear differential equations. A large amount of literature reports a great variety of numerical ODE solvers. Numerical methods can give very accurate results.

Table 12.4 gives explicit ODE solvers provided by MATLAB. The success of a numerical ODE solver depends on whether or not the ODE is a **stiff ODE**. A stiff ODE is an ODE that has a solution that can change drastically over a small change in the independent variable. Like Euler's integration method, all numerical integration methods segment a time range into small segments, called the **step size**. In Euler's integration method, the step size is fixed. However, to solve stiff ODEs a numerical integration method must include a means with which to assess how rapidly the solution is changing, and adaptively reduce the step size to maintain accurate results. Then, when the solution tends to change less rapidly, the step size must be automatically increased to reduce computing time. Stiff ODE solvers can require much more computing time than nonstiff solvers.

Table 12.4   MATLAB ODE solvers

Solver	Problem type	Accuray	Usage
ode45	nonstiff	medium	Should be the first solver you try
ode23	nonstiff	low	Problems with low error tolerances or for solving moderately stiff problems
ode113	nonstiff	low to high	Problems with stringent error tolerances or for solving computationally intensive problems
ode15s	stiff	low to medium	If ode45 is slow because the problem is stiff
ode23s	stiff	low	Low error tolerances to solve stiff systems
ode23t	moderately stiff	low	Moderately stiff problems
ode23tb	stiff	low	Low error tolerances to solve stiff systems

The ODE solvers provided by MATLAB solve the system of first-order differential equations written as

$$\frac{dx}{dt} = F(t, x) \tag{12.45}$$

where $x(t)$ is an $N \times 1$ vector, dx/dt means $[dx_1/dt \quad dx_2/dt \quad \cdots \quad dx_N/dt]'$, and $F(t, x)$ is a vector function. The syntax is given by

```
[Tout,Xout]  =  ode_solver(@odefun,tspan,x0)

[Tout,Xout]  =  ode_solver(@odefun,tspan,x0,options)
```

where @odefun is the handle of a function that evaluates $F(t, x)$, tspan is $[t_{\text{initial}} \ t_{\text{final}}]$, x_0 is a vector that holds the initial state $x(t_{\text{initial}}^-)$. For a scalar t and a vector x, odefun(t,x) must return a column vector given by $F(t, x)$. Each row in the solution matrix Xout is the state vector at a time given in the corresponding row of the column vector Tout. To obtain solutions at specific times, set the vector tspan to a time point sequence. With the second syntax, you can use the function **odeset** to set parameter values concerned with the integration method. Use **doc ode45** for details.

To assess the behavior of an LTI CTS described by (12.17) or (12.37) and (12.38), it is useful to see its unit step or impulse response. The **impulse** (11.10) (also called the **Dirac delta function**) function was introduced in Chapter 11, and its definition, repeated here for convenience, is given by

$$\delta(t \quad t_0) = \begin{cases} \infty, & t = t_0 \\ 0, & t \neq t_0 \end{cases}$$

where t_0 is an arbitrary time point. It was also shown in Chapter 11 that

$$\int_{-\infty}^{t} \delta(\tau - t_0) \, d\tau = u(t - t_0), \quad \rightarrow \quad \frac{d \, u(t - t_0)}{d \, t} = \delta(t - t_0)$$

The impulse function is nonzero only when the argument is zero. Therefore

$$\int_{-\infty}^{+\infty} v(t) \, \delta(t - t_0) \, dt = \int_{t_0-\varepsilon}^{t_0+\varepsilon} v(t) \, \delta(t - t_0) \, dt = \int_{t_0-\varepsilon}^{t_0+\varepsilon} v(t_0) \, \delta(t - t_0) \, dt$$

$$= v(t_0) \int_{t_0-\varepsilon}^{t_0+\varepsilon} \delta(t - t_0) \, dt = v(t_0)$$

where ε is an arbitrarily small positive number and $v(t)$ is a function of t that is continuous at t_0. This is called the **sifting property** of the impulse function. For example, a sample of the time function $v(t) = e^{-2t}u(t)$ at the time $t = 5$ can be obtained with

$$\int_{-\infty}^{\infty} e^{-2t}u(t) \, \delta(t - 5) \, dt = \int_{0}^{\infty} e^{-2t} \, \delta(t - 5) \, dt = e^{-10}$$

Example 12.12 (continued) ──────────────────────────────

Let us apply a MATLAB ODE solver to find the zero-input, zero-state, and complete responses of the low-pass filter shown in Fig. 6.13, which is described by (12.32). Prog. 12.11 gives the ODE function. It receives the matrices A and B and the input r as global variables.

```
function x_dot  =  state_function(t,x)
% function to compute: Ax+Br, where r is a constant
global A B r
x_dot  =  A*x + B*r;
end
```

Program 12.11   Function to compute the vector state function.

The statement to find x_dot looks simple. However, since most of the elements in the matrix A are zero, it is more efficient to find x_dot with

```
x_dot  =  [A(1,:)*x + r; x(1); x(2)];
```

Prog. 12.12 converts the $N = 3$ order differential equation (12.32) into a state variable model. Then, it finds the observability matrix to convert the given ICs into the initial state, which is used by ode45 to find $v_{zi}(t)$. The input is $r(t) = u(t)$, which is used by ode45 to find $v_{zs}(t)$.

```
clear all; clc
global A B r
a  =  [1 2 2 1]; b  =  [0 0 0 1];  % specify coefficients
[A B C D]  =  tf2ss(b,a); % get state variable description
O  =  obsv(A,C); % get observability matrix
tspan  =  0:0.05:15; % specify time points
r  =  0; % set input to zero for zero-input response
S  =  [1; 1; 0] % specify initial conditions
x0  =  inv(O)*S % get initial state
[t x] = ode45(@state_function,tspan,x0) % get state vector versus time
% each column of x' is the state vector at the corresponding time in t
v_zi  =  C*x' + D*r; % get zero-input response
plot(t,v_zi,'b'); grid on; hold on
xlabel('time - seconds'); ylabel('volts')
r  =  1; % specify input, a unit step
```

```
x0  =  [0; 0; 0] % specify initial state
[t x]  =  ode45(@state_function,tspan,x0);
v_zs  =  C*x' + D*r; % get zero-state response
plot(t,v_zs,'r')
v  =  v_zi + v_zs;
plot(t,v,'k')
title('Zero-Input, Zero-State and Complete Response')
```

Program 12.12   Program to find the complete response of a low-pass filter.

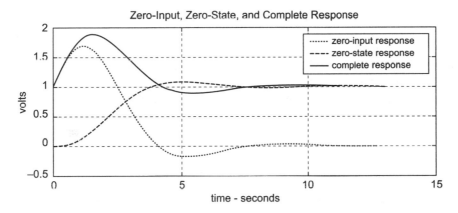

Figure 12.27   Complete unit step response.

The zero-input response has been found in two different ways. Fig. 12.26 shows the analytical solution obtained with (12.29), and Fig. 12.27 shows the numerical solution obtained with the ODE solver ode45. The results are substantially the same.

Let us examine Fig. 12.27 a little more. Notice that $v_{zi}(t)$ goes to zero because the natural frequencies have negative real parts. Until $v_{zi}(t)$ is nearly zero, it goes through transient behavior (exponential or oscillatory), which is determined by the natural frequencies. The input is a constant, and as $t \to \infty$, $v_{zs}(t)$ becomes a constant. This part of $v_{zs}(t)$ is called the **steady-state response**. The steady-state response is a function like the input. Until $v_{zs}(t)$ is nearly the same kind of function as the input, it goes through a transient behavior, which is also determined by the natural frequencies. The complete response $v(t)$ goes through transient behavior due to ICs and the input, and in steady state it behaves like the input. All LTI CTS, which have natural frequencies with negative real parts, behave in this manner.

If any natural frequency has a positive real part, then $v(t)$ will grow without bound, in which case, a physical system that behaves like this model will self-destruct. If any distinct natural frequency has a zero real part and a nonzero imaginary part, then (see Table 12.3) $v(t)$ will undergo a sustained oscillation, regardless of the input or the ICs. This can be useful if we want to build an oscillator. However, it would not be useful if, for example, the LTI CTS is involved in the control of the front wheels of your car, as they react to action by the steering mechanism.

The state variable description used in Example 12.12 is not unique. There are many other useful possibilities.

Example 12.13

Let us again apply a MATLAB ODE solver to find the unit pulse and unit step responses of the low-pass filter shown in Fig. 6.13, which is repeated for convenience in Fig. 12.28.

While we can work with (12.20), a third-order ODE, which can be converted into three first order differential equations, let us use a state variable description where the inductor

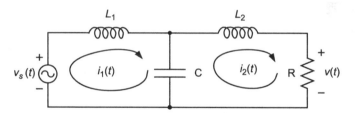

Figure 12.28   Circuit of a third-order Butterworth low-pass filter.

currents and capacitor voltage become the elements of the **state vector**. Let $x_1(t) = i_{L_1}(t) = i_1(t)$, $x_2(t) = i_{L_2}(t) = i_2(t)$, and $x_3(t) = v_C(t)$. Applying KVL to the two meshes and using the notation $\dot{x}(t) = dx/dt$ give

$$-v_s(t) + L_1 \frac{di_{L_1}}{dt} + v_C(t) = 0 \quad \rightarrow \quad \dot{x}_1(t) = -\frac{1}{L_1}x_3(t) + \frac{1}{L_1}v_s(t)$$

$$-v_C(t) + L_2 \frac{di_{L_2}}{dt} + Ri_{L_2}(t) = 0 \quad \rightarrow \quad \dot{x}_2(t) = -\frac{R}{L_2}x_2(t) + \frac{1}{L_2}x_3(t) \tag{12.46}$$

Applying KCL to the middle node gives

$$-i_{L_1}(t) + C\frac{dv_C}{dt} + i_{L_2}(t) = 0 \quad \rightarrow \quad \dot{x}_3(t) = \frac{1}{C}x_1(t) - \frac{1}{C}x_2(t) \tag{12.47}$$

Equations (12.46) and (12.47) are $N = 3$ first-order differential equations that model the behavior of the circuit. For linear systems these equations can be written using matrix notation, and become

$$\dot{x}(t) = \begin{bmatrix} 0 & 0 & -1/L_1 \\ 0 & -R/L_2 & 1/L_2 \\ 1/C & -1/C & 0 \end{bmatrix} x(t) + \begin{bmatrix} 1/L_1 \\ 0 \\ 0 \end{bmatrix} v_s(t) \tag{12.48}$$

where $x(t) = [x_1(t) \quad x_2(t) \quad x_3(t)]\,'$, a column vector, is the state vector, and (12.48) is the **state equation** of the circuit. To obtain the output, we have $v(t) = R\,x_2(t)$, or

$$v(t) = [0 \quad R \quad 0]x(t) + [0]v_s(t) \tag{12.49}$$

which is the **output equation**. Together, (12.48) and (12.49) are a **state variable description** of the circuit.

Let us apply the MATLAB ODE solver **ode45** to (12.48). Using the same component values as in Example 12.12, Prog. 12.13 gives the ODE function, where the input is a unit pulse, and Prog. 12.14 obtains the zero-state response. The unit pulse response is shown in Fig. 12.29. Notice that even though the input is nonzero for a very small time duration, the output does not become very small until around $t - 10$ sec. This is **transient behavior**. Fig. 12.29 also shows the unit step response, for which $v_s = 1$ for $t \geq 0$ was used.

```
function dxdt  =  low_pass_pulse(t,x)
% evaluate the state equation of a low-pass filter
global A13 A22 A23 A31 A32 B1
% specify a unit pulse
vs = 0; if t == 0, vs = 1; end; % input is one for one time step
% for the unit step response, replace the line above with: vs = 1;
% evaluate the derivative of each component of the state vector
x1dot  =  A13*x(3)+B1*vs;
x2dot  =  A22*x(2)+A23*x(3);
x3dot  =  A31*x(1)+A32*x(2);
dxdt  =  [x1dot; x2dot; x3dot]; % state vector derivative
end
```

Program 12.13   ODE function for the low-pass filter.

```
% Program to solve the ODE of a Butterworth third order low-pass filter
clear all; clc
global A13 A22 A23 A31 A32 B1
L1  =  3/2; L2  =  1/2; C  =  4/3; R  =  1; % circuit component values
A13  =  -1/L1; A22  =  -R/L2; A23  =  1/L2; A31  =  1/C; A32  =  -1/C; B1  =  1/L1;
```

```
tspan  =  linspace(0,20,1001); % time points
x0  =  [0; 0; 0]; % use zero initial state for the unit pulse response
[t x]  =  ode45(@low_pass_pulse,tspan,x0); % x is the state vector
v_zs  =  R*x(:,2); v_zs  =  v_zs/max(abs(v_zs)); % get output and normalize it
plot(t,v_zs); grid on
xlabel('time - seconds'); ylabel('output - volts')
title('Low-Pass Filter Unit Pulse and Unit Step Response')
[t x]  =  ode45(@low_pass_step,tspan,x0); % x is the state vector
v_zs  =  R*x(:,2); % get output
hold on; plot(t,v_zs)
```

Program 12.14   Program to find the unit pulse and unit step responses.

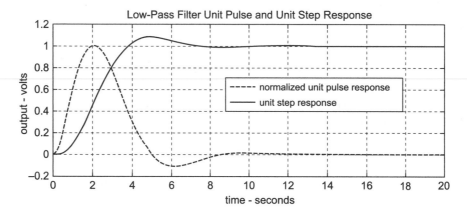

Figure 12.29   Unit step response and normalized unit pulse response.

Regardless of which state variable description you use, the input to output relationship is preserved. The state variables used here have physical meaning. The eigenvalues of the A matrix in (12.48) are given by

```
>> L1 = 3/2; L2 = 1/2; C = 4/3; R = 1; % circuit component values
>> A  =  [0 0 -1/L1; 0 -R/L2 1/L2; 1/C -1/C 0];
>> E  =  eig(A)
E  =
  -0.5000 + 0.8660i
  -0.5000 - 0.8660i
  -1.0000
```

which are the natural frequencies found in Example 12.12.

To better understand the impact of transient behavior on the complete response, let us apply a sinusoidal input given by $v_s(t) = \sin(\omega_0 t)\, u(t)$, where $\omega_0 = 1.0$ rad/sec. For this $v_s(t)$, the line in Prog. 12.13 that defines vs was replaced by vs = sin(t). The response is shown in Fig. 12.30, which was obtained by replacing the plotting statements in Prog. 12.14 with the following statements:

```
subplot(1,2,1); plot(t,v_zs); grid on
xlabel('time - seconds'); ylabel('output')
title('Low-Pass Filter Response')
subplot(1,2,2); plot3(x(:,1),x(:,2),x(:,3)); grid on
xlabel('L1 current'),ylabel('L2 current'),zlabel('C voltage')
title('State Vector Phase Diagram')
```

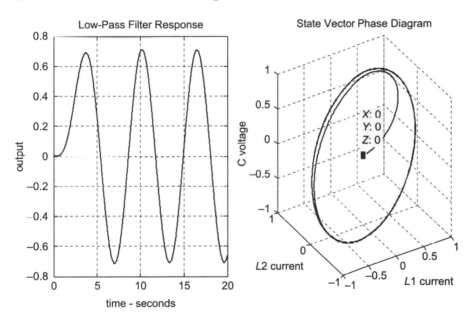

Figure 12.30   Sinusoidal response and phase diagram of the state vector.

Since the bandwidth of the filter is BW = 1 rad/sec and the input frequency is 1 rad/sec, the amplitude of the steady-state sinusoidal response is $1/\sqrt{2} = 0.707$. We see in Fig. 12.30 that this amplitude is not reached until around 10 sec, where the **transient behavior** has become negligible. The plot on the right of Fig. 12.30 shows the trajectory of the state vector, called a **phase diagram**, in a 3-D space from its initial state, through its transient and then **steady-state** oscillatory behavior.

To demonstrate the need of a stiff ODE solver, let us work with a nonlinear system.

Example 12.14

Solve the ODE that describes the AC to DC converter shown in Fig. 12.25. First, we must write the ODE in the form required by MATLAB ODE solvers, which is the first-order ($N = 1$) equation given by

$$\frac{dv}{dt} = F(t, v) = \frac{I_s}{C}(e^{(v_s(t)-v(t))/V_T} - 1) - \frac{v(t)}{RC}$$

A MATLAB function of the ODE function $F(t, v)$ is given in Prog. 12.15. Prog. 12.16 uses the ODE solver **ode15s**, which is specifically designed to solve stiff problems. The function **odeset** is used to set ODE solver parameters to values that achieve more accurate results than with default parameter values.

```
function dvdt  =  AC_DC(t,v)
% function of the ODE for the AC to DC converter
global R C w A  % parameters from user program
Is  =  1e-12; VT  =  25.85e-3; % specify diode parameters
vs  =  A*sin(w*t); % input sinusoid
dvdt  =  (Is/C)*(exp((vs-v)/VT)-1)-v/(R*C); % ODE function
end
```

Program 12.15   ODE function.

```
% Program to solve the ODE of an AC to DC converter
clear all; clc
global R C w A
f  =  60; w  =  2*pi*f; A  =  6.3; % sinusoid frequency is 60 Hz
R  =  100; C  =  1000e-6; % load resistance (Ohms), capacitor (Farads)
T0  =  1/f; tspan  =  linspace(0,4*T0,1001); % time points
v0  =  0; % initial condition
options = odeset('RelTol',1e-6,'AbsTol',1e-12); % stiff ODE parameter values
[tout vout]  =  ode15s(@AC_DC,tspan,v0,options);
vin  =  A*sin(w*tout);
plot(tout,vin,'k'); hold on; plot(tout,vout,'r'); grid on
xlabel('time - seconds'); ylabel('volts')
title('Input and Output of an AC to DC Converter')
R  =  33; % smaller load resistance
[tout vout]  =  ode15s(@AC_DC,tspan,v0,options);
plot(tout,vout,'b');
```

Program 12.16   Program to find the AC to DC converter output.

The input and output of the AC to DC converter are shown in Fig. 12.31. Starting at the IC $v(t = 0^-) = 0$ volts, the input charges the capacitor until the input decreases below the capacitor voltage. Then, the diode stops conducting and the capacitor starts to discharge through the resistor (called the load) until the input again becomes large enough to recharge the capacitor. The output voltage is not perfectly constant, and ripples more when the resistance of the load is decreased. We can increase the capacitor value to decrease the ripple in the output voltage. A better converter design is needed to achieve a more constant output voltage and tolerate load resistance variations. This depends on the needs of an application that is powered by an AC to DC converter.

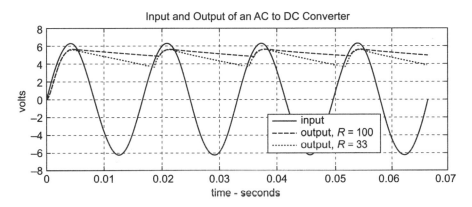

Figure 12.31   Input and output of the AC to DC converter.

Here, with an AC input amplitude of 6.3 volts, we get a DC output voltage of about 5 volts.

12.3.4   Impulse Response

A very useful function to know about an LTI CTS is its impulse response. The impulse response is found with $r(t) = \delta(t)$ and with all ICs set to zero. Then, the solution $y(t)$ of (12.17) or (12.37) and (12.38) is given the special notation $h(t) = y_{zs}(t)$. The function $h(t)$ is called the **impulse response**, and it is the solution of

$$h^{(N)}(t) + a_{N-1}\, h^{(N-1)}(t) + \cdots + a_0\, h^{(0)}(t) = b_N \delta^{(N)}(t) + \cdots + b_1\, \delta^{(1)}(t) + b_0\, \delta^{(0)}(t)$$

$$(12.50)$$

where for now, let $M = N$. By its definition, we have $h^{(n)}(0^-) = 0$, $n = 0, 1, \ldots, N - 1$. Soon, we will see that with the impulse response $h(t)$, the response $y_{zs}(t)$ to any input $r(t)$ can be found.

The right side of (12.50) is nonzero only at $t = 0$. At $t = 0$, the right side of (12.50) includes $b_N \delta^{(N)}(t)$. Therefore, to maintain equality, the left side of (12.50) must also include $b_N \delta^{(N)}(t)$. This can only happen if at $t = 0$, $h(t) = b_N \delta(t)$, which then gives $h^{(N)}(t) = b_N \delta^{(N)}(t)$, as required. However, if $M < N$ ($b_N = 0$), then the highest derivative of $\delta(t)$ in the right of (12.50) is due to the term $b_M \delta^{(M)}(t)$, and to maintain equality, the highest derivative of $\delta(t)$ that can appear in the left side of (12.50) is $\delta^{(M)}(t)$. Therefore, if $M < N$, then $h(t)$ cannot include an impulse $t = 0$.

Like we did with (12.33), let us solve the simpler problem given by

$$w^{(N)}(t) + a_{N-1}\, w^{(N-1)}(t) + a_{N-2}\, w^{(N-2)}(t) + \cdots + a_1\, w^{(1)}(t) + a_0\, w^{(0)}(t) = \delta\,(t)$$

$$(12.51)$$

and then, as with (12.36), for $t \geq 0$

$$h(t) = \begin{cases} b_N \delta(t) + b_N w^{(N)}(t) + \cdots + b_1\, w^{(1)}(t) + b_0\, w^{(0)}(t), & M = N \\ b_M w^{(M)}(t) + \cdots + b_1\, w^{(1)}(t) + b_0\, w^{(0)}(t), & M < N \end{cases} \qquad (12.52)$$

The N ICs of (12.51) are $w^{(n)}(0^-) = 0$, $n = 0, 1, \cdots, N-1$. To maintain the equality of (12.51) at $t = 0$, $w^{(N-1)}(t)$ must go through a step change at $t = 0$ to make $w^{(N)}(t) = \delta(t)$ at $t = 0$, which matches the right side of (12.51). The derivatives $w^{(n)}(t)$, $n = 0, 1, \ldots, N-2$ cannot go through a change at $t = 0$, because if, for example, $w^{(N-2)}(t)$ goes through a step change at $t = 0$, then at $t = 0$, $w^{(N-1)}(t) = \delta(t)$ and $w^{(N)}(t) = d\delta(t)/dt$, and there is no $d\delta(t)/dt$ on the right side of (12.51) at $t = 0$.

For $t > 0$, (12.51) becomes

$$w^{(N)}(t) + a_{N-1}\, w^{(N-1)}(t) + \cdots + a_0\, w^{(0)}(t) = 0, \quad t > 0 \qquad (12.53)$$

with ICs given by $w^{(n)}(0^+) = w^{(n)}(0^-) = 0$, $n = 0, 1, \ldots, N-2$ and, since $w^{(N-1)}(t)$ must go through a step change at $t = 0$, $w^{(N-1)}(0^+) = 1$. Since (12.53) is a homogeneous equation, its solution can be found in the same way that was used to solve (12.25), and therefore

$$w(t) = \sum_{n=1}^{N} K_n\, e^{\lambda_n t}, \quad t > 0 \qquad (12.54)$$

To find the constants K_n, $n = 0, 1, \ldots, N-1$, apply the ICs at $t = 0^+$ to get $\Lambda K = S$, where Λ and K are defined in (12.31), and $S = [0 \ 0 \ \cdots \ 0 \ 1]'$. With this $w(t)$, we can find the impulse response of (12.17) using (12.52). Notice that for $t > 0$, $w(t)$ is a linear combination of the characteristic modes of the system, and in view of (12.52), for $t > 0$, the impulse response of an LTI system is also a linear combination of its characteristic modes.

Example 12.15

Find the impulse response of the LTI CTS described by

$$y^{(2)}(t) + 7y^{(1)}(t) + 12y(t) = 2r^{(2)}(t) + 6r^{(1)}(t) + 4r(t)$$

Here, $M = N = 2$. To find $h(t)$, let $r(t) = \delta(t)$, and set $y^{(1)}(0^-) = 0$ and $y(0^-) = 0$. First, according to (12.53), we must solve

$$w^{(2)}(t) + 7w^{(1)}(t) + 12w(t) = 0, \quad t > 0, \quad w^{(1)}(0^+) = 1, \quad w(0^+) = 0$$

The characteristic equation is

$$\lambda^2 + 7\lambda + 12 = 0 \quad \rightarrow \quad (\lambda + 3)(\lambda + 4) = 0$$

and the natural frequencies are $\lambda_1 = -3$ and $\lambda_2 = -4$. The function $w(t)$ is given by

$$w(t) = K_1 e^{-3t} + K_2 e^{-4t}, \quad t > 0$$

The constants K_1 and K_2 are the solution of

$$\begin{bmatrix} 1 & 1 \\ -3 & -4 \end{bmatrix} \begin{bmatrix} K_1 \\ K_2 \end{bmatrix} = \begin{bmatrix} 0 \\ 1 \end{bmatrix} \quad \rightarrow \quad \begin{bmatrix} K_1 \\ K_2 \end{bmatrix} = \frac{1}{-1} \begin{bmatrix} -4 & -1 \\ 3 & 1 \end{bmatrix} \begin{bmatrix} 0 \\ 1 \end{bmatrix} = \begin{bmatrix} 1 \\ -1 \end{bmatrix}$$

and $w(t) = e^{-3t} - e^{-4t}$. For $t \geq 0$, the impulse response is given by

$$\begin{aligned}
h(t) &= 2\delta(t) + 2w^{(2)}(t) + 6w^{(1)}(t) + 4w(t) \\
&= 2\delta(t) + 2\frac{d}{dt}(-3e^{-3t} + 4e^{-4t}) + 6(-3e^{-3t} + 4e^{-4t}) + 4(e^{-3t} - e^{-4t}) \\
&= 2\delta(t) + 2(9e^{-3t} - 16e^{-4t}) - 14e^{-3t} + 20e^{-4t} = 2\delta(t) + 4e^{-3t} - 12e^{-4t}, \quad t \geq 0
\end{aligned}$$

and, since the CTS is causal, we write

$$h(t) = 2\delta(t) + (4e^{-3t} - 12e^{-4t})u(t)$$

For $N > 3$ or complex natural frequencies it can be challenging to manually find the impulse response. Then, we must use computing methods. Example 12.13 demonstrates one possible numerical method by finding the unit pulse response.

For an impulse response, the input $r(t) = \delta(t)$ is applied at $t = 0$. Therefore, if a CTS is a causal system, then

$$h(t)|_{t<0} = 0$$

MATLAB has the built-in function called **impulse** that finds, starting at $t = 0$, the impulse response $h(t)$ of an LTI CTS given its state variable model. If D in the state variable model is not zero ($M = N$ in (12.17)), then the impulse function (see (12.52).) in $h(t)$ is ignored. To obtain a plot of $h(t)$, use one of the following syntax options.

```
impulse(LTI_sys)
```

where LTI_sys is the object returned by the function **ss**. The time range and number of plotted points are chosen automatically.

```
impulse(LTI_sys,t_final)
```

where $h(t)$ is plotted over the range $0 \leq t \leq t_{final}$.

```
impulse(LTI_sys,t)
```

where the time range t is specified with t = t_start:dt:t_end.
You can specify line color, line style, and marker type with, for example

```
impulse(LTI_sys1, 'r',LTI_sys2,'y-',LTI_sys3,'gx')
```

To obtain a vector of impulse response values and no plot, the syntax options are

```
[h,t]   =  impulse(LTI_sys)
[h,t,x] =  impulse(LTI_sys)
```

where the elements of h correspond to the time points given in t. The second syntax option also returns the state vector x versus time. Use **doc impulse** for more details.

Two related functions are the built-in function **step**, which finds the step response of an LTI CTS under zero ICs, and the built-in function **initial**, which finds the response due to ICs. The functions step and impulse have the same syntax. The syntax for the function initial is similar to the function impulse syntax, but must include an initial state in the argument list. Some syntax options are

```
initial(LTI_sys,x0)
initial(LTI_sys,x0,t)
[yzi,t,x] = initial(LTI_sys,x0)
```

Example 12.16 ———————————————————————————————

Obtain the impulse and step responses of the low-pass filter given in Example 12.12. Prog. 12.17 uses the functions impulse and step to plot $h(t)$ and the step response, which are shown in Fig. 12.32.

```
clear all; clc
a = [1 2 2 1]; b = [0 0 0 1]; % specify coefficients of differential eq.
[A B C D] = tf2ss(b,a); % convert to state variable description
LTI_sys = ss(A,B,C,D); % create an object
save('state_space_model.mat','LTI_sys'); % to be used later
subplot(1,2,1); impulse(LTI_sys,'k'); grid on
subplot(1,2,2); step(LTI_sys,'k'); grid on
```

Program 12.17   Program to find the impulse and step responses of an LTI system.

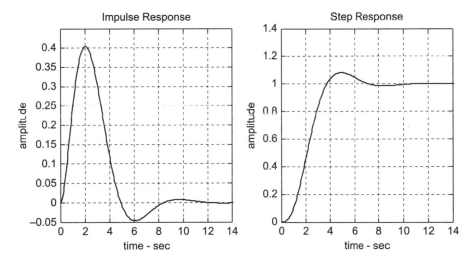

Figure 12.32   Impulse and step responses.

12.3.5   Convolution

Given the impulse response $h(t)$ of an LTI CTS, we can find the zero-state response of the system to an arbitrary input $r(t)$. To see how this can be done, we approximate the input with a sum of weighted pulses as shown in Fig. 12.33. The single pulse shown on the left of Fig. 12.33 has the property that $\lim_{T \to 0} p(t) = \delta(t)$. The height of the pulse $T\,p(t)$ is one.

On the right of Fig. 12.33, the pulse over $0 \leq t < T$ is $r(0)p(t)T$. Over $T \leq t < 2T$ the pulse is $r(T)p(t - T)T$, where $p(t - T)$ is $p(t)$ shifted to the right by T sec. If $r(t)$ is not zero for $t < 0$, then the pulse over $-T \leq t < 0$ is $r(-T)p(t + T)\, T$. In terms of weighted pulses, an approximation $\hat{r}(t)$ of $r(t)$ is given by

$$\hat{r}(t) = \sum_{k=-\infty}^{k=\infty} r(kT)\, p(t - kT)\, T \tag{12.55}$$

As T goes to zero, the summation in (12.55) becomes an integral, where $r(kT)$ becomes a function of a continuous variable, which will be denoted by τ, and we get

$$\hat{r}(t) = \sum_{k=-\infty}^{k=\infty} r(kT)\, p(t - kT)\, T \quad \rightarrow \quad \int_{-\infty}^{\infty} r(\tau)\delta(t - \tau)\, d\tau = r(t) \tag{12.56}$$

Let us consider the response of an LTI CTS to any one of the pulses in the sum of pulses

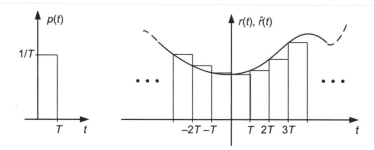

Figure 12.33   Approximation of $r(t)$ with a sum of weighted pulses.

in (12.55). Assume that for small T, the response to $p(t)$ is approximately $h(t)$. Since the system is time invariant, the response to $p(t - kT)$ is approximately $h(t - kT)$. Since the system is also linear, the response $\hat{y}_{zs}(t)$ to $\hat{r}(t)$ given by (12.55) is

$$\hat{y}_{zs}(t) = \sum_{k=-\infty}^{\infty} r(kT)\, h(t - kT)\, T \tag{12.57}$$

and as $T \rightarrow 0$, the summation in (12.57) becomes an integral given by

$$\hat{y}_{zs}(t) = \sum_{k=-\infty}^{\infty} r(kT)\, h(t - kT)T \quad \rightarrow \quad \int_{-\infty}^{\infty} h(t - \tau)\, r(\tau)\, d\tau = y_{zs}(t) \tag{12.58}$$

The integral in (12.58) is called the **convolution integral**. If we assume that the ICs of an LTI CTS are zero, then the complete response to an input $r(t)$ is $y(t) = y_{zs}(t)$ given by

$$y(t) = h(t) * r(t) = \int_{-\infty}^{\infty} h(t - \tau) \; r(\tau) \, d\tau \tag{12.59}$$

where the asterisk between $h(t)$ and $r(t)$ denotes the **convolution operation**. The convolution of any two signals $s_1(t)$ and $s_2(t)$ is commutative, that is, $s_1(t) * s_2(t) = s_2(t) * s_1(t)$, which means that

$$\int_{-\infty}^{\infty} s_1(t - \tau) \; s_2(\tau) \, d\tau = \int_{-\infty}^{\infty} s_1(\tau) \; s_2(t - \tau) \, d\tau$$

Example 12.17

The impulse response of an LTI CTS is $h(t) = b \, e^{-at} u(t)$. Manually, find the response $y(t)$ to a unit step input, $r(t) = u(t)$. The response is $y(t) = h(t) * r(t)$ given by

$$y(t) = \int_{-\infty}^{\infty} h(t - \tau) r(\tau) d\tau = \int_{-\infty}^{\infty} b \, e^{-a(t-\tau)} u(t - \tau) \, u(\tau) \, d\tau = b \int_{0}^{\infty} e^{-a(t-\tau)} u(t - \tau) d\tau$$

where $u(\tau)|_{\tau < 0} = 0$. Since τ in the rightmost integral is now nonnegative, then for $t < 0$, $u(t - \tau) = 0$, and therefore, $y(t)|_{t<0} = 0$. For $t > 0$, $u(t - \tau) = 0$ for $\tau > t$, and the integral for $y(t)$ becomes

$$y(t)|_{t>0} = b \int_{0}^{t} e^{-a(t-\tau)} d\tau = b \, e^{-at} \int_{0}^{t} e^{a\tau} d\tau = e^{-at} \frac{b}{a} (e^{a\tau}) \Big|_{0}^{t}$$

$$= e^{-at} \frac{b}{a} (e^{at} - 1) = \frac{b}{a} (1 - e^{-at})$$

and for all t, $y(t)$ is given by

$$y(t) = \frac{b}{a} (1 - e^{-at}) u(t)$$

In practical situations an expression for the input of an LTI CTS can be complicated, in which case, the convolution integral can be challenging to evaluate. Or, only samples of an input are available. Assume the input is zero for $t < 0$, and suppose we need to know $y(t)$

only for $t = nT$, $n = 0, 1, \ldots$, where T is some time increment. Then, (12.57) can be used to approximate $y(nT)$ with

$$y(nT) = \sum_{k=0}^{n} h(nT - kT)\, r(kT)T, \quad n = 0, 1, 2, \ldots$$

To simplify the notation, write

$$y(n) = T\sum_{k=0}^{n} h(n - k)\, r(k), \quad n = 0, 1, 2, \ldots \tag{12.60}$$

The built-in function **conv** computes the summation in (12.60), which is called **discrete convolution**. The syntax of the function conv is

```
c = conv(s1, s2)
```

where $s1$ is a vector that contains $N1$ samples of some signal $s_1(t)$, which could be the impulse response of some LTI CTS, and $s2$ is a vector that contains $N2$ samples of another signal $s_2(t)$, which could be the input of the LTI CTS. The length of the returned vector c is $N1+N2-1$.

Example 12.18

Find the response of the RC circuit to the pulse shown in Fig. 12.34. Summing the currents leaving the node at the output plus terminal gives

$$\frac{v(t) - v_s(t)}{R} + C\frac{dv}{dt} = 0 \quad \rightarrow \quad \frac{dv}{dt} + \frac{1}{RC}v(t) = \frac{1}{RC}v_s(t) \quad \rightarrow \quad \frac{dv}{dt} + a_0 v(t) = b_0 v_s(t)$$

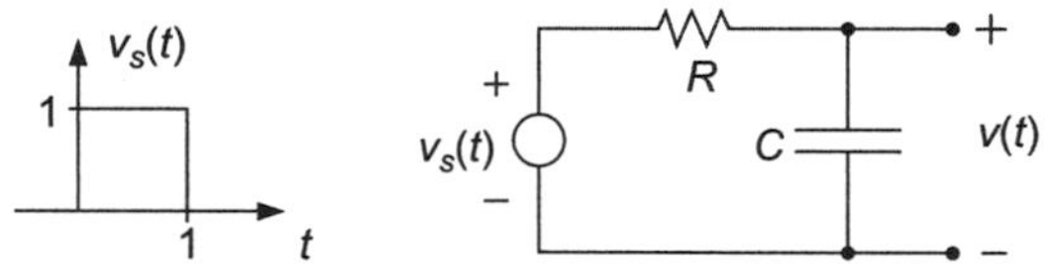

Figure 12.34 Pulse applied to an RC circuit.

Let $RC = 0.1$ sec. With (12.53), the impulse response is given by $h(t) = b_0\, e^{-a_0 t}\, u(t)$. Let $f_s = 2000$ samples/sec, and obtain samples of $v_s(t)$ over 2 sec and samples of $h(t)$ over 1 sec. Prog. 12.18 uses the function **conv** to find $v(t)$.

```
% Pulse response of an RC circuit
clear all; clc
fs  =  2000; % sampling rate
T   =  1/fs; % time increment
```

```
RC  =  0.1; a0  =  1/RC; b0  =  1/RC;  % time constant and coefficients
T2  =  1; N2  =  floor(T2/T)-1; % impulse response time range
t   =  [0:N2]*T;  h  =  b0*exp(-a0*t); % impulse response samples
T1  =  2; N1  =  floor(T1/T)-1; % input time range and number of samples
Nv  =  N1 + N2 -1; t  =  [0:Nv]*T; % convolution result time range
T_pulse  =  1; N_pulse  =  floor(T_pulse/T)-1; % pulse duration is 1 sec
vs  =  [ones(1,N_pulse),zeros(1,N1-N_pulse)]; % input samples
v   =  T*conv(vs,h); % obtain convolution approximation
subplot(1,3,1); plot(t,v,'k'); grid on; axis([0 3 -0.2 1.2])
xlabel('time - secs'); ylabel('volts'); title('Pulse Response')
v   =  v + 0.05*randn(1,length(v));  % introduce additive noise
subplot(1,3,2); plot(t,v,'k'); grid on; axis([0 3 -0.2 1.2])
xlabel('time - secs'); ylabel('volts'); title('Pulse Response')
T_pulse  =  0.1; N_pulse  =  floor(T_pulse/T)-1; % pulse duration is 0.1 secs
vs  =  [ones(1,N_pulse),zeros(1,N1-N_pulse)]; % input samples
v   =  T*conv(vs,h); % obtain convolution approximation
v   =  v + 0.05*randn(1,length(v));  % introduce additive noise
subplot(1,3,3); plot(t,v,'k'); grid on; axis([0 3 -0.2 1.2])
xlabel('time - secs'); ylabel('volts'); title('Pulse Response')
```

Program 12.18 Application of discrete convolution to approximate continuous convolution.

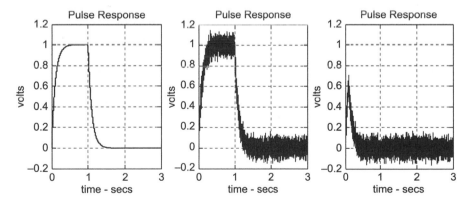

Figure 12.35   RC circuit pulse response.

The plot on the left in Fig. 12.35 shows the RC circuit output when the input is a pulse of duration 1 sec. Suppose this input pulse is applied to a long pair of wires. The RC circuit can be a model of the long pair of wires, where R is the series resistance of the wires and C accounts for the effect of two wires in close proximity. If over a long distance noise is introduced, then the plot in the center is the output at the other end of the pair of wires. The presence of a pulse is apparent. If the pulse duration is reduced, where, for example, it is 0.1 sec in the plot on the right, the presence of a pulse may not be so easily detected. If the objective is to communicate binary data over a long pair of wires, then Fig. 12.35 illustrates that the time duration of a bit cannot be arbitrarily small, which means that the bit rate is limited by the RC effect of the transmission wires.

12.3.6 Stability

From a practical viewpoint, the stability of a CTS is extremely important. Stability is defined in terms of the system's response to a bounded input, which means that for some positive and finite constant B, the input satisfies $|r(t)| \leq B$ for all t. A CTS is said to be **BIBO (bounded-input-bounded-output)** stable if and only if for a bounded input, the output is bounded.

For an LTI CTS, the output is given by (12.59), and to be bounded we consider

$$|y(t)| = \left| \int_{-\infty}^{\infty} h(t-\tau)\ r(\tau)\ d\tau \right| = \left| \int_{-\infty}^{\infty} h(\tau)\ r(t-\tau)\ d\tau \right|$$

$$\leq \left| \int_{-\infty}^{\infty} h(\tau)\ B\ d\tau \right| = B \left| \int_{-\infty}^{\infty} h(\tau)\ d\tau \right| \leq B \int_{-\infty}^{\infty} |h(\tau)| d\tau$$

Therefore, an LTI CTS is BIBO stable if the integral of the absolute value of the impulse response is finite. For an LTI CTS, the impulse response is a linear combination of the characteristic modes of the system (and possibly an impulse function). A characteristic mode is absolutely integrable if the real part of the corresponding natural frequency is negative. An LTI CTS is BIBO stable if and only if the real parts of all of its natural frequencies are negative.

12.3.7 Steady-State Response

Suppose an input, which is given by

$$r(t) = R\ e^{st} \tag{12.61}$$

where R and s are any real or complex numbers, has been applied to a BIBO stable LTI CTS described by (12.17) for a very long time. This means that all transient behavior incurred due to applying an input and due to ICs has become negligible. Substituting $r(t)$ into (12.17) gives

$$y^{(N)}(t) + a_{N-1}\ y^{(N-1)}(t) + \cdots + a_1\ y^{(1)}(t) + a_0\ y^{(0)}(t)$$
$$= b_M\ s^M R\ e^{st} + \cdots + b_1\ s\ R\ e^{st} + b_0\ R\ e^{st} = (b_M\ s^M + \cdots + b_1\ s\ + b_0\ )R\ e^{st} \tag{12.62}$$

Since the right side of (12.62) is an exponential function with frequency s, the left side must also be an exponential function with frequency s. Let us try $y(t) = Y e^{st}$, and (12.62) becomes

$$(s^N + a_{N-1} s^{N-1} + \cdots + a_1 s + a_0) Y e^{st} = (b_M s^M + \cdots + b_1 s + b_0) R e^{st}$$

and therefore

$$Y = \frac{b_M s^M + \cdots + b_1 s + b_0}{s^N + a_{N-1} s^{N-1} + \cdots + a_1 s + a_0} R = \frac{P(s)}{Q(s)} R = \frac{b_M \prod_{k=1}^{M}(s - z_k)}{\prod_{k=1}^{N}(s - p_k)} R = H(s) R$$

(12.63)

where $H(s) = P(s)/Q(s)$ is called the **transfer function** of the LTI CTS. Note that $H(s)$ can be found by inspection of (12.17). If $M < N$, then $H(s)$ is called a **strictly proper function**. If $M = N(M > N)$, then $H(s)$ is called a **proper (improper) function**.

Recall from Example 9.8 that the roots of $Q(s)$, p_k, $k = 1, \ldots, N$, and of $P(s)$, z_k, $k = 1, \ldots, M$, are called the **poles** and **zeros**, respectively, of $H(s)$. Notice too that $Q(s)$ is the characteristic polynomial, which means that the poles of the transfer function are the natural frequencies of the LTI CTS. The relationship between BIBO stability and the poles of the transfer function can be depicted with a pole-zero plot in a complex plane, called the **s-plane**. The built-in function **pzplot** produces a pole-zero plot given an object of a state variable description of an LTI CTS. A syntax option is given by

```
pzplot(LTI_sys)
```

Example 12.19 ───

Prog. 12.19 uses the function pzplot to plot in the s-plane the poles and zeros of the LTI CTS given in the program. The plot is shown in Fig. 12.36.

```
% pole-zero plot of an LTI CTS
clc; clear all;
% specify transfer function numerator and denominator coefficients
b  =  [1.0000 0.3000 1.0281 0.0508];
a  =  [1.0 0.4 74.1 14.8 1225.7];
[A,B,C,D]  =  tf2ss(b,a) % convert to state variable description
LTI_sys  =  ss(A,B,C,D); % create object of LTI CTS
pzplot(LTI_sys); axis([-0.3 0.1 -10 10]); % plot poles and zeros
```

Program 12.19   Program to obtain a pole-zero plot.

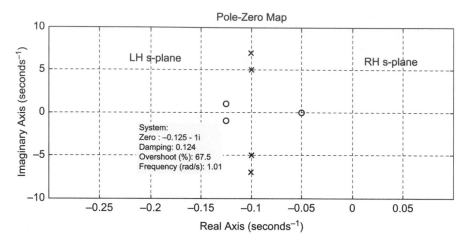

Figure 12.36   Pole-zero plot.

The pole-zero plot shows that all of the poles are located in the left-half (LH) of the
s-plane, which means that the poles have negative real parts, and therefore the LTI CTS is
BIBO stable.

Suppose the input to output relationship of a CTS is simply

$$\dot{y}(t) = r(t), \quad \rightarrow \quad y(t) = \int r(t)\, dt$$

If $r(t) = R\, e^{st}$, then

$$y(t) = \frac{1}{s} R\, e^{st}, \quad \rightarrow \quad Y = \frac{1}{s} R$$

and the transfer function of an **integrator** is $H(s) = 1/s$. If the input to output relationship is
$y(t) = \dot{r}(t)$, then $y(t) = sRe^{st}$, and the transfer function of a **differentiator** is $H(s) = s$.

   In Section 6.7 the concept of the frequency response of a LTI circuit was introduced.
Here, this concept will be extended to LTI systems described by (12.17). Let us write s as
$s = \sigma + j\,\omega$. Depending on the values of $\sigma = \text{Re}(s)$ and $\omega = \text{Im}(s)$, $r(t) = e^{st}$ can be anyone
of several different kinds of signals. Suppose the input $r(t)$ is given by

$$r(t) = A \cos(\omega\, t + \theta) = \frac{A}{2} e^{j\theta} e^{j\omega t} + \frac{A}{2} e^{-j\theta} e^{-j\omega t} = \frac{1}{2} R\, e^{j\omega t} + \frac{1}{2} R^{*} e^{-j\omega t} \qquad (12.64)$$

where $R = A e^{j\theta}$ is the phasor of $r(t)$. With $s = j\omega$, the response to $R e^{j\omega t}$ $(R^* e^{-j\omega t})$ is $Y e^{j\omega t}$ $(Y^* e^{-j\omega t})$, and with (12.63) we get $Y = H(j\omega)R$, the phasor of $y(t)$. Since (12.17) describes a linear system, the steady-state response $y(t)$ to $r(t)$, given in (12.64), is

$$y(t) = \frac{1}{2} H(j\,\omega) \, R \, e^{j\omega t} + \frac{1}{2} (H(j\,\omega) \, R)^* e^{-j\omega t} \tag{12.65}$$

Here, the transfer function $H(j\omega)$ is a complex function of the real variable ω, the frequency of the input. Writing $H(j\omega)$ in polar form changes (12.65) into

$$
\begin{aligned}
y(t) &= \frac{1}{2} \|H(j\,\omega)\| e^{j\angle H(j\omega)} \, A \, e^{j\theta} \, e^{j\omega t} + \frac{1}{2} \|H(j\,\omega)\| e^{-j\angle H(j\omega)} \, A \, e^{-j\theta} \, e^{-j\omega t} \\
&= A\|H(j\,\omega)\| \left(\frac{e^{j(\omega t + \theta + \angle H(j\omega))} + e^{-j(\omega t + \theta + \angle H(j\omega))}}{2} \right) \\
&= A\|H(j\,\omega)\| \cos\left(\omega\,t + \theta + \angle H(j\omega)\right) \tag{12.66}
\end{aligned}
$$

which shows how the output amplitude and phase depend on the transfer function $H(j\omega)$, a function of the frequency of the input.

For example, suppose the input of an integrator is $r(t)$ given by (12.64), where $R = A e^{j\theta}$ and $s = j\,\omega$. The integrator transfer function gives $H(s = j\omega) = 1/j\omega$, and the phasor of the output of the integrator is $Y = H(j\omega)R = Ae^{j\theta}/j\omega = (A/\omega)e^{j(\theta - \pi/2)}$. Therefore, according to (12.66), $y(t)$ is given by $y(t) = (A/\omega)\cos(\omega\,t + \theta - \pi/2)$.

Table 12.5 Functions concerned with the frequency response of an LTI CTS

Function	Brief description
bode	Plots magnitude and phase of $H(j\omega)$ versus ω
bodemag(LTI_sys)	Plots the magnitude of $H(j\omega)$ versus ω
bodeplot(LTI_sys)	Same as bode, but allows for more plot customizing
H = freqresp(LTI_sys,w, units)	Returns the frequency response using the specified units, where the string units can be any of the following:'rad/sec','cycles/sec','Hz','kHz','MHz','GHz', or'rpm'
ltiview(LTI_sys)	Graphical user interface (GUI) to obtain a variety of plots, including frequency response and impulse response; use doc ltiview for details
nichols(LTI_sys)	Plots magnitude versus phase of $H(j\omega)$, where ω is a parameter along the curve
nyquist(LTI_sys)	Plots imaginary part versus real part of $H(j\omega)$, where ω is a parameter along the curve

For a frequency response of an LTI CTS, the transfer function $H(s)$ is evaluated for values of s along the $j\omega$-axis ($\sigma = 0$) in the s-plane. See Fig. 9.5 in Example 9.4 for an example of plotting the frequency response of an LTI CTS. Also, Fig. 9.18 in Example 9.8 shows that a frequency response plot is a line over the $j\omega$-axis on the surface of $H(s)$ plotted over the *s*-plane. Plots of $\|H(j\omega)\|$ and $\angle H(j\omega)$ versus ω, which together are called a **Bode plot**, show the frequency selective behavior of an LTI CTS.

MATLAB provides several built-in functions, listed in Table 12.5, that are concerned with the frequency response of an LTI CTS. These frequency response plotting functions have the same syntax options as the function **bode**. To obtain a Bode plot, use one of the following syntax options.

```
bode(LTI_sys)
```

where LTI_sys is the object returned by the function **ss**. The frequency range and number of plotted points are chosen automatically.

```
bode(LTI_sys,{wmin,wmax})
```

where wmin and wmax specify the plot frequency range in rad/sec.

```
bode(LTI_sys,w)
```

where *w* is a vector of frequencies to be used in the plots.
You can specify line color, line style, and marker type with, for example

```
bode(LTI_sys1, 'r',LTI_sys2,'y--',LTI_sys3,'gx',w)
```

To obtain vectors of magnitude and phase and no plots, the syntax options are

```
[mag,phase] = bode(LTI_sys,w)
[mag,phase,w] = bode(LTI_sys)
```

The second syntax option also returns a vector of frequencies. Use **doc bode** for more details.

Example 12.20

```
% Obtain frequency response plots of system used in Example 12.16
clear all; clc
load('state_space_model.mat','LTI_sys') % get model
bode(LTI_sys); grid on; % Bode plot
figure;
nyquist(LTI_sys); grid on; % Nyquist plot
```

Program 12.20   Demonstration of using the functions bode and nyquist.

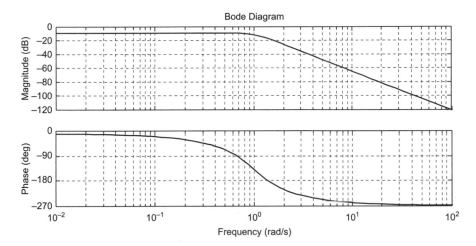

Figure 12.37 Bode plot of the frequency response.

Fig. 12.37 shows that in a Bode plot, a log frequency scale is used, and the magnitude is given in decibels.

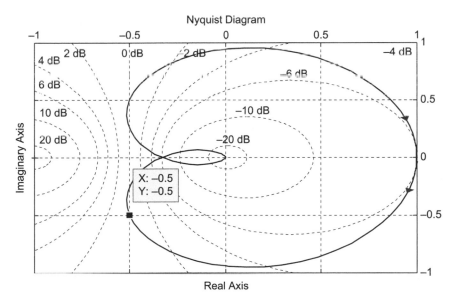

Figure 12.38 Nyquist plot of the frequency response.

Fig. 12.38 shows that in a Nyquist plot, the frequency is a parameter along the curve. Here, the plot starts at $\omega = 0$, where the real part of $H(j\omega)$ is one, and the imaginary part is zero. The downward arrow goes in the direction of an increasing frequency until $\omega \to +\infty$

at the origin. As the curve leaves the origin, $\omega = -\infty$, and increases until $\omega = 0$ again. Execute Prog. 9.8 in Example 9.7 to see a 3-D animation of a Nyquist plot.

12.4   Discrete Time Systems

A **discrete time system (DTS)** is an algorithm that operates on a number sequence $r(n)$, the input, and produces a number sequence $y(n)$, the output. In many practical applications, the number sequence $r(n)$ comes from a data acquisition system, as shown in Fig. 12.2, where the input is a continuous time signal $r(t)$. Recall that n is the discrete time index, and $r(n)$ and $y(n)$ are only defined at the sample time points given by $t = nT$, where $f_s = 1/T$ Hz is the sampling rate.

A DTS is modeled by a **difference equation**, for example

$$y(n) + a_1 y(n-1) + \cdots + a_{N-1} y(n-(N-1)) + a_N y(n-N)$$
$$= b_0 \, r(n) + b_1 \, r(n-1) + \cdots + b_{M-1} \, r(n-(M-1)) + b_M \, r(n-M) \tag{12.67}$$

where the a_k, $k = 1, \ldots, N$ and b_k, $k = 0, 1, \ldots, M$ are constants, $y(n)$ is the **present output**, $y(n-k)$, $k = 1, 2, \ldots, N$ are past outputs, $r(n)$ is the **present input** and $r(n-k)$, $k = 1, 2, \ldots, M$ are past inputs. Equation (12.67) is an Nth order difference equation. To operate (run) this DTS, reorganize (12.67) to express the present output in terms of past outputs and present and past inputs, and we get

$$y(n) = \sum_{k=0}^{M} b_k \, r(n-k) - \sum_{k=1}^{N} a_k \, y(n-k) \tag{12.68}$$

While an input can be applied at any time, it is conventional to apply inputs starting at a reference time given by $n = 0$. Unless stated otherwise, assume that $r(n < 0) = 0$. Given an input $r(n)$, then all of the terms in the right side of (12.67) can be combined into one function of time $g(n)$, called the **forcing function**. If $g(n) = 0$, then (12.67) is called a **homogeneous difference equation**.

At $n = 0$ the present input is $r(0)$, and to find the present output $y(0)$ with (12.68) requires the **initial conditions** given by

$$y(-1) = S_1, \qquad y(-2) = S_2, \ldots, y(-N) = S_N \tag{12.69}$$

for some given constants S_k, $k = 1, 2, \ldots, N$. Given the ICs and the present input $r(0)$, the present output $y(0)$ can be computed with (12.68). Then, when $n = 1$ and the present input is $r(1)$, the present output $y(1)$ can be computed. This can be continued for as long as we want to operate the DTS. The DTS described by (12.67) is a **causal** system, because a present output does not depend on future inputs. To operate a DTS in **real-time** (also called online), it must be possible, given an input $r(n)$, to compute an output $y(n)$ in an amount of time less than T sec. This requires that the DTS is a causal system.

If in (12.68) the right side includes terms such as $b_{-1} r(n+1)$, $b_{-2} r(n+2)$, ... , then to compute a present output $y(n)$, knowledge of future inputs is required, and the DTS cannot be operated in real time. Such a system is said to be **acausal**. However, when the entire history of an input is known (prerecorded), then at any time within this history, future values of the input are known, and an acausal system can be operated. Both a causal and an acausal system can be operated this way, which is referred to as **offline processing**. If the coefficients a_k and b_k and the ICs are real numbers, then a real input number sequence $r(n)$ causes a real output number sequence $y(n)$.

Suppose all ICs given in (12.69) are zero. Let $y(n) = y_1(n)$ denote the response when the input is $r(n) = r_1(n)$, and similarly for $y_2(n)$ and $r_2(n)$. Then the DTS is a **linear system** if when the input is $r(n) = c_1 r_1(n) + c_2 r_2(n)$, then the response is given by $y(n) = c_1 y_1(n) + c_2 y_2(n)$ for any constants c_1 and c_2. A DTS is a time invariant system if when $y(n) = y_1(n)$ is the response to $r(n) = r_1(n)$, then $y(n) = y_2(n) = y_1(n-k)$ is the response to $r(n) = r_2(n) = r_1(n-k)$, for any time shift k. The DTS described by (12.67) is a **linear** and **time invariant (LTI)** DTS.

We have already studied several discrete time systems. For example, Euler's method and the trapezoidal rule for numerical integration, which are described by (4.17) and (4.19), respectively, are LTI discrete time systems. The median filter presented in Section 4.4.2 and applied in Example 4.12 is a nonlinear discrete time system.

12.5 Response of LTI Discrete Time Systems

The complete response of the LTI DTS described by (12.67) can be written as

$$y(n) = y_{zi}(n) + y_{zs}(n)$$

where $y_{zi}(n)$ is the response to ICs, called the **zero-input response**, and $y_{zs}(n)$ is the response to the input $r(n)$ under zero ICs, called the **zero-state response**.

A special $y_{zs}(n)$ is the response when the input $r(n)$ is the unit pulse function $\delta(n)$, where $\delta(n < 0) = 0$, $\delta(0) = 1$, and $\delta(n > 0) = 0$. The unit pulse function is the **Kronecker delta function** defined by

$$\delta(n-k) = \begin{cases} 1, & n-k = 0 \\ 0, & n-k \neq 0 \end{cases}$$

where k is an arbitrary time point. The response $y_{zs}(n)$ to $r(n) = \delta(n)$ is called the **impulse response**, and it is denoted by $h(n)$. The impulse response $h(n)$ is the response of (12.67) under zero ICs when $r(n) = \delta(n)$. Soon you will see how the response to any input $r(n)$ can be found with $h(n)$.

The built-in function **filter** finds the offline response $y(n)$ to any input $r(n)$ given the difference equation coefficients. For zero ICs, a syntax option is given by

```
y  =  filter(b,a,r)
```

where r is a vector that holds all input samples, $b = [b_0 \; b_1 \; \ldots \; b_M]$, $a = [1 \; a_1 \; a_2 \; \ldots \; a_N]$ and the vector y holds the output samples.

To obtain a complete response by running the DTS, let

$$b = \begin{bmatrix} b_0 \\ b_1 \\ \vdots \\ b_M \end{bmatrix}, \quad R = \begin{bmatrix} r(n) \\ r(n-1) \\ \vdots \\ r(n-M) \end{bmatrix}, \quad a = \begin{bmatrix} a_1 \\ a_2 \\ \vdots \\ a_N \end{bmatrix}, \quad Y = \begin{bmatrix} y(n-1) \\ y(n-2) \\ \vdots \\ y(n-N) \end{bmatrix}$$

Then, execute the following steps.

```
1) R  =  zeros(M+1,1); n  =  0; Y  =  initial conditions;
2) get r; R(1)  =  r;   % input r(nT)
3) y  =  b' R - a' Y;   % output y(nT)
4) R  =  [0; R(1:M)]; Y  =  [y; Y(1:N-1)]; % shift down and
   prepare for next input sample
5) n  =  n+1; loop to step (2);
```

Discrete time systems are separated into two very different categories. Suppose that in (12.67) the $a_k = 0$, $k = 1, 2, \ldots, N$, which means that the present output does not depend on past outputs. Then, (12.68) becomes

$$y(n) = \sum_{k=0}^{M} b_k \, r(n-k) \tag{12.70}$$

and the present output depends only on present and past inputs. If the input has a finite time duration, then the response will have a finite time duration. For example, if $r(n) = \delta(n)$, which is nonzero only for one sample time, then the response is given by

$$y(n) = \sum_{k=0}^{M} b_k \, \delta(n-k) \tag{12.71}$$

and $y(n)$ is nonzero over a finite time range (has a finite time duration). An LTI DTS described by (12.70) is called a **finite impulse response (FIR)** DTS, because its unit pulse response, given by (12.71), has a finite time duration. For an FIR DTS, (12.71) gives the impulse response $h(n) = y(n)$.

Example 12.21 _____

By running the DTS, obtain the response of the FIR DTS described by

$$y(n) = \frac{1}{21}[-2r(n) + 3r(n-1) + 6r(n-2) + 7r(n-3) + 6r(n-4) + 3r(n-5) - 2r(n-6)]$$

Let the input $r(n)$ be the result of sampling $r(t) = \sin(2\pi f_1 t) + \sin(2\pi f_2 t)$ at the rate $f_s = 8000$ Hz. To avoid aliasing error, for the given f_s, $r(t)$ can have sinusoidal components with frequencies restricted to the range $0 \le f < 4000$ Hz. Let $f_1 = 200$ Hz and $f_2 = 2400$ Hz.

To operate the DTS, let

$$R' = [r(n) \ \ r(n-1) \ \ r(n-2) \ \ r(n-3) \ \ r(n-4) \ \ r(n-5) \ \ r(n-6)]$$
$$b' = [-2 \ 3 \ 6 \ 7 \ 6 \ 3 - 2]/21$$

Then, when $n = 0$, $R' = [r(0) \ 0 \ 0 \ 0 \ 0 \ 0 \ 0]$, and $y(0) = b'R$. When $n = 1$, $R' = [r(1) \ r(0) \ 0 \ 0 \ 0 \ 0 \ 0]$, and $y(1) = b'R$, and so on. Prog. 12.21 samples and plots the input, shown in Fig. 12.39, runs the DTS for $2/f_1$ sec, and plots the output, also shown in Fig. 12.39.

```
% Example of running an FIR digital filter
clear all; clc
fs  =  8000; % specify sampling frequency
T   =  1.0/fs; % sampling time increment
f1  =  200.0; w1  =  2*pi*f1; f2  =  2400; w2  =  2*pi*f2; % specify frequencies
N_samples  =  floor((2/f1)/T); % number of input samples to be processed
b   =  [-2; 3; 6; 7; 6; 3; -2]/21.0; % FIR filter coefficients
M   =  length(b)-1; % number of past inputs
R   =  zeros(M+1,1); % initialize vector of inputs
% for loop to go from one time point to the next time point
for n = 1:N_samples
    t   =  (n-1)*T; time(n)  =  t; % storing t in the vector time
    r(n)  =  cos(w1*t)+cos(w2*t); % get present input sample
    R(1)  =  r(n); % store present input as element R(1)
    y(n)  =  b'*R; % inner product of b and R
    R   =  [0; R(1:M)]; % shift down input samples
end
subplot(2,1,1); plot(time,r,'-o','MarkerSize',4); grid on; % plot input
title('FIR filter input'); ylabel('input');
```

```
subplot(2,1,2); plot(time,y,'-o','MarkerSize',4); grid on; % plot output
title('FIR filter output'); ylabel('output'); xlabel('time - seconds');
save('filter_input.mat','fs','r'); % used in another example
```

Program 12.21   Program to run an FIR discrete time system.

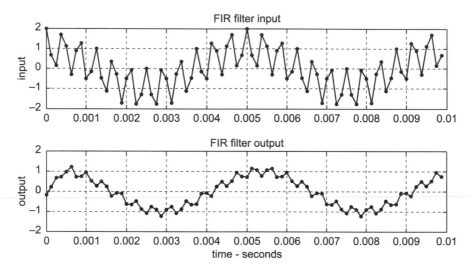

Figure 12.39   FIR DTS input and output.

From the input and output plots, we see that the high-frequency component in the input has been reduced in the output. This FIR DTS is a digital low-pass filter.

If at least one a_k in (12.68) is not zero, then the present output will depend on past outputs, which means that there is **feedback** in the system. In this case, the response to any input can never become and remain zero indefinitely. The response to $r(n) = \delta(n)$ will have an infinite time duration, in which case the DTS is called an **infinite impulse response (IIR)** DTS.

Example 12.22 _____

Some IIR DTS is described by

$$y(n) - 2.8135\, y(n-1) + 2.6728\, y(n-2) - 0.8574 y(n-3)$$

$$= r(n) + 2.8135\, r(n-1) + 2.6728\, r(n-2) + 0.8574 r(n-3)$$

By running the DTS, obtain its impulse response $h(n)$ and its response to the input used in Example 12.21. To operate the DTS, let

$$R' = [r(n) \ r(n-1) \ r(n-2) \ r(n-3)], \quad b' = [1 \ 2.8135 \ 2.6728 \ 0.8574]$$
$$Y' = [y(n-1) \ y(n-2) \ y(n-3)], \quad a' = [-2.8135 \ 2.6728 - 0.8574]$$

Then, when $n = 0$, Y holds the ICs, $R' = [r(0) \ 0 \ 0 \ 0]$ and $y(0) = b'R - a'Y$.

Prog. 12.22 uses the input that was used in Example 12.21, and the output is shown in Fig. 12.40. Then, the impulse response is obtained, which is also shown in Fig. 12.40.

```
% Example of running an IIR digital filter
clear all; clc
load('filter_input.mat','fs','r'); % input used in another example
N_samples  =  length(r); T  =  1/fs;
% specify IIR filter
b  =  [1; 2.8135; 2.6728; 0.8574]; a  =  [-2.8135; 2.6728; -0.8574];
R  =  [1; 0; 0; 0]; Y  =  [0; 0; 0]; % initialize for impulse response
M  =  length(b)-1; N  =  length(a);
% run DTS to get the impulse response
for n = 1:N_samples;
    h(n)  =  b'*R - a'*Y; % present output
    R  =  [0; R(1:M)]; % shift elements of R
    Y  =  [h(n); Y(1:N-1)]; % shift elements of Y and store present output
end
R  =  [0; 0; 0; 0]; Y  =  [0; 0; 0]; % using zero initial conditions
% run DTS to get response to sinusoidal input
for n = 1:N_samples;
    time(n)  =  (n-1)*T;
    R(1)  =  r(n); % store present input as element R(1)
    y(n)  =  b'*R - a'*Y;
    R  =  [0; R(1:M)]; % shift elements of R
    Y  =  [y(n); Y(1:N-1)]; % shift elements of Y and store present output
end
subplot(2,1,1); plot(time,h,'-o','MarkerSize',4); grid on; % plot input
title('IIR Filter Impulse Response'); ylabel('output');
subplot(2,1,2); plot(time,y,'-o','MarkerSize',4); grid on; % plot output
title('IIR Filter Response to Sinusoidal Input');
ylabel('output'); xlabel('time - seconds');
```

Program 12.22   Program to run an IIR discrete time system.

While the impulse response becomes very small, it never becomes and then remains exactly zero. Here, the impulse response starts to become negligible around $t = 0.01$ sec.

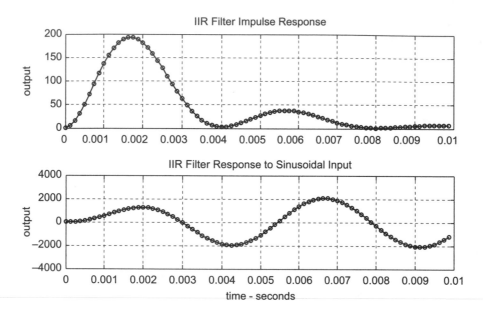

Figure 12.40   Impulse and sinusoidal response of an IIR low-pass filter.

By the time $t = 0.01$ sec, the sinusoidal response has almost completed its transient behavior. A comparison of the sinusoidal outputs of Figs. 12.39 and 12.40 shows that the IIR DTS does a much better job of removing the high-frequency component of the input than the FIR DTS. Nevertheless, each kind of DTS can be designed to have a near ideal low-pass filter frequency response.

In addition to the impulse response, it is also useful to assess the transient behavior of an LTI DTS based on its unit step response. The discrete time **unit step function** is defined by

$$u(n - k) = \begin{cases} 1, & n - k \geq 0 \\ 0, & n - k < 0 \end{cases}$$

where k is an arbitrary time point. Notice that the unit step function is one when the argument is nonnegative. The unit step function is commonly used to start (or stop) a given signal. For example, a sinusoidal pulse can be expressed as $\sin(n\pi/4)(u(n) - u(n - 8))$, which is one cycle of the sinusoid.

The **exponential function** z^n, where z is some real or complex number, also plays an important role in the study of LTI discrete time systems. For example, if we sample $r(t) = Ae^{\sigma t}\cos(\omega t + \theta)$ we get

$$
r(nT) = Ae^{\sigma nT}\frac{e^{j(\omega nT+\theta)} + e^{-j(\omega nT+\theta)}}{2} = \frac{A}{2}e^{j\theta}(e^{(\sigma+j\omega)T})^n + \frac{A}{2}e^{-j\theta}(e^{(\sigma-j\omega)T})^n
$$

$$
= \frac{K}{2}(z)^n + \frac{K^*}{2}(z^*)^n, \quad K = A\,e^{j\theta}, \quad z = e^{(\sigma+j\omega)T} = e^{\sigma T}e^{j\omega T}
$$

(12.72)

With (12.72) and the value of the complex number z we can represent a variety of discrete time signals that occur in the study of LTI discrete time systems.

12.5.1 Zero-Input Response

The **zero-input response** is caused by ICs, and it is the solution of the homogeneous equation given by

$$
y_{zi}(n) + a_1 y_{zi}(n-1) + \cdots + a_{N-1}y_{zi}(n-(N-1)) + a_N y_{zi}(n-N) = 0 \quad (12.73)
$$

The solution $y_{zi}(n)$ must be a function such that it and a linear combination of delayed $y_{zi}(n)$ can be zero for all n. Let us try $y_{zi}(n) = K\gamma^n$, and we get

$$
K\gamma^n + a_1 K\gamma^{n-1} + \cdots + a_{N-1}K\gamma^{n-(N-1)} + a_N K\gamma^{n-N} = 0
$$

Since K and γ^n cannot be zero, we can divide by K and γ^n to get

$$
Q(\gamma^{-1}) = 1 + a_1\gamma^{-1} + a_2\gamma^{-2} + \cdots + a_{N-1}\gamma^{-(N-1)} + a_N\gamma^{-N} = \prod_{k=1}^{N}(1 - \gamma_k\gamma^{-1}) = 0
$$

(12.74)

which is called the **characteristic equation** of the LTI DTS, and $Q(\gamma^{-1})$ is called the **characteristic polynomial**. Notice that the characteristic polynomial can be found by inspection of (12.67). The characteristic equation has N roots, γ_k, $k = 1, 2, \ldots, N$, which can be real or complex. Assume the roots are distinct.

For each distinct root γ_k there is a solution of (12.73) given by $K_k\gamma_k^n$, where we must still find K_k. Since (12.73) is a linear equation, then

$$
y_{zi}(n) = \sum_{k=1}^{N} K_k\gamma_k^n \quad (12.75)
$$

is also a solution of (12.73). The application of the N ICs gives

$$y(-1) \rightarrow \gamma_1^{-1} K_1 + \gamma_2^{-1} K_2 + \cdots + \gamma_N^{-1} K_N = S_1$$
$$y(-2) \rightarrow \gamma_1^{-2} K_1 + \gamma_2^{-2} K_2 + \cdots + \gamma_N^{-2} K_N = S_2$$
$$\vdots$$
$$y(-N) \rightarrow \gamma_1^{-N} K_1 + \gamma_2^{-N} K_2 + \cdots + \gamma_N^{-N} K_N = S_N$$

which are N equations in the N unknowns K_k, $k = 1, 2, \ldots, N$ that can be written as $\Lambda K = S$, where

$$\Lambda = \begin{bmatrix} \gamma_1^{-1} & \gamma_2^{-1} & \cdots & \gamma_N^{-1} \\ \gamma_1^{-2} & \gamma_2^{-2} & \cdots & \gamma_N^{-2} \\ \vdots & \vdots & \vdots & \vdots \\ \gamma_1^{-N} & \gamma_2^{-N} & \cdots & \gamma_N^{-N} \end{bmatrix}, \qquad K = \begin{bmatrix} K_1 \\ K_2 \\ \vdots \\ K_N \end{bmatrix}, \qquad S = \begin{bmatrix} S_1 \\ S_2 \\ \vdots \\ S_N \end{bmatrix}$$

and $K = \Lambda^{-1} S$.

The time domain behavior of the response to ICs depends on the γ_k, $k = 1, 2, \ldots, N$, and each function γ_k^n is called a **characteristic mode** of the LTI DTS. If a root γ_k occurs with multiplicity m, then m of the characteristic modes are given by $n^i \gamma_k^n$, $i = 0, \ldots, m - 1$. For all possible kinds of distinct values of γ, Table 12.6 gives the kind of characteristic mode that contributes to $y_{zi}(n)$.

Table 12.6   Characteristic modes

Distinct root of (12.74)	Contributions to the response
$\gamma = 1$	K, a constant
$\gamma = $ a real number	$K\gamma^n$, increasing, $\|\gamma\| > 1$, or decreasing, $\|\gamma\| < 1$ exponential function
$\gamma = $ complex conjugate pair	$2\|K\|\|\gamma\|^n \cos(\angle\gamma\, n + \angle K)$, exponentially increasing, $\|\gamma\| > 1$, or decreasing, $\|\gamma\| < 1$, sinusoidal function
$\gamma = $ complex conjugate pair, $\|\gamma\| = 1$	$2\|K\| \cos(\angle\gamma\, n + \angle K)$, sinusoidal function

For example, suppose two values of γ are $\gamma = \|\gamma\| e^{\pm j\angle\gamma}$, and, since $y_{zi}(n)$ is a real discrete time function, the contribution to $y_{zi}(n)$ is

$$K\gamma^n + K^*(\gamma^*)^n = \|K\| e^{j\angle K} \|\gamma\|^n e^{j\, n\angle\gamma} + \|K\| e^{-j\angle K} \|\gamma\|^n e^{-j\, n\angle\gamma}$$
$$= \|K\| \, \|\gamma\|^n (e^{j(n\angle\gamma + \angle K)} + e^{-j(n\angle\gamma + \angle K)}) = 2\|K\| \, \|\gamma\|^n \cos(\angle\gamma n + \angle K)$$

Table 12.6 shows that $\lim_{n \to \infty} y_{zi}(n) = 0$, if and only if all of the roots of the characteristic equation have magnitudes less than one. The angle $\angle\gamma$ is called the **angular frequency**. The frequency of oscillation ω is found with $\omega T = \angle\gamma$.

12.5.2 Zero-State Response

To find the **zero-state response** $y_{zs}(n)$ of (12.67), let us first consider a simpler problem given by

$$w(n) + a_1\,w(n-1) + \cdots + a_N w(n-N) = r\,(n) \tag{12.76}$$

where the ICs are zero. Assume that for a given $r(n)$, we can solve (12.76) for $w(n)$. If $g(n) = b_0 r(n)$, the response is $b_0 w(n)$. Time shifting each term in (12.76) gives

$$w(n-1) + a_1 w(n-2) + \cdots + a_N w(n-N-1) = r(n-1) \tag{12.77}$$

Therefore, if $g(n) = b_1 r(n-1)$, the response is $b_1 w(n-1)$. Since (12.67) is a linear equation, then if the forcing function is

$$g(n) = b_0 r(n) + b_1 r(n-1) + \cdots + b_{M-1} r(n-(M-1)) + b_M r(n-M)$$

which is the right side of (12.67), the zero-state solution of (12.67) is

$$y_{zs}(n) = b_0 w(n) + \cdots + b_{M-1} w(n-(M-1)) + b_M w(n-M), \quad n \geq 0 \tag{12.78}$$

Thus, if we can find $w(n)$, then with (12.78) we can find $y_{zs}(n)$. Soon, this method will be applied to obtain $y_{zs}(n)$ for a particular $r(n)$.

12.5.3 State Variables

Any IIR N^{th} order difference equation can be converted into a system of N first-order difference equations, and an LTI DTS can be described by

$$y(n) = Cx(n) + Dr(n) \tag{12.79}$$

$$x(n+1) = Ax(n) + Br(n) \tag{12.80}$$

where $x(n) = [x_1(n)\; x_2(n)\; \ldots\; x_N(n)]'$ is an $N \times 1$ column vector, called the **state vector,** $x(n+1)$ means $x(n+1) = [x_1(n+1)\; x_2(n+1)\; \ldots\; x_N(n+1)]'$, and the dimensions of the constant matrices A, B, C, and D are $N \times N$, $N \times 1$, $1 \times N$, and 1×1, respectively. Equation (12.80) is called the **state equation,** and (12.79) is called the **output equation.** Together, (12.80) and (12.79) are called a **state variable description** of an LTI DTS.

To solve (12.80) for $x(n)$ and obtain $y(n)$ with (12.79) requires the initial state $x(0)$, which we can find given the ICs of (12.69). Using (12.79) gives

$$y(n) = Cx(n) + Dr(n), \quad \rightarrow \quad y(-1) = S_1 = Cx(-1) + Dr(-1), \quad \rightarrow \quad Cx(-1) = S_1$$

where $r(n < 0) = 0$ was used. Using (12.79) again gives

$$y(-2) = Cx(-2) + Dr(-2) = CA^{-1}x(-1), \quad \rightarrow \quad CA^{-1}x(-1) = S_2$$

Continue with delays of $y(n)$ until we get

$$CA^{-(N-1)}x(-1) = S_N$$

Using matrix notation gives

$$\begin{bmatrix} C \\ CA^{-1} \\ \vdots \\ CA^{-(N-1)} \end{bmatrix} x(-1) = S, \quad S = \begin{bmatrix} S_1 \\ S_2 \\ \vdots \\ S_N \end{bmatrix} \quad \rightarrow \quad O\,x(-1) = S$$

where O, which is called the **observability matrix**, is an $N \times N$ matrix. Use the built-in function **obsv** to get O with

```
O  =  obsv(inv(A),C)
```

If O is nonsingular, then $x(0) = Ax(-1) = AO^{-1}S$. With $S = 0$ or $x(0) = 0$ (a vector of zeros), we get $y_{zs}(n)$ with (12.79).

The matrices A, B, C, and D must be found such that the state variable description gives the same input $r(n)$ to output $y(n)$ relationship as (12.67). There are an infinite number of possibilities. A commonly used conversion method starts with (12.76) by defining

$$
\begin{aligned}
x_N(n) &= w(n-N), & &\rightarrow x_N(n+1) = x_{N-1}(n) \\
x_{N-1}(n) &= w(n-(N-1)), & &\rightarrow x_{N-1}(n+1) = x_{N-2}(n) \\
x_{N-2}(n) &= w(n-(N-2)), & &\rightarrow x_{N-2}(n+1) = x_{N-3}(n)
\end{aligned}
$$

$$\vdots$$

$$
\begin{aligned}
x_2(n) &= w(n-2), & &\rightarrow x_2(n+1) = x_1(n) \\
x_1(n) &= w(n-1), & &\rightarrow x_1(n+1) = w(n) = r(n) - (a_1\,w(n-1) + \cdots + a_N\,w(n-N)) \\
& & & \qquad\qquad\quad = r(n) - (a_1\,x_1(n) + \cdots + a_N\,x_N(n))
\end{aligned}
$$

$$(12.81)$$

Using matrix notation, the equations of (12.81) become

$$
x(n+1) = \begin{bmatrix}
-a_1 & -a_2 & \cdots & \cdots & -a_{N-1} & -a_N \\
1 & 0 & \cdots & 0 & 0 & 0 \\
0 & 1 & 0 & 0 & 0 & 0 \\
\vdots & 0 & \ddots & \ddots & \vdots & \vdots \\
0 & \vdots & 0 & 1 & 0 & 0 \\
0 & 0 & \cdots & 0 & 1 & 0
\end{bmatrix} x(n) + \begin{bmatrix} 1 \\ 0 \\ 0 \\ \vdots \\ 0 \\ 0 \end{bmatrix} r(n) \qquad (12.82)
$$

which defines the matrices A and B in (12.80). Using (12.76) in (12.78) gives

$$
\begin{aligned}
y_{zs}(n) &= b_0(r(n) - (a_1 w(n-1) + \cdots + a_N w(n-N))) \\
&\quad + b_1 w(n-1) + \cdots + b_{M-1} w(n - (M-1)) + b_M w(n-M) \\
&= b_0 r(n) + (b_1 - b_0 a_1) x_1(n) + \cdots + (b_M - b_0 a_M) x_M(n) \\
&\quad - b_0 a_{M+1} x_{M+1}(n) - \cdots - b_0 a_N x_N(n) \\
&= [(b_1 - b_0 a_1) \cdots (b_M - b_0 a_M) \; -b_0 a_{M+1} \cdots \\
&\quad - b_0 a_N] x(n) + [b_0] r(n), \quad M < N
\end{aligned} \tag{12.83}
$$

which defines the matrices C and D in (12.79). If we set $x(0) = AO^{-1}S$, then (12.83) and (12.82) give $y(n)$, the complete solution of (12.67). For an LTI DTS, the matrices A, B, C, and D can be found with

$$
\text{[A, B, C, D]} \quad = \quad \text{dtf2ss(b,a)}
$$

where $b = [b_0 \; b_1 \ldots b_M]$ and $a = [1 \; a_1 \; a_2 \ldots a_N]$.

Example 12.23

Convert the DTS description given in Example 12.22 into a state variable description. Let the ICs be $y(-1) = 500, y(-2) = 0$, and $y(-3) = 0$. Then, with each description obtain the complete response to a unit step function input. Prog. 12.23 operates the two descriptions of the DTS, and plots the unit step responses shown in Fig. 12.41. The state variable description is given after the program.

```
% Step response of a DTS
clear all; clc;
% specify IIR filter difference equation
b  =  [1; 2.8135; 2.6728; 0.8574]; a  =  [-2.8135; 2.6728; -0.8574];
M  =  length(b)-1; N  =  length(a);
R  =  zeros(M+1,1); y_init  =  [500; 0; 0]; % output initial conditions
R(1)  =  1; Y  =  y_init; N_samples  =  101; t  =  0:N_samples-1; % time index
% run DTS to get response to a step input
for n = 1:N_samples;
    y(n)  =  b'*R - a'*Y;
    R  =  [1; R(1:M)]; % shift elements of R and get next input
    Y  =  [y(n); Y(1:N-1)]; % shift elements of Y and store present output
end
subplot(2,1,1); plot(t,y); grid on
ylabel('step response');title('Complete Response of Difference Equation')
a  =  [1; a]'; b  =  b'; % convert to row vectors
[A B C D]  =  dtf2ss(b,a) % obtain state variable description
O  =  obsv(inv(A),C); % obtain observability matrix
```

```
x_init  =  A*inv(O)*y_init; % get initial state
r  =  1; x  =  x_init; % input is a unit step
for n = 1:N_samples
    y(n) = C*x+D*r; % get output
    x = A*x+B*r; % update state
end
subplot(2,1,2); plot(t,y); grid on
xlabel('discrete time index'); ylabel('step response');
title('Complete Response of State Variable Description')
```

Program 12.23   Program to operate a DTS using two different descriptions.

```
A  =
    2.8135 -2.6728 0.8574
    1.0000    0       0
       0    1.0000    0
B  =
    1
    0
    0
C  =    5.6270  0   1.7148
D  =      1
x_init  =
  250.0000
  100.2923
    0.0000
```

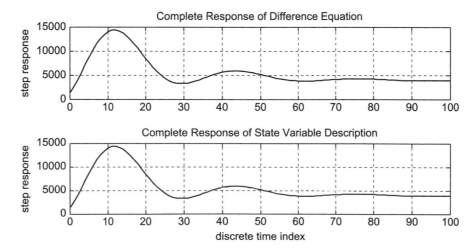

Figure 12.41   Unit step response.

Since the steady-state response is a constant, denote it by K, which must satisfy the difference equation in steady state. In steady state, the difference equation becomes

$$K - 2.8135\,K + 2.6728\,K - 0.8574\,K = 1 + 2.8135 + 2.6728 + 0.8574$$

giving $K = 3865.1$, which appears to be in agreement with the responses shown in Fig. 12.41.

12.5.4 Impulse Response

A very useful function to know about an LTI DTS is its **unit pulse response** $h(n)$ (also called the **impulse response**). Recall that $h(n)$ is the solution of (12.67) under zero ICs, when $r(n) = \delta(n)$. If the system is causal, then we must have $h(n < 0) = 0$. To find a function expression for $h(n)$, we first solve a simpler problem given by

$$w(n) + a_1 w(n-1) + \cdots + a_N w(n-N) = \delta(n) \tag{12.84}$$

Since ICs are zero, setting $n = 0$ in (12.84) gives $w(0) = 1$. If we can determine the solution $w(n)$ of (12.84), then, since the DTS is LTI and in view of the right side of (12.67), $h(n)$ is given by

$$h(n) = b_0 w(n) + b_1 w(n-1) + \cdots + b_M w(n-M), \quad n \geq 0 \tag{12.85}$$

For $n > 0$, (12.84) becomes

$$w(n) + a_1 w(n-1) + \cdots + a_N w(n-N) = 0, \quad n > 0 \tag{12.86}$$

which is a homogeneous equation. Therefore, we can instead solve (12.86) for $w(n)$ with N ICs given by $w(0) = 1$, $w(-1) = 0$, ... , $w(-(N-1)) = 0$. We want to find a function that can satisfy (12.86). Let us try $w(n) = K\gamma^n$, for some nonzero constants K and γ. Substituting this $w(n)$ into (12.86) gives

$$K\gamma^n + a_1 K\gamma^{n-1} + a_2 K\gamma^{n-2} + \cdots + a_N K\gamma^{n-N} = 0$$

and after multiplying by K^{-1} and γ^{-n} we get

$$Q(\gamma^{-1}) = 1 + a_1\gamma^{-1} + a_2\gamma^{-2} + \cdots + a_{N-1}\gamma^{-(N-1)} + a_N\gamma^{-N} = \prod_{k=1}^{N}(1 - \gamma_k\gamma^{-1}) = 0 \tag{12.87}$$

which is the **characteristic equation** of the DTS, and $Q(\gamma^{-1})$ is the **characteristic polynomial**. Let us assume that the roots are distinct. Since each function γ_k^n can satisfy (12.86), the linear combination given by

$$w(n) = \sum_{k=1}^{N} K_k \gamma_k^n \tag{12.88}$$

is also a solution of (12.86). To find the N constants K_k, $k = 1, \ldots, N$, set up N equations in N unknowns by applying the ICs $w(0) = 1$, $w(-1) = 0, \ldots, w(-(N-1)) = 0$. With $w(n)$ and (12.85) we can obtain the unit pulse response $h(n)$, which is a linear combination of the characteristic modes of the DTS.

Note that $\lim_{n \to \infty} h(n) = 0$, if and only if all roots of the characteristic equation satisfy

$$\|\gamma_k\| < 1, \; k = 1, 2, \ldots, N$$

As a special case, consider the second-order difference equation given by

$$y(n) + a_1 y(n-1) - y(n-2) = \delta(n), \quad y(-1) = 0, \quad y(-2) = 0 \tag{12.89}$$

Here, $y(n) = h(n)$, which can be found by running $h(n) = \delta(n) - a_1 h(n-1) - h(n-2)$. The characteristic ploynomial is given by

$$1 + a_1 \gamma^{-1} + \gamma^{-2} = (1 - \gamma_1 \gamma^{-1})(1 - \gamma_1^* \gamma^{-1}) = 1 - (\gamma_1 + \gamma_1^*)\gamma^{-1} + \|\gamma_1\|^2 \gamma^{-2}, \;\; \rightarrow \;\; \|\gamma_1\|^2 = 1$$

Therefore, $a_1 = -2\,\mathrm{Re}\,(\gamma_1) = -2\cos(\angle\gamma_1)$. The solution of (12.89) is given by

$$h(n) = \|K\|e^{j\angle K}e^{j\,n\angle\gamma_1}u(n) + \|K\|e^{-j\angle K}e^{-j\,n\angle\gamma_1}u(n) = 2\|K\|\cos(n\angle\gamma_1 + \angle K)\,u(n)$$

where the unit step function was introduced to ensure that $h(n < 0) = 0$. The constant K can be found by applying the ICs $h(0) = 1$ and $h(-1) = 0$. The solution shows that running (12.89) produces a sinusoidal signal with frequency determined by $\angle\gamma_1 = \omega T$. Equation (12.89) is a digital oscillator that we can operate at any frequency in the range $0 < \omega < \omega_s/2$ by setting the value of a_1. For example, assume $f_s = 8\mathrm{K}$ Hz, and we want to generate a 1K Hz digital sinusoid. Then, $\angle\gamma_1 = 2000\,\pi/8000 = \pi/4$ rad, and $a_1 = -\sqrt{2}$.

12.5.5 Convolution

The unit pulse function can be used to position a given value of a signal at a particular discrete time point. For example, $r(n) = -5\,\delta(n-2)$ positions the value -5 of $r(n)$ to occur at the time $n = 2$, while for all other n, r has zero value. We can write any discrete time signal $r(n)$ as a linear combination of unit pulse functions given by

$$r(n) = \sum_{k=-\infty}^{+\infty} r(k)\,\delta(n-k) \tag{12.90}$$

The zero-state response of an LTI DTS to any input can be found by writing the input as in (12.90). Since the DTS is LTI, the response to $r(k)\,\delta(n-k)$ is $r(k)\,h(n-k)$, and therefore, the response to $r(n)$ is given by

$$y(n) = \sum_{k=-\infty}^{+\infty} h(n-k)\,r(k) = h(n) * r(n) = \sum_{i=-\infty}^{+\infty} h(k)\,r(n-k) \tag{12.91}$$

which is called the **discrete linear convolution** operation. Since the DTS is time invariant, the response to $r(n - n_0)$ is $y(n - n_0) = h(n) * r(n - n_0)$. If the DTS is causal, and $r(n < 0) = 0$, then (12.91) becomes

$$y(n) = \sum_{k=0}^{n} h(n - k)r(k), \quad n = 0, 1, \ldots \tag{12.92}$$

which can be found with the built-in function **conv** (see (12.60)).

12.5.6   Stability

A DTS is said to be BIBO stable if for a bounded input the output is bounded. If the input is bounded, then for some real and positive number B, we have $|r(n)| \leq B$ for all n. For a bounded output, (12.91) becomes

$$|y(n)| = \left| \sum_{k=-\infty}^{+\infty} h(k)\, r(n - k) \right| \leq \sum_{k=-\infty}^{+\infty} |h(k)\, r(n - k)|$$

$$= \sum_{k=-\infty}^{+\infty} |h(k)|\, |r(n - k)| \leq B \sum_{k=-\infty}^{+\infty} |h(k)| < \infty$$

A sufficient condition for BIBO stability of an LTI DTS is that the unit pulse response must be absolutely summable.

Recall that the nature of the unit pulse response is determined by the roots of the characteristic equation. If all roots satisfy the condition $\|\gamma_i\| < 1$, then $h(n)$ is absolutely summable. Moreover, we then have $\lim_{n \to \infty} h(n) = 0$. This means that if we apply a finite duration input to a DTS, then eventually, the system output will approach zero. An important distinction between FIR and IIR systems is that an FIR system is unconditionally stable, while an IIR system can be stable or unstable, depending on the roots of the characteristic equation.

For example, to determine the stability of the DTS of Example 12.22, use the statements

```
Q = [1 -2.8135 2.6728 -0.8574]; % coefficients of characteristic polynomial
% check magnitudes of roots of characteristic polynomial
if isempty(find(abs(roots(Q))  >=  1))
      disp('LTI DTS is stable')
else
      disp('LTI DTS is not stable')
end

LTI DTS is stable
```

12.5.7   Steady-State Response

To assess the frequency selective behavior of a DTS, we investigate its response to an input given by $r(n) = A\cos(\omega n T + \phi)$. If the LTI DTS is stable, then, in steady state, the response will be a sinusoid. Of interest is the dependence of the response amplitude and phase on the frequency ω of the input (see Examples 12.21 and 12.22). It will be more convenient to determine the steady-state response if we work with an input $r(n)$ that comes from sampling the signal given by (12.61) to get

$$r(n) = R\,e^{snT} = R\big(e^{sT}\big)^n = R\,z^n, \quad z = e^{sT} = e^{(\sigma+j\omega)T} = e^{\sigma T}e^{j\omega T} \tag{12.93}$$

Like in (12.72), depending on the value of z (or s), (12.93) can be used to represent a variety of signals.

Substituting $r(n)$ into (12.67) gives

$$
\begin{aligned}
y(n) + a_1 y(n-1) &+ \cdots + a_{N-1}\,y(n-(N-1)) + a_N y(n-N) \\
&= \big(b_0 + b_1\,z^{-1} + \cdots + b_{M-1}\,z^{-(M-1)} + b_M\,z^{-M}\big)\,R\,z^n
\end{aligned}
\tag{12.94}
$$

Since the right side of (12.94) is an exponential function, the left side must also be an exponential function. Let us try $y(n) = Y\,z^n$, and (12.94) becomes

$$Y\,z^n + a_1 Y\,z^{n-1} + \cdots + a_{N-1}Y\,z^{n-(N-1)} + a_N Y\,z^{n-N}$$

$$= \big(1 + a_1\,z^{-1} + \cdots + a_{N-1}\,z^{-(N-1)} + a_N\,z^{-N}\big)Y\,z^n$$

$$= \big(b_0 + b_1\,z^{-1} + \cdots + b_{M-1}\,z^{-(M-1)} + b_M\,z^{-M}\big)\,R\,z^n$$

and therefore

$$Y = \frac{b_0 + b_1\,z^{-1} + \cdots + b_{M-1}\,z^{-(M-1)} + b_M\,z^{-M}}{1 + a_1\,z^{-1} + \cdots + a_{N-1}\,z^{-(N-1)} + a_N\,z^{-N}}R = \frac{P(z^{-1})}{Q(z^{-1})}\,R$$

$$= \frac{b_0\displaystyle\prod_{k=1}^{M}\big(1 - z_k\,z^{-1}\big)}{\displaystyle\prod_{k=1}^{N}\big(1 - p_k\,z^{-1}\big)}R = H(z^{-1})\,R \tag{12.95}$$

where $H(z^{-1}) = P(z^{-1})/Q(z^{-1})$ is called the **transfer function** of the LTI DTS. The roots of $Q(z^{-1})$ and $P(z^{-1})$ are called the **poles** and **zeros**, respectively, of $H(z^{-1})$. Note that $H(z^{-1})$ can be found by inspection of (12.67).

Notice that $Q(z^{-1})$ is the characteristic polynomial, which means that an LTI DTS is BIBO stable if and only if the poles of the transfer function all have magnitudes less than one. The relationship between BIBO stability and the poles of the transfer function can be depicted with a pole-zero plot in a complex plane, called the z-**plane**. The built-in function **zplane** produces a pole-zero plot given vectors b and a of the coefficients of $P(z^{-1})$ and $Q(z^{-1})$, respectively. A syntax option is

```
zplane(b,a)
```

Example 12.24 _____

Prog. 12.24 specifies an LTI DTS to have 2 zeros and 6 poles. With the coefficients in the vectors a and b, we can write the difference equation that relates the input $r(n)$ to the output $y(n)$. The function zplane produces the pole-zero plot shown in Fig. 12.42.

```
% Pole-zero description of an LTI IIR DTS
z(1) = 1; z(2) = -1; % specify zeros
p(1) = 0.9*exp(j*0.7*pi/2); p(2) = conj(p(1)); % specify poles
p(3) = 0.81*exp(j*0.9*pi/2); p(4) = conj(p(3));
p(5) = 0.9*exp(j*1.1*pi/2); p(6) = conj(p(5));
b = poly(z); % numerator polynomial of the transfer function
a = poly(p); % denominator polynomial of the transfer function
figure('Color',[1 1 1]) % make figure background white
zplane(b,a); grid on % get pole-zero plot in the z-plane
```

Program 12.24   Program to obtain a pole-zero plot given the transfer function.

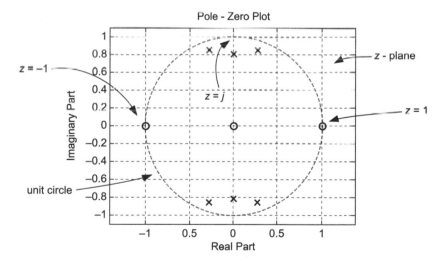

Figure 12.42   Pole-zero plot in the z-plane.

The function zplane also draws a circle with a radius equal to one, called the **unit circle**, in the z-plane, and then it places the poles (x mark) and zeros (o mark). In Fig. 12.42 we see that all of the poles of the transfer function are inside the unit circle in the z-plane, and therefore the LTI DTS is BIBO stable. The transfer function has 6 poles with magnitudes less than one, 2 zeros on the unit circle, and 4 zeros located at the origin in the z-plane. This stems from writing $H(z^{-1})$ as

$$H(z^{-1}) = \frac{\displaystyle\prod_{k=1}^{2}(1 - z_k z^{-1})}{\displaystyle\prod_{k=1}^{6}(1 - p_k z^{-1})} = \frac{z^4 \displaystyle\prod_{k=1}^{2}(z - z_k)}{\displaystyle\prod_{k=1}^{6}(z - p_k)}$$

Suppose the input to output relationship of an LTI DTS is simply

$$y(n) = r(n - 1)$$

This means that the present output $y(n)$ is the signal $r(n)$ at the previous sample time. For example, $y(0) = r(-1)$, $y(1) = r(0)$, and so on. Such an activity is achieved with a device called a **delayor**. If $r(n) = R\,z^n$, then $y(n) = Y\,z^n = R\,z^{n-1} = z^{-1}\,R\,z^n$, and $Y = z^{-1}R$. The transfer function of a delayor is $H(z^{-1}) = z^{-1}$. A computer memory register can be a delayor, where at a time n the register content is the data that was placed in the register at the previous sample time $n - 1$.

Examples 12.21 and 12.22 demonstrate that with three basic operations: **addition**, **multiplication**, and **delay**, digital filters (discrete time systems) can be implemented with a computer.

We know how to assess the frequency response of an LTI CTS. Here, this concept will be developed for discrete time systems. Recall (12.72), which shows that with Euler's identity we can write

$$r(n) = A \cos(\omega\,nT + \theta) = \frac{R}{2}(z)^n + \frac{R^*}{2}(z^*)^n, \quad R = A\,e^{j\theta}, \ z = z^{(\sigma + j\omega)T}\big|_{\sigma=0} = e^{j\omega T}$$

$$(12.96)$$

where R is the phasor of $r(n)$. As the frequency ω of $r(n)$ varies, $z = e^{j\omega T}$, of which the magnitude is one, takes on values along the unit circle in the z-plane. With $z = e^{j\omega T}$, the response to $R\,z^n$ $(R^* z^{-n})$ is $Y\,z^n$ $(Y^* z^{-n})$, and with (12.95) we get $Y = H(e^{-j\omega T})\,R$. Since (12.67) describes a linear system, the steady-state response $y(n)$ to $r(n)$, given in (12.96), is

$$y(n) = \frac{1}{2}H(e^{-j\omega T})\,R\,z^n + \frac{1}{2}(H(e^{-j\omega T})\,R)^*\,z^{-n} \qquad (12.97)$$

Here, the transfer function $H(e^{-j\omega T})$ is a complex function of the real variable ω, the frequency of the input. Writing $H(e^{-j\omega T})$ in polar form and applying Euler's identity changes (12.97) into

$$y(n) = \frac{1}{2}\|H(e^{-j\omega T})\|e^{j\angle H(e^{-j\omega T})} A e^{j\theta} e^{j\omega nT} + \frac{1}{2}\|H(e^{-j\omega T})\|e^{-j\angle H(e^{-j\omega T})} A e^{-j\theta} e^{-j\omega nT}$$

$$= A\|H(e^{-j\omega T})\|\cos{(\omega nT + \theta + \angle H(e^{-j\omega T}))}$$

(12.98)

which shows how the output amplitude and phase depend on the transfer function $H(e^{-j\omega T})$, a complex function of the input frequency ω.

The frequency ω appears in the argument of $z = e^{j\omega T} = \cos(\omega T) + j\sin(\omega T)$. When $\omega = 0$, then $z = 1$, and when $\omega = \omega_s/2$, $\omega T = \omega_s T/2 = (2\pi/T)(T/2) = \pi$, and $z = -1$. Therefore, as ω varies from $\omega = 0$ to $\omega = \omega_s/2$, z varies along the unit circle from $z = 1$ to $z = -1$, with an angle given by $\angle z = \omega T$. As ω is increased further from $\omega = \omega_s/2$ to ω_s, z continues to follow the unit circle from $z = -1$ back to $z = 1$. Therefore, for $\omega + k\omega_s$, for any integer k, we get the same value of z. This means that $H(e^{-j\omega T})$ is a **periodic function** of ω, with period ω_s. To avoid aliasing error when an analog signal $r(t)$ is processed by digital means, a digital filter, the signal $r(t)$ is restricted to have a bandwidth (BW) less than $\omega_s/2$.

Example 12.25 _____

Let us obtain a plot of the magnitude squared of $H(e^{-j\omega T})$ of the LTI DTS with the poles and zeros given in Example 12.24. Prog. 12.25 uses 3-D plot functions to show how the magnitude squared of $H(e^{-j\omega T})$ varies as z follows the unit circle. Fig. 12.43 shows the plot.

```
% Program to plot the magnitude squared of a transfer function H(z^-1)
% for values of z along the unit circle in the z-plane
clear all; clc;
N_pts  =  1000; k  =  [0:N_pts-1];
wT  =  k*2*pi/N_pts; % angles around unit circle
Z  =  exp(j*wT); Zm1  =  1./Z; % z^-1 on the unit circle
z(1)  =  1; z(2)  =  -1; % specify zeros
num  =  (1-z(1)*Zm1).*(1-z(2)*Zm1); % frequency response numerator
p(1)  =  0.9*exp(j*0.8*pi/2); p(2)  =  conj(p(1)); % specify poles
p(3)  =  0.81*exp(j*pi/2); p(4)  =  conj(p(3));
p(5)  =  0.9*exp(j*1.2*pi/2); p(6)  =  conj(p(5));
den = ones(1,N_pts); % initialize the denominator
for i = 1:6
    den  =  den.*(1-p(i)*Zm1); % frequency response denominator
end
H  =  num./den; % frequency response
H_mag_2  =  H.*conj(H); % magnitude squared
```

```
H_mag_2  =  H_mag_2/max(H_mag_2); % normalize
x  =  real(Z); y  =  imag(Z); % rectangular coordinates of the unit circle
zero  =  zeros(1,N_pts); % values for drawing unit circle
figure(1)
plot3(x,y,zero,'LineWidth',2.0); hold on; % plot unit circle
plot3(x,y,H_mag_2); grid on; % plot magnitude squared
xlabel('real axis'); ylabel('imaginary axis'); zlabel('magnitude squared')
```

Program 12.25   Program to obtain the frequency response of a digital filter.

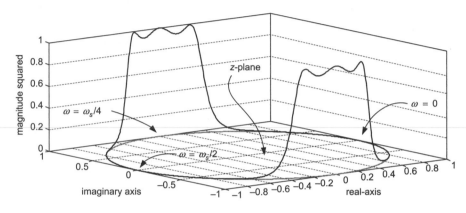

Figure 12.43   Magnitude squared frequency response of a digital band-pass filter.

Notice how the magnitude squared frequency response peaks in the vicinity of the poles of $H(z^{-1})$ (see Fig. 12.42) and becomes zero for $\omega = 0$ and $\omega = \omega_s/2$ due to the zeros of $H(z^{-1})$ located on the unit circle at $z = 1$ and $z = -1$.

While a plot of $H(z^{-1})$ versus z as z varies along the unit circle reveals how $H(e^{-j\omega T})$ is a periodic function of ω, to assess the frequency selective behavior of an LTI DTS, it is conventional to plot $\|H(e^{-j\omega T})\|^2$ versus ωT for $0 \le \omega T \le 2\pi$, which is shown in Fig. 12.44. The following MATLAB statements were appended to Prog. 12.25 to produce this plot.

```
figure(2)
plot(wT,H_mag_2); grid on; title('Frequency Response of an LTI DTS')
set(gca,'XTick',linspace(0,2*pi,5))
set(gca,'XTickLabel',{'0','pi/2','pi','3pi/2','2pi'})
axis([0 2*pi 0 1.1])
xlabel('\omegaT - radians'); ylabel('magnitude squared')
```

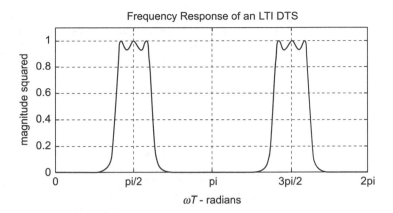

Figure 12.44 Magnitude squared frequency response of a digital band-pass filter.

The Signal Processing Toolbox contains many functions concerned with the time and frequency domain analysis of discrete time systems. Some of these functions are given in Table 12.7.

Table 12.7 Built in functions for the analysis of discrete time systems

Function	Brief description
freqz	Plots a frequency response given the transfer function
impz	Finds the impulse response given the transfer function
stepz	Finds the step response given the transfer function
dstep	Finds the step response given a state variable description
dimpulse	Finds the impulse response given a state variable description
dinitial	Finds the response to ICs given a state variable description

12.6 Ideal Digital Filters

We can design the coefficients of the digital filter described by (12.67) to achieve filter performance that comes arbitrarily close to the frequency response of an ideal filter. Fig. 12.45 shows the ideal performance of several standard digital filter types. The frequency response for each filter is specified over the frequency range, $-\omega_s/2 \leq \omega \leq \omega_s/2$, which is the frequency range of the filter input. Then, this frequency response is extended periodically to integer multiples of ω_s.

To justify an ideal linear phase characteristic, let us consider an input $x(t) = A_1\cos(\omega_1 t + \phi_1) + A_2\cos(\omega_2 t + \phi_2)$ of an analog filter, where ω_1 and ω_2 are within the filter passband. To preserve the input wave shape, the output $y(t)$ can only be a delayed version of the input. Now we have

$$y(t) = x(t - t_0) = A_1\cos\left(\omega_1(t - t_0) + \phi_1\right) + A_2\cos\left(\omega_2(t - t_0) + \phi_2\right)$$
$$= A_1\cos\left(\omega_1 t + \phi_1 - t_0\omega_1\right) + A_2\cos\left(\omega_2 t + \phi_2 - t_0\omega_2\right)$$

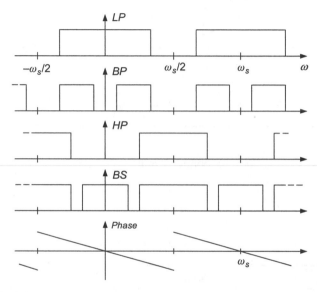

Figure 12.45 *Ideal frequency response of a low-pass (LP), band-pass (BP), high-pass (HP), and band-stop (BS) digital filter.*

Table 12.8 *Built-in functions for digital filter design*

Function	Brief description
fir1	Design of a linear phase FIR filter using the window method
fir2	Design of a linear phase FIR filter that has an arbitrary shaped magnitude frequency response using the frequency sampling method
firls	Design of a linear phase FIR filter by minimizing the error squared
fircls	Design of a linear phase FIR filter by constrained least squares
firpm	Design of a Parks–McClellan optimal equiripple FIR filter
butter	Design of Butterworth analog and IIR digital filters
cheby1	Chebyshev type I IIR digital and analog filter design
chevy2	Chebyshev type II IIR digital and analog filter design
ellip	Design of elliptic or Cauer IIR digital and analog filters

and to preserve the input wave shape, the input to output phase change must be proportional to frequency.

The MATLAB Signal Processing Toolbox includes many functions for the design of digital filters. Some of these are listed in Table 12.8. Use the **doc** help facility to learn how to use these functions to design a digital filter.

12.7 Conclusion

In the previous chapters we focused mainly on the features of the MATLAB programming language. MATLAB evolves continuously to incorporate new functions that implement algorithms to solve fundamental problems and apply methods for contemporary system design. In this chapter we investigated some of the fundamental mathematics of the many fields in which MATLAB has been applied. You should now know

- about the DFT and some of its properties
- how to apply the function fft for spectral analysis of stationary and nonstationary signals
- the importance of the sampling theorem
- about the kinds of errors that occur in Fourier-based spectral analysis by digital means
- that window functions can be used to reduce certain errors incurred in spectral analysis
- how to use MATLAB to analyze LTI continuous time and discrete time systems to find
  - the natural frequencies
  - characteristic modes
  - the response to ICs
  - impulse and step responses
  - the response to any input by convolution
  - the transfer function
  - pole-zero plots
  - stability condition
  - a state variable description
  - steady-state response
  - frequency response
- how to apply MATLAB ODE solvers
- how to operate a discrete time system both online and offline

Table 12.9 gives the additional built-in functions that were introduced in this chapter. Use the MATLAB help facility to learn more about these built-in functions.

With MATLAB tools to solve for the dynamic behavior of models of devices and systems, the next chapter introduces a GUI environment to conveniently build complex models of systems and simulate them.

Table 12.9   Built-in functions introduced in this chapter

Function	Brief description
fft	Returns the DFT using a fft algorithm
ifft	Returns the inverse DFT using a fft algorithm
tf2ss	Converts a transfer function of a CTS into a state variable description
dtf2ss	Converts a transfer function of a DTS into a state variable description
ss2tf	Converts a state variable description of a CTS into a transfer funstion
obsv	Returns the observability matrix
conv	Computes discrete convolution
ss	Creates an object of a CTS or DTS state variable description
pzplot	Plots a pole-zero plot of an LTI CTS given a transfer function
odeset	Sets parameters of ODE solvers
impulse	Returns the impulse response of an LTI CTS or DTS
filter	Performs offline operation of an LTI DTS
step	Returns the step response of a CTS
initial	Returns the response of a CTS to ICs
dtf2ss	Converts a DTS transfer function into a state variables description
zplane	Plots a pole-zero plot of an LTI DTS given H(z)

Further reading

Ambardar, A., 2007, Digital Signal Processing: *A Modern Introduction*, Thomson, Stamford, Conn.

Mandal, M. and Asif, A., 2007, *Continuous and Discrete Time Signals and Systems*, Cambridge University Press, Cambridge UK

Oppenheim, A.V. and Schafer, R.W., 2010, *Discrete Time Signal Processing*, Prentice-Hall, Englewood Cliffs, NJ

Orfanitis, S.J., 1996, *Introduction to Signal Processing*, Prentice-Hall, Englewood Cliffs, NJ

Priemer, R. 1991, *Introductory Digital Signal Processing*, World-Scientific, Singapore

Proakis, J.G., Manolakis, D.G., 1996, *Digital Signal Processing: Principles, Algorithms and Applications*, Prentice-Hall, Englewood Cliffs, NJ

Rabiner, L.R., Gold, B., 1974, *Theory and Application of Digital Signal Processing*, Prentice-Hall, Englewood Cliffs, NJ

Roberts, M.J., 2012, *Signals and Systems: Analysis Using Transform Methods and MATLAB*, McGraw-Hill, New York, NY

Tan, T., 2008, *Digital Signal Processing*, Fundamentals and Applications, Academic Press, Burlington, MA

Problems

Section 12.1

1) Manually, find the bandwidth, period, and complex Fourier series coefficients, and sketch versus frequency in hertz a stem plot of the magnitude spectrum of

(a) $x(t) = 6\sin(120\pi t + \pi/4) - 4\cos(30\pi t - \pi/3)$
(b) $y(t) = 3\cos(28\pi t - \pi/4) + 5\sin(112\pi t) - 2\cos(42\pi t)$

2) (a) Starting at $t = 0$, sketch $x(t)$ over three periods, where one period is given by

$$x(t) = \begin{cases} 2, -1 \leq t < 1 \\ -1, 1 < t < 4 \end{cases}$$

(b) Manually, find the Fourier series coefficients X_k. Write a MATLAB script to reconstruct $x(t)$. Use a program like Prog. 6.5. Does the reconstruction exhibit Gibbs' oscillation?

(c) Starting at $t = 0$, sketch $x(t)$ over three periods, where one period is given by

$$x(t) = |\sin(120\,\pi t)|, 0 \leq t < 1/120 \text{ sec}$$

(d) Repeat part (b) for the signal given in part (c).

3) One period of a discrete time signal $x(n)$ is given by

$$x(n) = \{x(-1) = 1, \ x(0) = 0, \ x(1) = -1, \ x(2) = 0\}$$

(a) What is N? Starting at $n = 0$, give a sketch of $x(n)$ versus n over three periods.
(b) Manually, find the DFT $X(k)$ of $x(n)$. Show details.

4) (a) Write a MATLAB script that obtains over one period $N = 20$ samples of the signal $x(t)$ given in part (a) of Prob. 12.1. What is the sampling frequency? Does the sampling frequency satisfy the sampling theorem? Use the **fft** function to obtain $X(k)$, $k = 0, 1, \ldots, N - 1$. Obtain a stem plot of the magnitude $\|X(k)\|$ versus k. What are the frequencies corresponding to $k = 0, 1, \ldots, N - 1$? How is $X(k)$ related to X_k?

(b) Repeat part (a), using $N = 6$ samples of $x(t)$ over one period. Is there aliasing error? If there is aliasing error, then what is(are) the frequency(ies) of the alias?

5) Repeat Prob. 12.4 for the signal given in part (b) of Prob. 12.1.

6) (a) Write a MATLAB script that samples the periodic signal given in Prob. 12.2, part (a) over one period at the rate $f_s = 10$ Hz (samples/sec). What is N? Use the fft function to find and stem plot $\|X(k)\|$ versus k. What is the frequency resolution? Is there aliasing error?

(b) Repeat part (a), but sample the signal over four periods. What is the frequency resolution? Is there aliasing error? Discuss the difference between the results and the results of part (a).

(c) Repeat part (a), but use $f_s = 100$ Hz. Discuss the difference between the results and those of part (a).

7) Given is $x(t) = \sin(t)/t$. Is this time function bandlimited? If you are not sure, then consider the following investigation.

(a) Write a MATLAB script that samples (use $f_s = 20$ Hz) and plots $x(t)$ over $-5 \le t \le 5$. Then, use the fft function to find and line plot $\|X(k)\|$ versus k.

(b) Repeat part (a), but sample $x(t)$ over $-20 \le t \le 20$.

(c) In view of the trend from the results of parts (a) and (b), is $x(t)$ bandlimited? If it is bandlimited, then what is the bandwidth?

8) (a) Given is $x(t) = 4\sin(8\pi t)$. Write a MATLAB script that samples the signal over one period T_x. Use a sampling frequency $f_s = 1/T$ that satisfies the sampling theorem. Make sure that T_x/T (T_x is the period of $x(t)$.) is an integer N. Use the fft function to find and plot $\|X(k)\|/N$ versus k. Also, output a table of $X(k)$ versus k.

(b) Instead of the samples $x(n)$ of $x(t)$, you have available $y(n) = x(n) + e(n)$, where $e(n)$ is additive noise found with the function randn. Apply your MATLAB script to $y(n)$. Give plots of $y(n)$ and $\|Y(k)\|/N$. From the spectrum plot, is it possible to estimate the frequency of $x(t)$?

(c) Repeat part (a), but sample the signal over a time range given by $1.25T_x$. What is N? Discuss the difference between the results and those of part (a).

(d) Repeat part (c), but apply the Hann window to the data before applying the fft function. Discuss the difference between the results and those of part (c).

9) One period of the DFT $X(k)$ of a periodic discrete time signal $x(n)$ is given by

$$X(-5) = 2e^{-j3\pi/4}, X(-4) = X(-3) = 0, X(-2) = 4e^{j2\pi/5}, X(-1) = 3,$$
$$X(0) = 1, X(k) = X^*(-k), k = 1, 2, \dots, 5$$

(a) What is N? Starting at $k = 0$, sketch a stem plot of two periods of $\|X(k)\|$.

(b) Write a MATLAB script that sets up an appropriate $X(k)$, and uses the **ifft** function to find $x(n)$. Then, give a stem plot of it.

(c) If $x(n)$ comes from sampling a continuous time signal $x(t)$ over one period, then what additional information must be known to know the bandwidth of $x(t)$? Assume that the period T_x is given.

10) In the DFT algorithm, let $W = e^{-j2\pi/N}$, and consider the expansion of the DFT given by

$$X(0) = x(0) + x(1) + \dots + x(N - 1)$$
$$X(1) = x(0) + x(1)W + \dots + x(N - 1)W^{N-1}$$
$$X(2) = x(0) + x(1)W^2 + \dots + x(N - 1)W^{2(N-1)}$$
$$\vdots$$
$$X(N - 1) = x(0) + x(1)W^{N-1} + \dots + x(N - 1)W^{(N-1)\,(N-1)}$$

Let $x' = [x(0)\ x(1)\ x(2) \dots x(N - 1)]$, and let $X' = [X(0)\ X(1)\ \dots\ X(N - 1)]$.

(a) Write a MATLAB script that finds the matrix D_N such that $X = D_N x$, for $N = 32$.

(b) Continue the script to obtain $N = 32$ samples of $x(t) = 4\cos(20\pi t + \pi/3)$ sampled at the rate $f_s = 1/T$ Hz, where $T = 2T_x/N$. Use $X = D_N x$ to find the DFT of

$x(n)$, $n = 0, 1, \ldots, N - 1$. Plot the magnitude spectrum. Are the results as expected?

11) Denote the DFT algorithm with $X = \text{DFT}(x)$, where the input is the vector $x' = [x(0)\ x(1)\ x(2) \ldots x(N-1)]$ and the output is the vector $X' = [X(0)\ X(1) \ldots X(N-1)]$. Or, let $C = x$, and then $X = \text{DFT}(C)$. The IDFT of X gives x. The IDFT of X can also be found with the DFT algorithm.

(a) Prove that $x = (1/N)\,\text{real}(\text{DFT}(C^*))$, where $C = X$. Therefore, only one program is needed to do signal analysis and signal reconstruction. Hint: take the conjugate of (12.14), and then, since $x(n)$ is real, use $x^*(n) = x(n)$.

(b) Write a MATLAB script that uses the function fft to test part(a). Sample $x(t) = 4\cos(20\,\pi\,t + \pi/3)$ at the rate $f_s = 1/T$ Hz, where $T = T_x/N$ and $N = 32$. Plot x and $\text{real}(\text{DFT}(X^*/N))$, and compare them.

12) Given is $x(t) = e^{-4t}\sin(16\,\pi t)\,u(t)$. Write a MATLAB script that samples the signal at a rate $f_s = 1\text{K Hz}$ to obtain $x(n)$, $n = 0, 1, \ldots, N - 1$, where $N = 2^{13}$. Use $x(t = 0^+)$ for $x(n = 0)$.

(a) If the $x(n)$ are considered to be the result of sampling a periodic signal over one period, then what is the period T_x?

(b) Use the function **spectrogram** to find an estimate of the power spectral density P of $x(t)$. Use a Bartlett window. Use $N_W = 2^7$ window points. Make the segment to segment overlap $N_o = N_W - 2^5$, and use N_W points for the fft. Use the function **surf** to obtain a 3-D plot of P using a decibel scale. You may have to rotate the plot for a good viewpoint. Does the plot show whether or not $x(t)$ is a stationary signal? Explain.

(c) Repeat part (b) for $N_W = 2^6$ and $N_W = 2^9$. Do these results change your opinion about the stationarity of $x(t)$?

13) The MATLAB statements

```
t = 0:0.001:6; f_start = 0; f_end = 100; t_cross = 1.5;
x = chirp(t, f_start, t_cross, f_end, 'linear');
```

use the function **chirp** to sample at a rate $f_s = 1\text{K Hz}$ a sweep frequency signal over 6 sec. Write a MATLAB script that uses the function spectrogram to find and plot the STFT power spectral density of the chirp signal. Try various values of N_W and N_o.

Sections 12.2 and 12.3

14) Find the differential equation that relates the current $i(t)$ to the voltage $v_s(t)$ of the circuit shown in P12.14. In terms of the initial voltage on the capacitor, give $i(0^-)$.

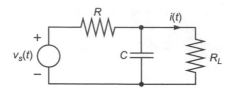

Figure P12.14   RCR circuit.

15) Find the differential equation that relates the voltage $v(t)$ to the voltage $v_s(t)$ of the circuit shown in P12.15. In terms of the initial voltage of the capacitor and the initial current in the inductor, give $v(0^-)$ and $\dot{v}(0^-)$. Assume that $v_s(0^-) = 0$.

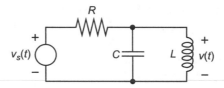

Figure P12.15   RLC circuit.

16) An LTI CTS is described by

$$y^4(t) + 6y^3(t) + 138y^2(t) + 462y^1(t) + 2929y^0(t) = 573r(t)$$

with initial conditions: $y^3(0^-) = 0$, $y^2(0^-) = 1$, $y^{(1)}(0^-) = 0$, $y^{(0)}(0^-) = 0$.
(a) Give the characteristic equation.
(b) Write a MATLAB script that finds and line plots the zero-input response.

17) An LTI CTS is described by

$$y^3(t) + 5y^2(t) + 33y^1(t) + 29y^0(t) = 2r^2(t) + 4r^1(t) + 202r^0(t)$$

with initial conditions: $y^2(0^-) = 0$, $y^{(1)}(0^-) = 1$, $y^{(0)}(0^-) = 0$. If $w(t)$ is the zero-state solution of $w^3(t) + 5w^2(t) + 33w^1(t) + 29w^0(t) = r(t)$
then in terms of $w(t)$ write $y_{zs}(t)$.

18) (a) Manually, find the matrices A, B, C, and D of a state variable description of the LTI CTS given in Prob. 12.17. Write a MATLAB script that sets up the observability matrix and finds the initial state.
   (b) Check your results by also using the function **tf2ss** to find A, B, C, and D.

19) (a) Write a MATLAB script that finds a state variable description of the LTI CTS given in Prob. 12.16.
   (b) Find the eigenvalues of A and the roots of the characteristic equation.
   (c) Create an object of the state variable description with the function **ss**. Using the object description, find the observability matrix and then the initial state.

20) For the LTI CTS given in Prob. 12.15, write a MATLAB function script and a script that applies an ODE solver to find and plot:
    (a) $y_{zi}(t)$. What happens to this response as t increases?
    (b) $y_{zs}(t)$, where $r(t) = u(t)$.
    (c) $y(t)$. Compare $y(t)$ to a plot of the sum of $y_{zi}(t)$ and $y_{zs}(t)$.
    (d) The steady-state response $y_{ss}(t)$ is a constant, and it is given by $y_{ss}(t) = 202/29$. Subtract the steady-state response from $y(t)$ to find and line plot the transient response. What happens to the transient response as t increases?

21) Repeat Prob. 12.20, parts (a), (b), and (c) for $r(t) = \sin(4\pi t)\,u(t)$. What is the amplitude of the steady-state response?

22) Repeat Prob. 12.20 for the LTI CTS given in Prob. 12.16.

23) Manually, find the impulse response $h(t)$ of the LTI CTS described by

$$\dot{y}(t) + 2y(t) = 3r(t), \quad y(0^-) = 1$$

24) Manually, find the impulse response $h(t)$ of the LTI CTS described by

$$\ddot{y}(t) + 2\dot{y}(t) + 2y(t) = \dot{r}(t) + r(t), \quad \dot{y}(0^-) = 1,\ y(0^-) = 1$$

25) (a) Write a MATLAB script that uses the function **impulse** to find and plot the impulse response $h(t)$ of the LTI CTS given in Prob. 12.17. What happens to $h(t)$ as t increases, and give an explanation?
    (b) Continue the script to use the function **step** to find and plot the step response. Describe how you can find the step response knowing the impulse response $h(t)$. Consider how the impulse function $\delta(t)$ and the unit step function $u(t)$ are related.
    (c) Continue the script to use the function **initial** to find and plot $y_{zi}(t)$.

26) Repeat Prob. 12.25 for the LTI CTS given in Prob. 12.16.

27) The impulse response of an LTI CTS is given by $h(t) = 2\,e^{-t}u(t)$.
    (a) Manually, find $y_{zs}(t)$ for the input $r(t) = 3e^{-2t}u(t)$ by convolving $h(t)$ and $r(t)$.
    (b) Repeat part (a) if the input is $r(t-1)$.
    (c) Repeat part (a) if the input is $r(t+1)$.

28) Given are two signals $s_1(t)$ and $s_2(t)$. Write a MATLAB script that uses the function **conv** to find and line plot $s_3(t) = s_1(t) * s_2(t)$. Obtain N_1 samples of $s_1(t)$ and N_2 samples of $s_2(t)$. Use $s_1(t) = u(t)$, $N_1 = 100$, $s_2(t) = e^{-t}u(t)$, $N_2 = 150$ and $f_s = 50$ Hz. How many samples N_3 of $s_3(t)$ are produced by the conv function? How is N_3 related to N_1 and N_2?

29) Repeat Prob. 12.28 for $s_2(t) = (e^{-10t} - e^{-20t})\sin(100\pi t)$. Here, you must pick appropriate values for N_1, N_2 and f_s to obtain a good approximation of $s_3(t)$ over $0 \le t \le 0.5$.

30) Let A_1 and A_2 denote the areas under the functions $x_1(t)$ and $x_2(t)$, respectively. Prove that $A_3 = A_1 A_2$, where A_3 is the area under the function $x_3(t) = x_1(t) * x_2(t)$.

Hint: The area $A_3 = \int_{-\infty}^{+\infty} x_3(t)\, dt$. Substitute in the convolution integral for $x_3(t)$, and proceed from there.

31) (a) For the LTI CTS given in Prob. 12.17 give the transfer function $H(s)$.

    (b) Write a MATLAB script that uses the function **pzplot** to obtain a pole-zero plot. Is the LTI system BIBO stable? Explain.

    (c) Using the transfer function $H(j\omega)$, continue the script to find the steady-state response $y_{ss}(t)$ to $r(t) = \sin(2\pi t)$. Plot $y_{ss}(t)$ and $r(t)$ on the same figure.

    (d) Obtain a Bode plot.

Sections 12.4 and 12.5

32) A DTS is described by

$$y(n) + 0.5\,y(n-1) + 0.25\,y(n-3) + 0.125\,y(n-5) = r(n) + 0.25\,r(n-1)$$

    (a) What order is this difference equation?

    (b) What initial conditions must be known to solve for $y(n)$, $n = 0,\ 1,\ldots$?

    (c) State whether or not this DTS is causal, time invariant and linear.

    (d) Give the homogeneous equation of this DTS.

    (e) If $y(-2) = 3$, and all other initial conditions are zero, manually find $y(n = 2)$ for $r(n < 0) = 0$ and $r(n \geq 0) = K = 1$.

    (f) Now assume that all initial conditions are zero. Find $y(n = 2)$ for $r(n < 0) = 0$, $r(n \geq 0) = K$ and $K = 1,\ 2$ and 3. Denote each result by $y_1(2)$, $y_2(2)$ and $y_3(2)$, respectively. What is the relationship between $y_1(2)$, $y_2(2)$ and $y_3(2)$? Explain. Do you expect this relationship to apply for all n? Does this relationship apply if some initial condition is not zero?

33) For the DTS given in Prob. 12.32 assume that all initial conditions are zero.

    (a) Manually find $y(n)$, $n = 0, 1, 2$, and 3 for $r(n) = \delta(n)$, and denote the result by $y_1(n)$, which is the unit pulse response.

    (b) Repeat part (a) for $r(n) = \delta(n-2)$, and denote the result by $y_2(n)$. What is the relationship between $y_1(n)$ and $y_2(n)$?

34) Let $y(t)$ be the integral of $r(t)$, and write

$$y(t) = \int_0^t r(\tau)\, d\tau + y(0^-)$$

    Assume that $r(t)$ is finite, and that it has a finite number of discontinuities. Recall from Section 4.7.2 that the trapezoidal rule gives an approximation of this integral with

$$y(n+1) = \frac{T}{2}\left(r(n+1) + r(n)\right) + y(n), \quad t = nT,\ n = 0, 1, \ldots$$

where $y(n = 0) = S_0$ is the initial condition given by $S_0 = y(t = 0^-)$, T is the time increment, and $f_s = 1/T$ Hz is the sampling rate. This is a first order difference equation. Write a MATLAB script that runs this difference equation to find $y(n)$, $n = 0, 1, \ldots, 99$ for $r(t) = u(t)$. Use $r(0) = r(t = 0^+)$, and let $f_s = 100$ Hz. Assume that $S_0 = 0$. Use a **for loop** to find $y(n)$ recursively. Provide a line plot of $y(n)$ versus t. Knowing that $y(t) = t\,u(t)$, discuss the error between $y(t = nT)$ and $y(n)$.

35) Repeat Prob. 12.34 for $r(t) = \sin(2\pi f\, t)$ and $f = 5$ and 40 Hz. Provide line plots for each case. For each f, assess the approximation error by computing $e(n) = y(t = nT) - y(n)$, $\overline{e(n)}$, which is the average error, $\varepsilon(n) = e(n) - \overline{e(n)}$ and σ^2, which is the average of $\varepsilon^2(n)$. What happens to σ^2 as f is increased?

36) Given is the LTI DTS

$$\sum_{i=0}^{8} a_i\, y(n - i) = \sum_{i=0}^{8} b_i\, r(n - i)$$

where a_i and b_i, $i = 0, 1, \cdots, 8$ are given by

$a = [1 \ -2.1878931\text{e-}1 \ -2.4620748 \ 3.0601433\text{e-}1 \ 2.6490803 \ -1.7792892\text{e-}1 \ \ldots$
$\qquad -1.3579514 \ \ 9.3413671\text{e-}3 \ \ 3.2604906\text{e-}1]$
$b = [4.5347167\text{e-}1 \ -9.4249009\text{e-}3 \ -1.7473316 \ 9.0243936\text{e-}3 \ 2.5892471 \ \ldots$
$\qquad 9.0243936\text{e-}3 \ -1.7473316 \ -9.4249009\text{e-}3 \ 4.5347167\text{e-}1]$

The input $r(n)$ comes from sampling $r(t) - \sin(\omega_1 t) + \sin(\omega_2 t) + \sin(\omega_3 t)$ at the rate $f_s = 8\text{K}$ Hz, where $\omega_1 = 2\pi(100)$, $\omega_2 = 2\pi(1500)$ and $\omega_3 = 2\pi(3800)$. Assume $r(n < 0) = 0$, and that initial conditions are zero. Write a MATLAB script that uses the built in function **filter** to obtain the response $y(n)$ to $r(n)$. Process enough samples to see the response over $0 \leq t \leq 0.01$ sec. Provide line plots of $y(n)$ and $r(n)$ versus t. What kind of a filter do you think this DTS is? Try other input frequencies to support your opinion.

37) Write a MATLAB script to run the filter given in Prob. 12.36, and find and line plot the unit pulse and unit step responses over $0 \leq t \leq 0.01$ sec.

38) (a) Write a MATLAB script that uses the function **dtf2ss** to obtain a state variable description of the DTS described in Prob. 12.36.
    (b) Continue the script by using a state variable description to obtain and line plot the unit step response over $0 \leq t \leq 0.01$ sec.

39) The coefficients of an FIR LTI DTS are given by

$$b_k = \frac{1}{(k - 128)\pi}(\sin((k - 128)3\pi/4) - \sin((k - 128)\pi/4)), \quad k = 0, 1, \ldots, 256$$

where $b_{128} = 1/2$. Using the same sampling frequency and input as described in Prob. 12.36, write a MATLAB script that runs this FIR filter over $0 \leq t \leq 0.01$ sec. Provide line plots of $y(n)$ and $r(n)$ versus t. What kind of a filter do you think this DTS is?

40) An LTI DTS is described by

$$y(n) + 0.81\, y(n-2) = r(n) - r(n-2), \quad y(-1) = 1, \quad y(-2) = 0$$

(a) Give the characteristic equation.
(b) What are the characteristic modes?
(c) Manually, find $y_{zi}(n)$.

41) Manually, find the unit pulse response $h(n)$ of the DTS given in Prob. 12.40.

42) (a) Give the transfer function $H(z^{-1})$ of the DTS described in Prob. 12.36.
(b) Write a MATLAB script that uses the function **zplane** to obtain a pole-zero plot. Is the DTS stable? Explain.
(c) Continue the script to plot the magnitude frequency response for $0 \leq \omega T \leq \pi$, which corresponds to $0 \leq \omega \leq \omega_s/2$.

43) Give the transfer function $H(z^{-1})$ of the LTI DTS described in Prob. 12.40.
(a) Write a MATLAB script that produces a 3-D plot of the magnitude frequency response over the unit circle.
(b) If $f_s = 1\text{K Hz}$, and $r(n)$ comes from sampling $r(t) = \cos(250\pi t)$, then what is the steady-state response $y_{ss}(n)$? Provide a plot of $r(n)$ and $y_{ss}(n)$.

Section 12.6

44) The sampling frequency is $f_s = 8\text{K Hz}$. Consider the design of a low-pass filter with a bandwidth of $f_c = 500$ Hz.
(a) Sketch the ideal magnitude frequency response of such a filter over $0 \leq f \leq 2f_s$.
(b) Write a MATLAB script that uses the function **fir1** to design the coefficients b_k, $k = 0, 1, \ldots, 256$ of an FIR filter. Use doc fir1 to find out how to use the function fir1.
(c) Continue the script to find and plot the magnitude frequency response of the designed filter over $0 \leq f \leq f_s$. Provide a copy of the plot.
(d) Continue the script to use the function **filter** to obtain the response $y(n)$ to samples of $r(t) = \sin(300\pi t) + \cos(2000\pi t)$ over $0 \leq t \leq 20$ msec. Provide plots of $y(n)$ and $r(n)$ versus t. Discuss the performance of the filter.

CHAPTER 13

Introduction to Simulink®

Simulink® is a facility for modeling and analyzing dynamic systems. With its graphical user interface (GUI), you can build a block diagram model of a dynamic system and then run the model. Simulink includes many libraries of ready to use blocks that represent both linear and nonlinear continuous and discrete time operations. There are libraries of blocks for synthesizing signals and for observing signals. You can also design your own blocks. A Simulink simulation can access MATLAB data files and functions, and output results to MATLAB for further analysis and visualization. While a simulation executes, you can view any part of model behavior as it changes with time.

In this chapter, you will learn how to

- build a model of a dynamic system using Simulink blocks
- design a block
- simulate a system and observe its behavior
- use MATLAB functions in a Simulink simulation

13.1 Simulink® Environment

To start Simulink, you must have already started MATLAB. While in MATLAB, specify a *Current Folder*, from which Simulink files, called **model files**, can be retrieved and in which model files can be saved. Then, click the Simulink icon in the MATLAB desktop toolbar, or you can use the menu sequence

```
start → Simulink → Library Browser
```

Or, in the MATLAB *Command Window*, enter the command **simulink** to open the *Simulink Library Browser* shown in Fig. 13.1. Here, Simulink is highlighted, and you can see all of the standard Simulink libraries, each of which contains blocks that you can use to build dynamic system models. If you only want to open a window of the standard Simulink libraries, then in the *Command Window* enter

```
open_system('simulink.mdl')
```

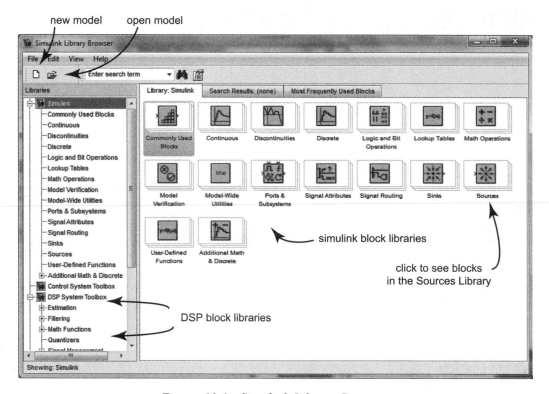

Figure 13.1   Simulink Library Browser.

Clicking on the Commonly Used Blocks Library opens a window that shows all of the blocks that it contains, as shown in Fig. 13.2.

From the Simulink Library Browser, you can create a new Simulink model with

File → New → Model

or you can click on the new model icon in the toolbar of the *Library Browser Window*. You can open an existing model with

File → Open

or you can click on the open model icon in the toolbar.

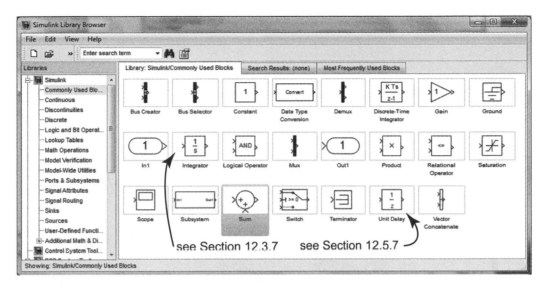

Figure 13.2 Window of blocks in the Commonly Used Blocks Library.

All Simulink model files end with the file extension .mdl. In the MATLAB desktop set a *Current Folder* for all of your Simulink model files, and use the *Current Folder* in Simulink in the same way as it is used in MATLAB. Then, from the MATLAB *Command Window* you need only enter the model name to open it.

Example 13.1

This example demonstrates building a model. We begin by clicking the new model icon in the *Library Browser Window* toolbar. Fig. 13.3 shows the *Model Editor Window*, where Simulink models are created. The *Model Browser Window* was anchored in the *Model Editor Window* with

View → Model Bowser Options → Model Browser

which shows all of the model files in the *Current Folder*.

Some blocks have already been dragged into the *Model Editor Window*. The Sine Wave blocks and the Clock block came from the Sources Library, the Summation block (converted to a Subtraction block) came from the Commonly Used Blocks Library and the XY Graph and To Workspace blocks came from the Sinks Library. The Clock block output is time. With the To Workspace block, we can make data obtained with Simulink available to programs in MATLAB. In this case, the data will be placed in a variable named sine_sum. It is also possible to write data to a .mat file (Use the To File block in the Sinks Library.). You should click on each of the libraries to see the great

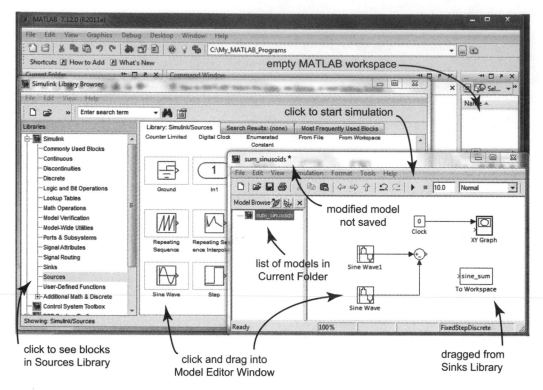

Figure 13.3   Desktop containing the Library Browser and Model Editor.

variety of blocks available to build dynamic systems. Right click a block to access information about it.

Generally, each block has one or more input ports, and/or one or more output ports. In Fig. 13.3, the output ports of the sine wave generators are already connected to the two input ports of the Subtraction block. See Summation block help for information about increasing the number of input ports and converting it to do subtraction. To connect an output port of one block to an input port of another block, start by clicking the output port, and then drag the mouse pointer to an input port. To go around a corner, release and depress the mouse button as you create the path you prefer. Or, you can click on a block icon, which makes it active and a source, and while depressing the keyboard Control key click on another block icon, which makes it active and a sink (destination). Simulink takes care of drawing the connection, which depends on how you place block icons. You can always move block icons around for another arrangement. As you do this, all connections are retained.

Fig. 13.3 does not show all desired connections. The Subtraction block output port must be connected to the Y input port of the XY Graph block and to the input port of the To Workspace block. To do this, first connect the Subtraction block output port to the Y input port. To also connect to the To Workspace block input port, we must add a **branch line** by

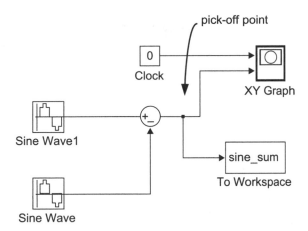

Figure 13.4   Completed model.

first left clicking the connection to the Y input port, which makes it active. Then, with the mouse pointer anywhere on the active connection, right click the mouse and drag the mouse pointer to the input port of the To Workspace block. Simulink will draw a pick-off point and a connection (branch line), as shown in Fig. 13.4. Or, you can start at the input port of the To Workspace block, and drag the mouse pointer to a desired location on the other connection.

Each block has parameters (left click icon twice) and properties (right click icon once), as shown in Fig. 13.5 for a Sine Wave block. In the *Block Parameters Window* you can set parameters such as frequency, phase, amplitude, sampling frequency, and more. You can also access this window from the Property Menu.

After any modification in a model, it must be saved for the change to be effective in a simulation invoked from the *Command Window*. However, you can run a model from the *Model Editor Window* without first saving the model. A summary of a model can be obtained with

View → Model Explorer

as shown in Fig. 13.6, where block parameters of the To Workspace block are given.

Before we can run the model, simulation options must be set by opening the Configuration Parameters dialog box with

Simulation → Configuration Parameters

A part of the dialog box is shown in Fig. 13.7, where all parameters have default values. The finish time was set to 10 sec, and the max step size was set to 0.01 sec.

To run the model, click the play button on the toolbar of the *Model Editor Window*. Or, you can use

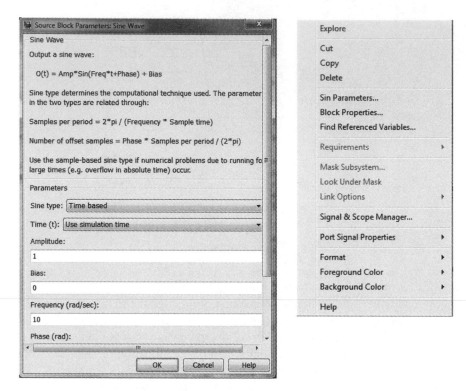

Figure 13.5   *Block Parameters Window (left) and Property Menu (right) of a Sine Wave block.*

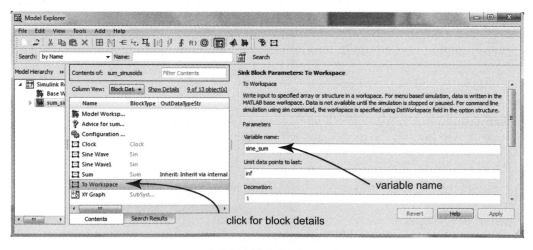

Figure 13.6   *Model Explorer Window.*

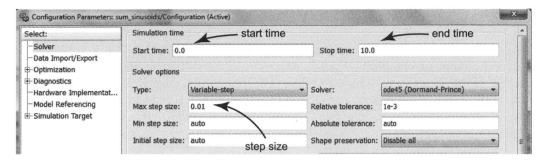

Figure 13.7   Simulation Configuration Parameters dialog box.

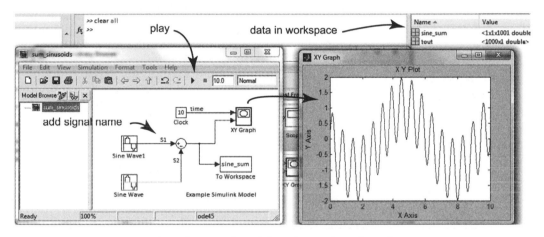

Figure 13.8   Simulation of model with graph and workspace output.

Simulation → Start

The output of interest is displayed by the XY Graph block, which is shown in Fig. 13.8. Notice that the simulation output is also in the MATLAB workspace, to which the graphics capabilities of MATLAB can be applied. In the *Command Window*, use the function **whos** to see more detail about the workspace.

For reference, it is useful to label signal lines (the connections between blocks). To do this, double click a signal line, which causes a blinking cursor to appear. You can then type in text to associate with the signal line. This text can be dragged to a preferred position along the signal line. To place information about the model, double click anywhere in the *Model Editor Window,* which also causes a blinking cursor to appear. You can then type in text, and drag it anywhere within the window. See Fig. 13.8, for example.

Example 13.2 _____

This example demonstrates the utility of the Fcn block. In Fig. 13.9, two Fcn blocks are used. The input of one block (Fcn) is time. The input to output function expression is defined in the Block Parameters dialog box. The input of the other Fcn block (Fcn1) is the output of a block (the Fcn block in this case). The Scope block input consists of two signals that are displayed on the scope screen. These signals are combined into a signal vector with a Mux (multiplexer) block, which can be set to have any number of inputs.

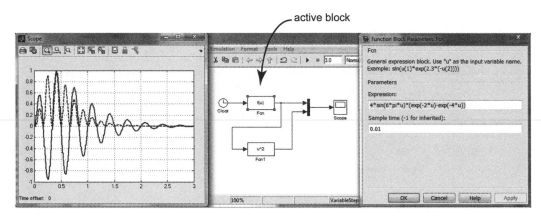

Figure 13.9   Demonstration of using Fcn blocks.

13.2   Dynamic Systems

A continuous time dynamic system is described by a differential equation. In Chapter 12 we solved both linear and nonlinear differential equations by writing programs in MATLAB. With Simulink, a GUI is used to build with blocks a model of a dynamic system. The model is converted by Simulink into a set of equations, which are converted into a program, the output of which is the solution of the set of equations.

Example 13.3 _____

Let us build a model of a dynamic system described by

$$\ddot{y}(t) + a_1 \dot{y}(t) + a_0 y(t) = b_2 \ddot{r}(t) + b_1 \dot{r}(t) + b_0 r(t)$$
$$\dot{y}(0^-) = S_1, \quad y(0^-) = S_0 \tag{13.1}$$

Equation (13.1) is a special case of (12.17), with $N = 2$, $M = 2$, and $M \leq N$. The transfer function is given by

$$H(s) = \frac{b_2 s^2 + b_1 s^1 + b_0 s^0}{s^2 + a_1 s^1 + a_0 s^0}$$

which is a proper function. Depending on the meaning of $r(t)$ and $y(t)$, (13.1) can describe the behavior of a wide variety of physical systems, the band-pass filter in Fig. 6.11, for example.

To build a Simulink model of (13.1), integrate it N times to get

$$y(t) = \left(b_2 r(t) + b_1 \int r(t)dt + b_0 \iint r(t)dt\, dt \right) - \left(a_1 \int y(t)dt + a_0 \iint y(t)dt\, dt \right) \quad (13.2)$$

In (13.2), $y(t)$ is equal to the difference between two main terms. The first term involves $r(t)$ and integrals of $r(t)$, and the second term involves integrals of $y(t)$. We can now draw a Simulink model of (13.2) using blocks given in the Simulink standard libraries.

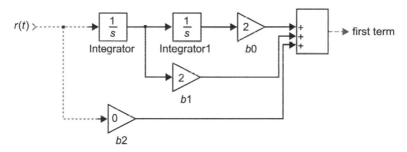

Figure 13.10   The first term of (13.2).

Fig. 13.10 uses two Integrator blocks, three Gain blocks (gains set to 0, 2, and 2) and a Sum block that are interconnected to duplicate the combination of parts in the first term of (13.2). The first term literally is a template to build this section of the model. Similarly, the second term of (13.2) is a template to complete the model, which is shown in Fig. 13.11, where the format block option was used to flip the gain feedback blocks.

The Integrator2 and Integrator3 blocks can be double clicked to open their parameter dialog boxes to set the initial conditions. If the output of Integrator2 is denoted by $x_1(t)$, then the input, which is $y(t)$, is also $\dot{x}_1(t) = y(t)$. Similarly, $\dot{x}_2(t) = x_1(t)$. Since $r(t < 0) = 0$,

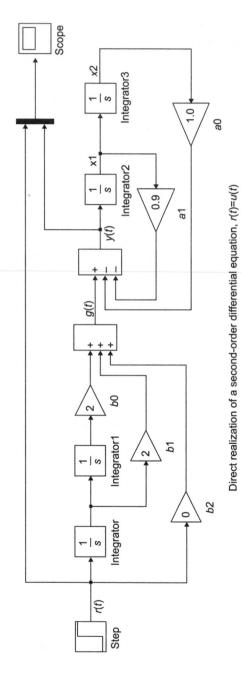

Direct realization of a second-order differential equation, $r(t) = u(t)$

Figure 13.11 Complete model of (13.1). (Gain3 = a_1, Gain4 = a_0).

then $g(t < 0) = 0$. With $y(t) = g(t) - a_1 x_1(t) - a_0 x_2(t)$, we can find the initial condition $x_1(0^-)$ of Integrator2 and $x_2(0^-)$ of Integrator3 using

$$y(0^-) = S_0 = g(0^-) - a_1 x_1(0^-) - a_0 x_2(0^-) \rightarrow -a_1 x_1(0^-) - a_0 x_2(0^-) = S_0$$
$$\dot{y}(0^-) = S_1 = \dot{g}(0^-) - a_1 \dot{x}_1(0^-) - a_0 \dot{x}_2(0^-) \rightarrow -a_1 S_0 - a_0 x_1(0^-) = S_1$$

This model structure is called a **direct realization** of (13.1). Notice how this realization method can be extended to realize higher order LTI differential equations. There are many other possible kinds of model structures.

The input in Fig. 13.11 is a unit step function, where in its parameter dialog box, the sample time was set to 0.01 sec. A Scope (oscilloscope) block is used to see the input and output. The signals $r(t)$ and $y(t)$ are combined into a signal vector with a Mux (multiplexer) block. In this case, the Scope block will display two signals. Fig. 13.12 shows the unit step response under zero initial conditions, where $a1 = 0.9$ and $a0 = 1.0$. Like the XY Graph block, the data displayed by the Scope block can also be made available to MATLAB by setting parameters in the Scope Parameters dialog box, as shown in Fig. 13.12. The array step_response contains the input, output, and time data.

With Simulink it is convenient to study the behavior of a dynamic system under various conditions, possibly followed by further analysis using MATLAB. For example, in Fig. 13.13 the unit step input in Fig. 13.11 was replaced by a sinusoidal input plus additive noise. The Scope block displays the input and output, which is also placed in the MATLAB workspace.

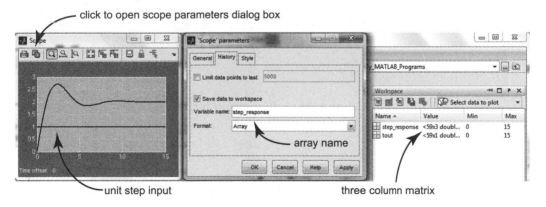

Figure 13.12   Unit step response of a second-order dynamic system.

There are many other input possibilities. For example, the input could be the output of a (1) From File block, a .mat file of a recorded speech signal, for example, (2) From Work-space block, (3) user defined Fcn block, and many more. The Fcn block is particularly

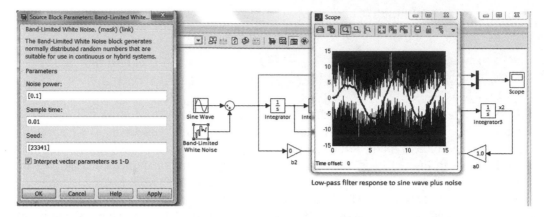

Figure 13.13   Low-pass filter response to signal plus noise input.

useful because its output can be the result of evaluating any valid MATLAB expression that operates on a scalar or vector input of the block.

Example 13.4

Let us build a Simulink model of the circuit shown in Fig. 13.14. We have studied this circuit in Examples 6.11, 11.4, 12.12, and 12.13. Each time, the method and objective of the analysis was different.

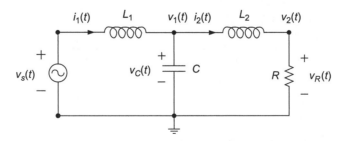

Figure 13.14   An RLC circuit.

Generally, the order of a differential equation that relates some input (current or voltage source) to some output (current through or voltage across a component) in an RLC circuit is the sum of the number of dynamic components (inductors and capacitors) in the circuit.

For example, a third-order differential equation relates the voltage $v_R(t)$ to the voltage source $v_s(t)$ (see Example 12.10). Starting with the third-order differential equation, the method followed in Example 13.3 can be applied to build a Simulink model. However, to find the initial conditions requires some intermediate calculations, and the given initial state of the inductors and capacitor is lost in the arithmetic. An alternative is to use the Transfer Fcn block in which case initial conditions must be zero.

Instead, we could apply KVL to obtain two mesh equations in terms of mesh currents, which are

$$L_1\frac{di_1}{dt} + \frac{1}{C}\int(i_1(t) - i_2(t))dt = v_s(t)$$

$$-\frac{1}{C}\int(i_1(t) - i_2(t))dt + L_2\frac{di_2}{dt} + Ri_2(t) = 0$$

(13.3)

If the equations in (13.3) are each integrated once, then we can express $i_1(t)$ and $i_2(t)$ in terms of integrals of $i_1(t)$ and $i_2(t)$. Like (13.2), the expressions for $i_1(t)$ and $i_2(t)$ are templates for building a Simulink model. However, the initial state of the inductors and capacitor are not directly involved.

Another possibility is to apply KCL to obtain two node equations in terms of node voltages, which are

$$\frac{1}{L_1}\int(v_1(t) - v_s(t))dt + C\frac{dv_1}{dt} + \frac{1}{L_2}\int(v_1(t) - v_2(t))dt = 0$$

$$\frac{1}{L_2}\int(v_2(t) - v_1(t))dt + \frac{v_1(t)}{R} = 0$$

(13.4)

If the equations in (13.4) are each integrated once, then we can express $v_1(t)$ and $v_2(t)$ in terms of integrals of $v_1(t)$ and $v_2(t)$. Like (13.2), the expressions for $v_1(t)$ and $v_2(t)$ are templates for building a Simulink model. Again, the initial state of the inductors and capacitor are not directly involved.

Another possibility is to apply KCL and KVL to obtain equations in terms of inductor currents and capacitor voltage, which are

$$-v_s(t) + L_1\frac{di_1}{dt} + v_C(t) = 0 \rightarrow i_1(t) = \frac{1}{L_1}\int(v_s(t) - v_C(t))dt$$

$$-v_C(t) + L_2\frac{di_2}{dt} + Ri_2(t) = 0 \rightarrow i_2(t) = \frac{1}{L_2}\int(v_C(t) - Ri_2(t))dt$$

(13.5)

$$-i_1(t) + C\frac{dv_C}{dt} + i_2(t) = 0 \rightarrow v_C(t) = \frac{1}{C}\int(i_1(t) - i_2(t))dt$$

where each equation is integrated once, as shown in (13.5), to obtain expressions for the inductor currents and capacitor voltage in terms of their integrals.

The three equations in (13.5) are templates for the three subsystems shown in Fig. 13.15. If a variable is not available within a subsystem, then a place holder is used.

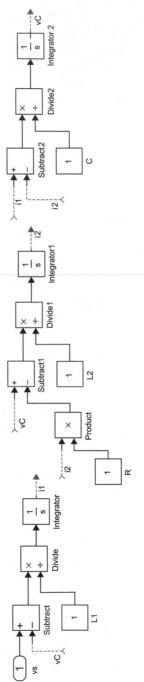

Figure 13.15 Subsystems for each of the equations in (13.5).

Once all subsystems have been designed, they can be connected to obtain the *RLC* circuit Simulink model shown in Fig. 13.16. With this model a variety of circuit behavior can be investigated. All circuit components are represented explicitly by Constant blocks. Also, the initial currents in inductors and initial voltage of the capacitor are the initial conditions of the integrators. Notice that Goto and From blocks are used to keep the connections uncluttered.

Recall that the transfer function of an Nth order differential equation is given by

$$H(s) = \frac{b_M s^M + \cdots + b_1 s + b_0}{s^N + a_{N-1} s^{N-1} + \cdots + a_1 s + a_0} = \frac{P(s)}{Q(s)} = \frac{b_M \prod_{k=1}^{M}(s - z_k)}{\prod_{k=1}^{N}(s - p_k)} \tag{13.6}$$

This transfer function can be modeled with the Transfer Fcn block in the Continuous Library. However, it is useful for you to see the details of another method, because the methodology of breaking a big problem into a set of smaller problems can be applied to other kinds of problems. Let $P = [b_M \ b_{M-1} \cdots b_1 \ b_0]$ and $Q = [1 \ a_{N-1} \cdots a_1 \ a_0]$. With the MATLAB function **roots**, the poles (roots(Q)) and zeros (roots(P)) of $H(s)$ can be found. Recombine into quadratics all complex conjugate pairs of poles and zeros. Then, $H(s)$ can be written as the product of functions given by

$$H(s) = H_1(s) H_2(s) \cdots H_L(s) \tag{13.7}$$

where each $H_i(s)$, $i = 1, \ldots, L$ is like one of the following proper or strictly proper transfer functions, all with real coefficients.

$$H_i(s) = \frac{c_2 s^2 + c_1 s^1 + c_0}{s^2 + d_1 s^1 + d_0}, \ \frac{c_1 s^1 + c_0}{s^2 + d_1 s^1 + d_0}, \ \frac{c_0}{s^2 + d_1 s^1 + d_0}, \ \frac{c_1 s^1 + c_0}{s^1 + d_0}, \ \text{or} \ \frac{c_0}{s^1 + d_0}$$

The differential equation (12.17), of which (13.6) is the transfer function, can be modeled with the block diagram shown in Fig. 13.17.

Each of the L subsystems in Fig. 13.17 is a special case of the model given in Fig. 13.11, as depicted in Fig. 13.18, where the input and output are Input Port and Output Port blocks, respectively. By a cascade connection of L models like the model shown in Fig. 13.18, a Simulink model of an Nth order LTI system can be created. Another method to model (12.17) is to use the State-Space block in the Continuous Library.

A Simulink model constructed with basic operations can easily become complicated. To reduce model complexity, it is useful to combine a set of basic operations that together do a particular activity into a subsystem block.

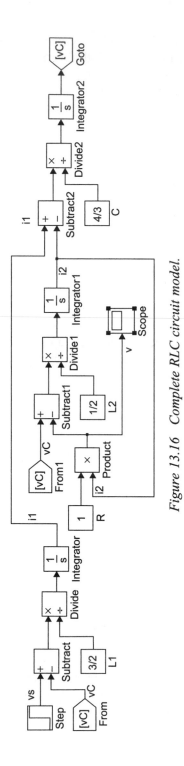

Figure 13.16 Complete RLC circuit model.

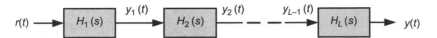

Figure 13.17   *Cascade connection of L subsystems.*

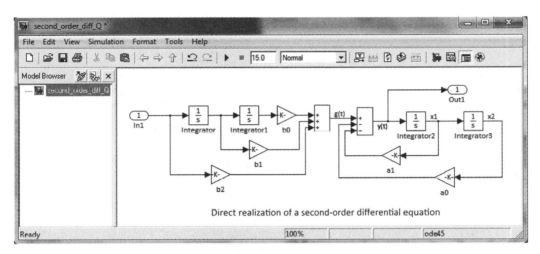

Figure 13.18   Simulink model of a proper or strictly proper second-order LTI system.

13.3   Custom Blocks

With Simulink you can create your own library of custom blocks.

Example 13.5 ————————————————————————————————

Let us make a subsystem block out of the model shown in Fig. 13.18. To do this, use the mouse to select and make active all elements (blocks and connections) in the model. You can also select multiple elements by depressing the Shift key as you click on each element. Or, click select all in the edit menu. Then, in the *Model Editor Window* use

```
Edit → create subsystem
```

to obtain the model shown in Fig. 13.19. You can click on the text below the 2nd Order Diff Eq block to change it. You can also add a title by double clicking anywhere in the window.

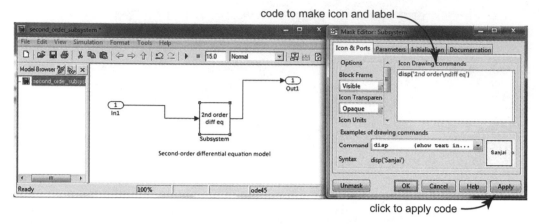

Figure 13.19   Subsystem block and mask editor.

To associate a label, icon, parameter values, and more with the subsystem, open the *Mask Editor Window* with

Edit → Mask Editor

Scroll through the Command list in the *Mask Editor Window* to see options for making a subsystem mask. For more details, see the Simulink help facility.

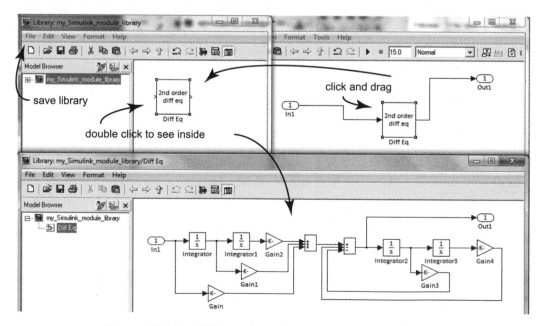

Figure 13.20   Addition of a subsystem to a custom library.

To create a custom library of custom blocks, use

File → New → Library

Then, drag the subsystem block into the library as shown in Fig. 13.20. Once you have created a custom library, you can open it to add more custom blocks.

Use the blow-up of the subsystem block to set the gains.

With the generic block that was made in Example 13.5, we can solve LTI differential equations, as suggested by Fig. 13.17, without the model becoming too complicated. It is also useful to create subsystem blocks of physical components that are often used.

Example 13.6

Fig. 13.21 shows an electro-mechanical model of a permanent magnet direct current (DC) motor. It converts electrical power provided by the voltage source $v(t)$ to mechanical power of the torque $T(t)$ applied to the armature and load with rotational moment of inertia J. Let us create a Simulink model of this motor, where the input is the voltage $v(t)$ and the output is the angular velocity $\omega(t)$.

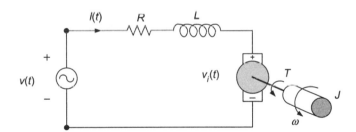

Figure 13.21   Electro-mechanical model of a permanent magnet DC motor.

On the rotor (armature) of the motor is wound a coil of wire with a resistance of R ohms and an inductance of L henrys. The stator of the motor is a permanent magnet. A current $i(t)$ in the armature coil causes an electromagnetic field. Magnetic repulsion and attraction between the magnetic poles of the electromagnetic field of the armature and the magnetic field of the stator force the armature to rotate. As the armature windings move through the magnetic field of the stator, an induced voltage $v_i(t)$, called the **back electromotive force** (emf), is generated across the terminals of the armature coil. Applying Kirchhoff's voltage law to the armature circuit gives

$$v(t) = Ri(t) + L\frac{di}{dt} + v_i(t) \tag{13.8}$$

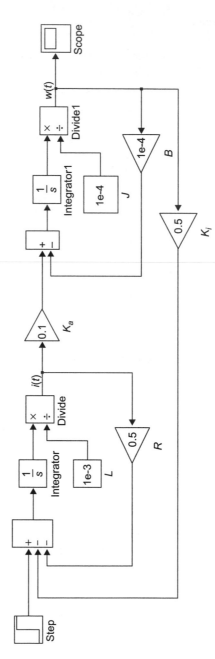

Figure 13.22 Simulink model of a permanent magnet DC motor.

The back emf is proportional to the angular velocity of the armature, and we write $v_i(t) = K_i\,\omega(t)$, where K_i is a proportionality constant.

The motor exerts a torque $T(t)$ (electromagnetic torque) that is proportional to the armature current, and we write $T(t) = K_a\,i(t)$, where K_a is a proportionality constant. Applying Newton's second law gives

$$T(t) = J\frac{d\omega}{dt} + B\omega \tag{13.9}$$

where $J\dot{\omega}$ is the torque due to accelerating the rotor and load, $B\omega$ is torque due to friction, and B is the angular coefficient of friction.

Integrating (13.8) gives

$$i(t) = \frac{1}{L}\int (v(t) - Ri(t) - K_i\omega(t))dt \tag{13.10}$$

and integrating (13.9) gives

$$\omega(t) = \frac{1}{J}\int (K_a i(t) - B\omega(t))dt \tag{13.11}$$

Like (13.2), equations (13.10) and (13.11) are templates for building the Simulink model of (13.8) and (13.9), which is shown in Fig. 13.22, where parameter values are made visible by sizing the blocks.

The Step and Scope blocks in Fig. 13.22 were replaced by Input and Output Port blocks to create a subsystem block that was added to the library of custom subsystem blocks shown in Fig. 13.23. Fig. 13.23 also shows the unit step response of the DC motor using its subsystem block.

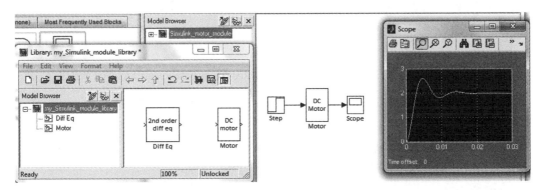

Figure 13.23   Demonstration of using a custom block in a Simulink model.

With the blocks in the standard Simulink libraries and blocks in a custom library, it becomes feasible to investigate interactively the behavior of many kinds of systems modeled with these blocks.

Example 13.7

Suppose the motor given in Example 13.6 must be used in some application. The input voltage controls the motor speed. This is called **open-loop** control because there is no provision to monitor the output (speed) and adjust the input (voltage) to achieve a particular output speed. There are some issues to consider. For example, a sudden input change causes the motor speed to oscillate (see Fig. 13.24) before it settles to a constant speed. Also, if the load changes, the motor speed can go through undesirable transients, as shown in Fig. 13.25 for a load increase, where (see left scope screen) the motor almost stalls for a while, and a load decrease, where (see right scope screen) the motor speed oscillates substantially.

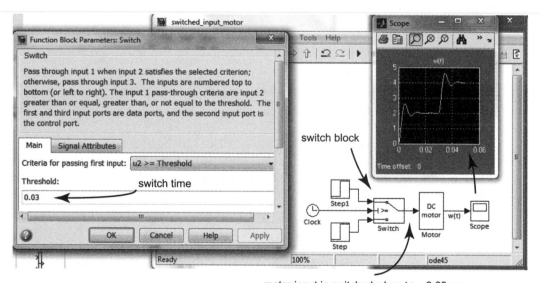

Figure 13.24   Oscillatory response when input (u(t) + u(t − 0.003)) changes suddenly.

To compensate for the kind of behavior shown in Figs. 13.23–13.25, let us introduce **feedback**, and use a controller, as shown in Fig. 13.26.

In Fig. 13.26, the input is no longer applied directly to the motor. Instead, the input (desired speed) is compared to the output (actual speed) as obtained with a sensor (tachometer, where the output voltage is proportional to the motor speed) to produce an error

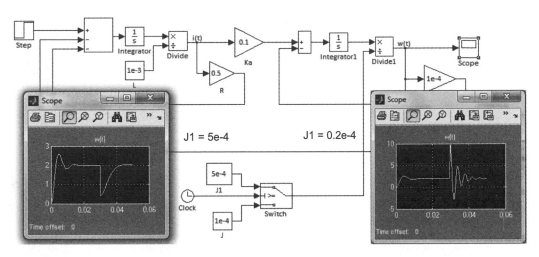

Figure 13.25 Motor speed response under load changes.

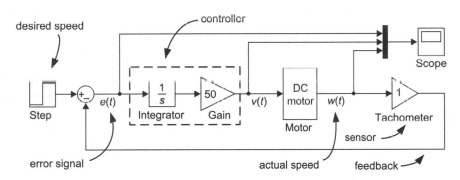

Figure 13.26 Closed-loop speed control system of a DC motor.

signal. If the actual speed is less than (greater than) the desired speed, the error signal is positive (negative), which increases (decreases) the controller output (power output of the amplifier) to increase (decrease) the motor speed. This is called **closed-loop** control. The objective of the controller is to make the motor speed track the desired speed. This is analogous to the cruise control of an automobile. It would be very undesirable if a cruise control system of an automobile behaved as shown in Figs. 13.24 or 13.25. Fig. 13.27 shows the behavior of the closed-loop motor speed control system for the range of loads: (a) $J = 5.0e-4$, (b) $J = 1.0e-4$, and (c) $J = 0.2e-4$, as used in Figs. 13.24 and 13.25. Each figure in Fig. 13.27 shows the error signal ($e(t)$), motor input ($v(t)$), and motor output ($\omega(t)$).

e(t), error signal   v(t), motor input   w(t), motor output

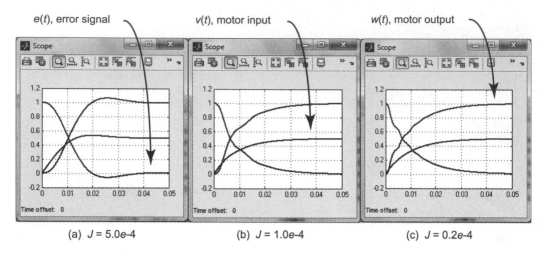

(a) J = 5.0e-4   (b) J = 1.0e-4   (c) J = 0.2e-4

Figure 13.27   Closed-loop system motor speed transient response.

Comparing Fig. 13.27(b) with Fig. 13.23, we see that the closed-loop controller has almost eliminated motor speed oscillation under its nominal load. The error signal smoothly becomes zero, as the motor speed matches the desired speed. We also see that when the load is decreased or increased from its nominal load by a factor of five, the motor speed is also well controlled without substantial oscillations. However, as is often the case, an advantage gained comes at a cost. Here we see that a smoother response requires more time to reach steady state. Such a trade-off and others can be a challenging problem for control system designers.

Examples 13.4, 13.6, and 13.7 demonstrate the utility of Simulink. Sometimes a model of a system can be readily constructed. Parameters can be changed to investigate various kinds of system behavior. By making a model that resembles how a physical system is put together, it can help to gain insight about system behavior. Included in the many Simulink demos is a model that illustrates this very well. In the MATLAB *Command Window*, enter at the prompt >> sldemo_househeat to see a demonstration of some of the possibilities.

13.4   Conclusion

Simulink has three functionalities that make it useful for problem solving. First, it is a GUI with which it is convenient to build a model of a dynamic system. Second, Simulink has a large set of standard and specialized libraries that contain a wide variety of blocks for building models, and you can create your own library of custom blocks. Third, Simulink, along with MATLAB, provides extensive computing and visualization capabilities to solve models of complex linear and nonlinear dynamic systems. You should now know how to

- use the Simulink GUI
- create a Simulink model with Simulink blocks
- set block and simulation parameters
- use Simulink to solve a differential equation
- export to and import from MATLAB data
- create custom blocks and a custom library of blocks

We looked at only a few of the many standard Simulink blocks. There is much more to learn about Simulink. There are many audio/video tutorials available both online and in the Simulink help facility. The demos are very helpful to learn more about how to use Simulink.

Problems

Section 13.1

1) In the MATLAB *Command Window*, enter the command to start Simulink, which opens the Simulink *Library Browser Window*. Open the Commonly Used Blocks Library by clicking its name in the libraries list.
   (a) Right click the Gain block, and select Help for the Gain block. Give a brief description of the purpose of the Gain block.
   (b) Repeat part (a) for the In1 block. What does it mean when the Sample Time is assigned a value of -1?
   (c) Repeat part (a) for the Saturation block. Explain how to set the lower and upper limits.
   (d) Repeat part (a) for Scope block. Explain the difference between a Scope block and a Floating Scope block. Explain how you can make the data displayed by a Scope block available to a MATLAB program.
2) Open the Continuous Library.
   (a) Give a brief description of the Transfer Fcn block.
   (b) Repeat part (a) for the State-Space block.
3) Open the Lookup Tables Library.
   (a) Give a brief description of the 1-D Lookup Table block. Explain how the number of independent variables is specified.
   (b) How is the number of data points in the Cosine block specified?
4) Open the Math Operations Library.
   (a) With a model sketch, explain the operation of the Divide block.
   (b) Manually, explain and give an example of the operation of the Polynomial block.
5) Open the Signal Routing Library. Explain the operation of the following blocks:
   (a) Mux, (b) Demux, (c) Switch, (d) Bus Creator, (e) Bus Selector, (f) Goto, (g) From.
6) Open the Discrete Library. Explain the operation of the following blocks:
   (a) Integer Delay. If the input is $x(n)=\sin(n\,\pi/4)$, then what is the output if the delay is set for two sample times?

(b) Discrete Transfer Fcn. If the transfer function is $H(z) = (z^2 - 1)/(z^2 + 0.5z + 0.75)$, then what is the difference equation that relates the output and input of the block?

(c) Discrete FIR Filter. What coefficients must be entered into the block if the input to output relationship must be $y(n) = (x(n) + 2x(n-1) + x(n-2))/4$?

7) Open the Sinks Library. Explain the operation of the following blocks: (a) Terminator, (b) To File, (c) To Workspace.

8) Open the Sources Library. Explain the operation of the following blocks:

(a) Signal Generator. What is the difference between the time-based and sample-based settings?

(b) What is the difference between the Random Number and Uniform Random Number blocks?

(c) Signal Builder. Explain how to build a signal.

9) Open the *New Model Window*, and make a Simulink model that includes a Sine Wave block, a Slider Gain block and a Scope block. In the sine wave parameters dialog box, set the frequency of the sine wave to 2 Hz, set the amplitude to 1 and set the sample time to 0.001 sec. In the simulation configuration parameters dialog box, set the stop time to two cycles of the sine wave. Save your model in the Current Folder.

(a) Run the simulation. In the scope display, click the autoscale icon. Provide a display of the scope screen.

(b) In the scope display, click the parameters button. The sampling decimation parameter should be set to 1. Set this parameter to 50, and repeat part (a). Explain the difference between this result and the result of part (a).

(c) Repeat part (a) for another setting of the slider gain.

10) Make a Simulink model that uses two Constant blocks, a Divide block and a Display block. Connect the blocks such that the Display block shows the result of dividing the Constant block by the Constant1 block. (a) Display 7/3, (b) Display 5/0.

11) Build a model that uses a Signal Builder block, a Product block, and an XY Graph block. The Signal Builder block must produce two signals. One signal is a square wave with a frequency of 1 Hz that oscillates between zero and one. Use the Product block to multiply the two outputs of the Signal Builder block and display the product. Set the stop time to 2 sec. Provide a display of the scope screen.

(a) The other signal is a sine wave with a frequency of 2 Hz using 20 samples per cycle.

(b) Repeat part (a) for a sine wave with frequency 100 Hz using 1000 samples per cycle.

12) Use a Fcn block, a Clock block, and a Scope block to generate and display the signal $x(t) = \exp(-t)\sin(5t)$ over a time range of 5 sec. Set the sample time to 0.01 sec. Provide a display of the scope screen.

13) Make a Simulink model that uses two Sine Wave blocks and a XY Graph block. For one of the Sine Wave blocks use a phase that converts the block to generating a cosine wave. Connect the Sine (Cosine) Wave block outputs to the X(Y) inputs of the XY Graph block. What does the XY Graph block display show? Provide a display of the graph.

14) Use Sine Wave, Integrator, Mux, and Scope blocks, and connect the Sine Wave block output to the Integrator block input. Use the Mux block to display on the scope the input and output of the Integrator block. Set the sine wave frequency to 0.25 Hz (pi/2 rad/sec), and use a sample time of 0.001 sec.

   (a) Run a simulation for 10 sec. Provide a display of the scope screen. Over what time range does the scope display signals?

   (b) In the *Scope Display Window*, click the parameters icon, set the decimation parameter to 2, and repeat part (a). What happened?

   (c) Click the Autoscale icon. Explain why the output is a raised cosine wave.

   (d) Set the decimation parameter back to 1, as in part(a). In the *Scope Display Window*, click the parameters icon. Click the History tab, and uncheck the limit data points to last check box. Run the simulation, and provide a display of the scope screen. How is this result different from the result of part (a)?

15) In a computer, all signals are number sequences, for example, a signal denoted by $x(n)$, where n is the discrete time index. Suppose $y(n)$ is a signal that is the result of decimating $x(n)$ by a factor of N. How is $y(n)$ related to $x(n)$?

16) Use Constant, Mux, and Integrator blocks to integrate a constant, and display the input and output of the integrator with a Scope block. Autoscale the display.

   (a) Set the Constant block to 0.5. Provide a display of the scope screen. What was the integrator initial condition?

   (b) Set the integrator initial condition to -2, and repeat part (a). Provide a display of the scope screen.

17) If the gain A of the op-amp is very large, then the gain of the circuit shown in Fig. P13.17 is approximately $-R_2/R_1$.

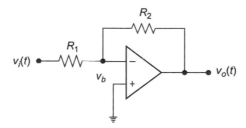

Figure P13.17   Op-amp used to make an inverting amplifier.

The equations that describe this circuit are

$$\frac{v_b - v_i}{R_1} + \frac{v_b - v_0}{R_2} = 0 \rightarrow v_0 = \frac{R_2}{R_1}(v_b - v_i) + v_b$$

$$v_0 = -Av_b \rightarrow v_b = \frac{1}{-A}v_0$$

Use Product, Divide, Constant, Subtract, Display, and $v_i = 3\, u(t)$ blocks to make a model of the two equations. With $R_1 = 10\text{K }\Omega$ and $R_2 = 20\text{K }\Omega$, run a simulation to display the circuit gain v_0/v_i for (a) $A = 1\text{e}2$, (b) $A = 1\text{e}4$, and (c) $A = 1\text{e}6$. Provide a copy of the model, where the display shows the circuit gain using a long format. What happens to the circuit gain as A is increased?

Section 13.2

18) Make a Simulink model of the circuit shown in Fig. P13.18.

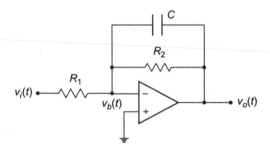

Figure P13.18   An op-amp circuit (active RC circuit).

Equation (12.22) gives the relationship between the input and output, which is

$$C\frac{dv_o(t)}{dt} + \frac{v_o(t)}{R_2} = -\frac{v_i(t)}{R_1}$$

Hint: This is an $N = 1$ order differential equation. To obtain a version of this equation that will serve as a template for designing the model, integrate the equation N times to get

$$v_o(t) = \frac{1}{C}\int\left(-\frac{v_o(t)}{R_2} - \frac{v_i(t)}{R_1}\right)dt$$

Design the model using $C = 0.1\ \mu F$, $R_1 = 10\text{K }\Omega$ and $R_2 = 30\text{K }\Omega$. Use a Scope block to display the response to (a) a unit step, $u(t)$, (b) $\sin(100\,\pi t)$, (c) $\sin(2000\,\pi t)$, and (d) what kind of filtering activity does this circuit do?

19) Make a Simulink model for the half-wave rectifier circuit in Fig. P13.19. The diode parameters are $I_s = 1\text{e-}12$, $V_T = 25.85\text{e-}3$. The input is $v_s(t) = 6.3\ \sin(120\,\pi t)$ volts. The equation that relates the output to the input is given by

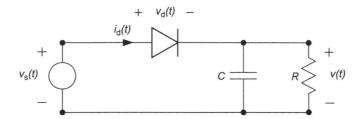

Figure P13.19   Half-wave rectifier circuit (AC/DC converter).

$$\frac{dv}{dt} = \frac{I_s}{C}\left(e^{(v_s(t)-v(t))/V_T} - 1\right) - \frac{v(t)}{RC}$$

Integrate this equation once to obtain an equation that will serve as a template for your model. Use a Fcn block to obtain $e^{v_d(t)/V_T}$, where $v_d(t) = v_s(t) - v(t)$. Use $C = 1000\ \mu\text{F}$ and $R = 100\ \Omega$. Run your model for 10 cycles of the input, and use Mux and Scope blocks to display the input and output. Set the sample time in the Sine Wave block parameter dialog box. Try a sample time of 0.0001 sec. You may have to adjust this.

To find the average value of the output, include in your model Clock, Switch, Integrator, and Divide blocks. One input to the switch is zero, and the other input is $v(t)$. The switch should connect $v(t)$ to the integrator after one cycle time, 1/60 sec, has elapsed. Divide the integrator output by the time duration of nine cycles, the value 9/60 sec in a Constant block, and use a Display block to show this result. After the simulation has finished, the Display block should show the average value of the output over nine cycles. Provide a copy of your model and the scope display.

20) By the direct realization method, build a Simulink model to find $i(t)$ and then $v(t)$ of the *RLC* circuit shown in Fig. P13.20.

The application of KVL results in

$$-v_s(t) + \frac{1}{C}\int i(t)dt + L\frac{di(t)}{dt} + Ri(t) = 0$$

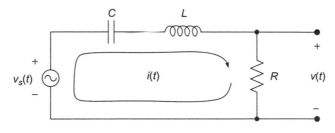

Figure P13.20   A series RLC circuit.

Let $R = 33\ \Omega$, $L = 11$ mH and $C = 0.01\ \mu$F. Include Mux and Scope blocks to display the input and output.

(a) Let the input be $v_s(t) = \sin(2\pi f t)$. Use a Constant block to set the value of the frequency f. Use Clock and Product blocks to find $u = ft$, and use a Fcn block to get $v_s(t) = \sin(2\pi u)$. By trial and error, find the frequency f_{max} for which the output has a maximum amplitude. You will have to adjust the total run time to see several cycles on the scope display. Provide a copy of your model and the scope display.

(b) Add Integrator, Derivative, Mux, and Scope blocks to your model and obtain a display of the capacitor and inductor voltages for $f = f_{max}$. Provide a copy of the scope display. When the capacitor (inductor) voltage is at a maximum, what is the inductor (capacitor) voltage? Does this happen at other frequencies?

21) Take the derivative of the KVL equation given in Prob. 13.20 to obtain a second-order differential equation. Use the component values given in Prob. 13.20, and obtain the transfer function $H(s)$ from the input $v_s(t)$ to the output $v(t)$. Use a Transfer Fcn block to make a Simulink model, and add a Scope block to display $v(t)$. Assuming zero initial conditions, find (a) the unit step response, (b) the response to $v_s(t) = \sin(2\pi f t)\, u(t)$, where $u(t)$ is the unit step function and $f = 10$K Hz. Provide copies of the model and scope displays.

22) The transfer function of an LTI CTS is given by

$$H(s) = \frac{s^2 + 3s + 2}{s^4 + 6s^3 + 138s^2 + 462s + 2929}$$

where the input is $r(t)$ and the output is $y(t)$. Assume zero initial conditions.

(a) Give the differential equation that relates $y(t)$ and $r(t)$.

(b) Build a Simulink model of this system using a Transfer Fcn block. Find and display with Scope block the step response. Use a run time long enough to see the step response settle. Provide copies of the model and the scope display.

(c) Use a Chirp Signal block for the input, and two Scope blocks, one to see the input and the other to see the output. Let the chirp signal vary in frequency from 0.1 to 50 Hz over a 10 sec time range. Run the simulation for 10 sec. Approximately, what is the frequency at which the output has a maximum amplitude? Provide copies of your model and scope displays.

(d) Repeat part (c), but let the chirp signal frequency vary from 20% below to 20% above your estimate of the peak output amplitude frequency found in part (c). Can you give a refined estimate of the output peak amplitude frequency?

23) The numerator and denominator of the transfer function given in Prob. 13.22 can be factored into

$$H(s) = \frac{(s+1)(s+2)}{(s+1+j10)(s+1-j10)(s+2+j5)(s+2-j5)}$$

$$= \frac{(s+1)(s+2)}{(s^2+4s+29)(s^2+2s+101)}$$

where complex conjugate pole pairs were recombined into quadratics with real coefficients. The transfer function can be regrouped into several different products of strictly proper and proper second-order transfer functions. For example

(a)

$$H(s) = \frac{(s+1)}{(s^2+4s+29)} \frac{(s+2)}{(s^2+2s+101)} = H_1(s)H_2(s)$$

where both $H_1(s)$ and $H_2(s)$ are strictly proper functions.

(b)

$$H(s) = \frac{(s^2+2s+3)}{(s^2+4s+29)} \frac{1}{(s^2+2s+101)} = H_1(s)H_2(s)$$

where $H_1(s)$ is a proper function and $H_2(s)$ is a strictly proper function.

(c) Give another possible break up of $H(s)$ into a product of two second order proper or strictly proper functions.

(d) For each case in parts (a), (b), and (c), use two Transfer Fcn blocks to build a Simulink model of a cascade connection of $H_1(s)$ and $H_2(s)$, and find the step response. Use XY Graph and Clock blocks to see the responses. Provide copies of the model and the displays. Are the responses in these three cases different? Discuss what will happen if in the cascade connection $H_1(s)$ and $H_2(s)$ are reversed.

Section 13.3

24) For the DC motor introduced in Example 13.6, find the transfer function from $v(t)$ to $\omega(t)$. You can do this by first finding the transfer functions of (13.8) and (13.9), which are

$$V = (Ls+R)I + K_i\Omega \rightarrow I = \frac{V}{(Ls+R)} - \frac{K_i}{(Ls+R)}\Omega$$

$$K_aI = (Js+B)\Omega \rightarrow \Omega = \frac{K_a}{(Js+B)}I$$

$$\therefore \quad H_{\text{motor}}(s) = \frac{K_a}{(Js+B)(Ls+R)+K_aK_i} = \frac{K_a/JL}{s^2+((JR+LB)/JL)s+K_aK_i/JL}$$

Using the parameter values given in Fig. 13.22, make a Simulink model of the DC motor. You can use the model given in Fig. 13.22, or use a Transfer Fcn block. Create a subsystem of the model. Create a custom library, and add the subsystem of the motor to it. Name this subsystem block Motor.

A DC motor can also be used to control the position of a load by integrating its output speed to obtain angular position. Make a Simulink model as shown in Fig. P13.24.

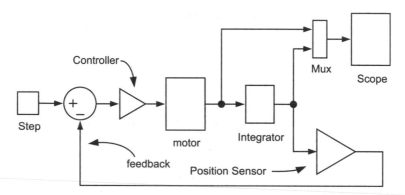

Figure P13.24   Closed-loop position control system.

The scope shows the motor speed and angular position of the armature. The output of the position sensor is a voltage proportional to input angular position. Assume that this device has been calibrated to produce a voltage in the range 0–10 volts as the angular position of the motor armature varies from 0 to 2π rad. This sensor could be variable resistor.

If this system works as it should, the motor speed should become zero, once the motor armature reaches a constant position. Try running this system for 0.2 sec with a sample time of 0.00001 sec. This system must be tuned, meaning that the controller gain must be adjusted for desired performance (a relatively smooth but quick change in position). Start with a controller gain of 10, and see what happens. Add a Scope block to monitor the error signal, the subtractor output. You may also want to introduce a scaling gain between the motor output and the scope to make the position and speed signals compatible on the display. Discuss what happens if the controller gain is too big. Provide copies of your model and scope displays, and a controller gain that you have found to work well.

25) A pair of wires can be surprisingly ineffective in the transmission of a binary data stream over a long distance. Suppose a binary voltage must be communicated over a pair of wires, as shown in Fig. P13.25(a). The resistance R_s is the source resistance, and R_L is the input resistance of the device that receives the signal. Every T sec, the data can be logic zero or logic one. Of interest is to communicate data at the highest possible bit rate that the destination device can detect without loss of data. Ideally, if $R_L = R_S$, then $v(t) = v_s(t)/2$, and there would be no data loss.

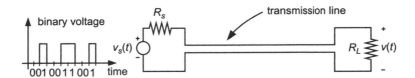

Figure P13.25(a)   Transmission line between a source and a load.

The problem is that the wires have an equivalent resistance, say $R = 0.05$ Ω/m and an equivalent capacitance, say $C = 10$ pF/m. There may also be an equivalent inductance, which will be ignored to keep the problem simple.

To study this problem a circuit model of the transmission line will be used. A commonly used model is shown in Fig. P13.25(b), which includes four RC stages. A more realistic model contains many more RC stages. Because of R_S, the first stage is different from the other stages.

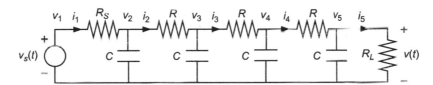

Figure P13.25(b)   Lumped component transmission line model.

To analyze this circuit, let us start with the first stage, and write

$$v_1 = R_S i_1 + v_2 \;\rightarrow\; i_1 = \frac{1}{R_S}(v_1 - v_2)$$

$$i_2 = i_1 - C\frac{dv_2}{dt} \;\rightarrow\; v_2 = \frac{1}{C}\int (i_1 - i_2)dt$$

$$v_2 = R i_2 + v_3 \;\rightarrow\; i_2 = \frac{1}{R}(v_2 - v_3)$$

$$i_3 = i_2 - C\frac{dv_3}{dt} \;\rightarrow\; v_3 = \frac{1}{C}\int (i_2 - i_3)dt$$

$$\vdots$$

$$v_4 = R i_4 + v_5 \;\rightarrow\; i_4 = \frac{1}{R}(v_4 - v_5)$$

$$i_5 = i_4 - C\frac{dv_5}{dt} \;\rightarrow\; v_5 = \frac{1}{C}\int (i_4 - i_5)dt$$

$$i_5 = \frac{v_5}{R_L}$$

The first two equations, where v_1 is the input and v_2 is the output, describe the first RC stage. The next two equations, where v_2 is the input and v_3 is the output, describe the second stage, and so on. A Simulink model of this circuit can become complicated. Let us make a subsystem block of one RC stage. Notice that for each stage, for example, the kth stage, v_k is the input, and v_{k+1} is the output. To find i_k we must feedback the output v_{k+1}, and to find the output v_{k+1}, we must feedback the current i_{k+1}. Fig. P13.25(c) gives a Simulink model of the kth RC stage. Check it for $k = 1$, where $R = R_S$, for $k = 2$, and for $k = 5$, where there is no next stage and $i_5 = v_5/R_L$ is used to feedback to the previous stage.

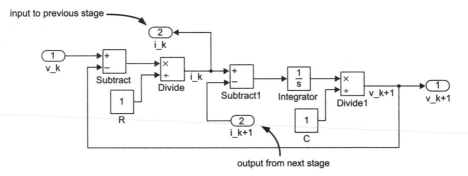

Figure P13.25(c)   Simulink model of the kth RC stage.

(a) Make a Simulink model, as shown in Fig. P13.25(c). Select all model elements, and make a subsystem of it. Add the subsystem to your library of subsystem blocks. Name the subsystem block RC Stage. Provide a copy of the subsystem model.

(b) To test your subsystem block, build a transmission line model using just one RC stage, where $v_1 = u(t)$, $R = R_S = 100\ \Omega$, $R_L = 100\ \Omega$, $C = 0.01\ \mu F$, and $i_2 = v_2/R_L$. To do this, add Divide and Constant blocks to obtain i_2. Attach a Scope block to display v_2. In the Simulation Configuration dialog box set the max step size to 1e-7 and the min step size to 1e-10. Run a simulation for 5e-2 msec. Autoscale the display. In view of the circuit you are simulating, does the response make sense? Provide a copy of your model and the scope display.

(c) Repeat part (b), but replace Step block input with a Pulse Generator block. Set the pulse generator to produce pulses with a period of 1e-2 msec and a duty cycle of 50%. Add a Mux block to display the input and output on the scope. Autoscale the display. Provide a copy of the scope display. What is the output like if the input data rate is increased by a factor of 10?

(d) Add three RC Stage blocks to the model of part (c), and connect them to simulate the transmission line model given in Fig. P13.25(b). Use $R = 10\ \Omega$. In this case, the Divide and Constant blocks must be used to find i_5. Use a Mux block to see the input and v_5. Run the simulation, and discuss the degree to which the input signal has been degraded by the transmission line. While realistic circuit parameter values were not used in this transmission line model, this simulation demonstrates the challenge of designing transmission lines for high bit rate data transfer.

Suggestions for Reporting Solutions to End of Chapter Problems

Many problems at the end of chapters require you to provide program and function listings and copies of alphanumeric and graphic output. This can be done in several ways, depending on your knowledge of your computer system and MATLAB®.

First, it is assumed that you have read Chapter 1. You can obtain a copy of the screen by depressing the Print Screen key on the keyboard. Whatever appeared on the screen at the time you depressed the Print Screen key is now stored on the Clipboard. If MATLAB is running, and you maximize the screen before depressing the Print Screen key, then you will get a copy of the entire MATLAB desktop. To obtain a copy of the *Command Window* or any other window, close the windows that you do not want to copy. The content of the Clipboard can be pasted directly into files of various other applications, including Microsoft Word®, Paint®, and Visio® (or other graphic applications).

By pasting a screen image, which is stored on the Clipboard, into Paint, you can edit colors and size, introduce various shapes and captions, and more. You can crop and cut out any portion of the image and paste the cut out into a Word doc file or into another open Paint screen for further editing. You can save the Paint image using a bmp, jpeg, or other file format and insert the saved file into a Word doc file as a picture.

You can paste an image on the Clipboard or cut out from Paint into an open file of a graphics application such as Visio. Within Visio, you can create and introduce a lot of other graphic and text material. From Visio, you can save the entire graphic in a variety of file formats, including an enhanced metafile (emf), which can be inserted into a Word doc file. You can also print a Paint or Visio image.

Now, it is assumed that you have read Chapter 2. If you have written a MATLAB script that produces a figure, then you can minimize all windows except the *Figure Window* and maximize it. From this point, you can work with the figure as described above.

If instead, you simply want a printed copy of a figure, then from within the *Figure Window* open the file menu and select print. From the file menu, you can also save the figure as a MATLAB fig file, and then open and print or edit the fig file at some other time. You can also get a print copy of the current figure by using the built-in function **print** from the *Command Window* or by invoking the print function in a program. Or, open the edit menu in the *Figure Window* to copy the figure and paste it into an open Word doc file or some other application. From the edit menu you can also elect to edit the figure or axes by selecting Figure Properties or Axes Properties. See Chapter 9 for details about using the figure property editor GUI.

To obtain a print copy of a script, open the file menu in the *Editor Window* and select print. To place a script into an open Word doc file, open the edit menu in the *Editor Window*, select all to highlight the entire script, right click the mouse, and copy and paste the script elsewhere. To print the contents of the *Command Window*, open the file menu and select print. You can also highlight any part of the text in the *Command Window*, and copy and paste the highlighted text into an open Word doc file or other application such as WordPad®.

To obtain a listing of any m-file, use the function **type** in the *Command Window*. For example, to list in the *Command Window* an m-file named DFT.m, which is in the current folder, enter type 'DFT.m' in the *Command Window*. Then, print the contents of the *Command Window*. Sometimes it is useful to include line numbers in an m-file listing. To do this, use the built-in function **dbtype**. For example, enter dbtype 'DFT.m' in the *Command Window*, and then print the contents of the *Command Window*.

Table of ASCII Codes

Table of ASCII (American Standard Code for Information Interchange) Codes
(7-bit codes (eighth bit = 0): 0–31, nonprintable; 32–127, printable)

Decimal	Octal	Hexadecimal	Binary	Symbol	Description
0	000	00	00000000	NUL	Null char
1	001	01	00000001	SOH	Start of Heading
2	002	02	00000010	STX	Start of Text
3	003	03	00000011	ETX	End of Text
4	004	04	00000100	EOT	End of Transmission
5	005	05	00000101	ENQ	Enquiry
6	006	06	00000110	ACK	Acknowledgment
7	007	07	00000111	BEL	Bell
8	010	08	00001000	BS	Back Space
9	011	09	00001001	HT	Horizontal Tab
10	012	0A	00001010	LF	Line Feed
11	013	0B	00001011	VT	Vertical Tab
12	014	0C	00001100	FF	Form Feed
13	015	0D	00001101	CR	Carriage Return
14	016	0E	00001110	SO	Shift Out/X-On
15	017	0F	00001111	SI	Shift In/X-Off
16	020	10	00010000	DLE	Data Line Escape
17	021	11	00010001	DC1	Device Control 1 (oft. XON)
18	022	12	00010010	DC2	Device Control 2
19	023	13	00010011	DC3	Device Control 3 (oft. XOFF)
20	024	14	00010100	DC4	Device Control 4
21	025	15	00010101	NAK	Negative Acknowledgment
22	026	16	00010110	SYN	Synchronous Idle
23	027	17	00010111	ETB	End of Transmit Block
24	030	18	00011000	CAN	Cancel

(Continues)

(Continued)

Decimal	Octal	Hexadecimal	Binary	Symbol	Description
25	031	19	00011001	EM	End of Medium
26	032	1A	00011010	SUB	Substitute
27	033	1B	00011011	ESC	Escape
28	034	1C	00011100	FS	File Separator
29	035	1D	00011101	GS	Group Separator
30	036	1E	00011110	RS	Record Separator
31	037	1F	00011111	US	Unit Separator
32	040	20	00100000		Space
33	041	21	00100001	!	Exclamation mark
34	042	22	00100010	"	Double quotes (or speech marks)
35	043	23	00100011	#	Number
36	044	24	00100100	$	Dollar
37	045	25	00100101	%	Procenttecken
38	046	26	00100110	&	Ampersand
39	047	27	00100111	'	Single quote
40	050	28	00101000	(	Open parenthesis (or open bracket)
41	051	29	00101001	)	Close parenthesis (or close bracket)
42	052	2A	00101010	*	Asterisk
43	053	2B	00101011	+	Plus
44	054	2C	00101100	,	Comma
45	055	2D	00101101	-	Hyphen
46	056	2E	00101110	.	Period, dot or full stop
47	057	2F	00101111	/	Slash or divide
48	060	30	00110000	0	Zero
49	061	31	00110001	1	One
50	062	32	00110010	2	Two
51	063	33	00110011	3	Three
52	064	34	00110100	4	Four
53	065	35	00110101	5	Five
54	066	36	00110110	6	Six
55	067	37	00110111	7	Seven
56	070	38	00111000	8	Eight
57	071	39	00111001	9	Nine
58	072	3A	00111010	:	Colon
59	073	3B	00111011	;	Semicolon
60	074	3C	00111100	<	Less than (or open angled bracket)
61	075	3D	00111101	=	Equals
62	076	3E	00111110	>	Greater than (or close angled bracket)
63	077	3F	00111111	?	Question mark
64	100	40	01000000	@	At symbol
65	101	41	01000001	A	Uppercase A
66	102	42	01000010	B	Uppercase B
67	103	43	01000011	C	Uppercase C
68	104	44	01000100	D	Uppercase D
69	105	45	01000101	E	Uppercase E

(Continues)

(*Continued*)

Decimal	Octal	Hexadecimal	Binary	Symbol	Description
70	106	46	01000110	F	Uppercase F
71	107	47	01000111	G	Uppercase G
72	110	48	01001000	H	Uppercase H
73	111	49	01001001	I	Uppercase I
74	112	4A	01001010	J	Uppercase J
75	113	4B	01001011	K	Uppercase K
76	114	4C	01001100	L	Uppercase L
77	115	4D	01001101	M	Uppercase M
78	116	4E	01001110	N	Uppercase N
79	117	4F	01001111	O	Uppercase O
80	120	50	01010000	P	Uppercase P
81	121	51	01010001	Q	Uppercase Q
82	122	52	01010010	R	Uppercase R
83	123	53	01010011	S	Uppercase S
84	124	54	01010100	T	Uppercase T
85	125	55	01010101	U	Uppercase U
86	126	56	01010110	V	Uppercase V
87	127	57	01010111	W	Uppercase W
88	130	58	01011000	X	Uppercase X
89	131	59	01011001	Y	Uppercase Y
90	132	5A	01011010	Z	Uppercase Z
91	133	5B	01011011	[	Opening bracket
92	134	5C	01011100	\	Backslash
93	135	5D	01011101	]	Closing bracket
94	136	5E	01011110	^	Caret - circumflex
95	137	5F	01011111	_	Underscore
96	140	60	01100000	`	Grave accent
97	141	61	01100001	a	Lowercase a
98	142	62	01100010	b	Lowercase b
99	143	63	01100011	c	Lowercase c
100	144	64	01100100	d	Lowercase d
101	145	65	01100101	e	Lowercase e
102	146	66	01100110	f	Lowercase f
103	147	67	01100111	g	Lowercase g
104	150	68	01101000	h	Lowercase h
105	151	69	01101001	i	Lowercase i
106	152	6A	01101010	j	Lowercase j
107	153	6B	01101011	k	Lowercase k
108	154	6C	01101100	l	Lowercase l
109	155	6D	01101101	m	Lowercase m
110	156	6E	01101110	n	Lowercase n
111	157	6F	01101111	o	Lowercase o
112	160	70	01110000	p	Lowercase p
113	161	71	01110001	q	Lowercase q
114	162	72	01110010	r	Lowercase r

(*Continues*)

(*Continued*)

Decimal	Octal	Hexadecimal	Binary	Symbol	Description
115	163	73	01110011	s	Lowercase s
116	164	74	01110100	t	Lowercase t
117	165	75	01110101	u	Lowercase u
118	166	76	01110110	v	Lowercase v
119	167	77	01110111	w	Lowercase w
120	170	78	01111000	x	Lowercase x
121	171	79	01111001	y	Lowercase y
122	172	7A	01111010	z	Lowercase z
123	173	7B	01111011	{	Opening brace
124	174	7C	01111100			Vertical bar
125	175	7D	01111101	}	Closing brace
126	176	7E	01111110	~	Equivalency sign–tilde
127	177	7F	01111111		Delete

There is also an extended 8-bit ASCII code (eighth bit $= 1$) ranging from 128 to 255, called ISO Latin-1, which includes codes of symbols, for example, $\pm$, 1/2, μ and more.

Answers to Selected Problems

Many problems at the end of chapters require you to provide answers that are MATLAB® statements, where you can refer to the chapter or the MATLAB help facilities and check your answer with MATLAB. To encourage you to develop programming and self-help skills, answers to these kinds of problems will not be given here. Instead, the answer given here will be: see MATLAB (**SM**), and for Simulink problems, the answer will be see Simulink (**SS**). Many problems require you to write a MATLAB script. Writing scripts is an essential part of learning MATLAB, and you can check your script with MATLAB. With some exceptions, the answer given here will be: your MATLAB script (**YMS**).

Chapter 1

1–14) SM; 15) f = 440, w = 2*pi*f, T0 = 1/f, t = T0/4, x = sin(w*t); 16–21) SM; 22d) mod(−7,−3) = x − n*y = −7 − n*(−3) = −1, n = floor((−7/−3) = 2; 23b) y = exp(−t/2) − exp(−2*t), x = y.* sin(w*t); 24–25) SM.

Chapter 2

1–4) SM; 5–6) YMS; 7a) One coulomb of charge moving through a 1 tesla magnetic field at the speed of 1 meter/sec and perpendicular to the magnetic field experiences a force of 1 newton; b) F (repulsive) = 0.020 newtons/meter; 8) 1.1234e9 newtons; 9) Energy required is: 1.348e-7 joules, and the voltage rise is: 134.813 volts; 10a) 12 mA, b) 144 mW; 10c) 518.4 joules; 11) 20 W, 7.2e4 joules; 12a) 3.6e6 joules; 12b) 1e-3 kWh; 13–18) SM; 19) Use: power = @(R,I) R*I^2, 20–22), and YMS; 23) Use: x_of_t = 'A*exp(-a*t).* cos(w*t+p)'; 24) Start your function as follows. Complete the function, save it and test it.

```
function sinusoid_plot(A,w,p,N)
% Function to plot a sinusoid over N cycles
T0=2*pi/w; T=T0/100; T_total=N*T0; t=0:T:T_total;
```

25–26) YMS; 27a) Since R_power is stored in a private folder, the m-file find_R_power cannot find the function R_power. MATLAB gives an error message. 27b) YMS; 28) SM; 29) The function evaluate is a function function. 30) SM; 31) YMS; 32) SM.

Chapter 3

1) A, 3×3; B, 3×2; C, 2×3; x, 3×1; 2) A = [2 −1 2; −2 1 0; −1 1 −2]; 3–4) SM; 5) I = eye(4), Z = zeros(3,2), T = 2*ones(100,1), R = rand(4,3), E = [], 6) z = −1.5:0.1:4.3, w = linspace(−1.5, 4.3, ceil((4.3−(−1.5))/0.1)+1); 7) SM; 8) Matrix multiplication is not commutative. 9) x′x is a scalar and xx′ is a 3×3 matrix. Dimension of CAB is $(2 \times 3)(3 \times 3)(3 \times 2) = 2 \times 2$. 10) SM; 11) Use given hint. 12) G = identity matrix; 13a) sqrt(1+q.^2); 13c) cos(2*q - pi/2); 14–19) YMS; 20) SM; 21) $|D| = 4$, $|E| = 1.4$; 22b) Let X = [z α], a row vector, and then A = [−4 2; 2 −3] and Y = [2 −3]. 23) Using row one, the three cofactors are $\alpha_{11} = -2$, $\alpha_{12} = 4$ and $\alpha_{13} = -1$. 24) Use given hint. 25) Start by forming the augmented matrix g. Apply elementary row operations to g. Do: 2 (row 1)+row 2 $\rightarrow$ row 2, row 1+row 3 $\rightarrow$ row 3 and 2(row 2)+row 3 $\rightarrow$ row 3. Now you can apply repeated backward substitution to find the solution. 26) After performing the elementary row operations given in the answer to Prob. 3.25, do: −1(row 2)+row 1 $\rightarrow$ row 1, −0.5(row 3)+row 2 $\rightarrow$ row 2 and 0.25(row 3)+row 1 $\rightarrow$ row 1. For X, element X(3) = 1.5. 27) SM; 28) SM; 29d) Do a search for sort in the Product Help Window. 29e) E = reshape(A′,10,2); 30) See Example 3.15. 31) $I_4 = -0.33258$ mA; 32) $i_1 = 53.2$ mA; 33) Power delivered to all components is zero. 34) $V_4 = 8.4615$ volts; 35) left circuit, $V_4 = 4.6154$ volts, right circuit, $V_4 = 3.8462$ volts; 36) same as for Prob. 3.35; 37) inconsistent; 38a) 0; 38b) $\sqrt{3}$; 38c) 1; 39) Use the p-norm definition, and pull out the common term k from the summation terms. 40) For example, see Fig. 3.5, and YMS; 41–44) YMS; 45) See Examples 3.20 and 3.21, and YMS. 46) See hint given in problem. 47) See Examples 3.22 and 3.23, and YMS. 48) YMS.

Chapter 4

1–5) SM; 6) One possible B is B = [0 0; 1 0]; 7–8) SM; 9–11) YMS; 12) To check if a number x is an integer, compare x to floor(x). To check if it is even, let y = x/2, and compare y to floor(y). 13) SM; 14) YMS; 15) Use suggestions given in problem. 16) $P = RV/(R_s+R)^2$, and YMS; 17–18) YMS; 19) The algorithm is an example of a digital low-pass filter. Use the suggested script given in the problem. 20–21) Use the suggested script given in the problem, and YMS. 22) You will need the derivative dP/dR = $(R_s^2 - R^2)/(R_s + R)^4$. 23–24) YMS; 25a) A = 0.5 − $e^{-5}(1 - 0.5e^{-5}) \cong 4.9328475e - 01$;

26) YMS; 27) Use x_of_t = @(t) exp(− t) − exp(−2*t), and then get the area with quad (x_of_t,0,5). 28 − 29) YMS.

Chapter 5

1–2) YMS; 3a) 1100011, 11111010; 3b) 255; 4a) 0.2 → 0.00110011, 0.21 → 0.00110101; 4b) 0.00390625; 5) SM; 6) YMS, and P = 1 to check even parity; 7) X = int16(X), SM; 8a) 2000 → 7d0; 8b) ab → 10101011 → 171, 7ff → 011111111111→ 2047; 9) SM; 10a) In hexadecimal, the number is: c18b0000. 10b) SM; 10c) SM; 10d) In hexadecimal, the rounded fraction is: 0.b2f0. 11) Use truth table for an AND gate. 12) Use truth table for an XOR gate. 13) See circuits. 14b) Yes, since a = b. 15) NOR gates can be used to make the basic gates, NOT, AND and OR. 16) YMS; 17) The additional term checks if $a_2 = b_2$, $a_1 = b_1$ and $a_0 = 1$ and $b_0 = 0$, and YMS. 18) See function in Prob 5.17, and realize it in a sum of products form. 19) See problem description. 20) See the MATLAB script in Example 5.6, as an example for making up a table of BCD codes for w, x, y and z, and then use an index to the table given by D+1 to get the BCD codes. 21) A+B = 1001111.010011101 = 4f.4e8 in hexadecimal, and YMS; 22–24) YMS; 25) K = 19, voltage resolution is: 3.815 uV; 26) Use Table 5.11.

Chapter 6

1a) $(x − 5)(x+3)$; 1b) $−(x − 2)(x+3)$; 1c) $(x+3)(x+1 − j)(x+1+j)$; 2a) min at x = 1; 2b) max at x = −1/2, 2c) max at x = −2 and min at x = −4/3; 3a) $−1+j5$; 3b) $1+j4$; 3c) $−9 − j7$; 3d) $(3 − j*2)/13$; 3e) $(− 3 − j11)/10$; 3f) $(9-j6)/13$; 4) SM; 5) See Fig. 6.3, for example. 6a) 2/13; 6b) −13; 6c) −6/13; 7–8) SM; 9) YMS; 10) Let $x_3 = x_1+x_2$. The three complex numbers x_1, x_2, and x_3 can be drawn as three vectors in a complex plane to form a triangle, where $\|x_1\|$, $\|x_2\|$ and $\|x_3\|$ are the lengths of the triangle sides. Now invoke the triangle inequality. 11) $c_1 = e^{j\pi/2}$, $c_2 = \sqrt{13}\, e^{j\tan^{-1}(3/2)}$, $c_3 = \sqrt{10}\, e^{j(\pi/2 + \tan^{-1}(3/1))} = \sqrt{10}\, e^{j(\pi − \tan^{-1}(1/3))}$, $c_4 = \sqrt{25}\, e^{j(\pi + \tan^{-1}(3/4))} = \sqrt{25}\, e^{j(3\pi/2 − \tan^{-1}(4/3))}$, $c_5 = 3\, e^{j3\pi/2}$, $c_6 = \sqrt{13}\, e^{j(3\pi/2+ \tan^{-1}(2/3))} = \sqrt{13}\, e^{j(2\pi − \tan^{-1}(3/2))}$; 12) SM; 13) $c_7 = \sqrt{2}+j\sqrt{2}$, $c_8 = j3$, $c_9 = −3\sqrt{2}/2 − j\,3\sqrt{2}/2$, $c_{10} = −3/2 − j\,3\sqrt{3}/2$. 14) $c_7 c_8 = 6e^{j3\pi/4}$, $c_8 + c_9 = −3\sqrt{2}/2 +j(3 − 3\sqrt{2}/2)$, $c_{10}^* = 3e^{−j7\pi/3}$, $c_7/c_9 = −j\,2/3$; 15) SM; 16) Use polar forms. 17) Let $f(x) = a_n x^n + a_{n−1}x^{n−1} + \cdots + a_1 x + a_0$, where the coefficients are real numbers. Use the property that the conjugate of a sum of complex numbers equals the sum of the conjugates, a property from Prob. 6.17. 18) Apply Euler's identity to each sinusoid on the right side of the equal sign, multiply out, cancel some terms and again apply Euler's identity to match the left side of the equal sign. 19) First, convert to a cosine function with a positive amplitude. 20) YMS; 21a) $X = 3e^{j7\pi/6}$; 21b) $X = 5e^{−j\pi/4}$; 21c) $X = 7e^{j\pi/3}$; 21d) $X = 2e^{−j\pi/2}$; 22a) $x(t) = 3\cos(10\pi t + \pi/3)$; 23) YMS; 24a) $f_0 = 0.25$ Hz; 24b) $a_0 = 1$, $a_k = (−4\sin(k\omega_0) +2\sin(3k\omega_0))/(k\omega_0)$, $b_k = (2\cos(k\omega_0) − 2\cos(3k\omega_0))/(k\omega_0) = 0$; 25) $X_k = (2e^{−j2k\omega_0} − 1)\sin(k\omega_0)/(k\omega_0)$, $X_0 = 1$; 26–28) YMS; 29a) $Z_L(\omega) = j\omega L$, $Z_L(0) = 0$,

$Z_L(\infty) = \infty$; 29b) $Z_C(\omega) = 1/j\omega C$, $Z_C(0) = \infty$, $Z_C(\infty) = 0$; 30) See, for example, Example 6.11. 31a) $- V_s + \frac{1}{j\omega\,(0.25 \times 10^{-6})}\,I + j\omega\,(10 \times 10^{-3})\,I + 100\,I = 0$, $V_s = 5\,e^{-j\pi/2}$, $V = 100\,I$, $\omega = 10^4$ rad/sec , and YMS; 31b) YMS; 31c) We can write $V = H(j\omega)\,V_s$. The denominator of $\|H(j\omega)\|$ has its smallest value when $\frac{-1}{\omega\,(0.25 \times 10^{-6})} + \omega\,(10 \times 10^{-3}) = 0$, which gives $-1 + \omega^2 LC = 0 \;\rightarrow\; \omega = 20000$ rad/sec. 32a) $H(j\omega) = \frac{1/j\omega C}{R + 1/j\omega C}$; 32b) YMS, a low-pass filter and the BW is the frequency at which $\|H(j\omega)\|^2$ has a value that is 1/2 its peak value. 33) YMS; 34a) $H(j\omega) = R/(R + 1/j\omega C)$; 34b) YMS and an inspection of the plot of the magnitude squared gives a BW of approximately 3600 Hz. 35a) Consider (6.27), which can be written as $b_k = \frac{2}{T_0}\int_{-T_0/2}^{T_0/2} x(t)\,\sin(k\omega_0\,t)\,dt$. Since the periodic function is given to be an even function, and the sine function is an odd function, then the integrand is an odd function. The integral of an odd function over a symmetric range is zero, which makes $b_k = 0$. According to (6.28), X_k is real. 35b) Start with (6.26). 36) Use $R_2/R_1 = 10$ for each amplifier. 37) The gain must be less than 45. 38) Start by applying KCL at the negative terminal of the op-amp.

Chapter 7

1–2) SM; 3) After making the assignments, use the function whos. 4) Use braces to define a cell array. 5) SM; 6) Use the function isletter, and also try isstrprop. 7) In part(b), use the function findstr. 8) In part (b), initialize digits with digits = '', using two single quotes, and with each new digit name concatenate digits with the name to obtain an updated digits. 9) YMS; 10) SM; 11) See Example 7.3, and YMS. 12) See Example 7.4. 13–14) SM.

Chapter 8

1–3) SM; 4b) See Example 8.2, or try using the function which. 5–10) YMS; 11) See Example 8.4. 12–13) SM; 14) See Example 8.4. 15) Use suggestion given in problem. 16–17) SM; 18) YMS; 19) SM; 20–21) Script is given in problem.

Chapter 9

1–4) SM; 5–6) YMS; 7) Script is described in the problem. 8) YMS; 9) Since f = 60 Hz, time for four cycles is total_time = 4/60 sec. 10–11) Define a vector of frequency values, and use a vectorized expression to obtain $H(j\omega)$. See Prog. 9.3. The filter is a low-pass filter. 12) The algorithm to obtain u from y can be implemented with a for loop. Before the for loop compute u(1) and u(2) as described in the problem. The for loop is: for n = 3:length(t); u(n) = (y(n)+ 2*y(n−1)+y(n−2))/4; end. To find v, compute v(1) = u(1)/4; v(2) = (u(2)+2*u(1))/4; for n = 3:length(t); v(n) = (u(n)+2*u(n−1)+u(n−2))/4; end. Like x and y, u and v are time functions versus t. Once plotted, see Prog. 9.3 for examples of customizing the plots. 13) YMS; 14) Define three vectors, x1, x2, and y, for example, x1 = [0 1 0 1 0 1 0 1 1 0 0].

To better see the logic signals, use axis([0 length(n) 0 1.1]), and turn the grid on. See Prog. 9.5. 15–18) SM; 19) YMS; 20) See Example 9.8. 21–22) YMS; 23c) low-pass filter; 24) See Example 9.10. 25–27) YMS; 28) See Examples 9.13 and 9.14. 29) See Example 9.11.

Chapter 10

1–8) SM; 9) YMS; 10–17) SM.

Chapter 11

1) SM; 2) SM, and also, use the function whos; 3–9) SM; 10) Start with (11.1), and multiply both sides by r to get $r\,S = \sum_{n=0}^{N-1} r^{n+1} = \sum_{n=1}^{N} r^n = r^N + \sum_{n=0}^{N-1} r^n - 1 = r^N + S - 1$, and solve for S. 11–12) SM; 13) The series converges if $\|z\| < 1$. 14–19) SM; 20a) The KCL equations are

$$\frac{V_R - V_S}{R_S} + \frac{V_R - V_C}{j\omega L} = 0$$

$$\frac{V_C - V_S}{R_S} + \frac{V_C - V_R}{j\omega L} + \frac{V_C}{R_C} + \frac{V_C}{1/j\omega C} = 0$$

20b–c) See Example 11.4. 21) SM; 22b) $x_p(t) = u(t) - 3u(t-1) + 5u(t-2) - 3u(t-4)$; 22c) $\dot{x}_p(t) = \delta(t) - 3\,\delta(t-1) + 5\,\delta(t-2) - 3\,\delta(t-4)$; 23a) Since $i(t) = C\,dv/dt$, a sudden increase (decrease) in $v(t)$ will cause a large positive (negative) capacitor current. 23b) Since $v(t) = L\,di/dt$, a sudden increase (decrease) in $i(t)$ will cause a large positive (negative) inductor voltage. 24–25) SM; 26a) $Y(j\omega) = 1/(j\omega + a)$; 26b) $Y(j\omega) = T\sin(\omega T/2)/(\omega T/2)$; 27a) YMS; 27b) $C = B$; 28a) $Y(s) = s/(s^2 + \omega^2)$; 28b) $Y(s) = (s+a)/((s+a)^2 + \omega^2)$.

Chapter 12

1a) BW $= 60$ Hz, $T_0 = 1/15$ sec, $X_4 = 3\,e^{-j\pi/4}$, $X_{-4} = X_4^*$, $X_1 = 2\,e^{j2\pi/3}$, $X_{-1} = X_1^*$; 1b) BW $= 56$ Hz, $T_0 = 1/7$ sec, $X_8 = (5/2)\,e^{-j\pi/2}$, $X_{-8} = X_8^*$, $X_3 = e^{j\pi}$, $X_{-3} = X_3^*$, $X_2 = (3/2)\,e^{-j\pi/4}$, $X_{-2} = X_2^*$; 2a) $T_0 = 6$ sec; 2b) $X_0 = 1/2$, $X_k = (e^{j\,k\pi/6} - 0.5\,e^{-j5\,k\pi/6})\sin(k\pi/2)/(k\pi/2)$. The reconstruction does exhibit Gibbs' oscillation. 2d) $X_0 = 2/\pi$, $X_k = 2/(\pi(1 - 4k^2))$. The reconstruction does not exhibit Gibbs' oscillation. 3a) $N = 4$; 3b) $X(0) = 0$, $X(1) = j\,2$, $X(2) = 0$, $X(3) = -j\,2$; 4a) See Example 12.3. The sampling frequency is $f_s = 300$ Hz (samples/sec), and it satisfies the sampling theorem. The frequencies are $f = k\,f_0$, where $f_0 = 15$ Hz. $X_k = X(k)/N$; 4b) $f_s = 80$ Hz, which does not satisfy the sampling theorem, and there will be aliasing error. 5) See Example 12.3. 6) YMS; 7) The given signal is bandlimited. 8a) The results give X_k without error. 8b) Yes, it should be possible to estimate the frequency of $x(t)$. 8c) In this case, the spectral lines occur at frequencies other than the frequency of $x(t)$. 8d) See Example 12.7. 9a) $N = 11$, where $X(N - 1 = 10) = X(-1)$, ...; 9b) See Prog. 12.6, and YMS. 9c) If the sampling frequency satisfies the sampling theorem, then $BW = 5/T_x$ Hz. 10a) Your script should start by assigning to N an integer, and $W = \exp(-j*2*pi/N)$. Then, initialize DN with $DN = ones(N,N)$. Now, the first column of DN holds the coefficients of $x(0)$ in each of the DFT equations. Then, define a column vector w with: for $k = 1:N$; $w(k) = W^\wedge(k-1)$; end; $w = w'$. To obtain DN, use: for $k = 2: N$; $DN(:, k) = DN(:, k-1).*w$; end. 10b) YMS; 11a) Apply suggestion given in problem. 11b) YMS; 12) YMS; 12a) $T_x = 8.192$ sec. See Examples 12.8 and 12.9. 13) YMS; 14) $\frac{di}{dt} + \left(\left(\frac{1}{RC}\right) + \left(\frac{1}{R_L C}\right)\right)i(t) = \frac{1}{RR_L C}v_s(t)$. Since

$v_C(t) = R_L i(t)$, $i(0^-) = v_C(0^-)/R$. 15) $\frac{d^2 v}{dt^2} + \frac{1}{RC}\frac{dv}{dt} + \frac{1}{LC} v(t) = \frac{1}{RC}\frac{dv_s}{dt}$. Since $v(t) = v_C(t)$,

$v(0^-) = v_C(0^-)$. Since $\frac{v(t)-v_s(t)}{R} + C\frac{dv}{dt} + i_L(t) = 0$, $\dot v(0^-) = (- i_L(0^-) - v_C(0^-)/R)/C$.

16a) $\lambda^4 + 6\lambda^3 + 138\lambda^2 + 462\lambda + 2929 = 0$; 16b) See Prog. 12.10, and YMS. 17) $y_{zs}(t) = 2\ddot w(t) + 4\dot w(t) + 202w(t)$; 18a) Apply (12.40) and (12.42), and YMS to find the initial state. 18b) SM; 19–22) YMS; 23) $h(t) = 3e^{-2t} u(t)$; 24) To find $h(t)$, follow the procedure given in Example 12.15 to get $h(t) = (K_1 e^{(-1+j)t} + K_2 e^{(-1-j)t})u(t)$, where $K_1 = 1/2$ and $K_2 = 1/2$. Using Euler's identity gives $h(t) = e^{-t}\cos(t)u(t)$. 25–26) YMS; 27a) $y_{z\,s}(t) = 6 (e^{-t} - e^{-2t}) u(t)$; 27b) $y_{zs}(t) = h(t) * r(t - 1) = 6 (e^{-(t-1)} - e^{-2(t-1)}) u(t - 1)$; 27c) $y_{zs}(t) = h(t) * r(t + 1) = 6 (e^{-(t+1)} - e^{-2(t+1)}) u(t + 1)$; 28) YMS, and $N_3 = N_1 + N_2 - 1 = 249$; 29) YMS; 30) Use suggestion given in problem. 31a) $H(s) = (2s^2 + 4s + 202)/(s^3 + 5s^2 + 33s + 29)$; 31b) YMS; 31c) $r(t) = \cos(2 \pi t -\pi/2)$, $H(j\omega) = (2(j\omega)^2 + 4(j\omega) + 202)/((j\omega)^3 + 5(j\omega)^2 + 33(j\omega) + 29)$, $\omega = 2\pi$ rad/sec, $H(j2\pi) = -0.7244 + 0.0259i = 0.7249 e^{j3.1059}$, $y_{ss}(t) = 0.7249\cos(2 \pi t - \pi/2 + 3.1059)$; 32a) 5th order difference equation; 32b) $y(-1)$, $y(-2)$, $y(-3)$, $y(-4)$ and $y(-5)$; 32c) causal, time invariant and linear DTS; 32d) $y(n) + 0.5 y(n - 1) +0.25 y(n - 3) + 0.125 y(n - 5) = 0$; 32e) $y(2) = 1.25$; 32f) The DTS is a linear system. Thus, if initial conditions are zero, then if $y(n) = y_1(n)$ is the response to $r(n) = r_1(n)$, then if $r(n) = r_2(n) = a r_1(n)$, for any constant a, the response is $y(n) = y_2(n) = a y_1(n)$. This is called the homogeneity property. Here, $r(n) = r_1(n) = 1$ gives $y(2) = y_1(2) = 0.875$, $r(n) = r_2(n) = 2 r_1(n)$ gives $y(2) = y_2(2) = 2y_1(2) = 1.75$ and $r(n) = r_3(n) = 3 r_1(n)$ gives $y(2) = y_3(2) = 3y_1(2) = 2.625$. The homogeneity property is not satisfied if initial conditions are not zero. 33a) $y_1(0) = 1.0$, $y_1(1) = -0.25$, $y_1(2) = 0.125$ and $y_1(3) = -0.3125$; 33b) $y_2(0) = 0$, $y_2(1) = 0$, $y_2(2) = 1.0$ and $y_2(3) = -0.25$. Since the DTS is a time invariant system, time shifting the input, time shifts the output, and $y_2(n) = y_1(n - 2)$. 34) YMS; 35) YMS, and as f is increased, σ^2 increases. 36) YMS; 37) See Examples 12.21 and 12.22 and YMS. 38-39) YMS; 40a) $1 + 0.81 \gamma^{-2} = 0$; 40b) $(0.9e^{j\pi/2})^n$ and $(0.9e^{-j\pi/2})^n$; 40c) $y_{zi}(n) = (0.9)^{n+1}\cos((n + 1) \pi/2)$; 41) $h(n) = (0.9)^n\cos(n \pi/2)$ $u(n) - (0.9)^{n-2}\cos((n - 2) \pi/2) u(n - 2)$; 42a) $H(z^{-1}) = (b_0 + b_1 z^{-1} + \cdots + b_8 z^{-8})/(a_0 + a_1 z^{-1} + \cdots + a_8 z^{-8})$; 42b-42c) YMS;

43) $H(z^{-1}) = (1 - z^{-2})/(1 + z^{-2})$;   43a)   See   Example   12.25,   and   YMS;
43b) $\omega = 250\,\pi$ rad/sec and $\omega T = \pi/4$; Evaluate $H(z^{-1})$ for $z = e^{j\pi/4}$, and apply (12.98).
44) YMS.

Chapter 13

1–5) SS; 6a) output is $y(n) = \sin((n-2)\pi/2)$; 6b) $y(n) + 0.5y(n\text{-}1) + 0.75y(n-2) = x(n) -$
$x(n-2)$; 6c) SS; 7–12) SS; 13) a circle; 14a-b) SS; 14c) The scope output is a raised cosine,
because the area under a sine wave is positive over the first half of its cycle, and becomes zero
when the cycle is completed. 14d) SS; 15) $y(n)$ is every N^{th} value of $x(n)$. 16-17) SS; 18) See
problem suggestion. 19) SS, and the average value should be approximately 5 volts. 20) SS;
21) $H(s) = RCs/(LCs^2 + RCs + 1)$; 22a) $y^{(4)}(t) + 6\,y^{(3)}(t) + 138\,y^{(2)}(t) + 462\,y^{(1)}(t) + 2929\,y(t)$
$= r^{(2)}(t) + 3\,r^{(1)}(t) + 2\,r(t)$; 22b–d) SS; 23c) $H(s) = \dfrac{(s+2)}{(s^2+4s+29)}\dfrac{(s+1)}{(s^2+2s+101)} = H_1(s)\,H_2(s)$;
23d) Ideally, the results should be the same. 24) Transfer function is given in problem. 24) SS,
25) Use Simulink model of an RC stage given in the problem.

Index